中国科学院科学出版基金资助出版

《国外数学名著系列》(影印版)专家委员会

(按姓氏笔画排序)

丁伟岳　王　元　石钟慈　冯克勤　严加安　李邦河
李大潜　张伟平　张继平　杨　乐　姜伯驹　郭　雷

项目策划

向安全　林　鹏　王春香　吕　虹　范庆奎　王　璐

执行编辑

范庆奎

国外数学名著系列(影印版) 6

Numerical Optimization

数值最优化

Jorge Nocedal   Stephen J. Wright

科学出版社
北京

图字:01-2005-6747

Jorge Nocedal, Stephen J. Wright: Numerical Optimization
© 1999 Springer Science+Business Media, Inc.

This reprint has been authorized by Springer-Verlag (Berlin/Heidelberg/New York) for sale in the People's Republic of China only and not for export therefrom.

本书英文影印版由德国施普林格出版公司授权出版。未经出版者书面许可,不得以任何方式复制或抄袭本书的任何部分。本书仅限在中华人民共和国销售,不得出口。版权所有,翻印必究。

图书在版编目(CIP)数据

数值最优化＝Numerical Optimization/(美)劳斯特(Nocedal, J.)等著.—影印版.—北京:科学出版社,2006

(国外数学名著系列)

ISBN 978-7-03-016675-3

Ⅰ.数… Ⅱ.劳… Ⅲ.最优化算法-研究生-教材-英文 Ⅳ.O224

中国版本图书馆 CIP 数据核字(2005)第 154405 号

科 学 出 版 社 出版
北京东黄城根北街 16 号
邮政编码:100717
http://www.sciencep.com

北京虎彩文化传播有限公司 印刷
科学出版社发行　各地新华书店经销

*

2006 年 1 月第 一 版　　开本:B5(720×1000)
2021 年 1 月第六次印刷　　印张:41 1/2
字数:779 000

定价:198.00 元
(如有印装质量问题,我社负责调换)

《国外数学名著系列》(影印版)序

要使我国的数学事业更好地发展起来,需要数学家淡泊名利并付出更艰苦地努力。另一方面,我们也要从客观上为数学家创造更有利的发展数学事业的外部环境,这主要是加强对数学事业的支持与投资力度,使数学家有较好的工作与生活条件,其中也包括改善与加强数学的出版工作。

从出版方面来讲,除了较好较快地出版我们自己的成果外,引进国外的先进出版物无疑也是十分重要与必不可少的。从数学来说,施普林格(Springer)出版社至今仍然是世界上最具权威的出版社。科学出版社影印一批他们出版的好的新书,使我国广大数学家能以较低的价格购买,特别是在边远地区工作的数学家能普遍见到这些书,无疑是对推动我国数学的科研与教学十分有益的事。

这次科学出版社购买了版权,一次影印了 23 本施普林格出版社出版的数学书,就是一件好事,也是值得继续做下去的事情。大体上分一下,这 23 本书中,包括基础数学书 5 本,应用数学书 6 本与计算数学书 12 本,其中有些书也具有交叉性质。这些书都是很新的,2000 年以后出版的占绝大部分,共计 16 本,其余的也是 1990 年以后出版的。这些书可以使读者较快地了解数学某方面的前沿,例如基础数学中的数论、代数与拓扑三本,都是由该领域大数学家编著的"数学百科全书"的分册。对从事这方面研究的数学家了解该领域的前沿与全貌很有帮助。按照学科的特点,基础数学类的书以"经典"为主,应用和计算数学类的书以"前沿"为主。这些书的作者多数是国际知名的大数学家,例如《拓扑学》一书的作者诺维科夫是俄罗斯科学院的院士,曾获"菲尔兹奖"和"沃尔夫数学奖"。这些大数学家的著作无疑将会对我国的科研人员起到非常好的指导作用。

当然,23 本书只能涵盖数学的一部分,所以,这项工作还应该继续做下去。更进一步,有些读者面较广的好书还应该翻译成中文出版,使之有更大的读者群。

总之,我对科学出版社影印施普林格出版社的部分数学著作这一举措表示热烈的支持,并盼望这一工作取得更大的成绩。

王 元

2005 年 12 月 3 日

To Our Parents:

Raúl and Concepción   Peter and Berenice

Preface

This is a book for people interested in solving optimization problems. Because of the wide (and growing) use of optimization in science, engineering, economics, and industry, it is essential for students and practitioners alike to develop an understanding of optimization algorithms. Knowledge of the capabilities and limitations of these algorithms leads to a better understanding of their impact on various applications, and points the way to future research on improving and extending optimization algorithms and software. Our goal in this book is to give a comprehensive description of the most powerful, state-of-the-art, techniques for solving continuous optimization problems. By presenting the motivating ideas for each algorithm, we try to stimulate the reader's intuition and make the technical details easier to follow. Formal mathematical requirements are kept to a minimum.

Because of our focus on continuous problems, we have omitted discussion of important optimization topics such as discrete and stochastic optimization. However, there are a great many applications that can be formulated as continuous optimization problems; for instance,

finding the optimal trajectory for an aircraft or a robot arm;

identifying the seismic properties of a piece of the earth's crust by fitting a model of the region under study to a set of readings from a network of recording stations;

designing a portfolio of investments to maximize expected return while maintaining an acceptable level of risk;

controlling a chemical process or a mechanical device to optimize performance or meet standards of robustness;

computing the optimal shape of an automobile or aircraft component.

Every year optimization algorithms are being called on to handle problems that are much larger and complex than in the past. Accordingly, the book emphasizes large-scale optimization techniques, such as interior-point methods, inexact Newton methods, limited-memory methods, and the role of partially separable functions and automatic differentiation. It treats important topics such as trust-region methods and sequential quadratic programming more thoroughly than existing texts, and includes comprehensive discussion of such "core curriculum" topics as constrained optimization theory, Newton and quasi-Newton methods, nonlinear least squares and nonlinear equations, the simplex method, and penalty and barrier methods for nonlinear programming.

THE AUDIENCE

We intend that this book will be used in graduate-level courses in optimization, as offered in engineering, operations research, computer science, and mathematics departments. There is enough material here for a two-semester (or three-quarter) sequence of courses. We hope, too, that this book will be used by practitioners in engineering, basic science, and industry, and our presentation style is intended to facilitate self-study. Since the book treats a number of new algorithms and ideas that have not been described in earlier textbooks, we hope that this book will also be a useful reference for optimization researchers.

Prerequisites for this book include some knowledge of linear algebra (including numerical linear algebra) and the standard sequence of calculus courses. To make the book as self-contained as possible, we have summarized much of the relevant material from these areas in the Appendix. Our experience in teaching engineering students has shown us that the material is best assimilated when combined with computer programming projects in which the student gains a good feeling for the algorithms—their complexity, memory demands, and elegance—and for the applications. In most chapters we provide simple computer exercises that require only minimal programming proficiency.

EMPHASIS AND WRITING STYLE

We have used a conversational style to motivate the ideas and present the numerical algorithms. Rather than being as concise as possible, our aim is to make the discussion flow in a natural way. As a result, the book is comparatively long, but we believe that it can be read relatively rapidly. The instructor can assign substantial reading assignments from the text and focus in class only on the main ideas.

A typical chapter begins with a nonrigorous discussion of the topic at hand, including figures and diagrams and excluding technical details as far as possible. In subsequent sections, the algorithms are motivated and discussed, and then stated explicitly. The major theoretical results are stated, and in many cases proved, in a rigorous fashion. These proofs can be skipped by readers who wish to avoid technical details.

The practice of optimization depends not only on efficient and robust algorithms, but also on good modeling techniques, careful interpretation of results, and user-friendly software. In this book we discuss the various aspects of the optimization process—modeling, optimality conditions, algorithms, implementation, and interpretation of results—but not with equal weight. Examples throughout the book show how practical problems are formulated as optimization problems, but our treatment of modeling is light and serves mainly to set the stage for algorithmic developments. We refer the reader to Dantzig [63] and Fourer, Gay, and Kernighan [92] for more comprehensive discussion of this issue. Our treatment of optimality conditions is thorough but not exhaustive; some concepts are discussed more extensively in Mangasarian [154] and Clarke [42]. As mentioned above, we are quite comprehensive in discussing optimization algorithms.

TOPICS NOT COVERED

We omit some important topics, such as network optimization, integer programming, stochastic programming, nonsmooth optimization, and global optimization. Network and integer optimization are described in some excellent texts: for instance, Ahuja, Magnanti, and Orlin [1] in the case of network optimization and Nemhauser and Wolsey [179], Papadimitriou and Steiglitz [190], and Wolsey [249] in the case of integer programming. Books on stochastic optimization are only now appearing; we mention those of Kall and Wallace [139], Birge and Louveaux [11]. Nonsmooth optimization comes in many flavors. The relatively simple structures that arise in robust data fitting (which is sometimes based on the ℓ_1 norm) are treated by Osborne [187] and Fletcher [83]. The latter book also discusses algorithms for nonsmooth penalty functions that arise in constrained optimization; we discuss these briefly, too, in Chapter 18. A more analytical treatment of nonsmooth optimization is given by Hiriart-Urruty and Lemaréchal [137]. We omit detailed treatment of some important topics that are the focus of intense current research, including interior-point methods for nonlinear programming and algorithms for complementarity problems.

ADDITIONAL RESOURCE

The material in the book is complemented by an online resource called the NEOS Guide, which can be found on the World-Wide Web at

```
http://www.mcs.anl.gov/otc/Guide/
```

The Guide contains information about most areas of optimization, and presents a number of case studies that describe applications of various optimization algorithms to real-world problems such as portfolio optimization and optimal dieting. Some of this material is interactive in nature and has been used extensively for class exercises.

For the most part, we have omitted detailed discussions of specific software packages, and refer the reader to Moré and Wright [173] or to the Software Guide section of the NEOS Guide, which can be found at

http://www.mcs.anl.gov/otc/Guide/SoftwareGuide/

Users of optimization software refer in great numbers to this web site, which is being constantly updated to reflect new packages and changes to existing software.

ACKNOWLEDGMENTS

We are most grateful to the following colleagues for their input and feedback on various sections of this work: Chris Bischof, Richard Byrd, George Corliss, Bob Fourer, David Gay, Jean-Charles Gilbert, Phillip Gill, Jean-Pierre Goux, Don Goldfarb, Nick Gould, Andreas Griewank, Matthias Heinkenschloss, Marcelo Marazzi, Hans Mittelmann, Jorge Moré, Will Naylor, Michael Overton, Bob Plemmons, Hugo Scolnik, David Stewart, Philippe Toint, Luis Vicente, Andreas Waechter, and Ya-xiang Yuan. We thank Guanghui Liu, who provided help with many of the exercises, and Jill Lavelle who assisted us in preparing the figures. We also express our gratitude to our sponsors at the Department of Energy and the National Science Foundation, who have strongly supported our research efforts in optimization over the years.

One of us (JN) would like to express his deep gratitude to Richard Byrd, who has taught him so much about optimization and who has helped him in very many ways throughout the course of his career.

FINAL REMARK

In the preface to his 1987 book [83], Roger Fletcher described the field of optimization as a "fascinating blend of theory and computation, heuristics and rigor." The ever-growing realm of applications and the explosion in computing power is driving optimization research in new and exciting directions, and the ingredients identified by Fletcher will continue to play important roles for many years to come.

Jorge Nocedal   Stephen J. Wright
Evanston, IL   *Argonne, IL*

Contents

Preface v

1 **Introduction**   1
    Mathematical Formulation 2
    Example: A Transportation Problem 4
    Continuous versus Discrete Optimization 4
    Constrained and Unconstrained Optimization 6
    Global and Local Optimization 6
    Stochastic and Deterministic Optimization 7
    Optimization Algorithms 7
    Convexity 8
  Notes and References 9

2 **Fundamentals of Unconstrained Optimization**   10
  2.1 What Is a Solution? 13
        Recognizing a Local Minimum 15
        Nonsmooth Problems 18

	2.2	Overview of Algorithms	19
		Two Strategies: Line Search and Trust Region	19
		Search Directions for Line Search Methods	21
		Models for Trust-Region Methods	26
		Scaling	27
		Rates of Convergence	28
		R-Rates of Convergence	29
	Notes and References		30
	Exercises		30
3	**Line Search Methods**		**34**
	3.1	Step Length	36
		The Wolfe Conditions	37
		The Goldstein Conditions	41
		Sufficient Decrease and Backtracking	41
	3.2	Convergence of Line Search Methods	43
	3.3	Rate of Convergence	46
		Convergence Rate of Steepest Descent	47
		Quasi-Newton Methods	49
		Newton's Method	51
		Coordinate Descent Methods	53
	3.4	Step-Length Selection Algorithms	55
		Interpolation	56
		The Initial Step Length	58
		A Line Search Algorithm for the Wolfe Conditions	58
	Notes and References		61
	Exercises		62
4	**Trust-Region Methods**		**64**
		Outline of the Algorithm	67
	4.1	The Cauchy Point and Related Algorithms	69
		The Cauchy Point	69
		Improving on the Cauchy Point	70
		The Dogleg Method	71
		Two-Dimensional Subspace Minimization	74
		Steihaug's Approach	75
	4.2	Using Nearly Exact Solutions to the Subproblem	77
		Characterizing Exact Solutions	77
		Calculating Nearly Exact Solutions	78
		The Hard Case	82
		Proof of Theorem 4.3	84
	4.3	Global Convergence	87

		Reduction Obtained by the Cauchy Point	87
		Convergence to Stationary Points	89
		Convergence of Algorithms Based on Nearly Exact Solutions	93
	4.4	Other Enhancements	94
		Scaling	94
		Non-Euclidean Trust Regions	96
	Notes and References	97
	Exercises	97

5 Conjugate Gradient Methods — 100

	5.1	The Linear Conjugate Gradient Method	102
		Conjugate Direction Methods	102
		Basic Properties of the Conjugate Gradient Method	107
		A Practical Form of the Conjugate Gradient Method	111
		Rate of Convergence	112
		Preconditioning	118
		Practical Preconditioners	119
	5.2	Nonlinear Conjugate Gradient Methods	120
		The Fletcher–Reeves Method	120
		The Polak–Ribière Method	121
		Quadratic Termination and Restarts	122
		Numerical Performance	124
		Behavior of the Fletcher–Reeves Method	124
		Global Convergence	127
	Notes and References	131
	Exercises	132

6 Practical Newton Methods — 134

	6.1	Inexact Newton Steps	136
	6.2	Line Search Newton Methods	139
		Line Search Newton–CG Method	139
		Modified Newton's Method	141
	6.3	Hessian Modifications	142
		Eigenvalue Modification	143
		Adding a Multiple of the Identity	144
		Modified Cholesky Factorization	145
		Gershgorin Modification	150
		Modified Symmetric Indefinite Factorization	151
	6.4	Trust-Region Newton Methods	154
		Newton–Dogleg and Subspace-Minimization Methods	154
		Accurate Solution of the Trust-Region Problem	155
		Trust-Region Newton–CG Method	156

        Preconditioning the Newton–CG Method 157
        Local Convergence of Trust-Region Newton Methods 159
Notes and References 162
Exercises 162

7 Calculating Derivatives   164

7.1   Finite-Difference Derivative Approximations 166
        Approximating the Gradient 166
        Approximating a Sparse Jacobian 169
        Approximating the Hessian 173
        Approximating a Sparse Hessian 174
7.2   Automatic Differentiation 176
        An Example 177
        The Forward Mode 178
        The Reverse Mode 179
        Vector Functions and Partial Separability 183
        Calculating Jacobians of Vector Functions 184
        Calculating Hessians: Forward Mode 185
        Calculating Hessians: Reverse Mode 187
        Current Limitations 188
Notes and References 189
Exercises 189

8 Quasi-Newton Methods   192

8.1   The BFGS Method 194
        Properties of the BFGS Method 199
        Implementation 200
8.2   The SR1 Method 202
        Properties of SR1 Updating 205
8.3   The Broyden Class 207
        Properties of the Broyden Class 209
8.4   Convergence Analysis 211
        Global Convergence of the BFGS Method 211
        Superlinear Convergence of BFGS 214
        Convergence Analysis of the SR1 Method 218
Notes and References 219
Exercises 220

9 Large-Scale Quasi-Newton and Partially Separable Optimization   222

9.1   Limited-Memory BFGS 224
        Relationship with Conjugate Gradient Methods 227
9.2   General Limited-Memory Updating 229

		Compact Representation of BFGS Updating	230
		SR1 Matrices	232
		Unrolling the Update	232
	9.3	Sparse Quasi-Newton Updates	233
	9.4	Partially Separable Functions	235
		A Simple Example	236
		Internal Variables	237
	9.5	Invariant Subspaces and Partial Separability	240
		Sparsity vs. Partial Separability	242
		Group Partial Separability	243
	9.6	Algorithms for Partially Separable Functions	244
		Exploiting Partial Separability in Newton's Method	244
		Quasi-Newton Methods for Partially Separable Functions	245
	Notes and References		247
	Exercises		248

10 Nonlinear Least-Squares Problems — 250

	10.1	Background	253
		Modeling, Regression, Statistics	253
		Linear Least-Squares Problems	256
	10.2	Algorithms for Nonlinear Least-Squares Problems	259
		The Gauss–Newton Method	259
		The Levenberg–Marquardt Method	262
		Implementation of the Levenberg–Marquardt Method	264
		Large-Residual Problems	266
		Large-Scale Problems	269
	10.3	Orthogonal Distance Regression	271
	Notes and References		273
	Exercises		274

11 Nonlinear Equations — 276

	11.1	Local Algorithms	281
		Newton's Method for Nonlinear Equations	281
		Inexact Newton Methods	284
		Broyden's Method	286
		Tensor Methods	290
	11.2	Practical Methods	292
		Merit Functions	292
		Line Search Methods	294
		Trust-Region Methods	298
	11.3	Continuation/Homotopy Methods	304
		Motivation	304

Practical Continuation Methods 306
Notes and References 310
Exercises 311

12 Theory of Constrained Optimization 314

Local and Global Solutions 316
Smoothness 317
12.1 Examples 319
A Single Equality Constraint 319
A Single Inequality Constraint 321
Two Inequality Constraints 324
12.2 First-Order Optimality Conditions 327
Statement of First-Order Necessary Conditions 327
Sensitivity 330
12.3 Derivation of the First-Order Conditions 331
Feasible Sequences 332
Characterizing Limiting Directions: Constraint Qualifications 336
Introducing Lagrange Multipliers 339
Proof of Theorem 12.1 341
12.4 Second-Order Conditions 342
Second-Order Conditions and Projected Hessians 348
Convex Programs 350
12.5 Other Constraint Qualifications 351
12.6 A Geometric Viewpoint 354
Notes and References 357
Exercises 358

13 Linear Programming: The Simplex Method 362

Linear Programming 364
13.1 Optimality and Duality 366
Optimality Conditions 366
The Dual Problem 367
13.2 Geometry of the Feasible Set 370
Basic Feasible Points 370
Vertices of the Feasible Polytope 372
13.3 The Simplex Method 374
Outline of the Method 374
Finite Termination of the Simplex Method 377
A Single Step of the Method 378
13.4 Linear Algebra in the Simplex Method 379
13.5 Other (Important) Details 383
Pricing and Selection of the Entering Index 383

	Starting the Simplex Method	386

        Starting the Simplex Method 386
        Degenerate Steps and Cycling 389
   13.6 Where Does the Simplex Method Fit? 391
   Notes and References 392
   Exercises 393

14 Linear Programming: Interior-Point Methods   394
   14.1 Primal–Dual Methods 396
         Outline 396
         The Central Path 399
         A Primal–Dual Framework 401
         Path-Following Methods 402
   14.2 A Practical Primal–Dual Algorithm 404
         Solving the Linear Systems 408
   14.3 Other Primal–Dual Algorithms and Extensions 409
         Other Path-Following Methods 409
         Potential-Reduction Methods 409
         Extensions 410
   14.4 Analysis of Algorithm 14.2 411
   Notes and References 416
   Exercises 417

15 Fundamentals of Algorithms for Nonlinear Constrained Optimization   420
         Initial Study of a Problem 422
   15.1 Categorizing Optimization Algorithms 423
   15.2 Elimination of Variables 426
         Simple Elimination for Linear Constraints 427
         General Reduction Strategies for Linear Constraints 430
         The Effect of Inequality Constraints 434
   15.3 Measuring Progress: Merit Functions 434
   Notes and References 437
   Exercises 438

16 Quadratic Programming   440
         An Example: Portfolio Optimization 442
   16.1 Equality–Constrained Quadratic Programs 443
         Properties of Equality-Constrained QPs 444
   16.2 Solving the KKT System 447
         Direct Solution of the KKT System 448
         Range-Space Method 449
         Null-Space Method 450
         A Method Based on Conjugacy 452

	16.3	Inequality-Constrained Problems	453
		Optimality Conditions for Inequality-Constrained Problems	454
		Degeneracy	455
	16.4	Active-Set Methods for Convex QP	457
		Specification of the Active-Set Method for Convex QP	461
		An Example	463
		Further Remarks on the Active-Set Method	465
		Finite Termination of the Convex QP Algorithm	466
		Updating Factorizations	467
	16.5	Active-Set Methods for Indefinite QP	470
		Illustration	472
		Choice of Starting Point	474
		Failure of the Active-Set Method	475
		Detecting Indefiniteness Using the LBL^T Factorization	475
	16.6	The Gradient–Projection Method	476
		Cauchy Point Computation	477
		Subspace Minimization	480
	16.7	Interior-Point Methods	481
		Extensions and Comparison with Active-Set Methods	484
	16.8	Duality	484
	Notes and References		485
	Exercises		486

17 Penalty, Barrier, and Augmented Lagrangian Methods 490

	17.1	The Quadratic Penalty Method	492
		Motivation	492
		Algorithmic Framework	494
		Convergence of the Quadratic Penalty Function	495
	17.2	The Logarithmic Barrier Method	500
		Properties of Logarithmic Barrier Functions	500
		Algorithms Based on the Log-Barrier Function	505
		Properties of the Log-Barrier Function and Framework 17.2	507
		Handling Equality Constraints	509
		Relationship to Primal–Dual Methods	510
	17.3	Exact Penalty Functions	512
	17.4	Augmented Lagrangian Method	513
		Motivation and Algorithm Framework	513
		Extension to Inequality Constraints	516
		Properties of the Augmented Lagrangian	518
		Practical Implementation	521
	17.5	Sequential Linearly Constrained Methods	523
	Notes and References		525

Exercises 526

18 Sequential Quadratic Programming   528
- 18.1 Local SQP Method 530
  - SQP Framework 531
  - Inequality Constraints 533
  - IQP vs. EQP 534
- 18.2 Preview of Practical SQP Methods 534
- 18.3 Step Computation 536
  - Equality Constraints 536
  - Inequality Constraints 538
- 18.4 The Hessian of the Quadratic Model 539
  - Full Quasi-Newton Approximations 540
  - Hessian of Augmented Lagrangian 541
  - Reduced-Hessian Approximations 542
- 18.5 Merit Functions and Descent 544
- 18.6 A Line Search SQP Method 547
- 18.7 Reduced-Hessian SQP Methods 548
  - Some Properties of Reduced-Hessian Methods 549
  - Update Criteria for Reduced-Hessian Updating 550
  - Changes of Bases 551
  - A Practical Reduced-Hessian Method 552
- 18.8 Trust-Region SQP Methods 553
  - Approach I: Shifting the Constraints 555
  - Approach II: Two Elliptical Constraints 556
  - Approach III: Sℓ_1QP (Sequential ℓ_1 Quadratic Programming) 557
- 18.9 A Practical Trust-Region SQP Algorithm 560
- 18.10 Rate of Convergence 563
  - Convergence Rate of Reduced-Hessian Methods 565
- 18.11 The Maratos Effect 567
  - Second-Order Correction 570
  - Watchdog (Nonmonotone) Strategy 571
- Notes and References 573
- Exercises 574

A Background Material   576
- A.1 Elements of Analysis, Geometry, Topology 577
  - Topology of the Euclidean Space $\mathbf{R}^n$ 577
  - Continuity and Limits 580
  - Derivatives 581
  - Directional Derivatives 583
  - Mean Value Theorem 584

	Implicit Function Theorem	585
	Geometry of Feasible Sets	586
	Order Notation	591
	Root-Finding for Scalar Equations	592
A.2	Elements of Linear Algebra	593
	Vectors and Matrices	593
	Norms	594
	Subspaces	597
	Eigenvalues, Eigenvectors, and the Singular-Value Decomposition	598
	Determinant and Trace	599
	Matrix Factorizations: Cholesky, LU, QR	600
	Sherman–Morrison–Woodbury Formula	605
	Interlacing Eigenvalue Theorem	605
	Error Analysis and Floating-Point Arithmetic	606
	Conditioning and Stability	608

References 611

Index 625

CHAPTER 1

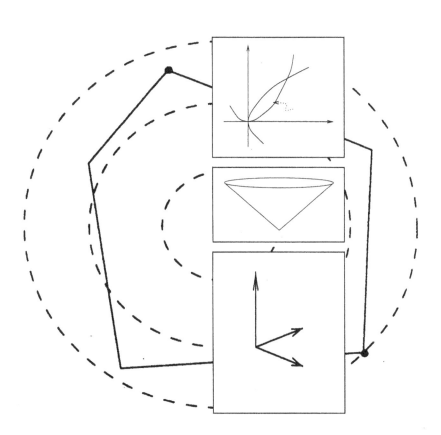

Introduction

People optimize. Airline companies schedule crews and aircraft to minimize cost. Investors seek to create portfolios that avoid excessive risks while achieving a high rate of return. Manufacturers aim for maximum efficiency in the design and operation of their production processes.

Nature optimizes. Physical systems tend to a state of minimum energy. The molecules in an isolated chemical system react with each other until the total potential energy of their electrons is minimized. Rays of light follow paths that minimize their travel time.

Optimization is an important tool in decision science and in the analysis of physical systems. To use it, we must first identify some *objective*, a quantitative measure of the performance of the system under study. This objective could be profit, time, potential energy, or any quantity or combination of quantities that can be represented by a single number. The objective depends on certain characteristics of the system, called *variables* or *unknowns*. Our goal is to find values of the variables that optimize the objective. Often the variables are restricted, or *constrained*, in some way. For instance, quantities such as electron density in a molecule and the interest rate on a loan cannot be negative.

The process of identifying objective, variables, and constraints for a given problem is known as *modeling*. Construction of an appropriate model is the first step—sometimes the most important step—in the optimization process. If the model is too simplistic, it will not give useful insights into the practical problem, but if it is too complex, it may become too difficult to solve.

Once the model has been formulated, an optimization algorithm can be used to find its solution. Usually, the algorithm and model are complicated enough that a computer is needed to implement this process. There is no universal optimization algorithm. Rather, there are numerous algorithms, each of which is tailored to a particular type of optimization problem. It is often the user's responsibility to choose an algorithm that is appropriate for their specific application. This choice is an important one; it may determine whether the problem is solved rapidly or slowly and, indeed, whether the solution is found at all.

After an optimization algorithm has been applied to the model, we must be able to recognize whether it has succeeded in its task of finding a solution. In many cases, there are elegant mathematical expressions known as *optimality conditions* for checking that the current set of variables is indeed the solution of the problem. If the optimality conditions are not satisfied, they may give useful information on how the current estimate of the solution can be improved. Finally, the model may be improved by applying techniques such as *sensitivity analysis*, which reveals the sensitivity of the solution to changes in the model and data.

MATHEMATICAL FORMULATION

Mathematically speaking, optimization is the minimization or maximization of a function subject to constraints on its variables. We use the following notation:

x is the vector of *variables*, also called *unknowns* or *parameters*;

f is the *objective function*, a function of x that we want to maximize or minimize;

c is the vector of *constraints* that the unknowns must satisfy. This is a vector function of the variables x. The number of components in c is the number of individual restrictions that we place on the variables.

The optimization problem can then be written as

$$\min_{x \in \mathbb{R}^n} f(x) \quad \text{subject to} \quad \begin{cases} c_i(x) = 0, & i \in \mathcal{E}, \\ c_i(x) \geq 0, & i \in \mathcal{I}. \end{cases} \quad (1.1)$$

Here f and each c_i are scalar-valued functions of the variables x, and $\mathcal{I}, \mathcal{E}$ are sets of indices. As a simple example, consider the problem

$$\min (x_1 - 2)^2 + (x_2 - 1)^2 \quad \text{subject to} \quad \begin{cases} x_1^2 - x_2 & \leq 0, \\ x_1 + x_2 & \leq 2. \end{cases} \quad (1.2)$$

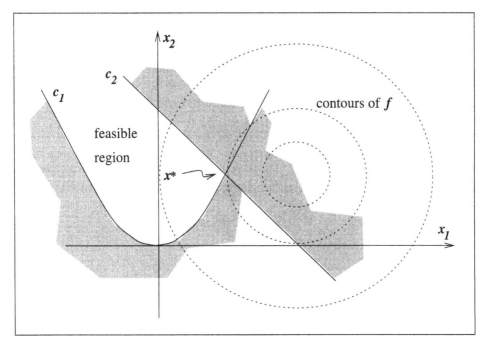

Figure 1.1 Geometrical representation of an optimization problem.

We can write this problem in the form (1.1) by defining

$$f(x) = (x_1 - 2)^2 + (x_2 - 1)^2, \qquad x = \begin{bmatrix} x_1 \\ x_2 \end{bmatrix},$$

$$c(x) = \begin{bmatrix} c_1(x) \\ c_2(x) \end{bmatrix} = \begin{bmatrix} -x_1^2 + x_2 \\ -x_1 - x_2 + 2 \end{bmatrix}, \qquad \mathcal{I} = \{1, 2\}, \qquad \mathcal{E} = \emptyset.$$

Figure 1.1 shows the contours of the objective function, i.e., the set of points for which $f(x)$ has a constant value. It also illustrates the *feasible region*, which is the set of points satisfying all the constraints, and the optimal point x^*, the solution of the problem. Note that the "infeasible side" of the inequality constraints is shaded.

The example above illustrates, too, that transformations are often necessary to express an optimization problem in the form (1.1). Often it is more natural or convenient to label the unknowns with two or three subscripts, or to refer to different variables by completely different names, so that relabeling is necessary to achieve the standard form. Another common difference is that we are required to *maximize* rather than minimize f, but we can accommodate this change easily by *minimizing* $-f$ in the formulation (1.1). Good software systems perform the conversion between the natural formulation and the standard form (1.1) transparently to the user.

CHAPTER 1. INTRODUCTION

EXAMPLE: A TRANSPORTATION PROBLEM

A chemical company has 2 factories F_1 and F_2 and a dozen retail outlets $R_1, \ldots, R_{12}$. Each factory F_i can produce a_i tons of a certain chemical product each week; a_i is called the *capacity* of the plant. Each retail outlet R_j has a known weekly *demand* of b_j tons of the product. The cost of shipping one ton of the product from factory F_i to retail outlet R_j is c_{ij}.

The problem is to determine how much of the product to ship from each factory to each outlet so as to satisfy all the requirements and minimize cost. The variables of the problem are x_{ij}, $i = 1, 2$, $j = 1, \ldots, 12$, where x_{ij} is the number of tons of the product shipped from factory F_i to retail outlet R_j; see Figure 1.2. We can write the problem as

$$\min \sum_{ij} c_{ij} x_{ij} \tag{1.3}$$

subject to

$$\sum_{j=1}^{12} x_{ij} \leq a_i, \quad i = 1, 2, \tag{1.4a}$$

$$\sum_{i=1}^{2} x_{ij} \geq b_j, \quad j = 1, \ldots, 12, \tag{1.4b}$$

$$x_{ij} \geq 0, \quad i = 1, 2, \ j = 1, \ldots, 12. \tag{1.4c}$$

In a practical model for this problem, we would also include costs associated with manufacturing and storing the product. This type of problem is known as a *linear programming* problem, since the objective function and the constraints are all linear functions.

CONTINUOUS VERSUS DISCRETE OPTIMIZATION

In some optimization problems the variables make sense only if they take on integer values. Suppose that in the transportation problem just mentioned, the factories produce tractors rather than chemicals. In this case, the x_{ij} would represent integers (that is, the number of tractors shipped) rather than real numbers. (It would not make much sense to advise the company to ship 5.4 tractors from factory 1 to outlet 12.) The obvious strategy of ignoring the integrality requirement, solving the problem with real variables, and then rounding all the components to the nearest integer is by no means guaranteed to give solutions that are close to optimal. Problems of this type should be handled using the tools of discrete optimization. The mathematical formulation is changed by adding the constraint

$$x_{ij} \in \mathbb{Z}, \quad \text{for all } i \text{ and } j,$$

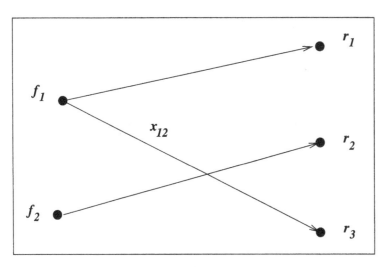

Figure 1.2 A transportation problem.

to the existing constraints (1.4), where **Z** is the set of all integers. The problem is then known as an *integer programming problem*.

The generic term *discrete optimization* usually refers to problems in which the solution we seek is one of a number of objects in a finite set. By contrast, *continuous optimization* problems—the class of problems studied in this book—find a solution from an uncountably infinite set—typically a set of vectors with real components. Continuous optimization problems are normally easier to solve, because the smoothness of the functions makes it possible to use objective and constraint information at a particular point x to deduce information about the function's behavior at all points close to x. The same statement cannot be made about discrete problems, where points that are "close" in some sense may have markedly different function values. Moreover, the set of possible solutions is too large to make an exhaustive search for the best value in this finite set.

Some models contain variables that are allowed to vary continuously and others that can attain only integer values; we refer to these as *mixed integer programming* problems.

Discrete optimization problems are not addressed directly in this book; we refer the reader to the texts by Papadimitriou and Steiglitz [190], Nemhauser and Wolsey [179], Cook et al. [56], and Wolsey [249] for comprehensive treatments of this subject. We point out, however, that the continuous optimization algorithms described here are important in discrete optimization, where a sequence of continuous subproblems are often solved. For instance, the branch-and-bound method for integer linear programming problems spends much of its time solving linear program "relaxations," in which all the variables are real. These subproblems are usually solved by the simplex method, which is discussed in Chapter 13 of this book.

CONSTRAINED AND UNCONSTRAINED OPTIMIZATION

Problems with the general form (1.1) can be classified according to the nature of the objective function and constraints (linear, nonlinear, convex), the number of variables (large or small), the smoothness of the functions (differentiable or nondifferentiable), and so on. Possibly the most important distinction is between problems that have constraints on the variables and those that do not. This book is divided into two parts according to this classification.

Unconstrained optimization problems arise directly in many practical applications. If there are natural constraints on the variables, it is sometimes safe to disregard them and to assume that they have no effect on the optimal solution. Unconstrained problems arise also as reformulations of constrained optimization problems, in which the constraints are replaced by penalization terms in the objective function that have the effect of discouraging constraint violations.

Constrained optimization problems arise from models that include explicit constraints on the variables. These constraints may be simple bounds such as $0 \leq x_1 \leq 100$, more general linear constraints such as $\sum_i x_i \leq 1$, or nonlinear inequalities that represent complex relationships among the variables.

When both the objective function and all the constraints are linear functions of x, the problem is a *linear programming* problem. Management sciences and operations research make extensive use of linear models. *Nonlinear programming* problems, in which at least some of the constraints or the objective are nonlinear functions, tend to arise naturally in the physical sciences and engineering, and are becoming more widely used in management and economic sciences.

GLOBAL AND LOCAL OPTIMIZATION

The fastest optimization algorithms seek only a local solution, a point at which the objective function is smaller than at all other feasible points in its vicinity. They do not always find the best of all such minima, that is, the *global solution*. Global solutions are necessary (or at least highly desirable) in some applications, but they are usually difficult to identify and even more difficult to locate. An important special case is *convex programming* (see below), in which all local solutions are also global solutions. Linear programming problems fall in the category of convex programming. However, general nonlinear problems, both constrained and unconstrained, may possess local solutions that are not global solutions.

In this book we treat global optimization only in passing, focusing instead on the computation and characterization of local solutions, issues that are central to the field of optimization. We note, however, that many successful global optimization algorithms proceed by solving a sequence of local optimization problems, to which the algorithms described in this book can be applied. A collection of recent research papers on global optimization can be found in Floudas and Pardalos [90].

STOCHASTIC AND DETERMINISTIC OPTIMIZATION

In some optimization problems, the model cannot be fully specified because it depends on quantities that are unknown at the time of formulation. In the transportation problem described above, for instance, the customer demands b_j at the retail outlets cannot be specified precisely in practice. This characteristic is shared by many economic and financial planning models, which often depend on the future movement of interest rates and the future behavior of the economy.

Frequently, however, modelers can predict or estimate the unknown quantities with some degree of confidence. They may, for instance, come up with a number of possible *scenarios* for the values of the unknown quantities and even assign a probability to each scenario. In the transportation problem, the manager of the retail outlet may be able to estimate demand patterns based on prior customer behavior, and there may be different scenarios for the demand that correspond to different seasonal factors or economic conditions. *Stochastic optimization* algorithms use these quantifications of the uncertainty to produce solutions that optimize the expected performance of the model.

We do not consider stochastic optimization problems further in this book, focusing instead on *deterministic optimization* problems, in which the model is fully specified. Many algorithms for stochastic optimization do, however, proceed by formulating one or more deterministic subproblems, each of which can be solved by the techniques outlined here. For further information on stochastic optimization, consult the books by Birge and Louveaux [11] and Kall and Wallace [139].

OPTIMIZATION ALGORITHMS

Optimization algorithms are iterative. They begin with an initial guess of the optimal values of the variables and generate a sequence of improved estimates until they reach a solution. The strategy used to move from one iterate to the next distinguishes one algorithm from another. Most strategies make use of the values of the objective function f, the constraints c, and possibly the first and second derivatives of these functions. Some algorithms accumulate information gathered at previous iterations, while others use only local information from the current point. Regardless of these specifics (which will receive plenty of attention in the rest of the book), all good algorithms should possess the following properties:

- Robustness. They should perform well on a wide variety of problems in their class, for all reasonable choices of the initial variables.

- Efficiency. They should not require too much computer time or storage.

- Accuracy. They should be able to identify a solution with precision, without being overly sensitive to errors in the data or to the arithmetic rounding errors that occur when the algorithm is implemented on a computer.

These goals may conflict. For example, a rapidly convergent method for nonlinear programming may require too much computer storage on large problems. On the other hand, a robust method may also be the slowest. Tradeoffs between convergence rate and storage requirements, and between robustness and speed, and so on, are central issues in numerical optimization. They receive careful consideration in this book.

The mathematical theory of optimization is used both to characterize optimal points and to provide the basis for most algorithms. It is not possible to have a good understanding of numerical optimization without a firm grasp of the supporting theory. Accordingly, this book gives a solid (though not comprehensive) treatment of optimality conditions, as well as convergence analysis that reveals the strengths and weaknesses of some of the most important algorithms.

CONVEXITY

The concept of convexity is fundamental in optimization; it implies that the problem is benign in several respects. The term *convex* can be applied both to sets and to functions.

$S \in \mathbf{R}^n$ is a *convex set* if the straight line segment connecting any two points in S lies entirely inside S. Formally, for any two points $x \in S$ and $y \in S$, we have $\alpha x + (1-\alpha) y \in S$ for all $\alpha \in [0, 1]$.

f is a *convex function* if its domain is a convex set and if for any two points x and y in this domain, the graph of f lies below the straight line connecting $(x, f(x))$ to $(y, f(y))$ in the space $\mathbf{R}^{n+1}$. That is, we have

$$f(\alpha x + (1-\alpha) y) \leq \alpha f(x) + (1-\alpha) f(y), \quad \text{for all } \alpha \in [0, 1].$$

When f is smooth as well as convex and the dimension n is 1 or 2, the graph of f is bowl-shaped (See Figure 1.3), and its contours define convex sets. A function f is said to be *concave* if $-f$ is convex.

As we will see in subsequent chapters, algorithms for unconstrained optimization are usually guaranteed to converge to a stationary point (maximizer, minimizer, or inflection point) of the objective function f. If we know that f is convex, then we can be sure that the algorithm has converged to a *global* minimizer. The term *convex programming* is used to describe a special case of the constrained optimization problem (1.1) in which

- the objective function is convex;
- the equality constraint functions $c_i(\cdot)$, $i \in \mathcal{E}$, are linear;
- the inequality constraint functions $c_i(\cdot)$, $i \in \mathcal{I}$, are concave.

As in the unconstrained case, convexity allows us to make stronger claims about the convergence of optimization algorithms than we can make for nonconvex problems.

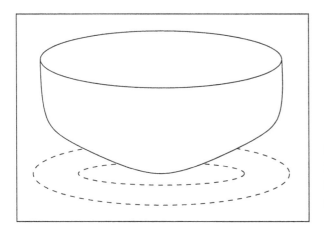

Figure 1.3
The convex function $f(x) = (x_1-6)^2 + \frac{1}{25}(x_2-4.5)^4$.

NOTES AND REFERENCES

Optimization traces its roots to the calculus of variations and the work of Euler and Lagrange. The development of linear programming in the 1940s broadened the field and stimulated much of the progress in modern optimization theory and practice during the last 50 years.

Optimization is often called *mathematical programming*, a term that is somewhat confusing because it suggests the writing of computer programs with a mathematical orientation. This term was coined in the 1940s, before the word "programming" became inextricably linked with computer software. The original meaning of this word (and the intended one in this context) was more inclusive, with connotations of problem formulation and algorithm design and analysis.

Modeling will not be treated extensively in the book. Information about modeling techniques for various application areas can be found in Dantzig [63], Ahuja, Magnanti, and Orlin [1], Fourer, Gay, and Kernighan [92], and Winston [246].

CHAPTER 2

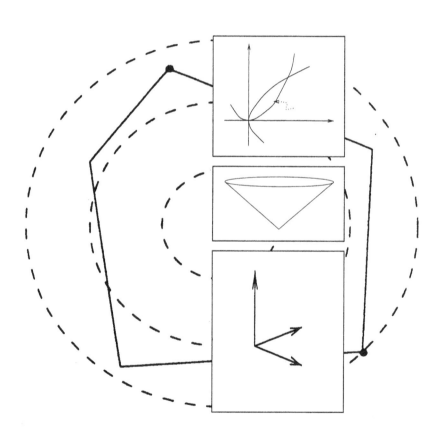

Fundamentals of Unconstrained Optimization

In unconstrained optimization, we minimize an objective function that depends on real variables, with no restrictions at all on the values of these variables. The mathematical formulation is

$$\min_x f(x), \qquad (2.1)$$

where $x \in \mathbf{R}^n$ is a real vector with $n \geq 1$ components and $f : \mathbf{R}^n \to \mathbf{R}$ is a smooth function.

Usually, we lack a global perspective on the function f. All we know are the values of f and maybe some of its derivatives at a set of points $x_0, x_1, x_2, \ldots$. Fortunately, our algorithms get to choose these points, and they try to do so in a way that identifies a solution reliably and without using too much computer time or storage. Often, the information about f does not come cheaply, so we usually prefer algorithms that do not call for this information unnecessarily.

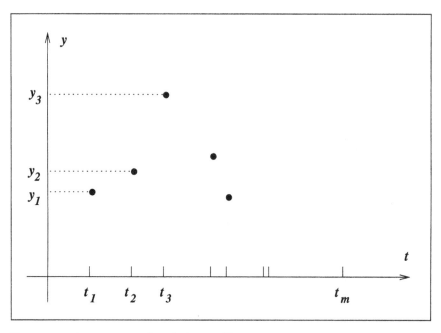

Figure 2.1 Least squares data fitting problem.

❏ EXAMPLE 2.1

Suppose that we are trying to find a curve that fits some experimental data. Figure 2.1 plots measurements $y_1, y_2, \ldots, y_m$ of a signal taken at times $t_1, t_2, \ldots, t_m$. From the data and our knowledge of the application, we deduce that the signal has exponential and oscillatory behavior of certain types, and we choose to model it by the function

$$\phi(t; x) = x_1 + x_2 e^{-(x_3-t)^2/x_4} + x_5 \cos(x_6 t).$$

The real numbers x_i, $i = 1, 2, \ldots, 6$, are the parameters of the model. We would like to choose them to make the model values $\phi(t_j; x)$ fit the observed data y_j as closely as possible. To state our objective as an optimization problem, we group the parameters x_i into a vector of unknowns $x = (x_1, x_2, \ldots, x_6)^T$, and define the residuals

$$r_j(x) = y_j - \phi(t_j; x), \qquad j = 1, \ldots, m, \tag{2.2}$$

which measure the discrepancy between the model and the observed data. Our estimate of x will be obtained by solving the problem

$$\min_{x \in \mathbb{R}^6} f(x) = r_1^2(x) + \cdots + r_m^2(x). \tag{2.3}$$

This is a *nonlinear least-squares problem*, a special case of unconstrained optimization. It illustrates that some objective functions can be expensive to evaluate even when the number of variables is small. Here we have $n = 6$, but if the number of measurements m is large (10^5, say), evaluation of $f(x)$ for a given parameter vector x is a significant computation.

□

Suppose that for the data given in Figure 2.1 the optimal solution of (2.3) is approximately $x^* = (1.1, 0.01, 1.2, 1.5, 2.0, 1.5)$ and the corresponding function value is $f(x^*) = 0.34$. Because the optimal objective is nonzero, there must be discrepancies between the observed measurements y_j and the model predictions $\phi(t_j, x^*)$ for some (usually most) values of j—the model has not reproduced all the data points exactly. How, then, can we verify that x^* is indeed a minimizer of f? To answer this question, we need to define the term "solution" and explain how to recognize solutions. Only then can we discuss algorithms for unconstrained optimization problems.

2.1 WHAT IS A SOLUTION?

Generally, we would be happiest if we found a *global minimizer* of f, a point where the function attains its least value. A formal definition is

> A point x^* is a *global minimizer* if $f(x^*) \leq f(x)$ for all x,

where x ranges over all of $\mathbf{R}^n$ (or at least over the domain of interest to the modeler). The global minimizer can be difficult to find, because our knowledge of f is usually only local. Since our algorithm does not visit many points (we hope!), we usually do not have a good picture of the overall shape of f, and we can never be sure that the function does not take a sharp dip in some region that has not been sampled by the algorithm. Most algorithms are able to find only a *local* minimizer, which is a point that achieves the smallest value of f in its neighborhood. Formally, we say:

> A point x^* is a *local minimizer* if there is a neighborhood $\mathcal{N}$ of x^* such that $f(x^*) \leq f(x)$ for $x \in \mathcal{N}$.

(Recall that a neighborhood of x^* is simply an open set that contains x^*.) A point that satisfies this definition is sometimes called a *weak local minimizer*. This terminology distinguishes it from a strict local minimizer, which is the outright winner in its neighborhood. Formally,

> A point x^* is a *strict local minimizer* (also called a *strong local minimizer*) if there is a neighborhood $\mathcal{N}$ of x^* such that $f(x^*) < f(x)$ for all $x \in \mathcal{N}$ with $x \neq x^*$.

For the constant function $f(x) = 2$, every point x is a weak local minimizer, while the function $f(x) = (x - 2)^4$ has a strict local minimizer at $x = 2$.

A slightly more exotic type of local minimizer is defined as follows.

A point x^* is an *isolated local minimizer* if there is a neighborhood $\mathcal{N}$ of x^* such that x^* is the only local minimizer in $\mathcal{N}$.

Some strict local minimizers are not isolated, as illustrated by the function

$$f(x) = x^4 \cos(1/x) + 2x^4, \qquad f(0) = 0,$$

which is twice continuously differentiable and has a strict local minimizer at $x^* = 0$. However, there are strict local minimizers at many nearby points x_n, and we can label these points so that $x_n \to 0$ as $n \to \infty$.

While strict local minimizers are not always isolated, it is true that all isolated local minimizers are strict.

Figure 2.2 illustrates a function with many local minimizers. It is usually difficult to find the global minimizer for such functions, because algorithms tend to be "trapped" at the local minimizers. This example is by no means pathological. In optimization problems associated with the determination of molecular conformation, the potential function to be minimized may have millions of local minima.

Sometimes we have additional "global" knowledge about f that may help in identifying global minima. An important special case is that of convex functions, for which every local minimizer is also a global minimizer.

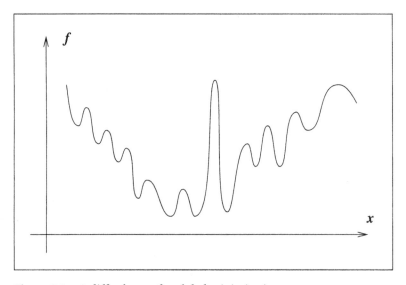

Figure 2.2 A difficult case for global minimization.

RECOGNIZING A LOCAL MINIMUM

From the definitions given above, it might seem that the only way to find out whether a point x^* is a local minimum is to examine all the points in its immediate vicinity, to make sure that none of them has a smaller function value. When the function f is *smooth*, however, there are much more efficient and practical ways to identify local minima. In particular, if f is twice continuously differentiable, we may be able to tell that x^* is a local minimizer (and possibly a strict local minimizer) by examining just the gradient $\nabla f(x^*)$ and the Hessian $\nabla^2 f(x^*)$.

The mathematical tool used to study minimizers of smooth functions is Taylor's theorem. Because this theorem is central to our analysis throughout the book, we state it now. Its proof can be found in any calculus textbook.

Theorem 2.1 (Taylor's Theorem).
    Suppose that $f : \mathbf{R}^n \to \mathbf{R}$ is continuously differentiable and that $p \in \mathbf{R}^n$. Then we have that

$$f(x+p) = f(x) + \nabla f(x+tp)^T p, \tag{2.4}$$

for some $t \in (0, 1)$. Moreover, if f is twice continuously differentiable, we have that

$$\nabla f(x+p) = \nabla f(x) + \int_0^1 \nabla^2 f(x+tp) p \, dt, \tag{2.5}$$

and that

$$f(x+p) = f(x) + \nabla f(x)^T p + \tfrac{1}{2} p^T \nabla^2 f(x+tp) p, \tag{2.6}$$

for some $t \in (0, 1)$.

Necessary conditions for optimality are derived by assuming that x^* is a local minimizer and then proving facts about $\nabla f(x^*)$ and $\nabla^2 f(x^*)$.

Theorem 2.2 (First-Order Necessary Conditions).
    If x^* is a local minimizer and f is continuously differentiable in an open neighborhood of x^*, then $\nabla f(x^*) = 0$.

PROOF. Suppose for contradiction that $\nabla f(x^*) \neq 0$. Define the vector $p = -\nabla f(x^*)$ and note that $p^T \nabla f(x^*) = -\|\nabla f(x^*)\|^2 < 0$. Because ∇f is continuous near x^*, there is a scalar $T > 0$ such that

$$p^T \nabla f(x^* + tp) < 0, \qquad \text{for all } t \in [0, T].$$

For any $\bar{t} \in (0, T]$, we have by Taylor's theorem that

$$f(x^* + \bar{t}p) = f(x^*) + \bar{t}p^T \nabla f(x^* + tp), \quad \text{for some } t \in (0, \bar{t}).$$

Therefore, $f(x^* + \bar{t}p) < f(x^*)$ for all $\bar{t} \in (0, T]$. We have found a direction leading away from x^* along which f decreases, so x^* is not a local minimizer, and we have a contradiction. □

We call x^* a *stationary point* if $\nabla f(x^*) = 0$. According to Theorem 2.2, any local minimizer must be a stationary point.

For the next result we recall that a matrix B is positive definite if $p^T B p > 0$ for all $p \neq 0$, and positive semidefinite if $p^T B p \geq 0$ for all p (see the Appendix).

Theorem 2.3 (Second-Order Necessary Conditions).
If x^* is a local minimizer of f and $\nabla^2 f$ is continuous in an open neighborhood of x^*, then $\nabla f(x^*) = 0$ and $\nabla^2 f(x^*)$ is positive semidefinite.

PROOF. We know from Theorem 2.2 that $\nabla f(x^*) = 0$. For contradiction, assume that $\nabla^2 f(x^*)$ is not positive semidefinite. Then we can choose a vector p such that $p^T \nabla^2 f(x^*) p < 0$, and because $\nabla^2 f$ is continuous near x^*, there is a scalar $T > 0$ such that $p^T \nabla^2 f(x^* + tp) p < 0$ for all $t \in [0, T]$.

By doing a Taylor series expansion around x^*, we have for all $\bar{t} \in (0, T]$ and some $t \in (0, \bar{t})$ that

$$f(x^* + \bar{t}p) = f(x^*) + \bar{t}p^T \nabla f(x^*) + \tfrac{1}{2}\bar{t}^2 p^T \nabla^2 f(x^* + tp) p < f(x^*).$$

As in Theorem 2.2, we have found a direction from x^* along which f is decreasing, and so again, x^* is not a local minimizer. □

We now describe *sufficient conditions*, which are conditions on the derivatives of f at the point z^* that guarantee that x^* is a local minimizer.

Theorem 2.4 (Second-Order Sufficient Conditions).
Suppose that $\nabla^2 f$ is continuous in an open neighborhood of x^* and that $\nabla f(x^*) = 0$ and $\nabla^2 f(x^*)$ is positive definite. Then x^* is a strict local minimizer of f.

PROOF. Because the Hessian is continuous and positive definite at x^*, we can choose a radius $r > 0$ so that $\nabla^2 f(x)$ remains positive definite for all x in the open ball $\mathcal{D} = \{z \mid \|z - x^*\| < r\}$. Taking any nonzero vector p with $\|p\| < r$, we have $x^* + p \in \mathcal{D}$ and so

$$f(x^* + p) = f(x^*) + p^T \nabla f(x^*) + \tfrac{1}{2} p^T \nabla^2 f(z) p$$
$$= f(x^*) + \tfrac{1}{2} p^T \nabla^2 f(z) p,$$

where $z = x^* + tp$ for some $t \in (0, 1)$. Since $z \in \mathcal{D}$, we have $p^T \nabla^2 f(z) p > 0$, and therefore $f(x^* + p) > f(x^*)$, giving the result. □

Note that the second-order sufficient conditions of Theorem 2.4 guarantee something stronger than the necessary conditions discussed earlier; namely, that the minimizer is a *strict* local minimizer. Note too that the second-order sufficient conditions are not necessary: A point x^* may be a strict local minimizer, and yet may fail to satisfy the sufficient conditions. A simple example is given by the function $f(x) = x^4$, for which the point $x^* = 0$ is a strict local minimizer at which the Hessian matrix vanishes (and is therefore not positive definite).

When the objective function is convex, local and global minimizers are simple to characterize.

Theorem 2.5.
When f is convex, any local minimizer x^* is a global minimizer of f. If in addition f is differentiable, then any stationary point x^* is a global minimizer of f.

PROOF. Suppose that x^* is a local but not a global minimizer. Then we can find a point $z \in \mathbb{R}^n$ with $f(z) < f(x^*)$. Consider the line segment that joins x^* to z, that is,

$$x = \lambda z + (1 - \lambda) x^*, \quad \text{for some } \lambda \in (0, 1]. \tag{2.7}$$

By the convexity property for f, we have

$$f(x) \leq \lambda f(z) + (1 - \lambda) f(x^*) < f(x^*). \tag{2.8}$$

Any neighborhood $\mathcal{N}$ of x^* contains a piece of the line segment (2.7), so there will always be points $x \in \mathcal{N}$ at which (2.8) is satisfied. Hence, x^* is not a local minimizer.

For the second part of the theorem, suppose that x^* is not a global minimizer and choose z as above. Then, from convexity, we have

$$\nabla f(x^*)^T (z - x^*) = \frac{d}{d\lambda} f(x^* + \lambda(z - x^*))|_{\lambda=0} \quad \text{(see the Appendix)}$$
$$= \lim_{\lambda \downarrow 0} \frac{f(x^* + \lambda(z - x^*)) - f(x^*)}{\lambda}$$
$$\leq \lim_{\lambda \downarrow 0} \frac{\lambda f(z) + (1 - \lambda) f(x^*) - f(x^*)}{\lambda}$$
$$= f(z) - f(x^*) < 0.$$

Therefore, $\nabla f(x^*) \neq 0$, and so x^* is not a stationary point. □

These results, which are based on elementary calculus, provide the foundations for unconstrained optimization algorithms. In one way or another, all algorithms seek a point where $\nabla f(\cdot)$ vanishes.

CHAPTER 2. FUNDAMENTALS OF UNCONSTRAINED OPTIMIZATION

NONSMOOTH PROBLEMS

This book focuses on smooth functions, by which we generally mean functions whose second derivatives exist and are continuous. We note, however, that there are interesting problems in which the functions involved may be nonsmooth and even discontinuous. It is not possible in general to identify a minimizer of a general discontinuous function. If, however, the function consists of a few smooth pieces, with discontinuities between the pieces, it may be possible to find the minimizer by minimizing each smooth piece individually.

If the function is continuous everywhere but nondifferentiable at certain points, as in Figure 2.3, we can identify a solution by examining the *subgradient*, or *generalized gradient*, which is a generalization of the concept of gradient to the nonsmooth case. Nonsmooth optimization is beyond the scope of this book; we refer instead to Hiriart-Urruty and Lemaréchal [137] for an extensive discussion of theory. Here, we mention only that the minimization of a function such as the one illustrated in Figure 2.3 (which contains a jump discontinuity in the first derivative $f'(x)$ at the minimum) is difficult because the behavior of f is not predictable near the point of nonsmoothness. That is, we cannot be sure that information about f obtained at one point can be used to infer anything about f at neighboring points, because points of nondifferentiability may intervene. However, certain special nondifferentiable functions, such as functions of the form

$$f(x) = \|r(x)\|_1, \qquad f(x) = \|r(x)\|_\infty$$

(where $r(x)$ is the residual vector refined in (2.2)), can be solved with the help of special-purpose algorithms; see, for example, Fletcher [83, Chapter 14].

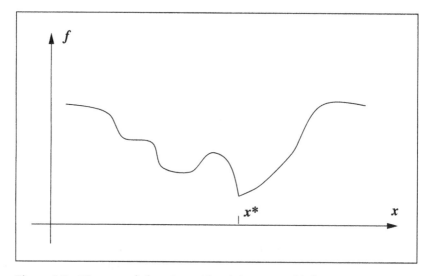

Figure 2.3 Nonsmooth function with minimum at a kink.

2.2 OVERVIEW OF ALGORITHMS

The last thirty years has seen the development of a powerful collection of algorithms for unconstrained optimization of smooth functions. We now give a broad description of their main properties, and we describe them in more detail in Chapters 3, 4, 5, 6, 8, and 9. All algorithms for unconstrained minimization require the user to supply a starting point, which we usually denote by x_0. The user with knowledge about the application and the data set may be in a good position to choose x_0 to be a reasonable estimate of the solution. Otherwise, the starting point must be chosen in some arbitrary manner.

Beginning at x_0, optimization algorithms generate a sequence of iterates $\{x_k\}_{k=0}^{\infty}$ that terminate when either no more progress can be made or when it seems that a solution point has been approximated with sufficient accuracy. In deciding how to move from one iterate x_k to the next, the algorithms use information about the function f at x_k, and possibly also information from earlier iterates $x_0, x_1, \ldots, x_{k-1}$. They use this information to find a new iterate x_{k+1} with a lower function value than x_k. (There exist *nonmonotone* algorithms that do not insist on a decrease in f at every step, but even these algorithms require f to be decreased after some prescribed number m of iterations. That is, they enforce $f(x_k) < f(x_{k-m})$.)

There are two fundamental strategies for moving from the current point x_k to a new iterate x_{k+1}. Most of the algorithms described in this book follow one of these approaches.

TWO STRATEGIES: LINE SEARCH AND TRUST REGION

In the *line search* strategy, the algorithm chooses a direction p_k and searches along this direction from the current iterate x_k for a new iterate with a lower function value. The distance to move along p_k can be found by approximately solving the following one-dimensional minimization problem to find a step length α:

$$\min_{\alpha>0} f(x_k + \alpha p_k). \tag{2.9}$$

By solving (2.9) exactly, we would derive the maximum benefit from the direction p_k, but an exact minimization is expensive and unnecessary. Instead, the line search algorithm generates a limited number of trial step lengths until it finds one that loosely approximates the minimum of (2.9). At the new point a new search direction and step length are computed, and the process is repeated.

In the second algorithmic strategy, known as *trust region*, the information gathered about f is used to construct a *model function* m_k whose behavior near the current point x_k is similar to that of the actual objective function f. Because the model m_k may not be a good approximation of f when x is far from x_k, we restrict the search for a minimizer of m_k to some region around x_k. In other words, we find the candidate step p by approximately solving the following subproblem:

$$\min_p m_k(x_k + p), \quad \text{where } x_k + p \text{ lies inside the trust region.} \tag{2.10}$$

If the candidate solution does not produce a sufficient decrease in f, we conclude that the trust region is too large, and we shrink it and re-solve (2.10). Usually, the trust region is a ball defined by $\|p\|_2 \leq \Delta$, where the scalar $\Delta > 0$ is called the trust-region radius. Elliptical and box-shaped trust regions may also be used.

The model m_k in (2.10) is usually defined to be a quadratic function of the form

$$m_k(x_k + p) = f_k + p^T \nabla f_k + \tfrac{1}{2} p^T B_k p, \tag{2.11}$$

where f_k, ∇f_k, and B_k are a scalar, vector, and matrix, respectively. As the notation indicates, f_k and ∇f_k are chosen to be the function and gradient values at the point x_k, so that m_k and f are in agreement to first order at the current iterate x_k. The matrix B_k is either the Hessian $\nabla^2 f_k$ or some approximation to it.

Suppose that the objective function is given by $f(x) = 10(x_2 - x_1^2)^2 + (1 - x_1)^2$. At the point $x_k = (0, 1)$ its gradient and Hessian are

$$\nabla f_k = \begin{bmatrix} -2 \\ 20 \end{bmatrix}, \quad \nabla^2 f_k = \begin{bmatrix} -38 & 0 \\ 0 & 20 \end{bmatrix}.$$

The contour lines of the quadratic model (2.11) with $B_k = \nabla^2 f_k$ are depicted in Figure 2.4, which also illustrates the contours of the objective function f and the trust region. We have indicated contour lines where the model m_k has values 1 and 12. Note from Figure 2.4 that each time we decrease the size of the trust region after failure of a candidate iterate, the step from x_k to the new candidate will be shorter, and it usually points in a different direction from the previous candidate. The trust-region strategy differs in this respect from line search, which stays with a single search direction.

In a sense, the line search and trust-region approaches differ in the order in which they choose the *direction* and *distance* of the move to the next iterate. Line search starts by fixing the direction p_k and then identifying an appropriate distance, namely the step length α_k. In trust region, we first choose a maximum distance—the trust-region radius Δ_k—and then seek a direction and step that attain the best improvement possible subject to this distance constraint. If this step proves to be unsatisfactory, we reduce the distance measure Δ_k and try again.

The line search approach is discussed in more detail in Chapter 3. Chapter 4 discusses the trust-region strategy, including techniques for choosing and adjusting the size of the region and for computing approximate solutions to the trust-region problems (2.10). We now preview two major issues: choice of the search direction p_k in line search methods, and choice of the Hessian B_k in trust-region methods. These issues are closely related, as we now observe.

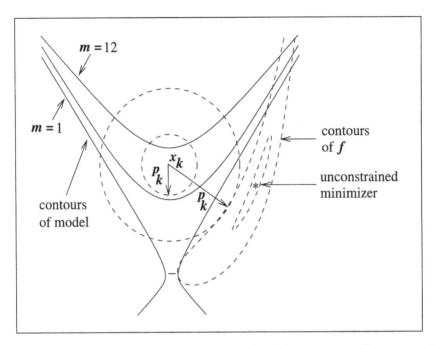

Figure 2.4 Two possible trust regions (circles) and their corresponding steps p_k. The solid lines are contours of the model function m_k.

SEARCH DIRECTIONS FOR LINE SEARCH METHODS

The steepest-descent direction $-\nabla f_k$ is the most obvious choice for search direction for a line search method. It is intuitive; among all the directions we could move from x_k, it is the one along which f decreases most rapidly. To verify this claim, we appeal again to Taylor's theorem (Theorem 2.1), which tells us that for any search direction p and step-length parameter α, we have

$$f(x_k + \alpha p) = f(x_k) + \alpha p^T \nabla f_k + \tfrac{1}{2}\alpha^2 p^T \nabla^2 f(x_k + tp)p, \quad \text{for some } t \in (0, \alpha)$$

(see (2.6)). The rate of change in f along the direction p at x_k is simply the coefficient of α, namely, $p^T \nabla f_k$. Hence, the unit direction p of most rapid decrease is the solution to the problem

$$\min_p \; p^T \nabla f_k, \quad \text{subject to } \|p\| = 1. \tag{2.12}$$

Since $p^T \nabla f_k = \|p\| \|\nabla f_k\| \cos\theta$, where θ is the angle between p and ∇f_k, we have from $\|p\| = 1$ that $p^T \nabla f_k = \|\nabla f_k\| \cos\theta$, so the objective in (2.12) is minimized when $\cos\theta$

22 Chapter 2. Fundamentals of Unconstrained Optimization

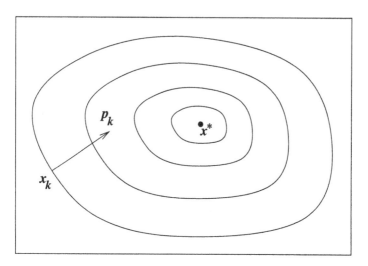

Figure 2.5 Steepest descent direction for a function of two variables.

takes on its minimum value of -1 at $\theta = \pi$ radians. In other words, the solution to (2.12) is

$$p = -\nabla f_k / \|\nabla f_k\|,$$

as claimed. As we show in Figure 2.5, this direction is orthogonal to the contours of the function.

The *steepest descent method* is a line search method that moves along $p_k = -\nabla f_k$ at every step. It can choose the step length α_k in a variety of ways, as we discuss in Chapter 3. One advantage of the steepest descent direction is that it requires calculation of the gradient ∇f_k but not of second derivatives. However, it can be excruciatingly slow on difficult problems.

Line search methods may use search directions other than the steepest descent direction. In general, any *descent* direction—one that makes an angle of strictly less than $\pi/2$ radians with $-\nabla f_k$—is guaranteed to produce a decrease in f, provided that the step length is sufficiently small (see Figure 2.6). We can verify this claim by using Taylor's theorem. From (2.6), we have that

$$f(x_k + \epsilon p_k) = f(x_k) + \epsilon p_k^T \nabla f_k + O(\epsilon^2).$$

When p_k is a downhill direction, the angle θ_k between p_k and ∇f_k has $\cos \theta_k < 0$, so that

$$p_k^T \nabla f_k = \|p_k\| \, \|\nabla f_k\| \cos \theta_k < 0.$$

It follows that $f(x_k + \epsilon p_k) < f(x_k)$ for all positive but sufficiently small values of ϵ.

Another important search direction—perhaps the most important one of all—is the *Newton direction*. This direction is derived from the second-order Taylor series

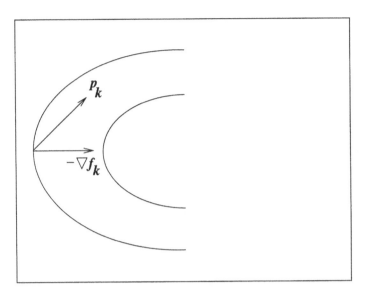

Figure 2.6 A downhill direction p_k

approximation to $f(x_k + p)$, which is

$$f(x_k + p) \approx f_k + p^T \nabla f_k + \tfrac{1}{2} p^T \nabla^2 f_k p \stackrel{\text{def}}{=} m_k(p). \tag{2.13}$$

Assuming for the moment that $\nabla^2 f_k$ is positive definite, we obtain the Newton direction by finding the vector p that minimizes $m_k(p)$. By simply setting the derivative of $m_k(p)$ to zero, we obtain the following explicit formula:

$$p_k^N = -\nabla^2 f_k^{-1} \nabla f_k. \tag{2.14}$$

The Newton direction is reliable when the difference between the true function $f(x_k + p)$ and its quadratic model $m_k(p)$ is not too large. By comparing (2.13) with (2.6), we see that the only difference between these functions is that the matrix $\nabla^2 f(x_k + tp)$ in the third term of the expansion has been replaced by $\nabla^2 f_k = \nabla^2 f(x_k)$. If $\nabla^2 f(\cdot)$ is sufficiently smooth, this difference introduces a perturbation of only $O(\|p\|^3)$ into the expansion, so that when $\|p\|$ is small, the approximation $f(x_k + p) \approx m_k(p)$ is very accurate indeed.

The Newton direction can be used in a line search method when $\nabla^2 f_k$ is positive definite, for in this case we have

$$\nabla f_k^T p_k^N = -p_k^{N^T} \nabla^2 f_k p_k^N \leq -\sigma_k \|p_k^N\|^2$$

for some $\sigma_k > 0$. Unless the gradient ∇f_k (and therefore the step p_k^N) is zero, we have that $\nabla f_k^T p_k^N < 0$, so the Newton direction is a descent direction. Unlike the steepest descent direction, there is a "natural" step length of 1 associated with the Newton direction. Most line search implementations of Newton's method use the unit step $\alpha = 1$ where possible and adjust this step length only when it does not produce a satisfactory reduction in the value of f.

When $\nabla^2 f_k$ is not positive definite, the Newton direction may not even be defined, since $\nabla^2 f_k^{-1}$ may not exist. Even when it *is* defined, it may not satisfy the descent property $\nabla f_k^T p_k^N < 0$, in which case it is unsuitable as a search direction. In these situations, line search methods modify the definition of p_k to make it satisfy the downhill condition while retaining the benefit of the second-order information contained in $\nabla^2 f_k$. We will describe these modifications in Chapter 6.

Methods that use the Newton direction have a fast rate of local convergence, typically quadratic. When a neighborhood of the solution is reached, convergence to high accuracy often occurs in just a few iterations. The main drawback of the Newton direction is the need for the Hessian $\nabla^2 f(x)$. Explicit computation of this matrix of second derivatives is sometimes, though not always, a cumbersome, error-prone, and expensive process.

Quasi-Newton search directions provide an attractive alternative in that they do not require computation of the Hessian and yet still attain a superlinear rate of convergence. In place of the true Hessian $\nabla^2 f_k$, they use an approximation B_k, which is updated after each step to take account of the additional knowledge gained during the step. The updates make use of the fact that changes in the gradient g provide information about the second derivative of f along the search direction. By using the expression (2.5) from our statement of Taylor's theorem, we have by adding and subtracting the term $\nabla^2 f(x) p$ that

$$\nabla f(x+p) = \nabla f(x) + \nabla^2 f(x) p + \int_0^1 \left[\nabla^2 f(x+tp) - \nabla^2 f(x)\right] p \, dt.$$

Because $\nabla f(\cdot)$ is continuous, the size of the final integral term is $o(\|p\|)$. By setting $x = x_k$ and $p = x_{k+1} - x_k$, we obtain

$$\nabla f_{k+1} = \nabla f_k + \nabla^2 f_{k+1}(x_{k+1} - x_k) + o(\|x_{k+1} - x_k\|).$$

When x_k and x_{k+1} lie in a region near the solution x^*, within which ∇f is positive definite, the final term in this expansion is eventually dominated by the $\nabla^2 f_k(x_{k+1} - x_k)$ term, and we can write

$$\nabla^2 f_{k+1}(x_{k+1} - x_k) \approx \nabla f_{k+1} - \nabla f_k. \tag{2.15}$$

We choose the new Hessian approximation B_{k+1} so that it mimics this property (2.15) of the true Hessian, that is, we require it to satisfy the following condition, known as the *secant equation*:

$$B_{k+1} s_k = y_k, \tag{2.16}$$

where

$$s_k = x_{k+1} - x_k, \qquad y_k = \nabla f_{k+1} - \nabla f_k.$$

Typically, we impose additional requirements on B_{k+1}, such as symmetry (motivated by symmetry of the exact Hessian), and a restriction that the difference between successive approximation B_k to B_{k+1} have low rank. The initial approximation B_0 must be chosen by the user.

Two of the most popular formulae for updating the Hessian approximation B_k are the *symmetric-rank-one* (SR1) formula, defined by

$$B_{k+1} = B_k + \frac{(y_k - B_k s_k)(y_k - B_k s_k)^T}{(y_k - B_k s_k)^T s_k}, \qquad (2.17)$$

and the *BFGS formula*, named after its inventors, Broyden, Fletcher, Goldfarb, and Shanno, which is defined by

$$B_{k+1} = B_k - \frac{B_k s_k s_k^T B_k}{s_k^T B_k s_k} + \frac{y_k y_k^T}{y_k^T s_k}. \qquad (2.18)$$

Note that the difference between the matrices B_k and B_{k+1} is a rank-one matrix in the case of (2.17), and a rank-two matrix in the case of (2.18). Both updates satisfy the secant equation and both maintain symmetry. One can show that BFGS update (2.18) generates positive definite approximations whenever the initial approximation B_0 is positive definite and $s_k^T y_k > 0$. We discuss these issues further in Chapter 8.

The quasi-Newton search direction is given by using B_k in place of the exact Hessian in the formula (2.14), that is,

$$p_k = -B_k^{-1} \nabla f_k. \qquad (2.19)$$

Some practical implementations of quasi-Newton methods avoid the need to factorize B_k at each iteration by updating the *inverse* of B_k, instead of B_k itself. In fact, the equivalent formula for (2.17) and (2.18), applied to the inverse approximation $H_k \stackrel{\text{def}}{=} B_k^{-1}$, is

$$H_{k+1} = \left(I - \rho_k s_k y_k^T\right) H_k \left(I - \rho_k y_k s_k^T\right) + \rho_k s_k s_k^T, \qquad \rho_k = \frac{1}{y_k^T s_k}. \qquad (2.20)$$

Calculation of p_k can then be performed by using the formula $p_k = -H_k \nabla f_k$. This can be implemented as a matrix–vector multiplication, which is typically simpler than the factorization/back-substitution procedure that is needed to implement the formula (2.19).

Two variants of quasi-Newton methods designed to solve large problems—partially separable and limited-memory updating—are described in Chapter 9.

The last class of search directions we preview here is that generated by *nonlinear conjugate gradient methods*. They have the form

$$p_k = -\nabla f(x_k) + \beta_k p_{k-1},$$

where β_k is a scalar that ensures that p_k and p_{k-1} are *conjugate*—an important concept in the minimization of quadratic functions that will be defined in Chapter 5. Conjugate gradient methods were originally designed to solve systems of linear equations $Ax = b$, where the coefficient matrix A is symmetric and positive definite. The problem of solving this linear system is equivalent to the problem of minimizing the convex quadratic function defined by

$$\phi(x) = \tfrac{1}{2} x^T A x + b^T x,$$

so it was natural to investigate extensions of these algorithms to more general types of unconstrained minimization problems. In general, nonlinear conjugate gradient directions are much more effective than the steepest descent direction and are almost as simple to compute. These methods do not attain the fast convergence rates of Newton or quasi-Newton methods, but they have the advantage of not requiring storage of matrices. An extensive discussion of nonlinear conjugate gradient methods is given in Chapter 5.

All of the search directions discussed so far can be used directly in a line search framework. They give rise to the steepest descent, Newton, quasi-Newton, and conjugate gradient line search methods. All except conjugate gradients have an analogue in the trust-region framework, as we now discuss.

MODELS FOR TRUST-REGION METHODS

If we set $B_k = 0$ in (2.11) and define the trust region using the Euclidean norm, the trust-region subproblem (2.10) becomes

$$\min_p \; f_k + p^T \nabla f_k \quad \text{subject to } \|p\|_2 \leq \Delta_k.$$

We can write the solution to this problem in closed form as

$$p_k = -\frac{\Delta_k \nabla f_k}{\|\nabla f_k\|}.$$

This is simply a steepest descent step in which the step length is determined by the trust-region radius; the trust-region and line search approaches are essentially the same in this case.

A more interesting trust-region algorithm is obtained by choosing B_k to be the exact Hessian $\nabla^2 f_k$ in the quadratic model (2.11). Because of the trust-region restriction $\|p\|_2 \leq \Delta_k$, there is no need to do anything special when $\nabla^2 f_k$ is not positive definite, since the subproblem (2.10) is guaranteed to have a solution p_k, as we see in Figure 2.4. The trust-region Newton method has proved to be highly effective in practice, as we discuss in Chapter 6.

If the matrix B_k in the quadratic model function m_k of (2.11) is defined by means of a quasi-Newton approximation, we obtain a trust-region quasi-Newton method.

SCALING

The performance of an algorithm may depend crucially on how the problem is formulated. One important issue in problem formulation is *scaling*. In unconstrained optimization, a problem is said to be *poorly scaled* if changes to x in a certain direction produce much larger variations in the value of f than do changes to x in another direction. A simple example is provided by the function $f(x) = 10^9 x_1^2 + x_2^2$, which is very sensitive to small changes in x_1 but not so sensitive to perturbations in x_2.

Poorly scaled functions arise, for example, in simulations of physical and chemical systems where different processes are taking place at very different rates. To be more specific, consider a chemical system in which four reactions occur. Associated with each reaction is a *rate constant* that describes the speed at which the reaction takes place. The optimization problem is to find values for these rate constants by observing the concentrations of each chemical in the system at different times. The four constants differ greatly in magnitude, since the reactions take place at vastly different speeds. Suppose we have the following rough estimates for the final values of the constants, each correct to within, say, an order of magnitude:

$$x_1 \approx 10^{-10}, \quad x_2 \approx x_3 \approx 1, \quad x_4 \approx 10^5.$$

Before solving this problem we could introduce a new variable z defined by

$$\begin{bmatrix} x_1 \\ x_2 \\ x_3 \\ x_4 \end{bmatrix} = \begin{bmatrix} 10^{-10} & 0 & 0 & 0 \\ 0 & 1 & 0 & 0 \\ 0 & 0 & 1 & 0 \\ 0 & 0 & 0 & 10^5 \end{bmatrix} \begin{bmatrix} z_1 \\ z_2 \\ z_3 \\ z_4 \end{bmatrix},$$

and then define and solve the optimization problem in terms of the new variable z. The optimal values of z will be within about an order of magnitude of 1, making the solution more balanced. This kind of scaling of the variables is known as *diagonal scaling*.

Scaling is performed (sometimes unintentionally) when the units used to represent variables are changed. During the modeling process, we may decide to change the units of some variables, say from meters to millimeters. If we do, the range of those variables and their size relative to the other variables will both change.

Some optimization algorithms, such as steepest descent, are sensitive to poor scaling, while others, such as Newton's method, are unaffected by it. Figure 2.7 shows the contours

28 CHAPTER 2. FUNDAMENTALS OF UNCONSTRAINED OPTIMIZATION

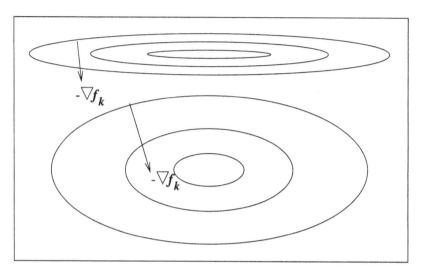

Figure 2.7 Poorly scaled and well-scaled problems, and performance of the steepest descent direction.

of two convex nearly quadratic functions, the first of which is poorly scaled, while the second is well scaled. For the poorly scaled problem, the one with highly elongated contours, the steepest descent direction (also shown on the graph) does not yield much reduction in the function, while for the well-scaled problem it performs much better. In both cases, Newton's method will produce a much better step, since the second-order quadratic model (m_k in (2.13)) happens to be a good approximation of f.

Algorithms that are not sensitive to scaling are preferable to those that are not, because they can handle poor problem formulations in a more robust fashion. In designing complete algorithms, we try to incorporate *scale invariance* into all aspects of the algorithm, including the line search or trust-region strategies and convergence tests. Generally speaking, it is easier to preserve scale invariance for line search algorithms than for trust-region algorithms.

RATES OF CONVERGENCE

One of the key measures of performance of an algorithm is its rate of convergence. We now define the terminology associated with different types of convergence, for reference in later chapters.

Let $\{x_k\}$ be a sequence in $\mathbf{R}^n$ that converges to x^*. We say that the convergence is *Q-linear* if there is a constant $r \in (0, 1)$ such that

$$\frac{\|x_{k+1} - x^*\|}{\|x_k - x^*\|} \leq r, \quad \text{for all } k \text{ sufficiently large.} \tag{2.21}$$

This means that the distance to the solution x^* decreases at each iteration by at least a constant factor. For example, the sequence $1 + (0.5)^k$ converges Q-linearly to 1. The prefix "Q" stands for "quotient," because this type of convergence is defined in terms of the quotient of successive errors.

The convergence is said to be *Q-superlinear* if

$$\lim_{k \to \infty} \frac{\|x_{k+1} - x^*\|}{\|x_k - x^*\|} = 0.$$

For example, the sequence $1 + k^{-k}$ converges superlinearly to 1. (Prove this statement!) *Q-quadratic* convergence, an even more rapid convergence rate, is obtained if

$$\frac{\|x_{k+1} - x^*\|}{\|x_k - x^*\|^2} \leq M, \quad \text{for all } k \text{ sufficiently large},$$

where M is a positive constant, not necessarily less than 1. An example is the sequence $1 + (0.5)^{2^k}$.

The speed of convergence depends on r and (more weakly) on M, whose values depend not only on the algorithm but also on the properties of the particular problem. Regardless of these values, however, a quadratically convergent sequence will always eventually converge faster than a linearly convergent sequence.

Obviously, any sequence that converges Q-quadratically also converges Q-superlinearly, and any sequence that converges Q-superlinearly also converges Q-linearly. We can also define higher rates of convergence (cubic, quartic, and so on), but these are less interesting in practical terms. In general, we say that the Q-order of convergence is p (with $p > 1$) if there is a positive constant M such that

$$\frac{\|x_{k+1} - x^*\|}{\|x_k - x^*\|^p} \leq M, \quad \text{for all } k \text{ sufficiently large}.$$

Quasi-Newton methods typically converge Q-superlinearly, whereas Newton's method converges Q-quadratically. In contrast, steepest descent algorithms converge only at a Q-linear rate, and when the problem is ill-conditioned the convergence constant r in (2.21) is close to 1.

Throughout the book we will normally omit the letter Q and simply talk about superlinear convergence, quadratic convergence, etc.

R-RATES OF CONVERGENCE

A slightly weaker form of convergence, characterized by the prefix "R" (for "root"), is concerned with the overall rate of decrease in the error, rather than the decrease over a single step of the algorithm. We say that convergence is *R-linear* if there is a sequence of nonnegative scalars $\{v_k\}$ such that

$$\|x_k - x^*\| \le v_k \text{ for all } k, \text{ and } \{v_k\} \text{ converges Q-linearly to zero.}$$

The sequence $\{\|x_k - x^*\|\}$ is said to be *dominated* by $\{v_k\}$. For instance, the sequence

$$x_k = \begin{cases} 1 + (0.5)^k, & k \text{ even,} \\ 1, & k \text{ odd,} \end{cases} \tag{2.22}$$

(the first few iterates are 2, 1, 1.25, 1, 1.03125, 1, ...) converges R-linearly to 1, because it is dominated by the sequence $1 + (0.5)^k$, which converges Q-linearly. Likewise, we say that $\{x_k\}$ converges R-superlinearly to x^* if $\{\|x_k - x^*\|\}$ is dominated by a Q-superlinear sequence, and $\{x_k\}$ converges R-quadratically to x^* if $\{\|x_k - x^*\|\}$ is dominated by a Q-quadratic sequence.

Note that in the R-linear sequence (2.22), the error actually increases at every second iteration! Such behavior occurs even in sequences whose R-rate of convergence is arbitrarily high, but it cannot occur for Q-linear sequences, which insist on a decrease at every step k, for k sufficiently large.

Most convergence analyses of optimization algorithms are concerned with Q-convergence.

NOTES AND REFERENCES

For an extensive discussion of convergence rates see Ortega and Rheinboldt [185].

EXERCISES

2.1 Compute the gradient $\nabla f(x)$ and Hessian $\nabla^2 f(x)$ of the Rosenbrock function

$$f(x) = 100(x_2 - x_1^2)^2 + (1 - x_1)^2. \tag{2.23}$$

Show that $x^* = (1, 1)^T$ is the only local minimizer of this function, and that the Hessian matrix at that point is positive definite.

2.2 Show that the function $f(x) = 8x_1 + 12x_2 + x_1^2 - 2x_2^2$ has only one stationary point, and that it is neither a maximum or minimum, but a saddle point. Sketch the contour lines of f.

2.3 Let a be a given n-vector, and A be a given $n \times n$ symmetric matrix. Compute the gradient and Hessian of $f_1(x) = a^T x$ and $f_2(x) = x^T A x$.

2.4 Write the second-order Taylor expansion (2.6) for the function $\cos(1/x)$ around a nonzero point x, and the third-order Taylor expansion of $\cos(x)$ around any point x. Evaluate the second expansion for the specific case of $x = 1$.

◈ **2.5** Consider the function $f : \mathbf{R}^2 \to \mathbf{R}$ defined by $f(x) = \|x\|^2$. Show that the sequence of iterates $\{x_k\}$ defined by

$$x_k = \left(1 + \frac{1}{2^k}\right) \begin{bmatrix} \cos k \\ \sin k \end{bmatrix}$$

satisfies $f(x_{k+1}) < f(x_k)$ for $k = 0, 1, 2, \ldots$. Show that every point on the unit circle $\{x \mid \|x\|^2 = 1\}$ is a limit point for $\{x_k\}$. Hint: Every value $\theta \in [0, 2\pi]$ is a limit point of the subsequence $\{\xi_k\}$ defined by

$$\xi_k = k(\bmod 2\pi) = k - 2\pi \left\lfloor \frac{k}{2\pi} \right\rfloor,$$

where the operator $\lfloor \cdot \rfloor$ denotes rounding down to the next integer.

◈ **2.6** Prove that all isolated local minimizers are strict. (Hint: Take an isolated local minimizer x^* and a neighborhood $\mathcal{N}$. Show that for any $x \in \mathcal{N}$, $x \neq x^*$ we must have $f(x) > f(x^*)$.)

◈ **2.7** Suppose that f is a convex function. Show that the set of global minimizers of f is a convex set.

◈ **2.8** Consider the function $f(x_1, x_2) = (x_1 + x_2^2)^2$. At the point $x^T = (1, 0)$ we consider the search direction $p^T = (-1, 1)$. Show that p is a descent direction and find all minimizers of the problem (2.9).

◈ **2.9** Suppose that $\tilde{f}(z) = f(x)$, where $x = Sz + s$ for some $S \in \mathbf{R}^{n \times n}$ and $s \in \mathbf{R}^n$. Show that

$$\nabla \tilde{f}(z) = S^T \nabla f(x), \qquad \nabla^2 \tilde{f}(z) = S^T \nabla^2 f(x) S.$$

(Hint: Use the chain rule to express $d\tilde{f}/dz_j$ in terms of df/dx_i and dx_i/dz_j for all $i, j = 1, 2, \ldots, n$.)

◈ **2.10** Show that the symmetric rank-one update (2.17) and the BFGS update (2.18) are scale-invariant if the initial Hessian approximations B_0 are chosen appropriately. That is, using the notation of the previous exercise, show that if these methods are applied to $f(x)$ starting from $x_0 = Sz_0 + s$ with initial Hessian B_0, and to $\tilde{f}(z)$ starting from z_0 with initial Hessian $S^T B_0 S$, then all iterates are related by $x_k = Sz_k + s$. (Assume for simplicity that the methods take unit step lengths.)

◈ **2.11** Suppose that a function f of two variables is poorly scaled at the solution x^*. Write two Taylor expansions of f around x^*—one along each coordinate direction—and use them to show that the Hessian $\nabla^2 f(x^*)$ is ill-conditioned.

✎ **2.12** Show that the sequence $x_k = 1/k$ is not Q-linearly convergent, though it does converge to zero. (This is called *sublinear convergence*.)

✎ **2.13** Show that the sequence $x_k = 1 + (0.5)^{2^k}$ is Q–quadratically convergent to 1.

✎ **2.14** Does the sequence $1/(k!)$ converge Q-superlinearly? Q-quadratically?

✎ * **2.15** Consider the sequence $\{x_k\}$ defined by

$$x_k = \begin{cases} \left(\frac{1}{4}\right)^{2^k}, & k \text{ even}, \\ (x_{k-1})/k, & k \text{ odd}. \end{cases}$$

Is this sequence Q-superlinearly convergent? Q-quadratically convergent? R-quadratically convergent?

Chapter 3

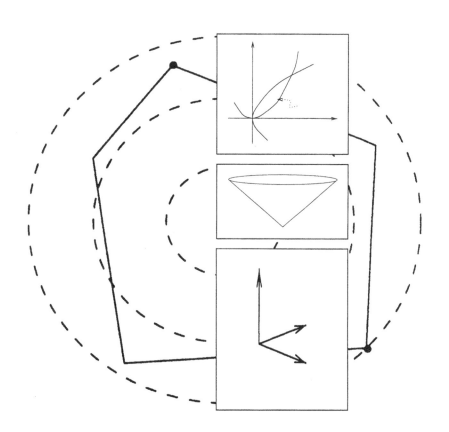

Line Search Methods

Each iteration of a line search method computes a search direction p_k and then decides how far to move along that direction. The iteration is given by

$$x_{k+1} = x_k + \alpha_k p_k, \tag{3.1}$$

where the positive scalar α_k is called the *step length*. The success of a line search method depends on effective choices of both the direction p_k and the step length α_k.

Most line search algorithms require p_k to be a *descent direction*—one for which $p_k^T \nabla f_k < 0$—because this property guarantees that the function f can be reduced along this direction, as discussed in the previous chapter. Moreover, the search direction often has the form

$$p_k = -B_k^{-1} \nabla f_k, \tag{3.2}$$

where B_k is a symmetric and nonsingular matrix. In the steepest descent method B_k is simply the identity matrix I, while in Newton's method B_k is the exact Hessian $\nabla^2 f(x_k)$. In quasi-Newton methods, B_k is an approximation to the Hessian that is updated at every

36 CHAPTER 3. LINE SEARCH METHODS

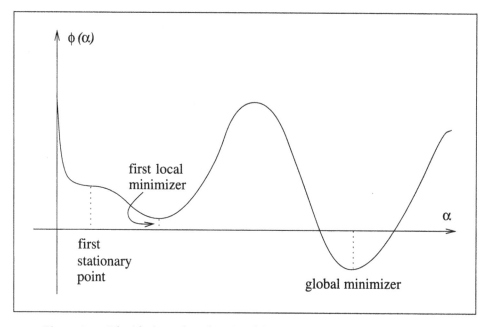

Figure 3.1 The ideal step length is the global minimizer.

iteration by means of a low-rank formula. When p_k is defined by (3.2) and B_k is positive definite, we have

$$p_k^T \nabla f_k = -\nabla f_k^T B_k^{-1} \nabla f_k < 0,$$

and therefore p_k is a descent direction.

In the next chapters we study how to choose the matrix B_k, or more generally, how to compute the search direction. We now give careful consideration to the choice of the step-length parameter α_k.

3.1 STEP LENGTH

In computing the step length α_k, we face a tradeoff. We would like to choose α_k to give a substantial reduction of f, but at the same time, we do not want to spend too much time making the choice. The ideal choice would be the global minimizer of the univariate function $\phi(\cdot)$ defined by

$$\phi(\alpha) = f(x_k + \alpha p_k), \quad \alpha > 0, \tag{3.3}$$

but in general, it is too expensive to identify this value (see Figure 3.1). To find even a local minimizer of ϕ to moderate precision generally requires too many evaluations of the objec-

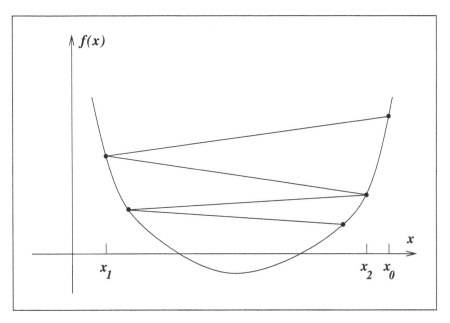

Figure 3.2 Insufficient reduction in f.

tive function f and possibly the gradient ∇f. More practical strategies perform an *inexact* line search to identify a step length that achieves adequate reductions in f at minimal cost.

Typical line search algorithms try out a sequence of candidate values for α, stopping to accept one of these values when certain conditions are satisfied. The line search is done in two stages: A bracketing phase finds an interval containing desirable step lengths, and a bisection or interpolation phase computes a good step length within this interval. Sophisticated line search algorithms can be quite complicated, so we defer a full description until the end of this chapter. We now discuss various termination conditions for the line search algorithm and show that effective step lengths need not lie near minimizers of the univariate function $\phi(\alpha)$ defined in (3.3).

A simple condition we could impose on α_k is that it provide a reduction in f, i.e., $f(x_k + \alpha_k p_k) < f(x_k)$. That this is not appropriate is illustrated in Figure 3.2, where the minimum is $f^* = -1$, but the sequence of function values $\{5/k\}$, $k = 0, 1, \ldots$, converges to zero. The difficulty is that we do not have sufficient reduction in f, a concept we discuss next.

THE WOLFE CONDITIONS

A popular inexact line search condition stipulates that α_k should first of all give *sufficient decrease* in the objective function f, as measured by the following inequality:

$$f(x_k + \alpha p_k) \leq f(x_k) + c_1 \alpha \nabla f_k^T p_k, \qquad (3.4)$$

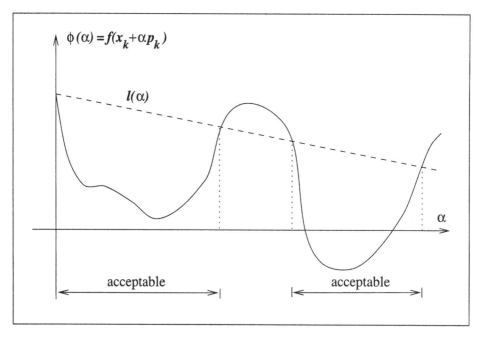

Figure 3.3 Sufficient decrease condition.

for some constant $c_1 \in (0, 1)$. In other words, the reduction in f should be proportional to both the step length α_k and the directional derivative $\nabla f_k^T p_k$. Inequality (3.4) is sometimes called the *Armijo condition*.

The sufficient decrease condition is illustrated in Figure 3.3. The right-hand-side of (3.4), which is a linear function, can be denoted by $l(\alpha)$. The function $l(\cdot)$ has negative slope $c_1 \nabla f_k^T p_k$, but because $c_1 \in (0, 1)$, it lies above the graph of ϕ for small positive values of α. The sufficient decrease condition states that α is acceptable only if $\phi(\alpha) \leq l(\alpha)$. The intervals on which this condition is satisfied are shown in Figure 3.3. In practice, c_1 is chosen to be quite small, say $c_1 = 10^{-4}$.

The sufficient decrease condition is not enough by itself to ensure that the algorithm makes reasonable progress, because as we see from Figure 3.3, it is satisfied for all sufficiently small values of α. To rule out unacceptably short steps we introduce a second requirement, called the *curvature condition*, which requires α_k to satisfy

$$\nabla f(x_k + \alpha_k p_k)^T p_k \geq c_2 \nabla f_k^T p_k, \tag{3.5}$$

for some constant $c_2 \in (c_1, 1)$, where c_1 is the constant from (3.4). Note that the left-hand-side is simply the derivative $\phi'(\alpha_k)$, so the curvature condition ensures that the slope of $\phi(\alpha_k)$ is greater than c_2 times the gradient $\phi'(0)$. This makes sense because if the slope $\phi'(\alpha)$ is strongly negative, we have an indication that we can reduce f significantly by moving further along the chosen direction. On the other hand, if the slope is only slightly negative or even positive, it is a sign that we cannot expect much more decrease in f in this direction,

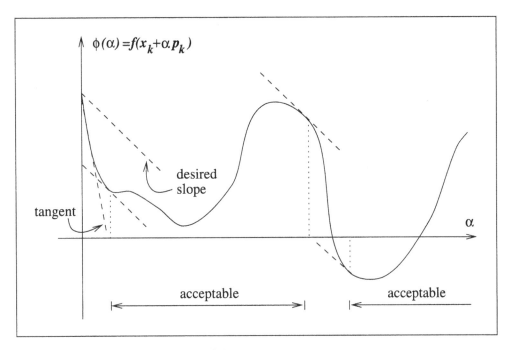

Figure 3.4 The curvature condition.

so it might make sense to terminate the line search. The curvature condition is illustrated in Figure 3.4. Typical values of c_2 are 0.9 when the search direction p_k is chosen by a Newton or quasi-Newton method, and 0.1 when p_k is obtained from a nonlinear conjugate gradient method.

The sufficient decrease and curvature conditions are known collectively as the *Wolfe conditions*. We illustrate them in Figure 3.5 and restate them here for future reference:

$$f(x_k + \alpha_k p_k) \leq f(x_k) + c_1 \alpha_k \nabla f_k^T p_k, \tag{3.6a}$$
$$\nabla f(x_k + \alpha_k p_k)^T p_k \geq c_2 \nabla f_k^T p_k, \tag{3.6b}$$

with $0 < c_1 < c_2 < 1$.

A step length may satisfy the Wolfe conditions without being particularly close to a minimizer of ϕ, as we show in Figure 3.5. We can, however, modify the curvature condition to force α_k to lie in at least a broad neighborhood of a local minimizer or stationary point of ϕ. The *strong Wolfe conditions* require α_k to satisfy

$$f(x_k + \alpha_k p_k) \leq f(x_k) + c_1 \alpha_k \nabla f_k^T p_k, \tag{3.7a}$$
$$|\nabla f(x_k + \alpha_k p_k)^T p_k| \leq c_2 |\nabla f_k^T p_k|, \tag{3.7b}$$

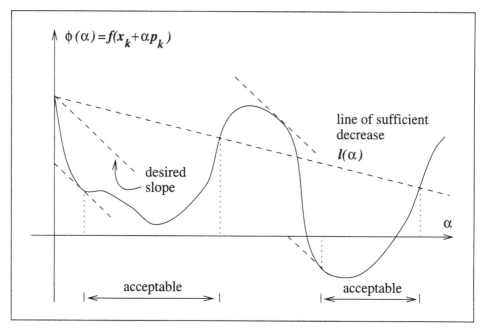

Figure 3.5 Step lengths satisfying the Wolfe conditions.

with $0 < c_1 < c_2 < 1$. The only difference with the Wolfe conditions is that we no longer allow the derivative $\phi'(\alpha_k)$ to be too positive. Hence, we exclude points that are far from stationary points of ϕ.

It is not difficult to prove that there exist step lengths that satisfy the Wolfe conditions for every function f that is smooth and bounded below.

Lemma 3.1.

Suppose that $f : \mathbb{R}^n \to \mathbb{R}$ is continuously differentiable. Let p_k be a descent direction at x_k, and assume that f is bounded below along the ray $\{x_k + \alpha p_k | \alpha > 0\}$. Then if $0 < c_1 < c_2 < 1$, there exist intervals of step lengths satisfying the Wolfe conditions (3.6) and the strong Wolfe conditions (3.7).

PROOF. Since $\phi(\alpha) = f(x_k + \alpha p_k)$ is bounded below for all $\alpha > 0$ and since $0 < c_1 < 1$, the line $l(\alpha) = f(x_k) + \alpha c_1 \nabla f_k^T p_k$ must intersect the graph of ϕ at least once. Let $\alpha' > 0$ be the smallest intersecting value of α, that is,

$$f(x_k + \alpha' p_k) = f(x_k) + \alpha' c_1 \nabla f_k^T p_k. \tag{3.8}$$

The sufficient decrease condition (3.6a) clearly holds for all step lengths less than α'.

By the mean value theorem, there exists $\alpha'' \in (0, \alpha')$ such that

$$f(x_k + \alpha' p_k) - f(x_k) = \alpha' \nabla f(x_k + \alpha'' p_k)^T p_k. \tag{3.9}$$

By combining (3.8) and (3.9), we obtain

$$\nabla f(x_k + \alpha'' p_k)^T p_k = c_1 \nabla f_k^T p_k > c_2 \nabla f_k^T p_k, \tag{3.10}$$

since $c_1 < c_2$ and $\nabla f_k^T p_k < 0$. Therefore, α'' satisfies the Wolfe conditions (3.6), and the inequalities hold strictly in both (3.6a) and (3.6b). Hence, by our smoothness assumption on f, there is an interval around α'' for which the Wolfe conditions hold. Moreover, since the term in the left-hand side of (3.10) is negative, the strong Wolfe conditions (3.7) hold in the same interval. □

The Wolfe conditions are scale-invariant in a broad sense: Multiplying the objective function by a constant or making an affine change of variables does not alter them. They can be used in most line search methods, and are particularly important in the implementation of quasi-Newton methods, as we see in Chapter 8.

THE GOLDSTEIN CONDITIONS

Like the Wolfe conditions, the *Goldstein conditions* also ensure that the step length α achieves sufficient decrease while preventing α from being too small. The Goldstein conditions can also be stated as a pair of inequalities, in the following way:

$$f(x_k) + (1 - c)\alpha_k \nabla f_k^T p_k \leq f(x_k + \alpha_k p_k) \leq f(x_k) + c\alpha_k \nabla f_k^T p_k, \tag{3.11}$$

with $0 < c < \frac{1}{2}$. The second inequality is the sufficient decrease condition (3.4), whereas the first inequality is introduced to control the step length from below; see Figure 3.6

A disadvantage of the Goldstein conditions vis-à-vis the Wolfe conditions is that the first inequality in (3.11) may exclude all minimizers of ϕ. However, the Goldstein and Wolfe conditions have much in common, and their convergence theories are quite similar. The Goldstein conditions are often used in Newton-type methods but are not well suited for quasi-Newton methods that maintain a positive definite Hessian approximation.

SUFFICIENT DECREASE AND BACKTRACKING

We have mentioned that the sufficient decrease condition (3.6a) alone is not sufficient to ensure that the algorithm makes reasonable progress along the given search direction. However, if the line search algorithm chooses its candidate step lengths appropriately, by using a so-called *backtracking* approach, we can dispense with the extra condition (3.6b) and use just the sufficient decrease condition to terminate the line search procedure. In its most basic form, backtracking proceeds as follows.

Procedure 3.1 (Backtracking Line Search).
 Choose $\bar{\alpha} > 0$, $\rho, c \in (0, 1)$; set $\alpha \leftarrow \bar{\alpha}$;
 repeat until $f(x_k + \alpha p_k) \leq f(x_k) + c\alpha \nabla f_k^T p_k$

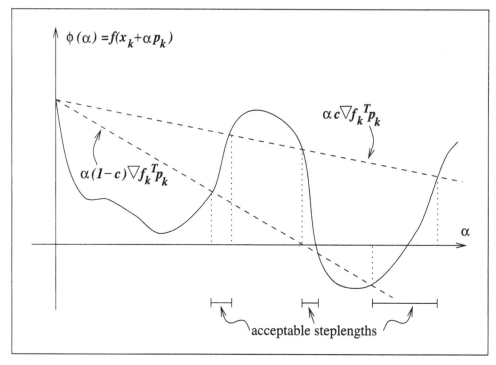

Figure 3.6 The Goldstein conditions.

$$\alpha \leftarrow \rho\alpha;$$
end (repeat)
Terminate with $\alpha_k = \alpha$.

In this procedure, the initial step length $\bar{\alpha}$ is chosen to be 1 in Newton and quasi-Newton methods, but can have different values in other algorithms such as steepest descent or conjugate gradient. An acceptable step length will be found after a finite number of trials because α_k will eventually become small enough that the sufficient decrease condition holds (see Figure 3.3). In practice, the contraction factor ρ is often allowed to vary at each iteration of the line search. For example, it can be chosen by safeguarded interpolation, as we describe later. We need ensure only that at each iteration we have $\rho \in [\rho_{lo}, \rho_{hi}]$, for some fixed constants $0 < \rho_{lo} < \rho_{hi} < 1$.

The backtracking approach ensures either that the selected step length α_k is some fixed value (the initial choice $\bar{\alpha}$), or else that it is short enough to satisfy the sufficient decrease condition but not *too* short. The latter claim holds because the accepted value α_k is within striking distance of the previous trial value, α_k/ρ, which was rejected for violating the sufficient decrease condition, that is, for being too long.

This simple and popular strategy for terminating a line search is well suited for Newton methods (see Chapter 6) but is less appropriate for quasi-Newton and conjugate gradient methods.

3.2 CONVERGENCE OF LINE SEARCH METHODS

To obtain global convergence, we must not only have well-chosen step lengths but also well-chosen search directions p_k. We discuss requirements on the search direction in this section, focusing on one key property: the angle θ_k between p_k and the steepest descent direction $-\nabla f_k$, defined by

$$\cos \theta_k = \frac{-\nabla f_k^T p_k}{\|\nabla f_k\| \|p_k\|}. \tag{3.12}$$

The following theorem, due to Zoutendijk, has far-reaching consequences. It shows, for example, that the steepest descent method is globally convergent. For other algorithms it describes how far p_k can deviate from the steepest descent direction and still give rise to a globally convergent iteration. Various line search termination conditions can be used to establish this result, but for concreteness we will consider only the Wolfe conditions (3.6). Though Zoutendijk's result appears, at first, to be technical and obscure, its power will soon become evident.

Theorem 3.2.
Consider any iteration of the form (3.1), where p_k is a descent direction and α_k satisfies the Wolfe conditions (3.6). Suppose that f is bounded below in $\mathbf{R}^n$ and that f is continuously differentiable in an open set $\mathcal{N}$ containing the level set $\mathcal{L} \stackrel{\text{def}}{=} \{x : f(x) \leq f(x_0)\}$, where x_0 is the starting point of the iteration. Assume also that the gradient ∇f is Lipschitz continuous on $\mathcal{N}$, that is, there exists a constant $L > 0$ such that

$$\|\nabla f(x) - \nabla f(\tilde{x})\| \leq L \|x - \tilde{x}\|, \quad \text{for all } x, \tilde{x} \in \mathcal{N}. \tag{3.13}$$

Then

$$\sum_{k \geq 0} \cos^2 \theta_k \|\nabla f_k\|^2 < \infty. \tag{3.14}$$

PROOF. From (3.6b) and (3.1) we have that

$$(\nabla f_{k+1} - \nabla f_k)^T p_k \geq (c_2 - 1) \nabla f_k^T p_k,$$

while the Lipschitz condition (3.13) implies that

$$(\nabla f_{k+1} - \nabla f_k)^T p_k \leq \alpha_k L \|p_k\|^2.$$

By combining these two relations, we obtain

$$\alpha_k \geq \frac{c_2 - 1}{L} \frac{\nabla f_k^T p_k}{\|p_k\|^2}.$$

By substituting this inequality into the first Wolfe condition (3.6a), we obtain

$$f_{k+1} \leq f_k - c_1 \frac{1 - c_2}{L} \frac{(\nabla f_k^T p_k)^2}{\|p_k\|^2}.$$

From the definition (3.12), we can write this relation as

$$f_{k+1} \leq f_k - c \cos^2 \theta_k \|\nabla f_k\|^2,$$

where $c = c_1(1 - c_2)/L$. By summing this expression over all indices less than or equal to k, we obtain

$$f_{k+1} \leq f_0 - c \sum_{j=0}^{k} \cos^2 \theta_j \|\nabla f_j\|^2. \tag{3.15}$$

Since f is bounded below, we have that $f_0 - f_{k+1}$ is less than some positive constant, for all k. Hence by taking limits in (3.15), we obtain

$$\sum_{k=0}^{\infty} \cos^2 \theta_k \|\nabla f_k\|^2 < \infty,$$

which concludes the proof. □

Similar results to this theorem hold when the Goldstein conditions (3.11) or strong Wolfe conditions (3.7) are used in place of the Wolfe conditions.

Note that the assumptions of Theorem 3.2 are not too restrictive. If the function f were not bounded below, the optimization problem would not be well-defined. The smoothness assumption—Lipschitz continuity of the gradient—is implied by many of the smoothness conditions that are used in local convergence theorems (see Chapters 6 and 8) and are often satisfied in practice.

Inequality (3.14), which we call the *Zoutendijk condition*, implies that

$$\cos^2 \theta_k \|\nabla f_k\|^2 \to 0. \tag{3.16}$$

3.2. CONVERGENCE OF LINE SEARCH METHODS

This limit can be used in turn to derive global convergence results for line search algorithms.

If our method for choosing the search direction p_k in the iteration (3.1) ensures that the angle θ_k defined by (3.12) is bounded away from 90°, there is a positive constant δ such that

$$\cos \theta_k \geq \delta > 0, \quad \text{for all } k. \tag{3.17}$$

It follows immediately from (3.16) that

$$\lim_{k \to \infty} \|\nabla f_k\| = 0. \tag{3.18}$$

In other words, we can be sure that the gradient norms $\|\nabla f_k\|$ converge to zero, provided that the search directions are never too close to orthogonality with the gradient. In particular, the method of steepest descent (for which the search direction p_k makes an angle of *zero* degrees with the negative gradient) produces a gradient sequence that converges to zero, provided that it uses a line search satisfying the Wolfe or Goldstein conditions.

We use the term *globally convergent* to refer to algorithms for which the property (3.18) is satisfied, but note that this term is sometimes used in other contexts to mean different things. For line search methods of the general form (3.1), the limit (3.18) is the strongest global convergence result that can be obtained: We cannot guarantee that the method converges to a minimizer, but only that it is attracted by stationary points. Only by making additional requirements on the search direction p_k—by introducing negative curvature information from the Hessian $\nabla^2 f(x_k)$, for example—can we strengthen these results to include convergence to a local minimum. See the Notes and References at the end of this chapter for further discussion of this point.

Consider now the Newton-like method (3.1), (3.2) and assume that the matrices B_k are positive definite with a uniformly bounded condition number. That is, there is a constant M such that

$$\|B_k\| \, \|B_k^{-1}\| \leq M, \quad \text{for all } k.$$

It is easy to show from the definition (3.12) that

$$\cos \theta_k \geq 1/M \tag{3.19}$$

(see Exercise 5). By combining this bound with (3.16) we find that

$$\lim_{k \to \infty} \|\nabla f_k\| = 0. \tag{3.20}$$

Therefore, we have shown that Newton and quasi-Newton methods are globally convergent if the matrices B_k have a bounded condition number and are positive definite (which is needed to ensure that p_k is a descent direction), and if the step lengths satisfy the Wolfe conditions.

For some algorithms, such as conjugate gradient methods, we will not be able to prove the limit (3.18), but only the weaker result

$$\liminf_{k \to \infty} \|\nabla f_k\| = 0. \tag{3.21}$$

In other words, just a subsequence of the gradient norms $\|\nabla f_{k_j}\|$ converges to zero, rather than the whole sequence (see Appendix A). This result, too, can be proved by using Zoutendijk's condition (3.14), but instead of a constructive proof, we outline a proof by contradiction. Suppose that (3.21) does not hold, so that the gradients remain bounded away from zero, that is, there exists $\gamma > 0$ such that

$$\|\nabla f_k\| \geq \gamma, \quad \text{for all } k \text{ sufficiently large}. \tag{3.22}$$

Then from (3.16) we conclude that

$$\cos \theta_k \to 0, \tag{3.23}$$

that is, the entire sequence $\{\cos \theta_k\}$ converges to 0. To establish (3.21), therefore, it is enough to show that a subsequence $\{\cos \theta_{k_j}\}$ is bounded away from zero. We will use this strategy in Chapter 5 to study the convergence of nonlinear conjugate gradient methods.

By applying this proof technique, we can prove global convergence in the sense of (3.20) or (3.21) for a general class of algorithms. Consider *any* algorithm for which (i) every iteration produces a decrease in the objective function, and (ii) every mth iteration is a steepest descent step, with step length chosen to satisfy the Wolfe or Goldstein conditions. Then since $\cos \theta_k = 1$ for the steepest descent steps, the result (3.20) holds. Of course, we would design the algorithm so that it does something "better" than steepest descent at the other $m - 1$ iterates; the occasional steepest descent steps may not make much progress, but they at least guarantee overall global convergence.

Note that throughout this section we have used only the fact that Zoutendijk's condition implies the limit (3.16). In later chapters we will make use of the bounded sum condition (3.14), which forces the sequence $\{\cos^2 \theta_k \|\nabla f_k\|^2\}$ to converge to zero at a sufficiently rapid rate.

3.3 RATE OF CONVERGENCE

It would seem that designing optimization algorithms with good convergence properties is easy, since all we need to ensure is that the search direction p_k does not tend to become orthogonal to the gradient ∇f_k, or that steepest descent steps are taken regularly. We could simply compute $\cos \theta_k$ at every iteration and turn p_k toward the steepest descent direction if $\cos \theta_k$ is smaller than some preselected constant $\delta > 0$. Angle tests of this type ensure global convergence, but they are undesirable for two reasons. First, they may impede a fast rate of convergence, because for problems with an ill-conditioned Hessian, it may be necessary to produce search directions that are almost orthogonal to the gradient, and an inappropriate choice of the parameter δ may prevent this. Second, angle tests destroy the invariance properties of quasi-Newton methods.

Algorithmic strategies that achieve rapid convergence can sometimes conflict with the requirements of global convergence, and vice versa. For example, the steepest descent method is the quintessential globally convergent algorithm, but it is quite slow in practice, as we shall see below. On the other hand, the pure Newton iteration converges rapidly when started close enough to a solution, but its steps may not even be descent directions away from the solution. The challenge is to design algorithms that incorporate both properties: good global convergence guarantees and a rapid rate of convergence.

We begin our study of convergence rates of line search methods by considering the most basic approach of all: the steepest descent method.

CONVERGENCE RATE OF STEEPEST DESCENT

We can learn much about the steepest descent method by considering the ideal case, in which the objective function is quadratic and the line searches are exact. Let us suppose that

$$f(x) = \tfrac{1}{2} x^T Q x - b^T x, \tag{3.24}$$

where Q is symmetric and positive definite. The gradient is given by $\nabla f(x) = Qx - b$, and the minimizer x^* is the unique solution of the linear system $Qx = b$.

Let us compute the step length α_k that minimizes $f(x_k - \alpha \nabla f_k)$. By differentiating

$$f(x_k - \alpha g_k) = \frac{1}{2}(x_k - \alpha g_k)^T Q (x_k - \alpha g_k) - b^T (x_k - \alpha g_k)$$

with respect to α, we obtain

$$\alpha_k = \frac{\nabla f_k^T \nabla f_k}{\nabla f_k^T Q \nabla f_k}. \tag{3.25}$$

If we use this exact minimizer α_k, the steepest descent iteration for (3.24) is given by

$$x_{k+1} = x_k - \left(\frac{\nabla f_k^T \nabla f_k}{\nabla f_k^T Q \nabla f_k} \right) \nabla f_k. \tag{3.26}$$

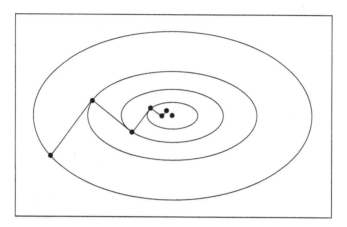

Figure 3.7 Steepest descent steps.

Since $\nabla f_k = Qx_k - b$, this equation yields a closed-form expression for x_{k+1} in terms of x_k. In Figure 3.7 we plot a typical sequence of iterates generated by the steepest descent method on a two-dimensional quadratic objective function. The contours of f are ellipsoids whose axes lie along the orthogonal eigenvectors of Q. Note that the iterates zigzag toward the solution.

To quantify the rate of convergence we introduce the weighted norm $\|x\|_Q^2 = x^T Q x$. By using the relation $Qx^* = b$, we can show easily that

$$\tfrac{1}{2}\|x - x^*\|_Q^2 = f(x) - f(x^*), \tag{3.27}$$

so that this norm measures the difference between the current objective value and the optimal value. By using the equality (3.26) and noting that $\nabla f_k = Q(x_k - x^*)$, we can derive the equality

$$\|x_{k+1} - x^*\|_Q^2 = \left\{1 - \frac{(\nabla f_k^T \nabla f_k)^2}{(\nabla f_k^T Q \nabla f_k)(\nabla f_k^T Q^{-1} \nabla f_k)}\right\} \|x_k - x^*\|_Q^2 \tag{3.28}$$

(see Exercise 7). This expression describes the exact decrease in f at each iteration, but since the term inside the brackets is difficult to interpret, it is more useful to bound it in terms of the condition number of the problem.

Theorem 3.3.

When the steepest descent method with exact line searches (3.26) is applied to the strongly convex quadratic function (3.24), the error norm (3.27) satisfies

$$\|x_{k+1} - x^*\|_Q^2 \leq \left(\frac{\lambda_n - \lambda_1}{\lambda_n + \lambda_1}\right)^2 \|x_k - x^*\|_Q^2, \tag{3.29}$$

where $0 < \lambda_1 \leq \cdots \leq \lambda_n$ are the eigenvalues of Q.

The proof of this result is given by Luenberger [152]. The inequalities (3.29) and (3.27) show that the function values f_k converge to the minimum f_* at a linear rate. As a special case of this result, we see that convergence is achieved in one iteration if all the eigenvalues are equal. In this case, Q is a multiple of the identity matrix, so the contours in Figure 3.7 are circles and the steepest descent direction always points at the solution. In general, as the condition number $\kappa(Q) = \lambda_n/\lambda_1$ increases, the contours of the quadratic become more elongated, the zigzagging in Figure 3.7 becomes more pronounced, and (3.29) implies that the convergence degrades. Even though (3.29) is a worst-case bound, it gives an accurate indication of the behavior of the algorithm when $n > 2$.

The rate-of-convergence behavior of the steepest descent method is essentially the same on general nonlinear objective functions. In the following result we assume that the step length is the global minimizer along the search direction.

Theorem 3.4.

Suppose that $f : \mathbf{R}^n \to \mathbf{R}$ is twice continuously differentiable, and that the iterates generated by the steepest descent method with exact line searches converge to a point x^* where the Hessian matrix $\nabla^2 f(x^*)$ is positive definite. Then

$$f(x_{k+1}) - f(x^*) \leq \left(\frac{\lambda_n - \lambda_1}{\lambda_n + \lambda_1}\right)^2 [f(x_k) - f(x^*)],$$

where $\lambda_1 \leq \cdots \leq \lambda_n$ are the eigenvalues of $\nabla^2 f(x^*)$.

In general, we cannot expect the rate of convergence to improve if an inexact line search is used. Therefore, Theorem 3.4 shows that the steepest descent method can have an unacceptably slow rate of convergence, even when the Hessian is reasonably well conditioned. For example, if $\kappa(Q) = 800$, $f(x_1) = 1$ and $f(x^*) = 0$, Theorem 3.4 suggests that the function value will still be about 0.08 after one thousand iterations of the steepest descent method.

QUASI-NEWTON METHODS

Let us now suppose that the search direction has the form

$$p_k = -B_k^{-1} \nabla f_k, \tag{3.30}$$

where the symmetric and positive definite matrix B_k is updated at every iteration by a quasi-Newton updating formula. We already encountered one quasi-Newton formula, the BFGS formula, in Chapter 2; others will be discussed in Chapter 8. We assume here that the step length α_k will be computed by an inexact line search that satisfies the Wolfe or strong Wolfe conditions, with one important proviso: The line search algorithm will always try the step length $\alpha = 1$ first, and will accept this value if it satisfies the Wolfe conditions. (We could enforce this condition by setting $\bar{\alpha} = 1$ in Procedure 3.1, for example.) This implementation detail turns out to be crucial in obtaining a fast rate of convergence.

The following result, due to Dennis and Moré, shows that if the search direction of a quasi-Newton method approximates the Newton direction well enough, then the unit step length will satisfy the Wolfe conditions as the iterates converge to the solution. It also specifies a condition that the search direction must satisfy in order to give rise to a superlinearly convergent iteration. To bring out the full generality of this result, we state it first in terms of a general descent iteration, and then examine its consequences for quasi-Newton and Newton methods.

Theorem 3.5.

Suppose that $f : \mathbb{R}^n \to \mathbb{R}$ is three times continuously differentiable. Consider the iteration $x_{k+1} = x_k + \alpha_k p_k$, where p_k is a descent direction and α_k satisfies the Wolfe conditions (3.6) with $c_1 \leq \frac{1}{2}$. If the sequence $\{x_k\}$ converges to a point x^* such that $\nabla f(x^*) = 0$ and $\nabla^2 f(x^*)$ is positive definite, and if the search direction satisfies

$$\lim_{k \to \infty} \frac{\|\nabla f_k + \nabla^2 f_k p_k\|}{\|p_k\|} = 0, \tag{3.31}$$

then

(i) the step length $\alpha_k = 1$ is admissible for all k greater than a certain index k_0; and

(ii) if $\alpha_k = 1$ for all $k > k_0$, $\{x_k\}$ converges to x^* superlinearly.

It is easy to see that if $c_1 > \frac{1}{2}$, then the line search would exclude the minimizer of a quadratic, and unit step lengths may not be admissible.

If p_k is a quasi-Newton search direction of the form (3.30), then (3.31) is equivalent to

$$\lim_{k \to \infty} \frac{\|(B_k - \nabla^2 f(x^*))p_k\|}{\|p_k\|} = 0. \tag{3.32}$$

Hence, we have the surprising (and delightful) result that a superlinear convergence rate can be attained even if the sequence of quasi-Newton matrices B_k does not converge to $\nabla^2 f(x^*)$; it suffices that the B_k become increasingly accurate approximations to $\nabla^2 f(x^*)$ *along the search directions* p_k.

An important remark is that condition (3.32) is both necessary and sufficient for the superlinear convergence of quasi-Newton methods.

Theorem 3.6.

Suppose that $f : \mathbf{R}^n \to \mathbf{R}$ is three times continuously differentiable. Consider the iteration $x_{k+1} = x_k + p_k$ (that is, the step length α_k is uniformly 1) and that p_k is given by (3.30). Let us also assume that $\{x_k\}$ converges to a point x^* such that $\nabla f(x^*) = 0$ and $\nabla^2 f(x^*)$ is positive definite. Then $\{x_k\}$ converges superlinearly if and only if (3.32) holds.

PROOF. We first show that (3.32) is equivalent to

$$p_k - p_k^N = o(\|p_k\|), \qquad (3.33)$$

where $p_k^N = -\nabla^2 f_k^{-1} \nabla f_k$ is the Newton step. Assuming that (3.32) holds, we have that

$$\begin{aligned} p_k - p_k^N &= \nabla^2 f_k^{-1}(\nabla^2 f_k p_k + \nabla f_k) \\ &= \nabla^2 f_k^{-1}(\nabla^2 f_k - B_k) p_k \\ &= O(\|(\nabla^2 f_k - B_k) p_k\|) \\ &= o(\|p_k\|), \end{aligned}$$

where we have used the fact that $\|\nabla^2 f_k^{-1}\|$ is bounded above for x_k sufficiently close to x^*, since the limiting Hessian $\nabla^2 f(x^*)$ is positive definite. The converse follows readily if we multiply both sides of (3.33) by $\nabla^2 f_k$ and recall (3.30).

For the remainder of the proof we need to look ahead to the proof of quadratic convergence of Newton's method and, in particular, the estimate (3.37). By using this inequality together with (3.33), we obtain that

$$\begin{aligned} \|x_k + p_k - x^*\| &\leq \|x_k + p_k^N - x^*\| + \|p_k - p_k^N\| \\ &= O(\|x_k - x^*\|^2) + o(\|p_k\|). \end{aligned}$$

A simple manipulation of this inequality reveals that $\|p_k\| = O(\|x_k - x^*\|)$, so we obtain

$$\|x_k + p_k - x^*\| \leq o(\|x_k - x^*\|),$$

giving the superlinear convergence result. □

We will see in Chapter 8 that quasi-Newton methods normally satisfy condition (3.32) and are superlinearly convergent.

NEWTON'S METHOD

Let us now consider the Newton iteration where the search direction is given by

$$p_k^N = -\nabla^2 f_k^{-1} \nabla f_k. \qquad (3.34)$$

Since the Hessian matrix $\nabla^2 f_k$ may not always be positive definite, p_k^N may not always be a descent direction, and many of the ideas discussed so far in this chapter no longer apply. In Chapter 6 we will describe two approaches for obtaining a globally convergent iteration based on the Newton step: a line search approach, in which the Hessian $\nabla^2 f_k$ is modified, if necessary, to make it positive definite and thereby yield descent, and a trust region approach, in which $\nabla^2 f_k$ is used to form a quadratic model that is minimized in a ball.

Here we discuss just the local rate-of-convergence properties of Newton's method. We know that for all x in the vicinity of a solution point x^* such that $\nabla^2 f(x^*)$ is positive definite, the Hessian $\nabla^2 f(x)$ will also be positive definite. Newton's method will be well-defined in this region and will converge quadratically, provided that the step lengths α_k are eventually always 1.

Theorem 3.7.

Suppose that f is twice differentiable and that the Hessian $\nabla^2 f(x)$ is Lipschitz continuous (see (A.8)) in a neighborhood of a solution x^* at which the sufficient conditions (Theorem 2.4) are satisfied. Consider the iteration $x_{k+1} = x_k + p_k$, where p_k is given by (3.34). Then

1. *if the starting point x_0 is sufficiently close to x^*, the sequence of iterates converges to x^*;*

2. *the rate of convergence of $\{x_k\}$ is quadratic; and*

3. *the sequence of gradient norms $\{\|\nabla f_k\|\}$ converges quadratically to zero.*

PROOF. From the definition of the Newton step and the optimality condition $\nabla f_* = 0$ we have that

$$x_k + p_k^N - x^* = x_k - x^* - \nabla^2 f_k^{-1} \nabla f_k$$
$$= \nabla^2 f_k^{-1} \left[\nabla^2 f_k (x_k - x^*) - (\nabla f_k - \nabla f_*) \right]. \tag{3.35}$$

Since

$$\nabla f_k - \nabla f_* = \int_0^1 \nabla^2 f(x_k + t(x^* - x_k))(x_k - x^*) \, dt,$$

we have

$$\left\| \nabla^2 f(x_k)(x_k - x^*) - (\nabla f_k - \nabla f(x^*)) \right\|$$
$$= \left\| \int_0^1 \left[\nabla^2 f(x_k) - \nabla^2 f(x_k + t(x^* - x_k)) \right] (x_k - x^*) \, dt \right\|$$
$$\leq \int_0^1 \left\| \nabla^2 f(x_k) - \nabla^2 f(x_k + t(x^* - x_k)) \right\| \|x_k - x^*\| \, dt$$
$$\leq \|x_k - x^*\|^2 \int_0^1 Lt \, dt = \tfrac{1}{2} L \|x_k - x^*\|^2, \tag{3.36}$$

where L is the Lipschitz constant for $\nabla^2 f(x)$ for x near x^*. Since $\nabla^2 f(x^*)$ is nonsingular, and since $\nabla^2 f_k \to \nabla^2 f(x^*)$, we have that $\|\nabla^2 f_k^{-1}\| \le 2\|\nabla^2 f(x^*)^{-1}\|$ for all k sufficiently large. By substituting in (3.35) and (3.36), we obtain

$$\|x_k + p_k^N - x^*\| \le L\|\nabla^2 f(x^*)^{-1}\|\|x_k - x^*\|^2 = \tilde{L}\|x_k - x^*\|^2, \qquad (3.37)$$

where $\tilde{L} = L\|\nabla^2 f(x^*)^{-1}\|$. Using this inequality inductively we deduce that if the starting point is sufficiently near x^*, then the sequence converges to x^*, and the rate of convergence is quadratic.

By using the relations $x_{k+1} - x_k = p_k^N$ and $\nabla f_k + \nabla^2 f_k p_k^N = 0$, we obtain that

$$\|\nabla f(x_{k+1})\| = \|\nabla f(x_{k+1}) - \nabla f_k - \nabla^2 f(x_k) p_k^N\|$$
$$= \left\|\int_0^1 \nabla^2 f(x_k + t p_k^N)(x_{k+1} - x_k)\, dt - \nabla^2 f(x_k) p_k^N\right\|$$
$$\le \int_0^1 \|\nabla^2 f(x_k + t p_k^N) - \nabla^2 f(x_k)\|\, \|p_k^N\|\, dt$$
$$\le \tfrac{1}{2} L \|p_k^N\|^2$$
$$\le \tfrac{1}{2} L \|\nabla^2 f(x_k)^{-1}\|^2 \|\nabla f_k\|^2$$
$$\le 2L \|\nabla^2 f(x^*)^{-1}\|^2 \|\nabla f_k\|^2,$$

proving that the gradient norms converge to zero quadratically. □

When the search direction is given by Newton's method, the limit (3.31) is satisfied (the ratio is zero for all k!), and Theorem 3.5 shows that the Wolfe conditions will accept the step length α_k for all large k. The same is true of the Goldstein conditions. Thus implementations of Newton's method using these conditions, and in which the line search always tries the unit step length first, will set $\alpha_k = 1$ for all large k and attain a local quadratic rate of convergence.

COORDINATE DESCENT METHODS

An approach that is frequently used in practice is to cycle through the n coordinate directions $e_1, e_2, \ldots, e_n$, using each in turn as a search direction. At the first iteration, we fix all except the first variable, and find a new value of this variable that minimizes (or at least reduces) the objective function. On the next iteration, we repeat the process with the *second* variable, and so on. After n iterations, we return to the first variable and repeat the cycle. The method is referred to as the *method of alternating variables* or the *coordinate descent method*. Though simple and somewhat intuitive, it can be quite inefficient in practice, as we illustrate in Figure 3.8 for a quadratic function in two variables. Note that after a few iterations, neither the vertical nor the horizontal move makes much progress toward the solution.

54 CHAPTER 3. LINE SEARCH METHODS

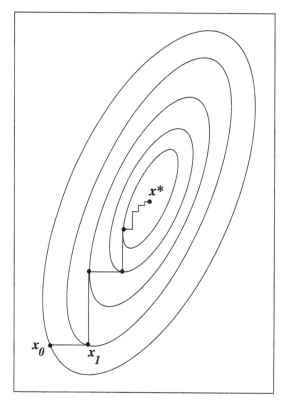

Figure 3.8
Coordinate descent.

In fact, the coordinate descent method with exact line searches can iterate infinitely without ever approaching a point where the gradient of the objective function vanishes. (By contrast, the steepest descent method produces a sequence for which $\|\nabla f_k\| \to 0$, as we showed earlier.) This observation can be generalized to show that a cyclic search along *any* set of linearly independent directions does not guarantee global convergence (Powell [198]). The difficulty that arises is that the gradient ∇f_k may become more and more perpendicular to the coordinate search direction, so that $\cos \theta_k$ approaches zero sufficiently rapidly that the Zoutendijk condition (3.14) is satisfied even when ∇f_k does not approach zero.

If the coordinate descent method converges to a solution, then its rate of convergence is often much slower than that of the steepest descent method, and the difference between them increases with the number of variables. However, the method may still be useful because

it does not require calculation of the gradient ∇f_k, and

the speed of convergence can be quite acceptable if the variables are loosely coupled.

Many variants of the coordinate descent method have been proposed, some of which are globally convergent. One simple variant is a "back-and-forth" approach in which we search along the sequence of directions

$$e_1, e_2, \ldots, e_{n-1}, e_n, e_{n-1}, \ldots, e_2, e_1, e_2, \ldots \quad \text{(repeats)}.$$

Another approach, suggested by Figure 3.8, is first to perform a sequence of coordinate descent steps and then search along the line joining the first and last points in the cycle. Several algorithms, such as that of Hooke and Jeeves, are based on these ideas; see [104, 83].

3.4 STEP-LENGTH SELECTION ALGORITHMS

We now consider techniques for finding a minimum of the one-dimensional function

$$\phi(\alpha) = f(x_k + \alpha p_k), \tag{3.38}$$

or for simply finding a step length α_k satisfying one of the termination conditions described in Section 3.1. We assume that p_k is a descent direction—that is, $\phi'(0) < 0$—so that our search can be confined to positive values of α.

If f is a convex quadratic, $f(x) = \frac{1}{2} x^T Q x + b^T x + c$, its one-dimensional minimizer along the ray $x_k + \alpha p_k$ can be computed analytically and is given by

$$\alpha_k = -\frac{\nabla f_k^T p_k}{p_k^T Q p_k}. \tag{3.39}$$

For general nonlinear functions, it is necessary to use an iterative procedure. Much attention must be given to this line search because it has a major impact on the robustness and efficiency of all nonlinear optimization methods.

Line search procedures can be classified according to the type of derivative information they use. Algorithms that use only function values can be inefficient, since to be theoretically sound, they need to continue iterating until the search for the minimizer is narrowed down to a small interval. In contrast, knowledge of gradient information allows us to determine whether a suitable step length has been located, as stipulated, for example, by the Wolfe conditions (3.6) or Goldstein conditions (3.11). Often, particularly when the iterates are close to the solution, the very first step satisfies these conditions, so the line search need not be invoked at all. In the rest of this section we will discuss only algorithms that make use of derivative information. More information on derivative-free procedures is given in the notes at the end of this chapter.

All line search procedures require an initial estimate α_0 and generate a sequence $\{\alpha_i\}$ that either terminates with a step length satisfying the conditions specified by the user (for example, the Wolfe conditions) or determines that such a step length does not exist. Typical procedures consist of two phases: a *bracketing phase* that finds an interval $[a, b]$ containing acceptable step lengths, and a *selection phase* that zooms in to locate the final step length. The selection phase usually reduces the bracketing interval during its search for the desired step length and interpolates some of the function and derivative information gathered on earlier steps to guess the location of the minimizer. We will first discuss how to perform this interpolation.

In the following discussion we let α_k and α_{k-1} denote the step lengths used at iterations k and $k-1$ of the optimization algorithm, respectively. On the other hand, we denote the trial step lengths generated during the line search by α_i and α_{i-1} and also α_j. We use α_0 to denote the initial guess.

INTERPOLATION

We begin by describing a line search procedure based on interpolation of known function and derivative values of the function ϕ. This procedure can be viewed as an enhancement of Procedure 3.1. The aim is to find a value of α that satisfies the sufficient decrease condition (3.6a), without being "too small." Accordingly, the procedures here generate a decreasing sequence of values α_i such that each value α_i is not too much smaller than its predecessor α_{i-1}.

Note that we can write the sufficient decrease condition in the notation of (3.38) as

$$\phi(\alpha_k) \le \phi(0) + c_1 \alpha_k \phi'(0), \tag{3.40}$$

and that since the constant c_1 is usually chosen to be small in practice ($c_1 = 10^{-4}$, say), this condition asks for little more than descent in f. We design the procedure to be "efficient" in the sense that it computes the derivative $\nabla f(x)$ as few times as possible.

Suppose that the initial guess α_0 is given. If we have

$$\phi(\alpha_0) \le \phi(0) + c_1 \alpha_0 \phi'(0),$$

this step length satisfies the condition, and we terminate the search. Otherwise, we know that the interval $[0, \alpha_0]$ contains acceptable step lengths (see Figure 3.3). We form a quadratic approximation $\phi_q(\alpha)$ to ϕ by interpolating the three pieces of information available—$\phi(0)$, $\phi'(0)$, and $\phi(\alpha_0)$—to obtain

$$\phi_q(\alpha) = \left(\frac{\phi(\alpha_0) - \phi(0) - \alpha_0 \phi'(0)}{\alpha_0^2} \right) \alpha^2 + \phi'(0)\alpha + \phi(0). \tag{3.41}$$

(Note that this function is constructed so that it satisfies the interpolation conditions $\phi_q(0) = \phi(0)$, $\phi_q'(0) = \phi'(0)$, and $\phi_q(\alpha_0) = \phi(\alpha_0)$.) The new trial value α_1 is defined as the minimizer of this quadratic, that is, we obtain

$$\alpha_1 = -\frac{\phi'(0)\alpha_0^2}{2[\phi(\alpha_0) - \phi(0) - \phi'(0)\alpha_0]}. \tag{3.42}$$

3.4. STEP-LENGTH SELECTION ALGORITHMS

If the sufficient decrease condition (3.40) is satisfied at α_1, we terminate the search. Otherwise, we construct a *cubic* function that interpolates the four pieces of information $\phi(0)$, $\phi'(0)$, $\phi(\alpha_0)$, and $\phi(\alpha_1)$, obtaining

$$\phi_c(\alpha) = a\alpha^3 + b\alpha^2 + \alpha\phi'(0) + \phi(0),$$

where

$$\begin{bmatrix} a \\ b \end{bmatrix} = \frac{1}{\alpha_0^2 \alpha_1^2 (\alpha_1 - \alpha_0)} \begin{bmatrix} \alpha_0^2 & -\alpha_1^2 \\ -\alpha_0^3 & \alpha_1^3 \end{bmatrix} \begin{bmatrix} \phi(\alpha_1) - \phi(0) - \phi'(0)\alpha_1 \\ \phi(\alpha_0) - \phi(0) - \phi'(0)\alpha_0 \end{bmatrix}.$$

By differentiating $\phi_c(x)$, we see that the minimizer α_2 of ϕ_c lies in the interval $[0, \alpha_1]$ and is given by

$$\alpha_2 = \frac{-b + \sqrt{b^2 - 3a\phi'(0)}}{3a}.$$

If necessary, this process is repeated, using a cubic interpolant of $\phi(0)$, $\phi'(0)$ and the two most recent values of ϕ, until an α that satisfies (3.40) is located. If any α_i is either too close to its predecessor α_{i-1} or else too much smaller than α_{i-1}, we reset $\alpha_i = \alpha_{i-1}/2$. This safeguard procedure ensures that we make reasonable progress on each iteration and that the final α is not too small.

The strategy just described assumes that derivative values are significantly more expensive to compute than function values. It is often possible, however, to compute the directional derivative simultaneously with the function, at little additional cost; see Chapter 7. Accordingly, we can design an alternative strategy based on cubic interpolation of the values of ϕ and ϕ' at the two most recent values of α.

Cubic interpolation provides a good model for functions with significant changes of curvature. Suppose we have an interval $[a, b]$ known to contain desirable step lengths, and two previous step length estimates α_{i-1} and α_i in this interval. We use a cubic function to interpolate $\phi(\alpha_{i-1})$, $\phi'(\alpha_{i-1})$, $\phi(\alpha_i)$, and $\phi'(\alpha_i)$. (This cubic function always exists and is unique; see, for example, Bulirsch and Stoer [29, p. 52].) The minimizer of this cubic in $[a, b]$ is either at one of the endpoints or else in the interior, in which case it is given by

$$\alpha_{i+1} = \alpha_i - (\alpha_i - \alpha_{i-1}) \left[\frac{\phi'(\alpha_i) + d_2 - d_1}{\phi'(\alpha_i) - \phi'(\alpha_{i-1}) + 2d_2} \right], \tag{3.43}$$

with

$$d_1 = \phi'(\alpha_{i-1}) + \phi'(\alpha_i) - 3\frac{\phi(\alpha_{i-1}) - \phi(\alpha_i)}{\alpha_{i-1} - \alpha_i},$$

$$d_2 = \left[d_1^2 - \phi'(\alpha_{i-1})\phi'(\alpha_i) \right]^{1/2}.$$

The interpolation process can be repeated by discarding the data at one of the step lengths α_{i-1} or α_i and replacing it by $\phi(\alpha_{i+1})$ and $\phi'(\alpha_{i+1})$. The decision on which of α_{i-1} and α_i should be kept and which discarded depends on the specific conditions used to terminate the line search; we discuss this issue further below in the context of the Wolfe conditions. Cubic interpolation is a powerful strategy, since it can produce a quadratic rate of convergence of the iteration (3.43) to the minimizing value of α.

THE INITIAL STEP LENGTH

For Newton and quasi-Newton methods the step $\alpha_0 = 1$ should always be used as the initial trial step length. This choice ensures that unit step lengths are taken whenever they satisfy the termination conditions and allows the rapid rate-of-convergence properties of these methods to take effect.

For methods that do not produce well-scaled search directions, such as the steepest descent and conjugate gradient methods, it is important to use current information about the problem and the algorithm to make the initial guess. A popular strategy is to assume that the first-order change in the function at iterate x_k will be the same as that obtained at the previous step. In other words, we choose the initial guess α_0 so that $\alpha_0 \nabla f_k^T p_k = \alpha_{k-1} \nabla f_{k-1}^T p_{k-1}$. We therefore have

$$\alpha_0 = \alpha_{k-1} \frac{\nabla f_{k-1}^T p_{k-1}}{\nabla f_k^T p_k}.$$

Another useful strategy is to interpolate a quadratic to the data $f(x_{k-1})$, $f(x_k)$, and $\phi'(0) = \nabla f_k^T p_k$ and to define α_0 to be its minimizer. This strategy yields

$$\alpha_0 = \frac{2(f_k - f_{k-1})}{\phi'(0)}. \tag{3.44}$$

It can be shown that if $x_k \to x^*$ superlinearly, then the ratio in this expression converges to 1. If we adjust the choice (3.44) by setting

$$\alpha_0 \leftarrow \min(1, 1.01\alpha_0),$$

we find that the unit step length $\alpha_0 = 1$ will eventually always be tried and accepted, and the superlinear convergence properties of Newton and quasi-Newton methods will be observed.

A LINE SEARCH ALGORITHM FOR THE WOLFE CONDITIONS

The Wolfe (or strong Wolfe) conditions are among the most widely applicable and useful termination conditions. We now describe in some detail a one-dimensional search procedure that is guaranteed to find a step length satisfying the *strong* Wolfe conditions (3.7)

3.4. STEP-LENGTH SELECTION ALGORITHMS

for any parameters c_1 and c_2 satisfying $0 < c_1 < c_2 < 1$. As before, we assume that p is a descent direction and that f is bounded below along the direction p.

The algorithm has two stages. This first stage begins with a trial estimate α_1, and keeps increasing it until it finds either an acceptable step length or an interval that brackets the desired step lengths. In the latter case, the second stage is invoked by calling a function called **zoom** (Algorithm 3.3 below), which successively decreases the size of the interval until an acceptable step length is identified.

A formal specification of the line search algorithm follows. We refer to (3.7a) as the *sufficient decrease condition* and to (3.7b) as the *curvature condition*. The parameter $\alpha_{\max}$ is a user-supplied bound on the maximum step length allowed. The line search algorithm terminates with α_* set to a step length that satisfies the strong Wolfe conditions.

Algorithm 3.2 (Line Search Algorithm).
    Set $\alpha_0 \leftarrow 0$, choose $\alpha_1 > 0$ and $\alpha_{\max}$;
    $i \leftarrow 1$;
    repeat
        Evaluate $\phi(\alpha_i)$;
        if $\phi(\alpha_i) > \phi(0) + c_1 \alpha_i \phi'(0)$ or $[\phi(\alpha_i) \geq \phi(\alpha_{i-1})$ and $i > 1]$
            $\alpha_* \leftarrow$ **zoom**(α_{i-1}, α_i) and **stop**;
        Evaluate $\phi'(\alpha_i)$;
        if $|\phi'(\alpha_i)| \leq -c_2 \phi'(0)$
            set $\alpha_* \leftarrow \alpha_i$ and **stop**;
        if $\phi'(\alpha_i) \geq 0$
            set $\alpha_* \leftarrow$ **zoom**(α_i, α_{i-1}) and **stop**;
        Choose $\alpha_{i+1} \in (\alpha_i, \alpha_{\max})$
        $i \leftarrow i + 1$;
    end (repeat)

Note that the sequence of trial step lengths $\{\alpha_i\}$ is monotonically increasing, but that the order of the arguments supplied to the **zoom** function may vary. The procedure uses the knowledge that the interval (α_{i-1}, α_i) contains step lengths satisfying the strong Wolfe conditions if one of the following three conditions is satisfied:

(i) α_i violates the sufficient decrease condition;

(ii) $\phi(\alpha_i) \geq \phi(\alpha_{i-1})$;

(iii) $\phi'(\alpha_i) \geq 0$.

The last step of the algorithm performs extrapolation to find the next trial value α_{i+1}. To implement this step we can use approaches like the interpolation procedures above, or we can simply set α_{i+1} to some constant multiple of α_i. Whichever strategy we use, it is important that the successive steps increase quickly enough to reach the upper limit $\alpha_{\max}$ in a finite number of iterations.

We now specify the function **zoom**, which requires a little explanation. The order of its input arguments is such that each call has the form **zoom**$(\alpha_{\text{lo}}, \alpha_{\text{hi}})$, where

(a) the interval bounded by α_{lo} and α_{hi} contains step lengths that satisfy the strong Wolfe conditions;

(b) α_{lo} is, among all step lengths generated so far and satisfying the sufficient decrease condition, the one giving the smallest function value; and

(c) α_{hi} is chosen so that $\phi'(\alpha_{\text{lo}})(\alpha_{\text{hi}} - \alpha_{\text{lo}}) < 0$.

Each iteration of **zoom** generates an iterate α_j between α_{lo} and α_{hi}, and then replaces one of these endpoints by α_j in such a way that the properties (a), (b), and (c) continue to hold.

Algorithm 3.3 (zoom).
    repeat
        Interpolate (using quadratic, cubic, or bisection) to find
            a trial step length α_j between α_{lo} and α_{hi};
        Evaluate $\phi(\alpha_j)$;
        if $\phi(\alpha_j) > \phi(0) + c_1 \alpha_j \phi'(0)$ or $\phi(\alpha_j) \geq \phi(\alpha_{\text{lo}})$
            $\alpha_{\text{hi}} \leftarrow \alpha_j$;
        else
            Evaluate $\phi'(\alpha_j)$;
            if $|\phi'(\alpha_j)| \leq -c_2 \phi'(0)$
                Set $\alpha_* \leftarrow \alpha_j$ and **stop**;
            if $\phi'(\alpha_j)(\alpha_{\text{hi}} - \alpha_{\text{lo}}) \geq 0$
                $\alpha_{\text{hi}} \leftarrow \alpha_{\text{lo}}$;
            $\alpha_{\text{lo}} \leftarrow \alpha_j$;
    end (repeat)

If the new estimate α_j happens to satisfy the strong Wolfe conditions, then **zoom** has served its purpose of identifying such a point, so it terminates with $\alpha_* = \alpha_j$. Otherwise, if α_j satisfies the sufficient decrease condition and has a lower function value than x_{lo}, then we set $\alpha_{\text{lo}} \leftarrow \alpha_j$ to maintain condition (b). If this results in a violation of condition (c), we remedy the situation by setting α_{hi} to the old value of α_{lo}. The reader should sketch some graphs to illustrate the workings of **zoom**!

As mentioned earlier, the interpolation step that determines α_j should be safeguarded to ensure that the new step length is not too close to the endpoints of the interval. Practical line search algorithms also make use of the properties of the interpolating polynomials to make educated guesses of where the next step length should lie; see [27, 172]. A problem that can arise in the implementation is that as the optimization algorithm approaches the solution, two consecutive function values $f(x_k)$ and $f(x_{k-1})$ may be indistinguishable in finite-precision arithmetic. Therefore, the line search must include a stopping test if it cannot attain a lower function value after a certain number (typically, ten) of trial step lengths. Some procedures also stop if the relative change in x is close to machine accuracy, or to some user-specified threshold.

A line search algorithm that incorporates all these features is difficult to code. We advocate the use of one of the several good software implementations available in the public domain. See Dennis and Schnabel [69], Lemaréchal [149], Fletcher [83], and in particular Moré and Thuente [172].

One may ask how much more expensive it is to require the strong Wolfe conditions instead of the regular Wolfe conditions. Our experience suggests that for a "loose" line search (with parameters such as $c_1 = 10^{-4}$ and $c_2 = 0.9$), both strategies require a similar amount of work. The strong Wolfe conditions have the advantage that by decreasing c_2 we can directly control the quality of the search by forcing the accepted value of α to lie closer to a local minimum. This feature is important in steepest descent or nonlinear conjugate gradient methods, and therefore a step selection routine that enforces the strong Wolfe conditions is of wider applicability.

NOTES AND REFERENCES

For an extensive discussion of line search termination conditions see Ortega and Rheinboldt [185]. Akaike [2] presents a probabilistic analysis of the steepest descent method with exact line searches on quadratic functions. He shows that when $n > 2$, the worst-case bound (3.28) can be expected to hold for most starting points. The case where $n = 2$ can be studied in closed form; see Bazaraa, Sherali, and Shetty [7].

Some line search methods (see Goldfarb [113] and Moré and Sorensen [169]) compute a direction of negative curvature, whenever it exists, to prevent the iteration from converging to nonminimizing stationary points. A direction of negative curvature p_- is one that satisfies $p_-^T \nabla^2 f(x_k) p_- < 0$. These algorithms generate a search direction by combining p_- with the steepest descent direction $-\nabla f$, and often perform a curvilinear backtracking line search. It is difficult to determine the relative contributions of the steepest descent and negative curvature directions, and due to this, this approach fell out of favor after the introduction of trust-region methods.

For a discussion on the rate of convergence of the coordinate descent method and for more references about this method see Luenberger [152].

Derivative-free line search algorithms include golden section and Fibonacci search. They share some of the features with the line search method given in this chapter. They typically store three trial points that determine an interval containing a one-dimensional minimizer. Golden section and Fibonacci differ in the way in which the trial step lengths are generated; see, for example, [58, 27].

Our discussion of interpolation follows Dennis and Schnabel [69], and the algorithm for finding a step length satisfying the strong Wolfe conditions can be found in Fletcher [83].

EXERCISES

3.1 Program the steepest descent and Newton algorithms using the backtracking line search, Procedure 3.1. Use them to minimize the Rosenbrock function (2.23). Set the initial step length $\alpha_0 = 1$ and print the step length used by each method at each iteration. First try the initial point $x_0 = (1.2, 1.2)$ and then the more difficult point $x_0 = (-1.2, 1)$.

3.2 Show that if $0 < c_2 < c_1 < 1$, then there may be no step lengths that satisfy the Wolfe conditions.

3.3 Show that the one-dimensional minimizer of a strongly convex quadratic function is given by (3.39).

3.4 Show that if $c \le \frac{1}{2}$, then the one-dimensional minimizer of a strongly convex quadratic function always satisfies the Goldstein conditions (3.11).

3.5 Prove that $\|Bx\| \ge \|x\|/\|B^{-1}\|$ for any nonsingular matrix B. Use this to establish (3.19).

3.6 Consider the steepest descent method with exact line searches applied to the convex quadratic function (3.24). Using the properties given in this chapter, show that if the initial point is such that $x_0 - x^*$ is parallel to an eigenvector of Q, then the steepest descent method will find the solution in one step.

3.7 Prove the result (3.28) by working through the following steps. First, use (3.26) to show that

$$\|x_k - x^*\|_Q^2 - \|x_{k+1} - x^*\|_Q^2 = 2\alpha_k \nabla f_k^T Q(x_k - x^*) - \alpha_k^2 \nabla f_k^T Q \nabla f_k,$$

where $\|\cdot\|_Q$ is defined by (3.27). Second, use the fact that $\nabla f_k = Q(x_k - x^*)$ to obtain

$$\|x_k - x^*\|_Q^2 - \|x_{k+1} - x^*\|_Q^2 = \frac{2(g_k^T g_k)^2}{(g_k^T Q g_k)} - \frac{(g_k^T g_k)^2}{(g_k^T Q g_k)}$$

and

$$\|x_k - x^*\|_Q^2 = \nabla f_k^T Q^{-1} \nabla f_k.$$

3.8 Let Q be a positive definite symmetric matrix. Prove that for any vector x,

$$\frac{(x^T x)^2}{(x^T Q x)(x^T Q^{-1} x)} \ge \frac{4 \lambda_n \lambda_1}{(\lambda_n + \lambda_1)^2},$$

where λ_n and λ_1 are, respectively, the largest and smallest eigenvalues of Q. (This relation, which is known as the Kantorovich inequality, can be used to deduce (3.29) from (3.28).

✏ **3.9** Program the BFGS algorithm using the line search algorithm described in this chapter that implements the strong Wolfe conditions. Have the code verify that $y_k^T s_k$ is always positive. Use it to minimize the Rosenbrock function using the starting points given in Exercise 1.

✏ **3.10** Show that the quadratic function that interpolates $\phi(0)$, $\phi'(0)$, and $\phi(\alpha_0)$ is given by (3.41). Then, make use of the fact that the sufficient decrease condition (3.6a) is not satisfied at α_0 to show that this quadratic has positive curvature and that the minimizer satisfies

$$\alpha_1 < \frac{1}{2(1-c_1)}.$$

Since c_1 is chosen to be quite small in practice, this indicates that α_1 cannot be much greater than $\frac{1}{2}$ (and may be smaller), which gives us an idea of the new step length.

✏ **3.11** If $\phi(\alpha_0)$ is large, (3.42) shows that α_1 can be quite small. Give an example of a function and a step length α_0 for which this situation arises. (Drastic changes to the estimate of the step length are not desirable, since they indicate that the current interpolant does not provide a good approximation to the function and that it should be modified before being trusted to produce a good step length estimate. In practice, one imposes a lower bound—typically, $\rho = 0.1$—and defines the new step length as $\alpha_i = \max(\rho\alpha_{i-1}, \hat{\alpha}_i)$, where $\hat{\alpha}_i$ is the minimizer of the interpolant.)

✏ **3.12** Suppose that the sufficient decrease condition (3.6a) is not satisfied at the step lengths α_0, and α_1, and consider the cubic interpolating $\phi(0)$, $\phi'(0)$, $\phi(\alpha_0)$ and $\phi(\alpha_1)$. By drawing graphs illustrating the two situations that can arise, show that the minimizer of the cubic lies in $[0, \alpha_1]$. Then show that if $\phi(0) < \phi(\alpha_1)$, the minimizer is less than $\frac{2}{3}\alpha_1$.

CHAPTER 4

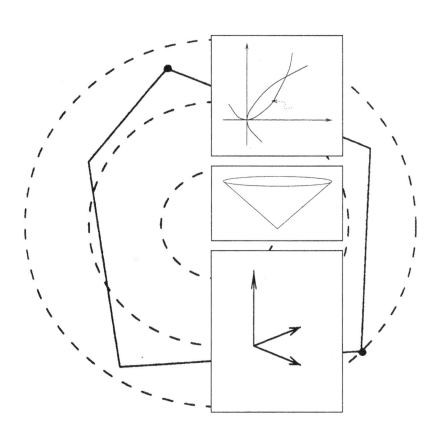

Trust-Region Methods

Line search methods and trust-region methods both generate steps with the help of a quadratic model of the objective function, but they use this model in different ways. Line search methods use it to generate a search direction, and then focus their efforts on finding a suitable step length α along this direction. Trust-region methods define a region around the current iterate within which they trust the model to be an adequate representation of the objective function, and then choose the step to be the approximate minimizer of the model in this trust region. In effect, they choose the direction and length of the step simultaneously. If a step is not acceptable, they reduce the size of the region and find a new minimizer. In general, the step direction changes whenever the size of the trust region is altered.

The size of the trust region is critical to the effectiveness of each step. If the region is too small, the algorithm misses an opportunity to take a substantial step that will move it much closer to the minimizer of the objective function. If too large, the minimizer of the model may be far from the minimizer of the objective function in the region, so we may have to reduce the size of the region and try again. In practical algorithms, we choose the size of the region according to the performance of the algorithm during previous iterations. If the model is generally reliable, producing good steps and accurately predicting the behavior of the objective function along these steps, the size of the trust region is steadily increased to

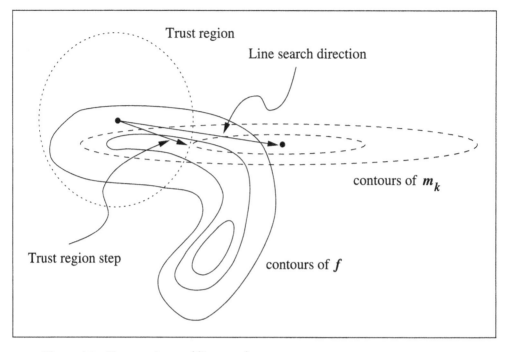

Figure 4.1 Trust-region and line search steps.

allow longer, more ambitious, steps to be taken. On the other hand, a failed step indicates that our model is an inadequate representation of the objective function over the current trust region, so we reduce the size of the region and try again.

Figure 4.1 illustrates the trust-region approach on a function f of two variables in which the current point lies at one end of a curved valley while the minimizer x^* lies at the other end. The quadratic model function m_k, whose elliptical contours are shown as dashed lines, is based on function and derivative information at x_k and possibly also on information accumulated from previous iterations and steps. A line search method based on this model searches along the step to the minimizer of m_k (shown), but this direction allows only a small reduction in f even if an optimal step is taken. A trust-region method, on the other hand, steps to the minimizer of m_k within the dotted circle, which yields a more significant reduction in f and a better step.

We will assume that the first two terms of the quadratic model functions m_k at each iterate x_k are identical to the first two terms of the Taylor-series expansion of f around x_k. Specifically, we have

$$m_k(p) = f_k + \nabla f_k^T p + \tfrac{1}{2} p^T B_k p, \qquad (4.1)$$

where $f_k = f(x_k)$, $\nabla f_k = \nabla f(x_k)$, and B_k is some symmetric matrix. Since by (2.6) we have

$$f(x_k + p) = f_k + \nabla f_k^T p + \tfrac{1}{2} p^T \nabla^2 f(x_k + tp) p, \tag{4.2}$$

for some scalar $t \in (0, 1)$, and since $m_k(p) = f_k + \nabla f_k^T p + O(\|p\|^2)$, the difference between $m_k(p)$ and $f(x_k + p)$ is $O(\|p\|^2)$, so the approximation error is small when p is small.

When B_k is equal to the true Hessian $\nabla^2 f(x_k)$, the model function actually agrees with the Taylor series to *three* terms. The approximation error is $O(\|p\|^3)$ in this case, so this model is especially accurate when $\|p\|$ is small. The algorithm based on setting $B_k = \nabla^2 f(x_k)$ is called the trust-region Newton method, and will be discussed further in Chapter 6. In the current chapter, we emphasize the generality of the trust-region approach by assuming little about B_k except symmetry and uniform boundedness in the index k.

To obtain each step, we seek a solution of the subproblem

$$\min_{p \in \mathbb{R}^n} m_k(p) = f_k + \nabla f_k^T p + \tfrac{1}{2} p^T B_k p \quad \text{s.t.} \ \|p\| \leq \Delta_k, \tag{4.3}$$

where $\Delta_k > 0$ is the trust-region radius. For the moment, we define $\|\cdot\|$ to be the Euclidean norm, so that the solution p_k^* of (4.3) is the minimizer of m_k in the ball of radius Δ_k. Thus, the trust-region approach requires us to solve a sequence of subproblems (4.3) in which the objective function and constraint (which can be written as $p^T p \leq \Delta_k^2$) are both quadratic. When B_k is positive definite and $\|B_k^{-1} \nabla f_k\| \leq \Delta_k$, the solution of (4.3) is easy to identify—it is simply the unconstrained minimum $p_k^B = -B_k^{-1} \nabla f_k$ of the quadratic $m_k(p)$. In this case, we call p_k^B the *full step*. The solution of (4.3) is not so obvious in other cases, but it can usually be found without too much expense. In any case, we need only an *approximate* solution to obtain convergence and good practical behavior.

OUTLINE OF THE ALGORITHM

The first issue to arise in defining a trust-region method is the strategy for choosing the trust-region radius Δ_k at each iteration. We base this choice on the agreement between the model function m_k and the objective function f at previous iterations. Given a step p_k we define the ratio

$$\rho_k = \frac{f(x_k) - f(x_k + p_k)}{m_k(0) - m_k(p_k)}; \tag{4.4}$$

the numerator is called the *actual reduction*, and the denominator is the *predicted reduction*. Note that since the step p_k is obtained by minimizing the model m_k over a region that includes the step $p = 0$, the predicted reduction will always be nonnegative. Thus if ρ_k is negative, the new objective value $f(x_k + p_k)$ is greater than the current value $f(x_k)$, so the step must be rejected.

On the other hand, if ρ_k is close to 1, there is good agreement between the model m_k and the function f over this step, so it is safe to expand the trust region for the next iteration. If ρ_k is positive but not close to 1, we do not alter the trust region, but if it is close to zero or negative, we shrink the trust region. The following algorithm describes the process.

Algorithm 4.1 (Trust Region).
Given $\bar{\Delta} > 0$, $\Delta_0 \in (0, \bar{\Delta})$, and $\eta \in [0, \frac{1}{4})$:
for $k = 0, 1, 2, \ldots$
    Obtain p_k by (approximately) solving (4.3);
    Evaluate ρ_k from (4.4);
    if $\rho_k < \frac{1}{4}$
        $\Delta_{k+1} = \frac{1}{4} \|p_k\|$
    else
        if $\rho_k > \frac{3}{4}$ and $\|p_k\| = \Delta_k$
            $\Delta_{k+1} = \min(2\Delta_k, \bar{\Delta})$
        else
            $\Delta_{k+1} = \Delta_k$;
    if $\rho_k > \eta$
        $x_{k+1} = x_k + p_k$
    else
        $x_{k+1} = x_k$;
end (for).

Here $\bar{\Delta}$ is an overall bound on the step lengths. Note that the radius is increased only if $\|p_k\|$ actually reaches the boundary of the trust region. If the step stays strictly inside the region, we infer that the current value of Δ_k is not interfering with the progress of the algorithm, so we leave its value unchanged for the next iteration.

To turn Algorithm 4.1 into a practical algorithm, we need to focus on solving (4.3). We first describe three strategies for finding *approximate* solutions, which achieve at least as much reduction in m_k as the reduction achieved by the so-called *Cauchy point*. This point is simply the minimizer of m_k along the steepest descent direction $-\nabla f_k$, subject to the trust-region bound. The first approximate strategy is the *dogleg method*, which is appropriate when the model Hessian B_k is positive definite. The second strategy, known as *two-dimensional subspace minimization*, can be applied when B_k is indefinite, though it requires an estimate of the most negative eigenvalue of this matrix. The third strategy, due to Steihaug, is most appropriate when B_k is the exact Hessian $\nabla^2 f(x_k)$ and when this matrix is large and sparse.

We also describe a strategy due to Moré and Sorensen that finds a "nearly exact" solution of (4.3). This strategy is based on the fact that the solution p satisfies $(B_k + \lambda I)p = -\nabla f_k$ for some positive value of $\lambda > 0$. This strategy seeks the value of λ that corresponds to

4.1 THE CAUCHY POINT AND RELATED ALGORITHMS

the trust-region radius Δ_k and performs additional calculations in the special case in which the resulting modified Hessian $(B_k + \lambda I)$ is nonsingular. Details are given below.

THE CAUCHY POINT

As we saw in the previous chapter, line search methods do not require optimal step lengths to be globally convergent. In fact, only a crude approximation to the optimal step length that satisfies certain loose criteria is needed. A similar situation applies in trust-region methods. Although in principle we are seeking the optimal solution of the subproblem (4.3), it is enough for global convergence purposes to find an approximate solution p_k that lies within the trust region and gives a *sufficient reduction* in the model. The sufficient reduction can be quantified in terms of the Cauchy point, which we denote by p_k^c and define in terms of the following simple procedure:

Algorithm 4.2 (Cauchy Point Calculation).
Find the vector p_k^s that solves a linear version of (4.3), that is,

$$p_k^s = \arg\min_{p \in \mathbb{R}^n} f_k + \nabla f_k^T p \quad \text{s.t.} \ \|p\| \leq \Delta_k; \qquad (4.5)$$

Calculate the scalar $\tau_k > 0$ that minimizes $m_k(\tau p_k^s)$ subject to satisfying the trust-region bound, that is,

$$\tau_k = \arg\min_{\tau > 0} m_k(\tau p_k^s) \quad \text{s.t.} \ \|\tau p_k^s\| \leq \Delta_k; \qquad (4.6)$$

Set $p_k^c = \tau_k p_k^s$.

In fact, it is easy to write down a closed-form definition of the Cauchy point. The solution of (4.5) is simply

$$p_k^s = -\frac{\Delta_k}{\|\nabla f_k\|} \nabla f_k.$$

To obtain τ_k explicitly, we consider the cases of $\nabla f_k^T B_k \nabla f_k \leq 0$ and $\nabla f_k^T B_k \nabla f_k > 0$ separately. For the former case, the function $m_k(\tau p_k^s)$ decreases monotonically with τ whenever $\nabla f_k \neq 0$, so τ_k is simply the largest value that satisfies the trust-region bound, namely, $\tau_k = 1$. For the case $\nabla f_k^T B_k \nabla f_k > 0$, $m_k(\tau p_k^s)$ is a convex quadratic in τ, so τ_k is either

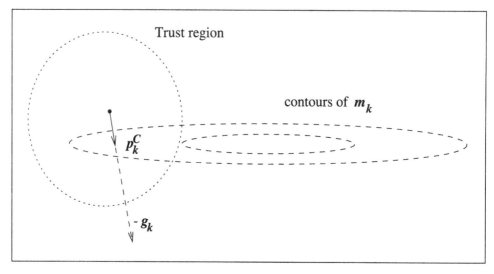

Figure 4.2 The Cauchy point.

the unconstrained minimizer of this quadratic, $\|\nabla f_k\|^3/(\Delta_k \nabla f_k^T B_k \nabla f_k)$, or the boundary value 1, whichever comes first. In summary, we have

$$p_k^C = -\tau_k \frac{\Delta_k}{\|\nabla f_k\|} \nabla f_k, \qquad (4.7)$$

where

$$\tau_k = \begin{cases} 1 & \text{if } \nabla f_k^T B_k \nabla f_k \leq 0; \\ \min\left(\|\nabla f_k\|^3/(\Delta_k \nabla f_k^T B_k \nabla f_k), 1\right) & \text{otherwise.} \end{cases} \qquad (4.8)$$

Figure 4.2 illustrates the Cauchy point for a subproblem in which B_k is positive definite. In this example, p_k^C lies strictly inside the trust region.

The Cauchy step p_k^C is inexpensive to calculate—no matrix factorizations are required—and is of crucial importance in deciding if an approximate solution of the trust-region subproblem is acceptable. Specifically, a trust-region method will be globally convergent if its steps p_k attain a sufficient reduction in m_k; that is, they give a reduction in the model m_k that is at least some fixed multiple of the decrease attained by the Cauchy step at each iteration.

IMPROVING ON THE CAUCHY POINT

Since the Cauchy point p_k^C provides sufficient reduction in the model function m_k to yield global convergence, and since the cost of calculating it is so small, why should we look

4.1. THE CAUCHY POINT AND RELATED ALGORITHMS

any further for a better approximate solution of (4.3)? The reason is that by always taking the Cauchy point as our step, we are simply implementing the steepest descent method with a particular choice of step length. As we have seen in Chapter 3, steepest descent performs poorly even if an *optimal* step length is used at each iteration.

The Cauchy point does not depend very strongly on the matrix B_k, which is used only in the calculation of the step length. Rapid convergence (superlinear, for instance) can be expected only if B_k plays a role in determining the *direction* of the step as well as its length.

A number of algorithms for generating approximate solutions p_k to the trust-region problem (4.3) start by computing the Cauchy point and then try to improve on it. The improvement strategy is often designed so that the full step $p_k^B = -B_k^{-1} \nabla f_k$ is chosen whenever B_k is positive definite and $\|p_k^B\| \leq \Delta_k$. When B_k is the exact Hessian $\nabla^2 f(x_k)$ or a quasi-Newton approximation, this strategy can be expected to yield superlinear convergence.

We now consider three methods for finding approximate solutions to (4.3) that have the features just described. Throughout this section we will be focusing on the internal workings of a single iteration, so we drop the subscript "k" from the quantities Δ_k, p_k, and m_k to simplify the notation. With this simplification, we restate the trust-region subproblem (4.3) as follows:

$$\min_{p \in \mathbb{R}^n} m(p) \stackrel{\text{def}}{=} f + g^T p + \tfrac{1}{2} p^T B p \quad \text{s.t.} \ \|p\| \leq \Delta. \tag{4.9}$$

We denote the solution of (4.9) by $p^*(\Delta)$, to emphasize the dependence on Δ.

THE DOGLEG METHOD

We start by examining the effect of the trust-region radius Δ on the solution $p^*(\Delta)$ of the subproblem (4.9). When B is positive definite, we have already noted that the unconstrained minimizer of m is the full step $p^B = -B^{-1}g$. When this point is feasible for (4.9), it is obviously a solution, so we have

$$p^*(\Delta) = p^B, \quad \text{when } \Delta \geq \|p^B\|. \tag{4.10}$$

When Δ is tiny, the restriction $\|p\| \leq \Delta$ ensures that the quadratic term in m has little effect on the solution of (4.9). The true solution $p(\Delta)$ is approximately the same as the solution we would obtain by minimizing the linear function $f + g^T p$ over $\|p\| \leq \Delta$, that is,

$$p^*(\Delta) \approx -\Delta \frac{g}{\|g\|}, \quad \text{when } \Delta \text{ is small.} \tag{4.11}$$

For intermediate values of Δ, the solution $p^*(\Delta)$ typically follows a curved trajectory like the one in Figure 4.3.

The *dogleg method* finds an approximate solution by replacing the curved trajectory for $p^*(\Delta)$ with a path consisting of two line segments. The first line segment runs from the

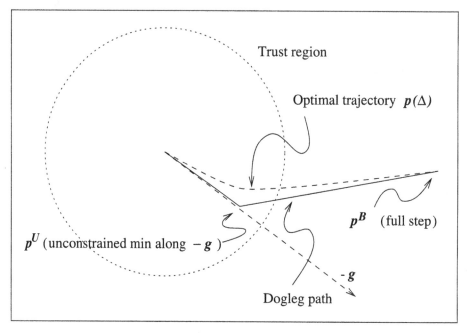

Figure 4.3 Exact trajectory and dogleg approximation.

origin to the unconstrained minimizer along the steepest descent direction defined by

$$p^{\text{U}} = -\frac{g^T g}{g^T B g} g, \tag{4.12}$$

while the second line segment runs from p^{U} to p^{B} (see Figure 4.3). Formally, we denote this trajectory by $\tilde{p}(\tau)$ for $\tau \in [0, 2]$, where

$$\tilde{p}(\tau) = \begin{cases} \tau p^{\text{U}}, & 0 \leq \tau \leq 1, \\ p^{\text{U}} + (\tau - 1)(p^{\text{B}} - p^{\text{U}}), & 1 \leq \tau \leq 2. \end{cases} \tag{4.13}$$

The dogleg method chooses p to minimize the model m along this path, subject to the trust-region bound. In fact, it is not even necessary to carry out a search, because the dogleg path intersects the trust-region boundary at most once and the intersection point can be computed analytically. We prove these claims in the following lemma.

Lemma 4.1.

Let B be positive definite. Then

(i) $\|\tilde{p}(\tau)\|$ *is an increasing function of τ, and*

(ii) $m(\tilde{p}(\tau))$ *is a decreasing function of τ.*

4.1. THE CAUCHY POINT AND RELATED ALGORITHMS

PROOF. It is easy to show that (i) and (ii) both hold for $\tau \in [0, 1]$, so we restrict our attention to the case of $\tau \in [1, 2]$.

For (i), define $h(\alpha)$ by

$$h(\alpha) = \tfrac{1}{2} \|\tilde{p}(1+\alpha)\|^2$$
$$= \tfrac{1}{2} \|p^{\scriptscriptstyle U} + \alpha(p^{\scriptscriptstyle B} - p^{\scriptscriptstyle U})\|^2$$
$$= \tfrac{1}{2} \|p^{\scriptscriptstyle U}\|^2 + \alpha p^{{\scriptscriptstyle U}T}(p^{\scriptscriptstyle B} - p^{\scriptscriptstyle U}) + \tfrac{1}{2}\alpha^2 \|p^{\scriptscriptstyle B} - p^{\scriptscriptstyle U}\|^2.$$

Our result is proved if we can show that $h'(\alpha) \geq 0$ for $\alpha \in (0, 1)$. Now,

$$h'(\alpha) = -p^{{\scriptscriptstyle U}T}(p^{\scriptscriptstyle U} - p^{\scriptscriptstyle B}) + \alpha \|p^{\scriptscriptstyle U} - p^{\scriptscriptstyle B}\|^2$$
$$\geq -p^{{\scriptscriptstyle U}T}(p^{\scriptscriptstyle U} - p^{\scriptscriptstyle B})$$
$$= \frac{g^T g}{g^T B g} g^T \left(-\frac{g^T g}{g^T B g} g + B^{-1} g \right)$$
$$= g^T g \frac{g B^{-1} g}{g^T B g} \left[1 - \frac{(g^T g)^2}{(g^T B g)(g^T B^{-1} g)} \right]$$
$$\geq 0,$$

where the final inequality follows from Exercise 3.

For (ii), we define $\hat{h}(\alpha) = m(\tilde{p}(1+\alpha))$ and show that $\hat{h}'(\alpha) \leq 0$ for $\alpha \in (0, 1)$. Substitution of (4.13) into (4.9) and differentiation with respect to the argument leads to

$$\hat{h}'(\alpha) = (p^{\scriptscriptstyle B} - p^{\scriptscriptstyle U})^T (g + B p^{\scriptscriptstyle U}) + \alpha (p^{\scriptscriptstyle B} - p^{\scriptscriptstyle U})^T B (p^{\scriptscriptstyle B} - p^{\scriptscriptstyle U})$$
$$\leq (p^{\scriptscriptstyle B} - p^{\scriptscriptstyle U})^T (g + B p^{\scriptscriptstyle U} + B(p^{\scriptscriptstyle B} - p^{\scriptscriptstyle U}))$$
$$= (p^{\scriptscriptstyle B} - p^{\scriptscriptstyle U})^T (g + B p^{\scriptscriptstyle B}) = 0,$$

giving the result. □

It follows from this lemma that the path $\tilde{p}(\tau)$ intersects the trust-region boundary $\|p\| = \Delta$ at exactly one point if $\|p^{\scriptscriptstyle B}\| \geq \Delta$, and nowhere otherwise. Since m is decreasing along the path, the chosen value of p will be at $p^{\scriptscriptstyle B}$ if $\|p^{\scriptscriptstyle B}\| \leq \Delta$, otherwise at the point of intersection of the dogleg and the trust-region boundary. In the latter case, we compute the appropriate value of τ by solving the following scalar quadratic equation:

$$\|p^{\scriptscriptstyle U} + (\tau - 1)(p^{\scriptscriptstyle B} - p^{\scriptscriptstyle U})\|^2 = \Delta^2.$$

The dogleg strategy can be adapted to handle indefinite B, but there is not much point in doing so because the full step $p^{\scriptscriptstyle B}$ is not the unconstrained minimizer of m in this case. Instead, we now describe another strategy, which aims to include directions of negative curvature (that is, directions d for which $d^T B d < 0$) in the space of candidate trust-region steps.

TWO-DIMENSIONAL SUBSPACE MINIMIZATION

When B is positive definite, the dogleg method strategy can be made slightly more sophisticated by widening the search for p to the entire two-dimensional subspace spanned by p^U and p^B (equivalently, g and $-B^{-1}g$). The subproblem (4.9) is replaced by

$$\min_p m(p) = f + g^T p + \tfrac{1}{2} p^T B p \quad \text{s.t.} \quad \|p\| \leq \Delta, \ p \in \text{span}[g, B^{-1}g]. \tag{4.14}$$

This is a problem in two variables that can be solved without much effort (see the exercises). Clearly, the Cauchy point p^c is feasible for (4.14), so the optimal solution of this subproblem yields at least as much reduction in m as the Cauchy point, resulting in global convergence of the algorithm. The two-dimensional subspace minimization strategy is obviously an extension of the dogleg method as well, since the entire dogleg path lies in $\text{span}[g, B^{-1}g]$.

An advantage of this strategy is that it can be modified to handle the case of indefinite B in a way that is intuitive, practical, and theoretically sound. We mention just the salient points of the handling of the indefiniteness here, and refer the reader to papers by Byrd, Schnabel, and Schultz (see [39] and [226]) for details. When B has negative eigenvalues, the two-dimensional subspace in (4.14) is changed to

$$\text{span}[g, (B + \alpha I)^{-1}g], \quad \text{for some } \alpha \in (-\lambda_1, -2\lambda_1], \tag{4.15}$$

where λ_1 denotes the most negative eigenvalue of B. (This choice of α ensures that $B + \alpha I$ is positive definite, and the flexibility in this definition allows us to use a numerical procedure such as the Lanczos method to compute an acceptable value of α.) When $\|(B+\alpha I)^{-1}g\| \leq \Delta$, we discard the subspace search of (4.14), (4.15) and instead define the step to be

$$p = -(B + \alpha I)^{-1}g + v, \tag{4.16}$$

where v is a vector that satisfies $v^T(B + \alpha I)^{-1}g \leq 0$. (This condition ensures that v does not move p back toward zero, but instead continues to move roughly in the direction of $-(B + \alpha I)^{-1}g$.)

When B has zero eigenvalues but no negative eigenvalues, the Cauchy step $p = p^c$ is used as the approximate solution of (4.9).

The reduction in model function m achieved by the two-dimensional minimization strategy often is close to the reduction achieved by the exact solution of (4.9). Most of the computational effort lies in a single factorization of B or $B + \alpha I$ (estimation of α and solution of (4.14) are less significant), while strategies that find nearly exact solutions of (4.9) typically require two or three such factorizations.

STEIHAUG'S APPROACH

Both methods described above require the solution of a single linear system involving B or $(B + \alpha I)$. When B is large, this operation may be quite costly, so we are motivated to consider other techniques for finding an approximate solution of (4.9) that do not require exact solution of a linear system but still produce an improvement on the Cauchy point. Steihaug [231] proposed a technique with these properties. Steihaug's implementation is based on the conjugate gradient algorithm, an iterative algorithm for solving linear systems with symmetric positive definite coefficient matrices. The conjugate gradient (CG) algorithm is the subject of Chapter 5, and the interested reader should look ahead to that chapter for further details. Our comments in this section focus on the differences between standard CG and Steihaug's approach, which are essentially that the algorithm terminates when it either exits the trust region $\|p\| \leq \Delta$ or when it encounters a direction of negative curvature in B.

Steihaug's approach can be stated formally as follows:

Algorithm 4.3 (CG–Steihaug).
Given $\epsilon > 0$;
Set $p_0 = 0, r_0 = g, d_0 = -r_0$;
if $\|r_0\| < \epsilon$
  return $p = p_0$;
for $j = 0, 1, 2, \ldots$
  if $d_j^T B d_j \leq 0$
    Find τ such that $p = p_j + \tau d_j$ minimizes $m(p)$ in (4.9)
    and satisfies $\|p\| = \Delta$;
    return p;
  Set $\alpha_j = r_j^T r_j / d_j^T B d_j$;
  Set $p_{j+1} = p_j + \alpha_j d_j$;
  if $\|p_{j+1}\| \geq \Delta$
    Find $\tau \geq 0$ such that $p = p_j + \tau d_j$ satisfies $\|p\| = \Delta$;
    return p;
  Set $r_{j+1} = r_j + \alpha_j B d_j$;
  if $\|r_{j+1}\| < \epsilon \|r_0\|$
    return $p = p_{j+1}$;
  Set $\beta_{j+1} = r_{j+1}^T r_{j+1} / r_j^T r_j$;
  Set $d_{j+1} = r_{j+1} + \beta_{j+1} d_j$;
end (for).

To connect this algorithm with Algorithm CG of Chapter 5, we note that $m(\cdot)$ takes the place of $\phi(\cdot)$, p takes the place of x, B takes the place of A, and $-g$ takes the place of b. The change of sign in the substitution $b \to -g$ propagates through the algorithm.

Algorithm 4.3 differs from standard CG in that two extra stopping criteria are present—the first two **if** statements inside the **for** loop. The first **if** statement stops the method if its current search direction d_j is a direction of zero curvature or negative curvature along B. The second one causes termination if p_{j+1} violates the trust-region bound. In both cases, a final point p is found by intersecting the current search direction with the trust-region boundary.

The initialization of p_0 to zero is a crucial feature of the algorithm. After the first iteration (assuming $\|r_0\|_2 \geq \epsilon$), we have

$$p_1 = \alpha_0 d_0 = \frac{r_0^T r_0}{d_0^T B d_0} d_0 = -\frac{g^T g}{g^T B g} g,$$

which is exactly the Cauchy point! Since each iteration of the conjugate gradient method reduces $m(\cdot)$, this algorithm fulfills the necessary condition for global convergence.

Another crucial property of the method is that each iterate p_j is larger in norm than its predecessor. This property is another consequence of the initialization $p_0 = 0$. Its main implication is that it is acceptable to stop iterating as soon as the trust-region boundary is reached, because no further iterates giving a lower value of ϕ will be inside the trust region. We state and prove this property formally in the following theorem. (The proof makes use of the expanding subspace property of the CG algorithm, which we do not describe until Chapter 5, so it can be skipped on the first pass.)

Theorem 4.2.
The sequence of vectors generated by Algorithm 4.3 satisfies

$$0 = \|p_0\|_2 < \cdots < \|p_j\|_2 < \|p_{j+1}\|_2 < \cdots < \|p\|_2 \leq \Delta.$$

PROOF. We first show that the sequences of vectors generated by Algorithm 4.3 satisfy $p_j^T r_j = 0$ for $j \geq 0$ and $p_j^T d_j > 0$ for $j \geq 1$.

Algorithm 4.3 computes p_{j+1} recursively in terms of p_j, but when all the terms of this recursion are written explicitly, we see that

$$p_j = p_0 + \sum_{i=0}^{j-1} \alpha_i d_i = \sum_{i=0}^{j-1} \alpha_i d_i,$$

since $p_0 = 0$. Multiplying by r_j and applying the expanding subspace property of CG gives

$$p_j^T r_j = \sum_{i=0}^{j-1} \alpha_i d_i^T r_j = 0.$$

An induction proof establishes the relation $p_j^T d_j > 0$. By applying the expanding subspace property again, we obtain

$$p_1^T d_1 = (\alpha_0 d_0)^T (r_1 + \beta_1 d_0) = \alpha_0 \beta_1 d_0^T d_0 > 0. \tag{4.17}$$

We now make the inductive hypothesis that $p_j^T d_j > 0$ and show that this implies $p_{j+1}^T d_{j+1} > 0$. From (4.17), we have $p_{j+1}^T r_{j+1} = 0$, and therefore we have

$$\begin{aligned} p_{j+1}^T d_{j+1} &= p_{j+1}^T (r_{j+1} + \beta_{j+1} d_j) \\ &= \beta_{j+1} p_{j+1}^T d_j \\ &= \beta_{j+1} (p_j + \alpha_j d_j)^T d_j \\ &= \beta_{j+1} p_j^T d_j + \alpha_j \beta_{j+1} d_j^T d_j. \end{aligned}$$

Because of the inductive hypothesis, the last expression is positive.

We now prove the theorem. If the algorithm stops because $d_j^T B d_j \leq 0$ or $\|p_{j+1}\|_2 \geq \Delta$, then the final point p is chosen to make $\|p\|_2 = \Delta$, which is the largest possible length any point can have. To cover all other possibilities in the algorithm we must show that $\|p_j\|_2 < \|p_{j+1}\|_2$ when $p_{j+1} = p_j + \alpha_j d_j$ and $j \geq 1$. Observe that

$$\|p_{j+1}\|_2^2 = (p_j + \alpha_j d_j)^T (p_j + \alpha_j d_j) = \|p_j\|_2^2 + 2\alpha_j p_j^T d_j + \alpha_j^2 \|d_j\|_2^2.$$

It follows from this expression and our intermediate result that $\|p_j\|_2 < \|p_{j+1}\|_2$, so our proof is complete. □

From this theorem we see that the iterates of Algorithm 4.3 sweep out points p_j that move on some interpolating path from p_1 to p, a path in which every step increases its total distance from the start point. When B is positive definite, this path may be compared to the path of the dogleg method, because both methods move from the Cauchy step p^C to the full step p^B, until the trust-region boundary intervenes.

A Newton trust-region method chooses B to be the exact Hessian $\nabla^2 f(x)$, which may be indefinite during the course of the iteration (hence our focus on the case of indefinite B). This method has excellent local and global convergence properties, as we see in Chapter 6.

4.2 USING NEARLY EXACT SOLUTIONS TO THE SUBPROBLEM

CHARACTERIZING EXACT SOLUTIONS

The methods discussed above make no serious attempt to seek the exact solution of the subproblem (4.9). They do, however, make some use of the information in the Hessian B, and they have advantages of low cost and global convergence, since they all generate a point that is at least as good as the Cauchy point.

When the problem is relatively small (that is, n is not too large), it may be worthwhile to exploit the model more fully by looking for a closer approximation to the solution of the subproblem. In the next few pages we describe an approach for finding a good approximation at the cost of about three factorizations of the matrix B, as compared with a single factorization for the dogleg and two-dimensional subspace minimization methods. This approach is based on a convenient characterization of the exact solution of (4.9) (we need to be able to recognize an exact solution when we see it, after all) and an ingenious application of Newton's method in one variable. Essentially, we see that a solution p of the trust-region problem satisfies the formula

$$(B + \lambda I)p^* = -g$$

for some $\lambda \geq 0$, and our algorithm for finding p^* aims to identify the appropriate value of λ.

The following theorem gives the precise characterization of the solution of (4.9).

Theorem 4.3.

The vector p^* is a global solution of the trust-region problem

$$\min_{p \in \mathbb{R}^n} m(p) = f + g^T p + \tfrac{1}{2} p^T B p, \quad \text{s.t. } \|p\| \leq \Delta, \quad (4.18)$$

if and only if p^* is feasible and there is a scalar $\lambda \geq 0$ such that the following conditions are satisfied:

$$(B + \lambda I)p^* = -g, \quad (4.19a)$$
$$\lambda(\Delta - \|p^*\|) = 0, \quad (4.19b)$$
$$(B + \lambda I) \quad \text{is positive semidefinite.} \quad (4.19c)$$

We delay the proof of this result until later in the chapter, and instead discuss just its key features here with the help of Figure 4.4. The condition (4.19b) is a complementarity condition that states that at least one of the nonnegative quantities λ and $(\Delta - \|p^*\|)$ must be zero. Hence, when the solution lies strictly inside the trust region (as it does when $\Delta = \Delta_1$ in Figure 4.4), we must have $\lambda = 0$ and so $Bp^* = -g$ with B positive semidefinite, from (4.19a) and (4.19c), respectively. In the other cases $\Delta = \Delta_2$ and $\Delta = \Delta_3$, we have $\|p^*\| = \Delta$, and so λ is allowed to take a positive value. Note from (4.19a) that

$$\lambda p^* = -Bp^* - g = -\nabla m(p^*),$$

that is, the solution p^* is collinear with the negative gradient of m and normal to its contours. These properties can be seen clearly in Figure 4.4.

CALCULATING NEARLY EXACT SOLUTIONS

The characterization of Theorem 4.3 suggests an algorithm for finding the solution p of (4.18). Either $\lambda = 0$ satisfies (4.19a) and (4.19c) with $\|p\| \leq \Delta$, or else we define

$$p(\lambda) = -(B + \lambda I)^{-1} g$$

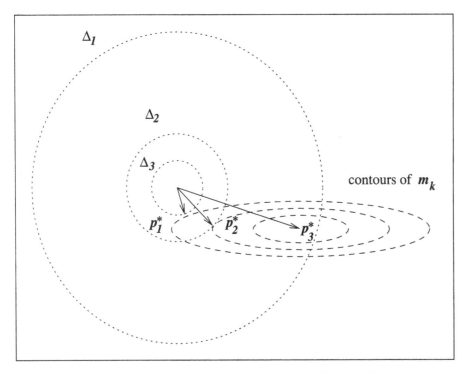

Figure 4.4 Solution of trust-region subproblem for different radii $\Delta_1, \Delta_2, \Delta_3$.

for λ sufficiently large that $B + \lambda I$ is positive definite (see the exercises), and seek a value $\lambda > 0$ such that

$$\|p(\lambda)\| = \Delta. \tag{4.20}$$

This problem is a one-dimensional root-finding problem in the variable λ.

To see that a value of λ with all the desired properties exists, we appeal to the eigendecomposition of B and use it to study the properties of $\|p(\lambda)\|$. Since B is symmetric, there is an orthogonal matrix Q and a diagonal matrix Λ such that $B = Q\Lambda Q^T$, where

$$\Lambda = \text{diag}(\lambda_1, \lambda_2, \ldots, \lambda_n),$$

and $\lambda_1 \leq \lambda_2 \leq \cdots \leq \lambda_n$ are the eigenvalues of B; see (A.46). Clearly, $B + \lambda I = Q(\Lambda + \lambda I)Q^T$, and for $\lambda \neq \lambda_j$, we have

$$p(\lambda) = -Q(\Lambda + \lambda I)^{-1} Q^T g = -\sum_{j=1}^{n} \frac{q_j^T g}{\lambda_j + \lambda} q_j, \tag{4.21}$$

where q_j denotes the jth column of Q. Therefore, by orthonormality of $q_1, q_2, \ldots, q_n$, we have

$$\|p(\lambda)\|^2 = \sum_{j=1}^{n} \frac{\left(q_j^T g\right)^2}{(\lambda_j + \lambda)^2}. \tag{4.22}$$

This expression tells us a lot about $\|p(\lambda)\|$. If $\lambda > -\lambda_1$, we have $\lambda_j + \lambda > 0$ for all $j = 1, 2, \ldots, n$, and so $\|p(\lambda)\|$ is a continuous, nonincreasing function of λ on the interval $(-\lambda_1, \infty)$. In fact, we have that

$$\lim_{\lambda \to \infty} \|p(\lambda)\| = 0. \tag{4.23}$$

Moreover, we have when $q_j^T g \neq 0$ that

$$\lim_{\lambda \to -\lambda_j} \|p(\lambda)\| = \infty. \tag{4.24}$$

These features can be seen in Figure 4.5. It is clear that the graph of $\|p(\lambda)\|$ attains the value Δ at exactly one point in the interval $(-\lambda_1, \infty)$, which is denoted by λ^* in the figure. For the case of B positive definite and $\|B^{-1}g\| \leq \Delta$, the value $\lambda = 0$ satisfies (4.19), so there

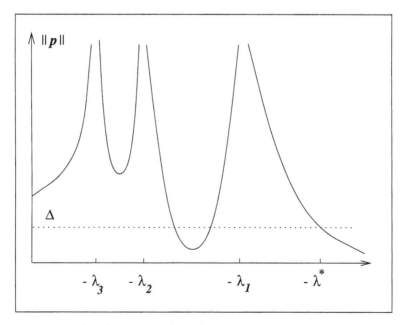

Figure 4.5 $\|p(\lambda)\|$ as a function of λ.

4.2. USING NEARLY EXACT SOLUTIONS TO THE SUBPROBLEM

is no need to carry out a search. When B is positive definite but $\|B^{-1}g\| > \Delta$, there is a strictly positive value of λ for which $\|p(\lambda)\| = \Delta$, so we search for the solution to (4.20) in the interval $(0, \infty)$.

For the case of B indefinite and $q_1^T g \neq 0$, (4.23) and (4.24) guarantee that we can find a solution in the interval $(-\lambda_1, \infty)$. We could use the root-finding Newton's method (see the Appendix) to find the value of $\lambda > \lambda_1$ that solves

$$\phi_1(\lambda) = \|p(\lambda)\| - \Delta = 0. \tag{4.25}$$

The disadvantage of this approach can be seen by considering the form of $\|p(\lambda)\|$ when λ is greater than, but close to, $-\lambda_1$. We then have

$$\phi_1(\lambda) \approx \frac{C_1}{\lambda + \lambda_1} + C_2,$$

where $C_1 > 0$ and C_2 are constants. For these values of λ the function is highly nonlinear, and therefore the root-finding Newton's method will be unreliable or slow. Better results will be obtained if we reformulate the problem (4.25) so that it is nearly linear near the optimal λ. By defining

$$\phi_2(\lambda) = \frac{1}{\Delta} - \frac{1}{\|p(\lambda)\|},$$

we see that for λ slightly greater than $-\lambda_1$ we have from (4.22) that

$$\phi_2(\lambda) \approx \frac{1}{\Delta} - \frac{\lambda + \lambda_1}{C_3}$$

for some $C_3 > 0$. Hence ϕ_2 is nearly linear in the range we consider, and the root-finding Newton's method will perform well, provided that it maintains $\lambda > -\lambda_1$ (see Figure 4.6). The root-finding Newton's method applied to ϕ_2 generates a sequence of iterates $\lambda^{(\ell)}$ by setting

$$\lambda^{(\ell+1)} = \lambda^{(\ell)} - \frac{\phi_2\left(\lambda^{(\ell)}\right)}{\phi_2'\left(\lambda^{(\ell)}\right)}. \tag{4.26}$$

After some elementary manipulation (see the exercises), this updating formula can be implemented in the following practical way.

Algorithm 4.4 (Exact Trust Region).
Given $\lambda^{(0)}$, $\Delta > 0$:
for $\ell = 0, 1, 2, \ldots$
    Factor $B + \lambda^{(\ell)} I = R^T R$;

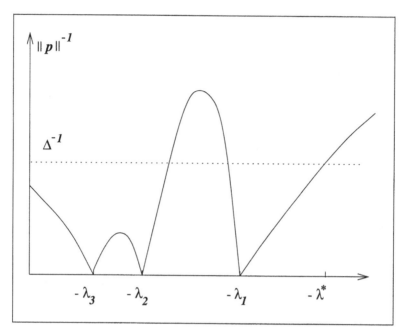

Figure 4.6 $1/\|p(\lambda)\|$ as a function of λ.

Solve $R^T R p_\ell = -g$, $R^T q_\ell = p_\ell$;
Set

$$\lambda^{(\ell+1)} = \lambda^{(\ell)} + \left(\frac{\|p_\ell\|}{\|q_\ell\|}\right)^2 \left(\frac{\|p_\ell\| - \Delta}{\Delta}\right); \qquad (4.27)$$

end (for).

Safeguards must be added to this algorithm to make it practical; for instance, when $\lambda^{(\ell)} < -\lambda_1$, the Cholesky factorization $B + \lambda^{(\ell)} I = R^T R$ will not exist. A slightly enhanced version of this algorithm does, however, converge to a solution of (4.20) in most cases.

The main work in each iteration of this method is, of course, the Cholesky factorization. Practical versions of this algorithm do not iterate until convergence to the optimal λ is obtained with high accuracy, but are content with an approximate solution that can be obtained in two or three iterations.

THE HARD CASE

Recall that in the discussion above, we assumed that $q_1^T g \neq 0$ in dealing with the case of indefinite B. In fact, the approach described above can be applied even when the most

4.2. USING NEARLY EXACT SOLUTIONS TO THE SUBPROBLEM

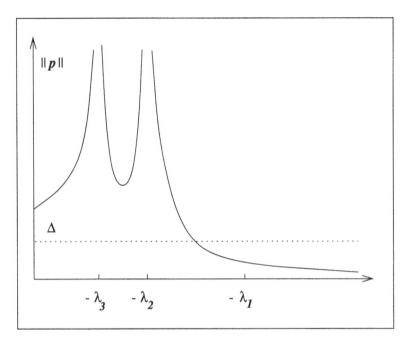

Figure 4.7 The hard case: $\|p(\lambda)\| < \Delta$ for all $\lambda \in (-\lambda_1, \infty)$.

negative eigenvalue is a multiple eigenvalue (that is, $0 > \lambda_1 = \lambda_2 = \cdots$), provided that $q_j^T g \neq 0$ for at least one of the indices j for which $\lambda_j = \lambda_1$. When this condition does not hold, the situation becomes a little complicated, because the limit (4.24) does not hold for $\lambda_j = \lambda_1$ and so there may not be a value $\lambda \in (-\lambda_1, \infty)$ such that $\|p(\lambda)\| = \Delta$ (see Figure 4.7). Moré and Sorensen [170] refer to this case as the *hard case*. At first glance, it is not clear how p and λ can be chosen to satisfy (4.19) in the hard case. Clearly, our root-finding technique will not work, since there is no solution for λ in the open interval $(-\lambda_1, \infty)$. But Theorem 4.3 assures us that the right value of λ lies in the interval $[-\lambda_1, \infty)$, so there is only one possibility: $\lambda = -\lambda_1$. To find p, it is not enough to delete the terms for which $\lambda_j = \lambda_1$ from the formula (4.21) and set

$$p = \sum_{j:\lambda_j \neq \lambda_1} \frac{q_j^T g}{\lambda_j + \lambda} q_j.$$

Instead, we note that $(B - \lambda_1 I)$ is singular, so there is a vector z such that $\|z\| = 1$ and $(B - \lambda_1 I)z = 0$. In fact, z is an eigenvector of B corresponding to the eigenvalue λ_1, so by orthogonality of Q we have $q_j^T z = 0$ for $\lambda_j \neq \lambda_1$. It follows from this property that if we set

$$p = \sum_{j:\lambda_j \neq \lambda_1} \frac{q_j^T g}{\lambda_j + \lambda} q_j + \tau z \qquad (4.28)$$

for any scalar τ, we have

$$\|p\|^2 = \sum_{j:\lambda_j \neq \lambda_1} \frac{\left(q_j^T g\right)^2}{(\lambda_j + \lambda)^2} + \tau^2,$$

so it is always possible to choose τ to ensure that $\|p\| = \Delta$. It is easy to check that (4.19) holds for this choice of p and $\lambda = -\lambda_1$.

PROOF OF THEOREM 4.3

We now give a formal proof of Theorem 4.3, the result that characterizes the exact solution of (4.9). The proof relies on the following technical lemma, which deals with the unconstrained minimizers of quadratics and is particularly interesting in the case where the Hessian is positive semidefinite.

Lemma 4.4.

Let m be the quadratic function defined by

$$m(p) = g^T p + \tfrac{1}{2} p^T B p, \qquad (4.29)$$

where B is any symmetric matrix. Then

(i) *m attains a minimum if and only if B is positive semidefinite and g is in the range of B;*

(ii) *m has a unique minimizer if and only if B is positive definite;*

(iii) *if B is positive semidefinite, then every p satisfying $Bp = -g$ is a global minimizer of m.*

PROOF. We prove each of the three claims in turn.

(i) We start by proving the "if" part. Since g is in the range of B, there is a p with $Bp = -g$. For all $w \in R^n$, we have

$$\begin{aligned}
m(p+w) &= g^T(p+w) + \tfrac{1}{2}(p+w)^T B(p+w) \\
&= (g^T p + \tfrac{1}{2} p^T B p) + g^T w + (Bp)^T w + \tfrac{1}{2} w^T B w \\
&= m(p) + \tfrac{1}{2} w^T B w \\
&\geq m(p),
\end{aligned}$$

since B is positive semidefinite. Hence p is a minimum of m.

For the "only if" part, let p be a minimizer of m. Since $\nabla m(p) = Bp + g = 0$, we have that g is in the range of B. Also, we have $\nabla^2 m(p) = B$ positive semidefinite, giving the result.

4.2. USING NEARLY EXACT SOLUTIONS TO THE SUBPROBLEM

(ii) For the "if" part, the same argument as in (i) suffices with the additional point that $w^T Bw > 0$ whenever $w \neq 0$. For the "only if" part, we proceed as in (i) to deduce that B is positive semidefinite. If B is not positive definite, there is a vector $w \neq 0$ such that $Bw = 0$. Hence from the logic above we have $m(p + w) = m(p)$, so the minimizer is not unique, giving a contradiction.

(iii) Follows from the proof of (i). □

To illustrate case (i), suppose that

$$B = \begin{bmatrix} 1 & 0 & 0 \\ 0 & 0 & 0 \\ 0 & 0 & 2 \end{bmatrix},$$

which has eigenvalues $0, 1, 2$ and is therefore singular. If g is any vector whose second component is zero, then g will be in the range of B, and the quadratic will attain a minimum. But if the second element in g is nonzero, we can decrease $m(\cdot)$ indefinitely by moving along the direction $\alpha(0, -g_2, 0)^T$ as $\alpha \uparrow \infty$.

We are now in a position to take account of the trust-region bound $\|p\| \leq \Delta$ and hence prove Theorem 4.3.

PROOF. (Theorem 4.3)
Assume first that there is $\lambda \geq 0$ such that the conditions (4.19) are satisfied. Lemma 4.4(iii) implies that p^* is a global minimum of the quadratic function

$$\hat{m}(p) = g^T p + \tfrac{1}{2} p^T (B + \lambda I) p = m(p) + \frac{\lambda}{2} p^T p. \tag{4.30}$$

Since $\hat{m}(p) \geq \hat{m}(p^*)$, we have

$$m(p) \geq m(p^*) + \frac{\lambda}{2}((p^*)^T p^* - p^T p). \tag{4.31}$$

Because $\lambda(\Delta - \|p^*\|) = 0$ and therefore $\lambda(\Delta^2 - (p^*)^T p^*) = 0$, we have

$$m(p) \geq m(p^*) + \frac{\lambda}{2}(\Delta^2 - p^T p).$$

Hence, from $\lambda \geq 0$, we have $m(p) \geq m(p^*)$ for all p with $\|p\| \leq \Delta$. Therefore, p^* is a global minimizer of (4.18).

For the converse, we assume that p^* is a global solution of (4.18) and show that there is a $\lambda \geq 0$ that satisfies (4.19).

In the case $\|p^*\| < \Delta$, p^* is an unconstrained minimizer of m, and so

$$\nabla m(p^*) = Bp^* + g = 0, \qquad \nabla^2 m(p^*) = B \text{ positive semidefinite},$$

and so the properties (4.19) hold for $\lambda = 0$.

Assume for the remainder of the proof that $\|p^*\| = \Delta$. Then (4.19b) is immediately satisfied, and p^* also solves the constrained problem

$$\min m(p) \quad \text{subject to } \|p\| = \Delta.$$

By applying optimality conditions for constrained optimization to this problem (see (12.30)), we find that there is a λ such that the Lagrangian function defined by

$$\mathcal{L}(p, \lambda) = m(p) + \frac{\lambda}{2}(p^T p - \Delta^2)$$

has a stationary point at p^*. By setting $\nabla_p \mathcal{L}(p^*, \lambda)$ to zero, we obtain

$$Bp^* + g + \lambda p^* = 0 \Rightarrow (B + \lambda I) p^* = -g, \qquad (4.32)$$

so that (4.19a) holds. Since $m(p) \geq m(p^*)$ for any p with $p^T p = (p^*)^T p^* = \Delta^2$, we have for such vectors p that

$$m(p) \geq m(p^*) + \frac{\lambda}{2}\left((p^*)^T p^* - p^T p\right).$$

If we substitute the expression for g from (4.32) into this expression, we obtain after some rearrangement that

$$\tfrac{1}{2}(p - p^*)^T (B + \lambda I)(p - p^*) \geq 0. \qquad (4.33)$$

Since the set of directions

$$\left\{ w : w = \pm \frac{p - p^*}{\|p - p^*\|}, \text{ for some } p \text{ with } \|p\| = \Delta \right\}$$

is dense on the unit sphere, (4.33) suffices to prove (4.19c).

It remains to show that $\lambda \geq 0$. Because (4.19a) and (4.19c) are satisfied by p^*, we have from Lemma 4.4(i) that p^* minimizes $\hat{m}$, so (4.31) holds. Suppose that there are only negative values of λ that satisfy (4.19a) and (4.19c). Then we have from (4.31) that $m(p) \geq m(p^*)$ whenever $\|p\| \geq \|p^*\| = \Delta$. Since we already know that p^* minimizes m for $\|p\| \leq \Delta$, it follows that m is in fact a global, unconstrained minimizer of m. From Lemma 4.4(i) it follows that $Bp = -g$ and B is positive semidefinite. Therefore conditions (4.19a) and

(4.19c) are satisfied by $\lambda = 0$, which contradicts our assumption that only negative values of λ can satisfy the conditions. We conclude that $\lambda \geq 0$, completing the proof. □

4.3 GLOBAL CONVERGENCE

REDUCTION OBTAINED BY THE CAUCHY POINT

In the preceding discussion of algorithms for approximately solving the trust-region subproblem, we have repeatedly emphasized that global convergence depends on the approximate solution obtaining at least as much decrease in the model function m as the Cauchy point. In fact, a fixed fraction of the Cauchy decrease suffices, as we show in the next few pages. We start by obtaining an estimate of the decrease in m achieved by the Cauchy point, and then use this estimate to prove that the sequence of gradients $\{\nabla f_k\}$ generated by Algorithm 4.1 either has an accumulation point at zero or else converges to zero, depending on whether we choose the parameter η to be zero or strictly positive in Algorithm 4.1. Finally, we state a convergence result for the version of Algorithm 4.1 that uses the nearly exact solutions calculated by Algorithm 4.4 above.

We start by proving that the dogleg and two-dimensional subspace minimization algorithms and Algorithm 4.3 produce approximate solutions p_k of the subproblem (4.3) that satisfy the estimate

$$m_k(0) - m_k(p_k) \geq c_1 \|\nabla f_k\| \min\left(\Delta_k, \frac{\|\nabla f_k\|}{\|B_k\|}\right), \quad (4.34)$$

for some constant $c_1 \in (0, 1]$. The presence of an alternative given by the minimum in (4.34) is typical of trust-region methods and arises because of the trust-region bound. The usefulness of this estimate will become clear in the following two sections. For now, we note that when Δ_k is the minimum value in (4.34), the condition is slightly reminiscent of the first Wolfe condition: The desired reduction in the model is proportional to the gradient and the size of the step.

We show now that the Cauchy point p_k^C satisfies (4.34), with $c_1 = \frac{1}{2}$.

Lemma 4.5.

The Cauchy point p_k^C satisfies (4.34) with $c_1 = \frac{1}{2}$, that is,

$$m_k(0) - m_k(p_k^C) \geq \frac{1}{2} \|\nabla f_k\| \min\left(\Delta_k, \frac{\|\nabla f_k\|}{\|B_k\|}\right). \quad (4.35)$$

PROOF. We consider first the case of $\nabla f_k^T B_k \nabla f_k \leq 0$. Here, we have

$$m_k(p_k^C) - m_k(0) = m_k(\Delta_k \nabla f_k / \|\nabla f_k\|)$$

$$= -\frac{\Delta_k}{\|\nabla f_k\|}\|\nabla f_k\|^2 + \tfrac{1}{2}\frac{\Delta_k^2}{\|\nabla f_k\|^2}\nabla f_k^T B_k \nabla f_k$$
$$\leq -\Delta_k \|\nabla f_k\|$$
$$\leq -\|\nabla f_k\| \min\left(\Delta_k, \frac{\|\nabla f_k\|}{\|B_k\|}\right),$$

and so (4.35) certainly holds.

For the next case, consider $\nabla f_k^T B_k \nabla f_k > 0$ and

$$\frac{\|\nabla f_k\|^3}{\Delta_k \nabla f_k^T B_k \nabla f_k} \leq 1. \tag{4.36}$$

We then have $\tau = \|\nabla f_k\|^3 / (\Delta_k \nabla f_k^T B_k \nabla f_k)$, and so

$$m_k(p_k^C) - m_k(0) = -\frac{\|\nabla f_k\|^4}{\nabla f_k^T B_k \nabla f_k} + \tfrac{1}{2}\nabla f_k^T B_k \nabla f_k \frac{\|\nabla f_k\|^4}{(\nabla f_k^T B_k \nabla f_k)^2}$$
$$= -\tfrac{1}{2}\frac{\|\nabla f_k\|^4}{\nabla f_k^T B_k \nabla f_k}$$
$$\leq -\tfrac{1}{2}\frac{\|\nabla f_k\|^4}{\|B_k\|\|\nabla f_k\|^2}$$
$$= -\tfrac{1}{2}\frac{\|\nabla f_k\|^2}{\|B_k\|}$$
$$\leq -\tfrac{1}{2}\|\nabla f_k\| \min\left(\Delta_k, \frac{\|\nabla f_k\|}{\|B_k\|}\right),$$

so (4.35) holds here too.

In the remaining case, (4.36) does not hold, and therefore

$$\nabla f_k^T B_k \nabla f_k < \frac{\|\nabla f_k\|^3}{\Delta_k}. \tag{4.37}$$

From the definition of p_k^C, we have $\tau = 1$, so using this fact together with (4.37), we obtain

$$m_k(p_k^C) - m_k(0) = -\frac{\Delta_k}{\|\nabla f_k\|}\|\nabla f_k\|^2 + \frac{1}{2}\frac{\Delta_k^2}{\|\nabla f_k\|^2}\nabla f_k^T B_k \nabla f_k$$
$$\leq -\Delta_k\|\nabla f_k\| + \frac{1}{2}\frac{\Delta_k^2}{\|\nabla f_k\|^2}\frac{\|\nabla f_k\|^3}{\Delta_k}$$
$$= -\tfrac{1}{2}\Delta_k\|\nabla f_k\|$$
$$\leq -\tfrac{1}{2}\|\nabla f_k\| \min\left(\Delta_k, \frac{\|\nabla f_k\|}{\|B_k\|}\right),$$

yielding the desired result (4.35) once more. □

To satisfy (4.34), our approximate solution p_k has only to achieve a reduction that is at least some fixed fraction c_2 of the reduction achieved by the Cauchy point. We state the observation formally as a theorem.

Theorem 4.6.
Let p_k be any vector such that $\|p_k\| \leq \Delta_k$ and $m_k(0) - m_k(p_k) \geq c_2 \left(m_k(0) - m_k(p_k^c)\right)$. Then p_k satisfies (4.34) with $c_1 = c_2/2$. In particular, if p_k is the exact solution p_k^* of (4.3), then it satisfies (4.34) with $c_1 = \frac{1}{2}$.

PROOF. Since $\|p_k\| \leq \Delta_k$, we have from (4.35) that

$$m_k(0) - m_k(p_k) \geq c_2 \left(m_k(0) - m_k(p_k^c)\right) \geq \tfrac{1}{2} c_2 \|\nabla f_k\| \min\left(\Delta_k, \frac{\|\nabla f_k\|}{\|B_k\|}\right),$$

giving the result. □

Note that the dogleg and two-dimensional subspace minimization algorithms and Algorithm 4.3 all satisfy (4.34) with $c_1 = \frac{1}{2}$, because they all produce approximate solutions p_k for which $m_k(p_k) \leq m_k(p_k^c)$.

CONVERGENCE TO STATIONARY POINTS

Global convergence results for trust-region methods come in two varieties, depending on whether we set the parameter η in Algorithm 4.1 to zero or to some small positive value. When $\eta = 0$ (that is, the step is taken whenever it produces a lower value of f), we can show that the sequence of gradients $\{\nabla f_k\}$ has a limit point at zero. For the more stringent acceptance test with $\eta > 0$, which requires the actual decrease in f to be at least some small fraction of the predicted decrease, we have the stronger result that $\nabla f_k \to 0$.

In this section we prove the global convergence results for both cases. We assume throughout that the approximate Hessians B_k are uniformly bounded in norm, and that the level set

$$\{x \mid f(x) \leq f(x_0)\} \tag{4.38}$$

is bounded. For generality, we also allow the length of the approximate solution p_k of (4.3) to exceed the trust-region bound, provided that it stays within some fixed multiple of the bound; that is,

$$\|p_k\| \leq \gamma \Delta_k, \quad \text{for some constant } \gamma \geq 1. \tag{4.39}$$

The first result deals with the case of $\eta = 0$.

Theorem 4.7.

Let $\eta = 0$ in Algorithm 4.1. Suppose that $\|B_k\| \leq \beta$ for some constant β, that f is continuously differentiable and bounded below on the level set (4.38), and that all approximate solutions of (4.3) satisfy the inequalities (4.34) and (4.39), for some positive constants c_1 and γ. We then have

$$\liminf_{k \to \infty} \|\nabla f_k\| = 0. \tag{4.40}$$

PROOF. We first perform some technical manipulation with the ratio ρ_k from (4.4). We have

$$|\rho_k - 1| = \left| \frac{(f(x_k) - f(x_k + p_k)) - (m_k(0) - m_k(p_k))}{m_k(0) - m_k(p_k)} \right|$$

$$= \left| \frac{m_k(p_k) - f(x_k + p_k)}{m_k(0) - m_k(p_k)} \right|.$$

Since from Taylor's theorem (Theorem 2.1) we have that

$$f(x_k + p_k) = f(x_k) + \nabla f(x_k)^T p_k + \int_0^1 [\nabla f(x_k + tp_k) - \nabla f(x_k)]^T p_k \, dt,$$

it follows from the definition (4.1) of m_k that

$$|m_k(p_k) - f(x_k + p_k)| = \left| \tfrac{1}{2} p_k^T B_k p_k - \int_0^1 [\nabla f(x_k + tp_k) - \nabla f(x_k)]^T p_k \, dt \right|$$

$$\leq (\beta/2)\|p_k\|^2 + C_4(p_k)\|p_k\|, \tag{4.41}$$

where we can make the scalar $C_4(p_k)$ arbitrarily small by restricting the size of p_k.

Suppose for contradiction that there is $\epsilon > 0$ and a positive index K such that

$$\|\nabla f_k\| \geq \epsilon \quad \text{for all } k \geq K. \tag{4.42}$$

From (4.34), we have for $k \geq K$ that

$$m_k(0) - m_k(p_k) \geq c_1 \|\nabla f_k\| \min\left(\Delta_k, \frac{\|\nabla f_k\|}{\|B_k\|}\right) \geq c_1 \epsilon \min\left(\Delta_k, \frac{\epsilon}{\beta}\right). \tag{4.43}$$

Using (4.43), (4.41), and the bound (4.39), we have

$$|\rho_k - 1| \leq \frac{\gamma \Delta_k (\beta \gamma \Delta_k / 2 + C_4(p_k))}{2 c_1 \epsilon \min(\Delta_k, \epsilon/\beta)}. \tag{4.44}$$

We now derive a bound on the right-hand-side that holds for all sufficiently small values of Δ_k, that is, for all $\Delta_k \leq \bar{\Delta}$, where $\bar{\Delta}$ is to be determined. By choosing $\bar{\Delta}$ to be small enough and noting that $\|p_k\| \leq \gamma \Delta_k \leq \gamma \bar{\Delta}$, we can ensure that the term in parentheses in the numerator of (4.44) satisfies the bound

$$\beta \gamma \Delta_k / 2 + C_4(p_k) \leq \frac{c_1 \epsilon}{2\gamma}. \tag{4.45}$$

By choosing $\bar{\Delta}$ even smaller, if necessary, to ensure that $\Delta_k \leq \bar{\Delta} \leq \epsilon/\beta$, we have from (4.44) that

$$|\rho_k - 1| \leq \frac{\gamma \Delta_k c_1 \epsilon / (2\gamma)}{2 c_1 \epsilon \Delta_k} = \frac{1}{4}.$$

Therefore, $\rho_k > \frac{3}{4}$, and so by the workings of Algorithm 4.1, we have $\Delta_{k+1} \geq \Delta_k$ whenever Δ_k falls below the threshold $\bar{\Delta}$. It follows that reduction of Δ_k (by a factor of $\frac{1}{4}$) can occur in our algorithm only if

$$\Delta_k \geq \bar{\Delta},$$

and therefore we conclude that

$$\Delta_k \geq \min\left(\Delta_K, \bar{\Delta}/4\right) \quad \text{for all } k \geq K. \tag{4.46}$$

Suppose now that there is an infinite subsequence $\mathcal{K}$ such that $\rho_k \geq \frac{1}{4}$ for $k \in \mathcal{K}$. If $k \in \mathcal{K}$ and $k \geq K$, we have from (4.43) that

$$\begin{aligned} f(x_k) - f(x_{k+1}) &= f(x_k) - f(x_k + p_k) \\ &\geq \tfrac{1}{4} [m_k(0) - m_k(p_k)] \\ &\geq \tfrac{1}{4} c_1 \epsilon \min(\Delta_k, \epsilon/\beta). \end{aligned}$$

Since f is bounded below, it follows from this inequality that

$$\lim_{k \in \mathcal{K}, k \to \infty} \Delta_k = 0,$$

contradicting (4.46). Hence no such infinite subsequence $\mathcal{K}$ can exist, and we must have $\rho_k < \frac{1}{4}$ for all k sufficiently large. In this case, Δ_k will eventually be reduced by a factor of $\frac{1}{4}$ at every iteration, and we have $\lim_{k \to \infty} \Delta_k = 0$, which again contradicts (4.46). Hence, our original assertion (4.42) must be false, giving (4.40). □

Our second global convergence result, for the case $\eta > 0$, borrows much of the analysis from the proof above. Our approach here follows that of Schultz, Schnabel, and Byrd [226].

Theorem 4.8.

Let $\eta \in \left(0, \frac{1}{4}\right)$ in Algorithm 4.1. Suppose that $\|B_k\| \leq \beta$ for some constant β, that f is Lipschitz continuously differentiable and bounded below on the level set (4.38), and that all approximate solutions p_k of (4.3) satisfy the inequalities (4.34) and (4.39) for some positive constants c_1 and γ. We then have

$$\lim_{k \to \infty} \nabla f_k = 0. \tag{4.47}$$

PROOF. Consider any index m such that $\nabla f_m \neq 0$. If we use β_1 to denote the Lipschitz constant for ∇f on the level set (4.38), we have

$$\|\nabla f(x) - \nabla f_m\| \leq \beta_1 \|x - x_m\|,$$

for all x in the level set. Hence, by defining the scalars

$$\epsilon = \tfrac{1}{2}\|\nabla f_m\|, \qquad R = \frac{\|\nabla f_m\|}{2\beta_1} = \frac{\epsilon}{\beta_1},$$

and the ball

$$\mathcal{B}(x_m, R) = \{x \mid \|x - x_m\| \leq R\},$$

we have

$$x \in \mathcal{B}(x_m, R) \Rightarrow \|\nabla f(x)\| \geq \|\nabla f_m\| - \|\nabla f(x) - \nabla f_m\| \geq \tfrac{1}{2}\|\nabla f_m\| = \epsilon.$$

If the entire sequence $\{x_k\}_{k \geq m}$ stays inside the ball $\mathcal{B}(x_m, R)$, we would have $\|\nabla f_k\| \geq \epsilon > 0$ for all $k \geq m$. The reasoning in the proof of Theorem 4.7 can be used to show that this scenario does not occur. Therefore, the sequence $\{x_k\}_{k \geq m}$ eventually leaves $\mathcal{B}(x_m, R)$.

Let the index $l \geq m$ be such that x_{l+1} is the first iterate after x_m outside $\mathcal{B}(x_m, R)$. Since $\|\nabla f_k\| \geq \epsilon$ for $k = m, m+1, \ldots, l$, we can use (4.43) to write

$$f(x_m) - f(x_{l+1}) = \sum_{k=m}^{l} f(x_k) - f(x_{k+1})$$

$$\geq \sum_{k=m, x_k \neq x_{k+1}}^{l} \eta[m_k(0) - m_k(p_k)]$$

$$\geq \sum_{k=m, x_k \neq x_{k+1}}^{l} \eta c_1 \epsilon \min\left(\Delta_k, \frac{\epsilon}{\beta}\right),$$

where we have limited the sum to the iterations k for which $x_k \neq x_{k+1}$, that is, those iterations on which a step was actually taken. If $\Delta_k \leq \epsilon/\beta$ for all $k = m, m+1, \ldots, l$, we have

$$f(x_m) - f(x_{l+1}) \geq \eta c_1 \epsilon \sum_{k=m, x_k \neq x_{k+1}}^{l} \Delta_k \geq \eta c_1 \epsilon R = \eta c_1 \epsilon^2 \frac{1}{\beta_1}. \tag{4.48}$$

Otherwise, we have $\Delta_k > \epsilon/\beta$ for some $k = m, m+1, \ldots, l$, and so

$$f(x_m) - f(x_{l+1}) \geq \eta c_1 \epsilon \frac{\epsilon}{\beta}. \tag{4.49}$$

Since the sequence $\{f(x_k)\}_{k=0}^{\infty}$ is decreasing and bounded below, we have that

$$f(x_k) \downarrow f^* \tag{4.50}$$

for some $f^* > -\infty$. Therefore, using (4.48) and (4.49), we can write

$$f(x_m) - f^* \geq f(x_m) - f(x_{l+1})$$
$$\geq \eta c_1 \epsilon^2 \min\left(\frac{1}{\beta}, \frac{1}{\beta_1}\right) = \frac{1}{4}\eta c_1 \min\left(\frac{1}{\beta}, \frac{1}{\beta_1}\right) \|\nabla f_m\|^2.$$

By rearranging this expression, we obtain

$$\|\nabla f_m\|^2 \leq \left(\frac{1}{4}\eta c_1 \min\left(\frac{1}{\beta}, \frac{1}{\beta_1}\right)\right)^{-1} (f(x_m) - f^*),$$

so from (4.50) we conclude that $\nabla f_m \to 0$, giving the result. □

CONVERGENCE OF ALGORITHMS BASED ON NEARLY EXACT SOLUTIONS

As we noted in the discussion of Algorithm 4.4, the loop to determine the optimal values of λ and p for the subproblem (4.9) does not iterate until high accuracy is achieved. Instead, it is terminated after two or three iterations with a fairly loose approximation to the true solution. The inexactness in this approximate solution is, however, measured in a different way from the dogleg and subspace minimization algorithms and Algorithm 4.3, and this difference affects the nature of the global convergence results that can be proved.

Moré and Sorensen [170] describe a safeguarded version of the root-finding Newton method that adds features for handling the hard case. Its termination criteria ensure that their approximate solution p satisfies the conditions

$$m(0) - m(p) \geq c_1(m(0) - m(p^*)), \tag{4.51a}$$
$$\|p\| \leq \gamma \Delta \tag{4.51b}$$

(where p^* is the exact solution of (4.3)), for some constants $c_1 \in (0, 1]$ and $\gamma > 0$. The condition (4.51a) ensures that the approximate solution achieves a significant fraction of the maximum decrease possible in the model function m. Of course, it is not necessary to know p^* to enforce this condition; it follows from practical termination criteria. One major difference between (4.51) and the earlier criterion (4.34) is that (4.51) makes better use of the second-order part of $m(\cdot)$, that is, the $p^T B p$ term. This difference is illustrated by the case in which $g = 0$ while B has negative eigenvalues, indicating that the current iterate x_k is a saddle point. Here, the right-hand-side of (4.34) is zero (indeed, the algorithms we described earlier would terminate at such a point). The right-hand-side of (4.51) is positive, indicating that decrease in the model function is still possible, so it forces the algorithm to move away from x_k.

The close attention that near-exact algorithms pay to the second-order term is warranted only if this term closely reflects the actual behavior of the function f—in fact, the trust-region Newton method, for which $B = \nabla^2 f(x)$, is the only case that has been treated in the literature. For purposes of global convergence analysis, the use of the exact Hessian allows us to say more about the limit points of the algorithm than merely that they are stationary points. In fact, second-order necessary conditions (Theorem 2.3) are satisfied at the limit points. The following result establishes this claim.

Theorem 4.9.

Suppose Algorithm 4.1 is applied with $B_k = \nabla^2 f(x_k)$, constant η in the open interval $\left(0, \frac{1}{4}\right)$, and the approximate solution p_k at each iteration satisfying (4.51) for some fixed $\gamma > 0$. Then $\lim_{k \to \infty} \|\nabla f_k\| = 0$.

If, in addition, the level set $\{x \mid f(x) \leq f(x_0)\}$ is compact, then either the algorithm terminates at a point x_k at which the second-order necessary conditions (Theorem 2.3) for a local minimum hold, or else $\{x_k\}$ has a limit point x^* in the level set at which the necessary conditions hold.

We omit the proof, which can be found in Moré and Sorensen [170, Section 4].

4.4 OTHER ENHANCEMENTS

SCALING

As we noted in Chapter 2, optimization problems are often posed with poor scaling—the objective function f is highly sensitive to small changes in certain components of the vector x and relatively insensitive to changes in other components. Topologically, a symptom of poor scaling is that the minimizer x^* lies in a narrow valley, so that the contours of the objective $f(\cdot)$ near x^* tend towards highly eccentric ellipses. Algorithms can perform poorly unless they compensate for poor scaling; see Figure 2.7 for an illustration of the poor performance of the steepest descent approach.

Recalling our definition of a trust region—a region around the current iterate within which the model $m_k(\cdot)$ is an adequate representation of the true objective $f(\cdot)$—it is easy to see that a *spherical* trust region is not appropriate to the case of poorly scaled functions. We can trust our model m_k to be reasonably accurate only for short distances along the highly sensitive directions, while it is reliable for longer distances along the less sensitive directions. Since the shape of our trust region should be such that our confidence in the model is more or less the same at all points on the boundary of the region, we are led naturally to consider

4.4. OTHER ENHANCEMENTS

elliptical trust regions in which the axes are short in the sensitive directions and longer in the less sensitive directions. Elliptical trust regions can be defined by

$$\|Dp\| \leq \Delta, \tag{4.52}$$

where D is a diagonal matrix with positive diagonal elements, yielding the following scaled trust-region subproblem:

$$\min_{p \in R^n} m_k(p) \stackrel{\text{def}}{=} f_k + \nabla f_k^T p + \tfrac{1}{2} p^T B_k p \quad \text{s.t.} \quad \|Dp\| \leq \Delta_k. \tag{4.53}$$

When $f(x)$ is highly sensitive to the value of the ith component x_i, we set the corresponding diagonal element d_{ii} of D to be large. The value of d_{ii} will be closer to zero for less-sensitive components x_i.

Information to construct the scaling matrix D can be derived reliably from the second derivatives $\partial^2 f/\partial x_i^2$. We can allow D to change from iteration to iteration, as long as all diagonal elements d_{ii} stay inside some predetermined range $[d_{\text{lo}}, d_{\text{hi}}]$, where $0 < d_{\text{lo}} \leq d_{\text{hi}} < \infty$. Of course, we do not need D to be a *precise* reflection of the scaling of the problem, so it is not necessary to devise elaborate heuristics or to spend a lot of computation to get it just right.

All algorithms discussed in this chapter can be modified for the case of elliptical trust regions, and the convergence theory continues to hold, with numerous superficial modifications. The Cauchy point calculation procedure (Algorithm 4.2), for instance, changes only in the specifications of the trust region in (4.5) and (4.6). We obtain the following generalized version.

Algorithm 4.5 (Generalized Cauchy Point Calculation).
Find the vector p_k^s that solves

$$p_k^s = \arg\min_{p \in R^n} f_k + \nabla f_k^T p \quad \text{s.t.} \quad \|Dp\| \leq \Delta_k; \tag{4.54}$$

Calculate the scalar $\tau_k > 0$ that minimizes $m_k(\tau p_k^s)$ subject to satisfying the trust-region bound, that is,

$$\tau_k = \arg\min_{\tau > 0} m_k(\tau p_k^s) \quad \text{s.t.} \quad \|\tau D p_k^s\| \leq \Delta_k; \tag{4.55}$$
$$p_k^c = \tau_k p_k^s.$$

For this scaled version, we find that

$$p_k^s = -\frac{\Delta_k}{\|D^{-1}\nabla f_k\|} D^{-2} \nabla f_k, \tag{4.56}$$

and that the step length τ_k is obtained from the following modification of (4.8):

$$\tau_k = \begin{cases} 1 & \text{if } \nabla f_k^T D^{-2} B_k D^{-2} \nabla f_k \leq 0 \\ \min\left(\dfrac{\|D^{-1}\nabla f_k\|^3}{\Delta_k \nabla f_k^T D^{-2} B_k D^{-2} \nabla f_k}, 1\right) & \text{otherwise.} \end{cases} \qquad (4.57)$$

(The details are left as an exercise.)

A simpler alternative for adjusting the definition of the Cauchy point and the various algorithms of this chapter to allow for the elliptical trust region is simply to rescale the variables p in the subproblem (4.53) so that the trust region is spherical in the scaled variables. By defining

$$\tilde{p} \stackrel{\text{def}}{=} Dp,$$

and by substituting into (4.53), we obtain

$$\min_{\tilde{p} \in \mathbb{R}^n} \tilde{m}_k(\tilde{p}) \stackrel{\text{def}}{=} f_k + \nabla f_k^T D^{-1} \tilde{p} + \tfrac{1}{2} \tilde{p}^T D^{-1} B_k D^{-1} \tilde{p} \qquad \text{s.t. } \|\tilde{p}\| \leq \Delta_k.$$

The theory and algorithms can now be derived in the usual way by substituting $\tilde{p}$ for p, $D^{-1}\nabla f_k$ for ∇f_k, $D^{-1} B_k D^{-1}$ for B_k, and so on.

NON-EUCLIDEAN TRUST REGIONS

Trust regions may also be defined in terms of norms other than the Euclidean norm. For instance, we may have

$$\|p\|_1 \leq \Delta_k \qquad \text{or} \qquad \|p\|_\infty \leq \Delta_k,$$

or their scaled counterparts

$$\|Dp\|_1 \leq \Delta_k \qquad \text{or} \qquad \|Dp\|_\infty \leq \Delta_k,$$

where D is a positive diagonal matrix as before. Norms such as these offer no obvious advantages for unconstrained optimization, but they may be useful for constrained problems. For instance, for the bound-constrained problem

$$\min_{x \in \mathbb{R}^n} f(x), \qquad \text{subject to } x \geq 0,$$

the trust-region subproblem may take the form

$$\min_{p \in \mathbb{R}^n} m_k(p) = f_k + \nabla f_k^T p + \tfrac{1}{2} p^T B_k p \qquad \text{s.t. } x_k + p \geq 0, \|p\| \leq \Delta_k. \qquad (4.58)$$

When the trust region is defined by a Euclidean norm, the feasible region for (4.58) consists of the intersection of a sphere and the nonnegative orthant—an awkward object, geometrically speaking. When the ∞-norm is used, however, the feasible region is simply the rectangular box defined by

$$x_k + p \geq 0, \quad p \geq -\Delta_k e, \quad p \leq \Delta_k e,$$

where $e = (1, 1, \ldots, 1)^T$, so the solution of the subproblem is easily calculated by standard techniques for quadratic programming.

NOTES AND REFERENCES

The influential paper of Powell [199] proves a result like Theorem 4.7 for the case of $\eta = 0$, where the algorithm takes a step whenever it decreases the function value. Powell uses a weaker assumption than ours on the matrices $\|B\|$, but his analysis is more complicated. Moré [167] summarizes developments in algorithms and software before 1982, paying particular attention to the importance of using a scaled trust-region norm. Much of the material in this chapter on methods that use nearly exact solutions to the subproblem (4.3) is drawn from the paper of Moré and Sorensen [170].

Byrd, Schnabel, and Schultz [226], [39] provide a general theory for inexact trust-region methods; they introduce the idea of two-dimensional subspace minimization and also focus on proper handling of the case of indefinite B to ensure stronger local convergence results than Theorems 4.7 and 4.8. Dennis and Schnabel [70] survey trust-region methods as part of their overview of unconstrained optimization, providing pointers to many important developments in the literature.

EXERCISES

4.1 Let $f(x) = 10(x_2 - x_1^2)^2 + (1 - x_1)^2$. At $x = (0, -1)$ draw the contour lines of the quadratic model (4.1) assuming that B is the Hessian of f. Draw the family of solutions of (4.3) as the trust region radius varies from $\Delta = 0$ to $\Delta = 2$. Repeat this at $x = (0, 0.5)$.

4.2 Write a program that implements the dogleg method. Choose B_k to be the exact Hessian. Apply it to solve Rosenbrock's function (2.23). Experiment with the update rule for the trust region by changing the constants in Algorithm 4.1, or by designing your own rules.

4.3 Program the trust region method based on Algorithm 4.3. Choose B_k to be the exact Hessian, and use it to minimize the function

$$\min\ f(x) = \sum_{i=1}^{n} \left[(1 - x_{2i-1})^2 + 10(x_{2i} - x_{2i-1}^2)^2\right]$$

with $n = 10$. Experiment with the starting point and the stopping test for the CG iteration. Repeat the computation with $n = 50$.

Your program should indicate, at every iteration, whether Algorithm 4.3 encountered negative curvature, reached the trust region boundary, or met the stopping test.

✐ **4.4** Theorem 4.7 shows that the sequence $\{\|g\|\}$ has an accumulation point at zero. Show that if the iterates x stay in a bounded set $\mathcal{B}$, then there is a limit point x_∞ of the sequence $\{x_k\}$ such that $\nabla f(x_\infty) = 0$.

✐ **4.5** Show that τ_k defined by (4.8) does indeed identify the minimizer of m_k along the direction $-\nabla f_k$.

✐ **4.6** The Cauchy–Schwarz inequality states that for any vectors u and v, we have

$$|u^T v|^2 \le (u^T u)(v^T v),$$

with equality only when u and v are parallel. When B is positive definite, use this inequality to show that

$$\gamma \stackrel{\text{def}}{=} \frac{\|g\|^4}{(g^T Bg)(g^T B^{-1}g)} \le 1,$$

with equality only if g and Bg (and $B^{-1}g$) are parallel.

✐ **4.7** When B is positive definite, the *double-dogleg method* constructs a path with three line segments from the origin to the full step. The four points that define the path are

- the origin;
- the unconstrained Cauchy step $p^C = -(g^T g)/(g^T Bg)g$;
- a fraction of the full step $\bar{\gamma} p^B = -\bar{\gamma} B^{-1}g$, for some $\bar{\gamma} \in (\gamma, 1]$, where γ is defined in the previous question; and
- the full step $p^B = -B^{-1}g$.

Show that $\|p\|$ increases monotonically along this path.

(Note: The double-dogleg method, as discussed in Dennis and Schnabel [69, Section 6.4.2], was for some time thought to be superior to the standard dogleg method, but later testing has not shown much difference in performance.)

✐ **4.8** Show that (4.26) and (4.27) are equivalent. Hints: Note that

$$\frac{d}{d\lambda}\left(\frac{1}{\|p(\lambda)\|}\right) = \frac{d}{d\lambda}\left(\|p(\lambda)\|^2\right)^{-1/2} = -\frac{1}{2}\left(\|p(\lambda)\|^2\right)^{-3/2} \frac{d}{d\lambda}\|p(\lambda)\|^2,$$

$$\frac{d}{d\lambda}\|p(\lambda)\|^2 = -2\sum_{j=1}^{n} \frac{(q_j^T g)^2}{(\lambda_j + \lambda)^3}.$$

(from (4.22)), and

$$\|q\|^2 = \|R^{-T}p\|^2 = p^T(B+\lambda I)^{-1}p = \sum_{j=1}^{n} \frac{(q_j^T g)^2}{(\lambda_j + \lambda)^3}.$$

◇ **4.9** Derive the solution of the two-dimensional subspace minimization problem in the case where B is positive definite.

◇ **4.10** Show that if B is any symmetric matrix, then there exists $\lambda \geq 0$ such that $B + \lambda I$ is positive definite.

◇ **4.11** Verify that the definitions (4.56) for p_k^s and (4.57) for τ_k are valid for the Cauchy point in the case of an elliptical trust region. (Hint: Using the theory of Chapter 12, we can show that the solution of (4.54) satisfies $\nabla f_k + \alpha D^2 p_k^s = 0$ for some scalar $\alpha \geq 0$.)

◇ **4.12** The following example shows that the reduction in the model function m achieved by the two-dimensional minimization strategy can be much smaller than that achieved by the exact solution of (4.9).
In (4.9), set

$$g = \left(-\frac{1}{\epsilon}, -1, -\epsilon^2\right)^T,$$

where ϵ is a small positive number. Set

$$B = \text{diag}\left(\frac{1}{\epsilon^3}, 1, \epsilon^3\right), \quad \Delta = 0.5.$$

Show that the solution of (4.9) has components $\left(O(\epsilon), \frac{1}{2} + O(\epsilon), O(\epsilon)\right)^T$ and that the reduction in the model m is $\frac{3}{8} + O(\epsilon)$. For the two-dimensional minimization strategy, show that the solution is a multiple of $B^{-1}g$ and that the reduction in m is $O(\epsilon)$.

CHAPTER 5

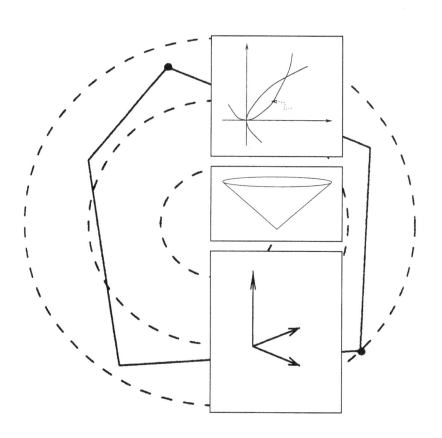

Conjugate Gradient Methods

Our interest in the conjugate gradient method is twofold. It is one of the most useful techniques for solving large linear systems of equations, and it can also be adapted to solve nonlinear optimization problems. These two variants of the fundamental approach, which we refer to as the *linear* and *nonlinear* conjugate gradient methods, respectively, have remarkable properties that will be described in this chapter.

The *linear* conjugate gradient method was proposed by Hestenes and Stiefel in the 1950s as an iterative method for solving linear systems with positive definite coefficient matrices. It is an alternative to Gaussian elimination that is very well suited for solving large problems. The performance of the linear conjugate gradient method is tied to the distribution of the eigenvalues of the coefficient matrix. By transforming, or *preconditioning*, the linear system, we can make this distribution more favorable and improve the convergence of the method significantly. Preconditioning plays a crucial role in the design of practical conjugate gradient strategies. Our treatment of the linear conjugate gradient method will highlight those properties of the method that are important in optimization.

The first *nonlinear* conjugate gradient method was introduced by Fletcher and Reeves in the 1960s. It is one of the earliest known techniques for solving large-scale nonlinear optimization problems. Over the years, many variants of this original scheme have been proposed, and some are widely used in practice. The key features of these algorithms are that they require no matrix storage and are faster than the steepest descent method.

5.1 THE LINEAR CONJUGATE GRADIENT METHOD

In this section we derive the linear conjugate gradient method and discuss its essential convergence properties. For simplicity, we drop the qualifier "linear" throughout.

The conjugate gradient method is an iterative method for solving a linear system of equations

$$Ax = b, \tag{5.1}$$

where A is an $n \times n$ matrix that is symmetric and positive definite. The problem (5.1) can be stated equivalently as the following minimization problem:

$$\phi(x) = \tfrac{1}{2} x^T A x - b^T x, \tag{5.2}$$

that is, both (5.1) and (5.2) have the same unique solution. This equivalence will allow us to interpret the conjugate gradient method either as an algorithm for solving linear systems or as a technique for minimization of convex quadratic functions. For future reference we note that the gradient of ϕ equals the residual of the linear system,

$$\nabla \phi(x) = Ax - b \stackrel{\text{def}}{=} r(x). \tag{5.3}$$

CONJUGATE DIRECTION METHODS

One of the remarkable properties of the conjugate gradient method is its ability to generate, in a very economical fashion, a set of vectors with a property known as *conjugacy*. A set of nonzero vectors $\{p_0, p_1, \ldots, p_l\}$ is said to be *conjugate* with respect to the symmetric positive definite matrix A if

$$p_i^T A p_j = 0, \quad \text{for all } i \neq j. \tag{5.4}$$

It is easy to show that any set of vectors satisfying this property is also linearly independent.

The importance of conjugacy lies in the fact that we can minimize $\phi(\cdot)$ in n steps by successively minimizing it along the individual directions in a conjugate set. To verify this claim, we consider the following *conjugate direction* method. (The distinction between the conjugate gradient method and the conjugate direction method will become clear as we proceed). Given a starting point $x_0 \in \mathbf{R}^n$ and a set of conjugate directions $\{p_0, p_1, \ldots, p_{n-1}\}$,

5.1. THE LINEAR CONJUGATE GRADIENT METHOD

let us generate the sequence $\{x_k\}$ by setting

$$x_{k+1} = x_k + \alpha_k p_k, \tag{5.5}$$

where α_k is the one-dimensional minimizer of the quadratic function $\phi(\cdot)$ along $x_k + \alpha p_k$, given explicitly by

$$\alpha_k = -\frac{r_k^T p_k}{p_k^T A p_k}; \tag{5.6}$$

see (3.38). We have the following result.

Theorem 5.1.

For any $x_0 \in \mathbf{R}^n$ the sequence $\{x_k\}$ generated by the conjugate direction algorithm (5.5), (5.6) converges to the solution x^* of the linear system (5.1) in at most n steps.

PROOF. Since the directions $\{p_i\}$ are linearly independent, they must span the whole space $\mathbf{R}^n$. Hence, we can write the difference between x_0 and the solution x^* in the following way:

$$x^* - x_0 = \sigma_0 p_0 + \sigma_1 p_1 + \cdots + \sigma_{n-1} p_{n-1},$$

for some choice of scalars σ_k. By premultiplying this expression by $p_k^T A$ and using the conjugacy property (5.4), we obtain

$$\sigma_k = \frac{p_k^T A(x^* - x_0)}{p_k^T A p_k}. \tag{5.7}$$

We now establish the result by showing that these coefficients σ_k coincide with the step lengths α_k generated by the formula (5.6).

If x_k is generated by algorithm (5.5), (5.6), then we have

$$x_k = x_0 + \alpha_0 p_0 + \alpha_1 p_1 + \cdots + \alpha_{k-1} p_{k-1}.$$

By premultiplying this expression by $p_k^T A$ and using the conjugacy property, we have that

$$p_k^T A(x_k - x_0) = 0,$$

and therefore

$$p_k^T A(x^* - x_0) = p_k^T A(x^* - x_k) = p_k^T (b - Ax_k) = -p_k^T r_k.$$

By comparing this relation with (5.6) and (5.7), we find that $\sigma_k = \alpha_k$, giving the result. □

Chapter 5. Conjugate Gradient Methods

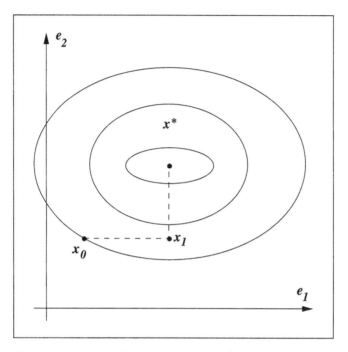

Figure 5.1 Successive minimizations along the coordinate directions find the minimizer of a quadratic with a diagonal Hessian in n iterations.

There is a simple interpretation of the properties of conjugate directions. If the matrix A in (5.2) is diagonal, the contours of the function $\phi(\cdot)$ are ellipses whose axes are aligned with the coordinate directions, as illustrated in Figure 5.1. We can find the minimizer of this function by performing one-dimensional minimizations along the coordinate directions $e_1, e_2, \ldots, e_n$ in turn.

When A is *not* diagonal, its contours are still elliptical, but they are usually no longer aligned with the coordinate directions. The strategy of successive minimization along these directions in turn no longer leads to the solution in n iterations (or even in a finite number of iterations). This phenomenon is illustrated in the two-dimensional example of Figure 5.2

We can recover the nice behavior of Figure 5.1 if we transform the problem to make A diagonal and then minimize along the coordinate directions. Suppose we transform the problem by defining new variables $\hat{x}$ as

$$\hat{x} = S^{-1}x, \tag{5.8}$$

where S is the $n \times n$ matrix defined by

$$S = [p_0 \; p_1 \; \cdots \; p_{n-1}],$$

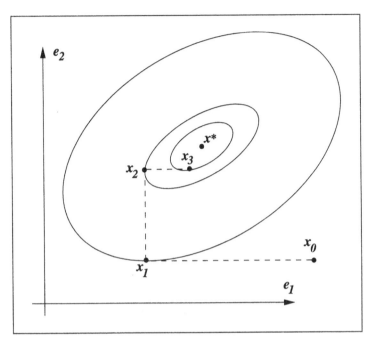

Figure 5.2 Successive minimization along coordinate axes does not find the solution in n iterations, for a general convex quadratic.

where $\{p_0, p_2, \ldots, p_{n-1}\}$ is the set of conjugate directions with respect to A. The quadratic ϕ defined by (5.2) now becomes

$$\hat{\phi}(\hat{x}) \stackrel{\text{def}}{=} \phi(S\hat{x}) = \tfrac{1}{2}\hat{x}^T (S^T A S)\hat{x} - (S^T b)^T \hat{x}.$$

By the conjugacy property (5.4), the matrix $S^T A S$ is diagonal, so we can find the minimizing value of $\hat{\phi}$ by performing n one-dimensional minimizations along the coordinate directions of $\hat{x}$. Because of the relation (5.8), however, each coordinate direction in $\hat{x}$-space corresponds to the direction p_i in x-space. Hence, the coordinate search strategy applied to $\hat{\phi}$ is equivalent to the conjugate direction algorithm (5.5), (5.6). We conclude, as in Theorem 5.1 that the conjugate direction algorithm terminates in at most n steps.

Returning to Figure 5.1, we note another interesting property: When the Hessian matrix is diagonal, each coordinate minimization correctly determines one of the components of the solution x^*. In other words, after k one-dimensional minimizations, the quadratic has been minimized on the subspace spanned by $e_1, e_2, \ldots, e_k$. The following theorem proves this important result for the general case in which the Hessian of the quadratic is not necessarily diagonal. (Here and later, we use the notation span$\{p_0, p_1, \ldots, p_k\}$ to denote the set of all linear combinations of $p_0, p_1, \ldots, p_k$.) In proving the result we will make use of the following expression, which is easily verified from the relations (5.3) and (5.5):

$$r_{k+1} = r_k + \alpha_k A p_k. \tag{5.9}$$

Theorem 5.2 (Expanding Subspace Minimization).
Let $x_0 \in \mathbb{R}^n$ be any starting point and suppose that the sequence $\{x_k\}$ is generated by the conjugate direction algorithm (5.5), (5.6). Then

$$r_k^T p_i = 0, \qquad \text{for } i = 0, \ldots, k-1, \tag{5.10}$$

and x_k is the minimizer of $\phi(x) = \frac{1}{2} x^T A x - b^T x$ over the set

$$\{x \mid x = x_0 + \operatorname{span}\{p_0, p_1, \ldots, p_{k-1}\}\}. \tag{5.11}$$

PROOF. We begin by showing that a point $\tilde{x}$ minimizes ϕ over the set (5.11) if and only if $r(\tilde{x})^T p_i = 0$, for each $i = 0, 1, \ldots, k-1$. Let us define $h(\sigma) = \phi(x_0 + \sigma_0 p_0 + \cdots + \sigma_{k-1} p_{k-1})$, where $\sigma = (\sigma_0, \sigma_1, \ldots, \sigma_{k-1})^T$. Since $h(\sigma)$ is a strictly convex quadratic, it has a unique minimizer σ^* that satisfies

$$\frac{\partial h(\sigma^*)}{\partial \sigma_i} = 0, \qquad i = 0, 1, \ldots, k-1.$$

By the chain rule, this implies that

$$\nabla \phi(x_0 + \sigma_0^* p_0 + \cdots + \sigma_{k-1}^* p_{k-1})^T p_i = 0, \qquad i = 0, 1, \ldots, k-1.$$

By recalling the definition (5.3), we obtain the desired result.

We now use induction to show that x_k satisfies (5.10). Since α_k is always the one-dimensional minimizer, we have immediately that $r_1^T p_0 = 0$. Let us now make the induction hypothesis, namely, that $r_{k-1}^T p_i = 0$ for $i = 0, \ldots, k-2$. By (5.9),

$$r_k = r_{k-1} + \alpha_{k-1} A p_{k-1},$$

we have

$$p_{k-1}^T r_k = p_{k-1}^T r_{k-1} + \alpha_{k-1} p_{k-1}^T A p_{k-1} = 0,$$

by the definition (5.6) of α_{k-1}. Meanwhile, for the other vectors $p_i, i = 0, 1, \ldots, k-2$, we have

$$p_i^T r_k = p_i^T r_{k-1} + \alpha_{k-1} p_i^T A p_{k-1} = 0$$

by the induction hypothesis and the conjugacy of the p_i. We conclude that $r_k^T p_i = 0$, for $i = 0, 1, \ldots, k-1$, so that the proof is complete. □

The fact that the current residual r_k is orthogonal to all previous search directions, as expressed in (5.10), is a property that will be used extensively in this chapter.

The discussion so far has been general, in that it applies to a conjugate direction method (5.5), (5.6) based on *any* choice of the conjugate direction set $\{p_0, p_1, \ldots, p_{n-1}\}$. There are many ways to choose the set of conjugate directions. For instance, the eigenvectors $v_1, v_2, \ldots, v_n$ of A are mutually orthogonal as well as conjugate with respect to A, so these could be used as the vectors $\{p_0, p_1, \ldots, p_{n-1}\}$. For large-scale applications, however, it is not practical to compute the complete set of eigenvectors, for this requires a large amount of computation. An alternative is to modify the Gram–Schmidt orthogonalization process to produce a set of conjugate directions rather than a set of orthogonal directions. (This modification is easy to produce, since the properties of conjugacy and orthogonality are closely related in spirit.) This approach is also expensive, since it requires us to store the entire direction set.

BASIC PROPERTIES OF THE CONJUGATE GRADIENT METHOD

The conjugate gradient method is a conjugate direction method with a very special property: In generating its set of conjugate vectors, it can compute a new vector p_k by using only the previous vector p_{k-1}. It does *not* need to know all the previous elements $p_0, p_1, \ldots, p_{k-2}$ of the conjugate set; p_k is automatically conjugate to these vectors. This remarkable property implies that the method requires little storage and computation.

Now for the details of the conjugate gradient method. Each direction p_k is chosen to be a linear combination of the steepest descent direction $-\nabla \phi(x_k)$ (which is the same as the negative residual $-r_k$, by (5.3)) and the previous direction p_{k-1}. We write

$$p_k = -r_k + \beta_k p_{k-1}, \qquad (5.12)$$

where the scalar β_k is to be determined by the requirement that p_{k-1} and p_k must be conjugate with respect to A. By premultiplying (5.12) by $p_{k-1}^T A$ and imposing the condition $p_{k-1}^T A p_k = 0$, we find that

$$\beta_k = \frac{r_k^T A p_{k-1}}{p_{k-1}^T A p_{k-1}}.$$

It makes intuitive sense to choose the first search direction p_0 to be the steepest descent direction at the initial point x_0. As in the general conjugate direction method, we perform successive one-dimensional minimizations along each of the search directions. We have thus specified a complete algorithm, which we express formally as follows:

Algorithm 5.1 (CG–Preliminary Version).
Given x_0;
Set $r_0 \leftarrow Ax_0 - b$, $p_0 \leftarrow -r_0$, $k \leftarrow 0$;
while $r_k \neq 0$

$$\alpha_k \leftarrow -\frac{r_k^T p_k}{p_k^T A p_k}; \tag{5.13a}$$

$$x_{k+1} \leftarrow x_k + \alpha_k p_k; \tag{5.13b}$$

$$r_{k+1} \leftarrow A x_{k+1} - b; \tag{5.13c}$$

$$\beta_{k+1} \leftarrow \frac{r_{k+1}^T A p_k}{p_k^T A p_k}; \tag{5.13d}$$

$$p_{k+1} \leftarrow -r_{k+1} + \beta_{k+1} p_k; \tag{5.13e}$$

$$k \leftarrow k + 1; \tag{5.13f}$$

end (while)

Later, we will present a more efficient version of the conjugate gradient method; the version above is useful for studying the essential properties of the method. We show first that the directions $p_0, p_1, \ldots, p_{n-1}$ are indeed conjugate, which by Theorem 5.1 implies termination in n steps. The theorem below establishes this property and two other important properties. First, the residuals r_i are mutually orthogonal. Second, each search direction p_k and residual r_k is contained in the *Krylov subspace of degree k for r_0*, defined as

$$\mathcal{K}(r_0; k) \stackrel{\text{def}}{=} \text{span}\{r_0, Ar_0, \ldots, A^k r_0\}. \tag{5.14}$$

Theorem 5.3.
Suppose that the kth iterate generated by the conjugate gradient method is not the solution point x^*. The following four properties hold:

$$r_k^T r_i = 0, \quad \text{for } i = 0, \ldots, k-1, \tag{5.15}$$

$$\text{span}\{r_0, r_1, \ldots, r_k\} = \text{span}\{r_0, Ar_0, \ldots, A^k r_0\}, \tag{5.16}$$

$$\text{span}\{p_0, p_1, \ldots, p_k\} = \text{span}\{r_0, Ar_0, \ldots, A^k r_0\}, \tag{5.17}$$

$$p_k^T A p_i = 0, \quad \text{for } i = 0, 1, \ldots, k-1. \tag{5.18}$$

Therefore, the sequence $\{x_k\}$ converges to x^* in at most n steps.

PROOF. The proof is by induction. The expressions (5.16) and (5.17) hold trivially for $k = 0$, while (5.18) holds by construction for $k = 1$. Assuming now that these three expressions are true for some k (the induction hypothesis), we show that they continue to hold for $k + 1$.

To prove (5.16), we show first that the set on the left-hand side is contained in the set on the right-hand side. Because of the induction hypothesis, we have from (5.16) and (5.17) that

$$r_k \in \text{span}\{r_0, Ar_0, \ldots, A^k r_0\}, \qquad p_k \in \text{span}\{r_0, Ar_0, \ldots, A^k r_0\},$$

while by multiplying the second of these expressions by A, we obtain

$$Ap_k \in \text{span}\{Ar_0, \ldots, A^{k+1} r_0\}. \tag{5.19}$$

By applying (5.9), we find that

$$r_{k+1} \in \text{span}\{r_0, Ar_0, \ldots, A^{k+1} r_0\}.$$

By combining this expression with the induction hypothesis for (5.16), we conclude that

$$\text{span}\{r_0, r_1, \ldots, r_k, r_{k+1}\} \in \text{span}\{r_0, Ar_0, \ldots, A^{k+1} r_0\}.$$

To prove that the reverse inclusion holds as well, we use the induction hypothesis on (5.17) to deduce that

$$A^{k+1} r_0 = A(A^k r_0) \in \text{span}\{Ap_0, Ap_1, \ldots, Ap_k\}.$$

Since by (5.9) we have $Ap_i = (r_{i+1} - r_i)/\alpha_i$ for $i = 0, 1, \ldots, k$, it follows that

$$A^{k+1} r_0 \in \text{span}\{r_0, r_1, \ldots, r_{k+1}\}.$$

By combining this expression with the induction hypothesis for (5.16), we find that

$$\text{span}\{r_0, Ar_0, \ldots, A^{k+1} r_0\} \subset \text{span}\{r_0, r_1, \ldots, r_k, r_{k+1}\}.$$

Therefore, the relation (5.16) continues to hold when k is replaced by $k+1$, as claimed.

We show that (5.17) continues to hold when k is replaced by $k+1$ by the following argument:

$$\begin{aligned}
&\text{span}\{p_0, p_1, \ldots, p_k, p_{k+1}\} \\
&= \text{span}\{p_0, p_1, \ldots, p_k, r_{k+1}\} && \text{by (5.13e)} \\
&= \text{span}\{r_0, Ar_0, \ldots, A^k r_0, r_{k+1}\} && \text{by induction hypothesis for (5.17)} \\
&= \text{span}\{r_0, r_1, \ldots, r_k, r_{k+1}\} && \text{by (5.16)} \\
&= \text{span}\{r_0, Ar_0, \ldots, A^{k+1} r_0\} && \text{by (5.16) for } k+1.
\end{aligned}$$

Next, we prove the conjugacy condition (5.18) with k replaced by $k+1$. By multiplying (5.13e) by Ap_i, $i = 0, 1, \ldots, k$, we obtain

$$p_{k+1}^T Ap_i = -r_{k+1}^T Ap_i + \beta_{k+1} p_k^T Ap_i. \tag{5.20}$$

By the definition (5.13d) of β_k, the right-hand-side of (5.20) vanishes when $i = k$. For $i \leq k - 1$ we need to collect a number of observations. Note first that our induction hypothesis for (5.18) implies that the directions $p_0, p_1, \ldots, p_k$ are conjugate, so we can apply Theorem 5.2 to deduce that

$$r_{k+1}^T p_i = 0, \qquad \text{for } i = 0, 1, \ldots, k. \tag{5.21}$$

Second, by repeatedly applying (5.17), we find that for $i = 0, 1, \ldots, k - 1$, the following inclusion holds:

$$Ap_i \in A \operatorname{span}\{r_0, Ar_0, \ldots, A^i r_0\} = \operatorname{span}\{Ar_0, A^2 r_0, \ldots, A^{i+1} r_0\}$$
$$\subset \operatorname{span}\{p_0, p_1, \ldots, p_{i+1}\}. \tag{5.22}$$

By combining (5.21) and (5.22), we deduce that

$$r_{k+1}^T Ap_i = 0, \qquad \text{for } i = 0, 1, \ldots, k - 1,$$

so the first term in the right-hand-side of (5.20) vanishes for $i = 0, 1, \ldots, k - 1$. Because of the induction hypothesis for (5.18), the second term vanishes as well, and we conclude that $p_{k+1}^T Ap_i = 0$, $i = 0, 1, \ldots, k$. Hence, the induction argument holds for (5.18) also.

It follows that the direction set generated by the conjugate gradient method is indeed a conjugate direction set, so Theorem 5.1 tells us that the algorithm terminates in at most n iterations.

Finally, we prove (5.15) by a noninductive argument. Because the direction set is conjugate, we have from (5.10) that $r_k^T p_i = 0$ for all $i = 0, 1, \ldots, k - 1$ and any $k = 1, 2, \ldots, n - 1$. By rearranging (5.13e), we find that

$$p_i = -r_i + \beta_i p_{i-1},$$

so that $r_i \in \operatorname{span}\{p_i, p_{i-1}\}$ for all $i = 1, \ldots, k - 1$. We conclude that $r_k^T r_i = 0$ for all $i = 1, \ldots, k - 1$, as claimed. □

The proof of this theorem relies on the fact that the first direction p_0 is the steepest descent direction $-r_0$; in fact, the result does not hold for other choices of p_0. Since the gradients r_k are mutually orthogonal, the term "conjugate gradient method" is actually a misnomer. It is the search directions, not the gradients, that are conjugate with respect to A.

5.1. THE LINEAR CONJUGATE GRADIENT METHOD

A PRACTICAL FORM OF THE CONJUGATE GRADIENT METHOD

We can derive a slightly more economical form of the conjugate gradient method by using the results of Theorems 5.2 and 5.3. First, we can use (5.13e) and (5.10) to replace the formula (5.13a) for α_k by

$$\alpha_k = \frac{r_k^T r_k}{p_k^T A p_k}.$$

Second, we have from (5.9) that $\alpha_k A p_k = r_{k+1} - r_k$, so by applying (5.13e) and (5.10) once again we can simplify the formula for β_{k+1} to

$$\beta_{k+1} = \frac{r_{k+1}^T r_{k+1}}{r_k^T r_k}.$$

By using these formulae together with (5.9), we obtain the following standard form of the conjugate gradient method.

Algorithm 5.2 (CG).
Given x_0;
Set $r_0 \leftarrow A x_0 - b$, $p_0 \leftarrow -r_0$, $k \leftarrow 0$;
while $r_k \neq 0$

$$\alpha_k \leftarrow \frac{r_k^T r_k}{p_k^T A p_k}; \quad (5.23a)$$
$$x_{k+1} \leftarrow x_k + \alpha_k p_k; \quad (5.23b)$$
$$r_{k+1} \leftarrow r_k + \alpha_k A p_k; \quad (5.23c)$$
$$\beta_{k+1} \leftarrow \frac{r_{k+1}^T r_{k+1}}{r_k^T r_k}; \quad (5.23d)$$
$$p_{k+1} \leftarrow -r_{k+1} + \beta_{k+1} p_k; \quad (5.23e)$$
$$k \leftarrow k + 1; \quad (5.23f)$$

end (while)

At any given point in Algorithm 5.2 we never need to know the vectors x, r, and p for more than the last two iterations. Accordingly, implementations of this algorithm overwrite old values of these vectors to save on storage. The major computational tasks to be performed at each step are computation of the matrix–vector product $A p_k$, calculation of the inner products $p_k^T (A p_k)$ and $r_{k+1}^T r_{k+1}$, and calculation of three vector sums. The inner product and vector sum operations can be performed in a small multiple of n floating-point operations, while the cost of the matrix–vector product is, of course, dependent on the problem. The CG method is recommended only for large problems; otherwise, Gaussian elimination or other factorization algorithms such as the singular value decomposition are to be preferred, since they are less sensitive to rounding errors. For large problems, the CG method has the advantage that it does not alter the coefficient matrix, and unlike factorization techniques, cannot produce fill in the arrays holding the matrix. The other key property is that the CG method sometimes approaches the solution very quickly, as we discuss next.

RATE OF CONVERGENCE

We have seen that in exact arithmetic the conjugate gradient method will terminate at the solution in at most n iterations. What is more remarkable is that when the distribution of the eigenvalues of A has certain favorable features, the algorithm will identify the solution in many fewer than n iterations. To show this we begin by viewing the expanding subspace minimization property proved in Theorem 5.2 in a slightly different way, using it to show that Algorithm 5.2 is optimal in a certain important sense.

From (5.23b) and (5.17), we have that

$$x_{k+1} = x_0 + \alpha_0 p_0 + \cdots + \alpha_k p_k$$
$$= x_0 + \gamma_0 r_0 + \gamma_1 A r_0 + \cdots + \gamma_k A^k r_0, \tag{5.24}$$

for some constants γ_i. We now define $P_k^*(\cdot)$ to be a polynomial of degree k with coefficients $\gamma_0, \gamma_1, \ldots, \gamma_k$. Like any polynomial, P_k^* can take either a scalar or a square matrix as its argument; for a matrix argument A, we have

$$P_k^*(A) = \gamma_0 I + \gamma_1 A + \cdots + \gamma_k A^k.$$

We can now write (5.24) as

$$x_{k+1} = x_0 + P_k^*(A) r_0. \tag{5.25}$$

We will now see that among all possible methods whose first k steps are restricted to the Krylov subspace $\mathcal{K}(r_0; k)$ given by (5.14), Algorithm 5.2 does the best job of minimizing the distance to the solution after k steps, when this distance is measured by the weighted norm measure $\|\cdot\|_A$ defined by

$$\|z\|_A^2 = z^T A z. \tag{5.26}$$

(Recall that this norm was used in the analysis of the steepest descent method of Chapter 3.)
Using this norm and the definition of ϕ (5.2), it is easy to show that

$$\tfrac{1}{2}\|x - x^*\|_A^2 = \tfrac{1}{2}(x - x^*)^T A (x - x^*) = \phi(x) - \phi(x^*). \tag{5.27}$$

5.1. THE LINEAR CONJUGATE GRADIENT METHOD

Theorem 5.2 states that x_{k+1} minimizes ϕ, and hence $\|x - x^*\|_A^2$, over the set $x_0 + \text{span}\{p_0, p_1, \ldots, p_k\}$. It follows from (5.25) that the polynomial P_k^* solves the following problem in which the minimum is taken over the space of all possible polynomials of degree k:

$$\min_{P_k} \|x_0 + P_k(A)r_0 - x^*\|_A. \tag{5.28}$$

We exploit this optimality property repeatedly in the remainder of the section. Since

$$r_0 = Ax_0 - b = Ax_0 - Ax^* = A(x_0 - x^*),$$

we have that

$$x_{k+1} - x^* = x_0 + P_k^*(A)r_0 - x^* = [I + P_k^*(A)A](x_0 - x^*). \tag{5.29}$$

Let $0 < \lambda_1 \leq \lambda_2 \leq \cdots \leq \lambda_n$ be the eigenvalues of A, and let $v_1, v_2, \ldots, v_n$ be the corresponding orthonormal eigenvectors. Since these eigenvectors span the whole space $\mathbf{R}^n$, we can write

$$x_0 - x^* = \sum_{i=1}^{n} \xi_i v_i, \tag{5.30}$$

for some coefficients ξ_i. It is easy to show that any eigenvector of A is also an eigenvector of $P_k(A)$ for any polynomial P_k. For our particular matrix A and its eigenvalues λ_i and eigenvectors v_i, we have

$$P_k(A)v_i = P_k(\lambda_i)v_i, \quad i = 1, 2, \ldots, n.$$

By substituting (5.30) into (5.29) we have

$$x_{k+1} - x^* = \sum_{i=1}^{n} [1 + \lambda_i P_k^*(\lambda_i)] \xi_i v_i,$$

and hence

$$\|x_{k+1} - x^*\|_A^2 = \sum_{i=1}^{n} \lambda_i [1 + \lambda_i P_k^*(\lambda_i)]^2 \xi_i^2. \tag{5.31}$$

Since the polynomial P_k^* generated by the CG method is optimal with respect to this norm, we have

$$\|x_{k+1} - x^*\|_A^2 = \min_{P_k} \sum_{i=1}^n \lambda_i [1 + \lambda_i P_k^*(\lambda_i)]^2 \xi_i^2.$$

By extracting the largest of the terms $[1 + \lambda_i P_k(\lambda_i)]^2$ from this expression, we obtain that

$$\|x_{k+1} - x^*\|_A^2 \leq \min_{P_k} \max_{1 \leq i \leq n} [1 + \lambda_i P_k(\lambda_i)]^2 \left(\sum_{j=1}^n \lambda_j \xi_j^2 \right)$$

$$= \min_{P_k} \max_{1 \leq i \leq n} [1 + \lambda_i P_k(\lambda_i)]^2 \|x_0 - x^*\|_A^2, \qquad (5.32)$$

where we have used the fact that $\|x_0 - x^*\|_A^2 = \sum_{j=1}^n \lambda_j \xi_j^2$.

The expression (5.32) allows us to quantify the convergence rate of the CG method by estimating the nonnegative scalar quantity

$$\min_{P_k} \max_{1 \leq i \leq n} [1 + \lambda_i P_k(\lambda_i)]^2. \qquad (5.33)$$

In other words, we search for a polynomial P_k that makes this expression as small as possible. In some practical cases, we can find this polynomial explicitly and draw some interesting conclusions about the properties of the CG method. The following result is an example.

Theorem 5.4.
If A has only r distinct eigenvalues, then the CG iteration will terminate at the solution in at most r iterations.

PROOF. Suppose that the eigenvalues $\lambda_1, \lambda_2, \ldots, \lambda_n$ take on the r distinct values $\tau_1 < \tau_2 < \cdots < \tau_r$. We define a polynomial $Q_r(\lambda)$ by

$$Q_r(\lambda) = \frac{(-1)^r}{\tau_1 \tau_2 \cdots \tau_r} (\lambda - \tau_1)(\lambda - \tau_2) \cdots (\lambda - \tau_r),$$

and note that $Q_r(\lambda_i) = 0$ for $i = 1, 2, \ldots, n$ and $Q_r(0) = 1$. From the latter observation, we deduce that $Q_r(\lambda) - 1$ is a polynomial of degree r with a root at $\lambda = 0$, so by polynomial division, the function $\bar{P}_{r-1}$ defined by

$$\bar{P}_{r-1}(\lambda) = (Q_r(\lambda) - 1)/\lambda$$

is a polynomial of degree $r - 1$. By setting $k = r - 1$ in (5.33), we have

$$0 \leq \min_{P_{r-1}} \max_{1 \leq i \leq n} [1 + \lambda_i P_{r-1}(\lambda_i)]^2 \leq \max_{1 \leq i \leq n} [1 + \lambda_i \bar{P}_{r-1}(\lambda_i)]^2 = \max_{1 \leq i \leq n} Q_r(\lambda_i) = 0.$$

Hence the constant in (5.33) is zero for the value $k = r - 1$, so we have by substituting into (5.32) that $\|x_r - x^*\|_A^2 = 0$, and therefore $x_r = x^*$, as claimed. □

By using similar reasoning, Luenberger [152] establishes the following estimate, which gives a useful characterization of the behavior of the CG method.

Theorem 5.5.
If A has eigenvalues $\lambda_1 \leq \lambda_2 \leq \cdots \leq \lambda_n$, we have that

$$\|x_{k+1} - x^*\|_A^2 \leq \left(\frac{\lambda_{n-k} - \lambda_1}{\lambda_{n-k} + \lambda_1}\right)^2 \|x_0 - x^*\|_A^2. \tag{5.34}$$

Without giving details of the proof, we describe how this result is obtained from (5.32). One selects a polynomial $\bar{P}_k$ of degree k such that the polynomial $Q_{k+1}(\lambda) = 1 + \lambda \bar{P}_k(\lambda)$ has roots at the k largest eigenvalues $\lambda_n, \lambda_{n-1}, \ldots, \lambda_{n-k+1}$, as well as at the midpoint between λ_1 and λ_{n-k}. It can be shown that the maximum value attained by Q_{k+1} on the remaining eigenvalues $\lambda_1, \lambda_2, \ldots, \lambda_{n-k}$ is precisely $(\lambda_{n-k} - \lambda_1)/(\lambda_{n-k} + \lambda_1)$.

We now illustrate how Theorem 5.5 can be used to predict the behavior of the CG method on specific problems. Suppose we have the situation plotted in Figure 5.3, where the eigenvalues of A consist of m large values, with the remaining $n - m$ smaller eigenvalues clustered around 1.

If we define $\epsilon = \lambda_{n-m} - \lambda_1$, Theorem 5.5 tells us that after $m + 1$ steps of the conjugate gradient algorithm, we have

$$\|x_{m+1} - x^*\| \approx \epsilon \|x_0 - x^*\|_A.$$

For a small value of ϵ, we conclude that the CG iterates will provide a good estimate of the solution after only $m + 1$ steps.

Figure 5.4 shows the behavior of CG on a problem of this type, which has five large eigenvalues with all the smaller eigenvalues clustered between 0.95 and 1.05, and compares this behavior with that of CG on a problem in which the eigenvalues satisfy some random distribution. In both cases, we plot the log of ϕ after each iteration.

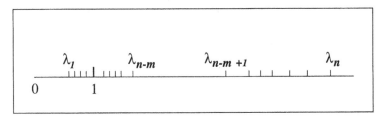

Figure 5.3 Two clusters of eigenvalues.

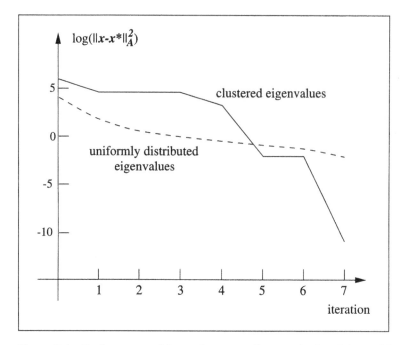

Figure 5.4 Performance of the conjugate gradient method on (a) a problem in which five of the eigenvalues are large and the remainder are clustered near 1, and (b) a matrix with uniformly distributed eigenvalues.

For the problem with clustered eigenvalues, Theorem 5.5 predicts a sharp decrease in the error measure at iteration 6. Note, however, that this decrease was achieved one iteration earlier, illustrating the fact that Theorem 5.5 gives only an upper bound, and that the rate of convergence can be faster. By contrast, we observe in Figure 5.4 that for the problem with randomly distributed eigenvalues the convergence rate is slower and more uniform.

Figure 5.4 illustrates another interesting feature: After one more iteration (a total of seven) on the problem with clustered eigenvalues, the error measure drops sharply. An extension of the arguments leading to Theorem 5.4 explains this behavior. It is *almost* true to say that the matrix A has just six distinct eigenvalues: the five large eigenvalues and 1. Then we would expect the error measure to be zero after six iterations. Because the eigenvalues near 1 are slightly spread out, however, the error does not become very small until the next iteration, i.e. iteration 7.

To state this more precisely, it is generally true that if the eigenvalues occur in r distinct clusters, the CG iterates will *approximately* solve the problem after r steps (see [115]). This result can be proved by constructing a polynomial $\bar{P}_{r-1}$ such that $(1 + \lambda \bar{P}_{r-1}(\lambda))$ has zeros inside each of the clusters. This polynomial may not vanish at the eigenvalues $\lambda_i, i = 1, 2, \ldots, n$, but its value will be small at these points, so the constant defined in (5.33) will be tiny for $k \geq r - 1$. We illustrate this behavior in Figure 5.5, which shows the

5.1. THE LINEAR CONJUGATE GRADIENT METHOD

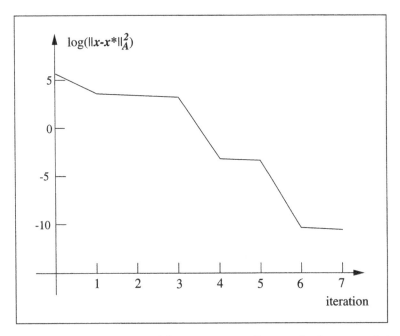

Figure 5.5 Performance of the conjugate gradient method on a matrix in which the eigenvalues occur in four distinct clusters.

performance of CG on a matrix of dimension $n = 14$ that has four clusters of eigenvalues: single eigenvalues at 140 and 120, a cluster of 10 eigenvalues very close to 10, with the remaining eigenvalues clustered between 0.95 and 1.05. After four iterations, the error norm is quite small. After six iterations, the solution is identified to good accuracy.

Another, more approximate, convergence expression for CG is based on the Euclidean condition number of A, which is defined by

$$\kappa(A) = \|A\|_2 \|A^{-1}\|_2 = \lambda_1/\lambda_n.$$

It can be shown that

$$\|x_k - x^*\|_A \leq \left(\frac{\sqrt{\kappa(A)} - 1}{\sqrt{\kappa(A)} + 1}\right)^{2k} \|x_0 - x^*\|_A. \tag{5.35}$$

This bound often gives a large overestimate of the error, but it can be useful in those cases where the only information we have about A is estimates of the extreme eigenvalues λ_1 and λ_n. This bound should be compared with that of the steepest descent method given by (3.28), which is identical in form but which depends on the condition number $\kappa(A)$, and not on its square root $\sqrt{\kappa(A)}$.

PRECONDITIONING

We can accelerate the conjugate gradient method by transforming the linear system to improve the eigenvalue distribution of A. The key to this process, which is known as *preconditioning*, is a change of variables from x to $\hat{x}$ via a nonsingular matrix C, that is,

$$\hat{x} = Cx. \tag{5.36}$$

The quadratic ϕ defined by (5.2) is transformed accordingly to

$$\hat{\phi}(\hat{x}) = \tfrac{1}{2}\hat{x}^T (C^{-T} A C^{-1})\hat{x} - (C^{-T} b)^T \hat{x}. \tag{5.37}$$

If we use Algorithm 5.2 to minimize $\hat{\phi}$ or, equivalently, to solve the linear system

$$(C^{-T} A C^{-1})\hat{x} = C^{-T} b,$$

then the convergence rate will depend on the eigenvalues of the matrix $C^{-T} A C^{-1}$ rather than those of A. Therefore, we aim to choose C such that the eigenvalues of $C^{-T} A C^{-1}$ are more favorable for the convergence theory discussed above. We can try to choose C such that the condition number of $C^{-T} A C^{-1}$ is much smaller than the original condition number of A, for instance, so that the constant in (5.35) is smaller. We could also try to choose C such that the eigenvalues of $C^{-T} A C^{-1}$ are clustered, which by the discussion of the previous section ensures that the number of iterates needed to find a good approximate solution is not much larger than the number of clusters.

It is not necessary to carry out the transformation (5.36) explicitly. Rather, we can apply Algorithm 5.2 to the problem (5.37), in terms of the variables $\hat{x}$, and then invert the transformations to reexpress all the equations in terms of x. This process of derivation results in Algorithm 5.3 (preconditioned conjugate gradient), which we now define. It happens that Algorithm 5.3 does not make use of C explicitly, but rather the matrix $M = C^T C$, which is symmetric and positive definite by construction.

Algorithm 5.3 (Preconditioned CG).
 Given x_0, preconditioner M;
 Set $r_0 \leftarrow Ax_0 - b$;
 Solve $My_0 = r_0$ for y_0;
 Set $p_0 = -r_0, k \leftarrow 0$;
 while $r_k \neq 0$

$$\alpha_k \leftarrow \frac{r_k^T y_k}{p_k^T A p_k}; \tag{5.38a}$$

$$x_{k+1} \leftarrow x_k + \alpha_k p_k; \tag{5.38b}$$

$$r_{k+1} \leftarrow r_k + \alpha_k A p_k; \qquad (5.38c)$$

$$M y_{k+1} \leftarrow r_{k+1}; \qquad (5.38d)$$

$$\beta_{k+1} \leftarrow \frac{r_{k+1}^T y_{k+1}}{r_k^T y_k}; \qquad (5.38e)$$

$$p_{k+1} \leftarrow -y_{k+1} + \beta_{k+1} p_k; \qquad (5.38f)$$

$$k \leftarrow k + 1; \qquad (5.38g)$$

end (while)

If we set $M = I$ in Algorithm 5.3, we recover the standard CG method, Algorithm 5.2. The properties of Algorithm 5.2 generalize to this case in interesting ways. In particular, the orthogonality property (5.15) of the successive residuals becomes

$$r_i^T M^{-1} r_j = 0 \quad \text{for all } i \neq j. \qquad (5.39)$$

In terms of computational effort, the main difference between the preconditioned and unpreconditioned CG methods is the need to solve systems of the form $My = r$.

PRACTICAL PRECONDITIONERS

No single preconditioning strategy is "best" for all conceivable types of matrices: The tradeoff between various objectives—effectiveness of M, inexpensive computation and storage of M, inexpensive solution of $My = r$—varies from problem to problem.

Good preconditioning strategies have been devised for specific types of matrices, in particular, those arising from discretizations of partial differential equations (PDEs). Often, the preconditioner is defined in such a way that the system $My = r$ amounts to a simplified version of the original system $Ax = b$. In the case of a PDE, $My = r$ could represent a coarser discretization of the underlying continuous problem than $Ax = b$. As in many other areas of optimization and numerical analysis, knowledge about the structure and origin of a problem (in this case, knowledge that the system $Ax = b$ is a finite-dimensional representation of a PDE) is the key to devising effective techniques for solving the problem.

General-purpose preconditioners have also been proposed, but their success varies greatly from problem to problem. The most important strategies of this type include symmetric successive overrelaxation (SSOR), incomplete Cholesky, and banded preconditioners. (See [220], [115], and [53] for discussions of these techniques.) *Incomplete Cholesky* is probably the most effective in general; we discussed it briefly in Chapter 6. The basic idea is simple: We follow the Cholesky procedure, but instead of computing the exact Cholesky factor L that satisfies $A = LL^T$, we compute an approximate factor $\tilde{L}$ that is sparser than L. (Usually, we require $\tilde{L}$ to be no denser, or not much denser, than the lower triangle of the original matrix A.) We then have $A \approx \tilde{L}\tilde{L}^T$, and by choosing $C = \tilde{L}^T$, we obtain $M = \tilde{L}\tilde{L}^T$

and

$$C^{-T}AC^{-1} = \tilde{L}^{-1}A\tilde{L}^{-T} \approx I,$$

so the eigenvalue distribution of $C^{-T}AC^{-1}$ is favorable. We do not compute M explicitly, but rather store the factor $\tilde{L}$ and solve the system $My = r$ by performing two triangular substitutions with $\tilde{L}$. Because the sparsity of $\tilde{L}$ is similar to that of A, the cost of solving $My = r$ is similar to the cost of computing the matrix–vector product Ap.

There are several possible pitfalls in the incomplete Cholesky approach. One is that the resulting matrix may not be (sufficiently) positive definite, and in this case one may need to increase the values of the diagonal elements to ensure that a value for $\tilde{L}$ can be found. Numerical instability or breakdown can occur during the incomplete factorization because of the sparsity conditions we impose on the factor $\tilde{L}$. This difficulty can be remedied by allowing additional fill-in in $\tilde{L}$, but the denser factor will be more expensive to compute and to apply at each iteration.

5.2 NONLINEAR CONJUGATE GRADIENT METHODS

We have noted that the CG method, Algorithm 5.2, can be viewed as a minimization algorithm for the convex quadratic function ϕ defined by (5.2). It is natural to ask whether we can adapt the approach to minimize general convex functions, or even general nonlinear functions f.

THE FLETCHER–REEVES METHOD

Fletcher and Reeves [88] showed that an extension of this kind is possible by making two simple changes in Algorithm 5.2. First, in place of the choice (5.23a) for the step length α_k (which minimizes ϕ along the search direction p_k), we need to perform a line search that identifies an approximate minimum of the nonlinear function f along p_k. Second, the residual r, which is simply the gradient of ϕ in Algorithm 5.2, must be replaced by the gradient of the nonlinear objective f. These changes give rise to the following algorithm for nonlinear optimization.

Algorithm 5.4 (FR-CG).
Given x_0;
Evaluate $f_0 = f(x_0)$, $\nabla f_0 = \nabla f(x_0)$;
Set $p_0 = -\nabla f_0, k \leftarrow 0$;
while $\nabla f_k \neq 0$
   Compute α_k and set $x_{k+1} = x_k + \alpha_k p_k$;
   Evaluate ∇f_{k+1};

5.2. NONLINEAR CONJUGATE GRADIENT METHODS

$$\beta_{k+1}^{FR} \leftarrow \frac{\nabla f_{k+1}^T \nabla f_{k+1}}{\nabla f_k^T \nabla f_k}; \tag{5.40a}$$

$$p_{k+1} \leftarrow -\nabla f_{k+1} + \beta_{k+1}^{FR} p_k; \tag{5.40b}$$

$$k \leftarrow k + 1; \tag{5.40c}$$

end (while)

If we choose f to be a strongly convex quadratic and α_k to be the exact minimizer, this algorithm reduces to the linear conjugate gradient method, Algorithm 5.2. Algorithm 5.4 is appealing for large nonlinear optimization problems because each iteration requires only evaluation of the objective function and its gradient. No matrix operations are performed, and just a few vectors of storage are required.

To make the specification of Algorithm 5.4 complete, we need to be more precise about the choice of line search parameter α_k. Because of the second term in (5.40b), the search direction p_k may fail to be a descent direction unless α_k satisfies certain conditions. By taking the inner product of (5.40b) (with k replacing $k+1$) with the gradient vector ∇f_k, we obtain

$$\nabla f_k^T p_k = -\|\nabla f_k\|^2 + \beta_k^{FR} \nabla f_k^T p_{k-1}. \tag{5.41}$$

If the line search is exact, so that α_{k-1} is a local minimizer of f along the direction p_{k-1}, we have that $\nabla f_k^T p_{k-1} = 0$. In this case we have from (5.41) that $\nabla f_k^T p_k < 0$, so that p_k is indeed a descent direction. But if the line search is not exact, the second term in (5.41) may dominate the first term, and we may have $\nabla f_k^T p_k > 0$, implying that p_k is actually a direction of ascent. Fortunately, we can avoid this situation by requiring the step length α_k to satisfy the *strong* Wolfe conditions, which we restate here:

$$f(x_k + \alpha_k p_k) \leq f(x_k) + c_1 \alpha_k \nabla f_k^T p_k, \tag{5.42a}$$

$$|\nabla f(x_k + \alpha_k p_k)^T p_k| \leq c_2 |\nabla f_k^T p_k|, \tag{5.42b}$$

where $0 < c_1 < c_2 < \frac{1}{2}$. (Note that we impose $c_2 < \frac{1}{2}$ here, in place of the looser condition $c_2 < 1$ that was used in the earlier statement (3.7).) By applying Lemma 5.6 below, we can show that condition (5.42b) implies that (5.41) is negative, and we conclude that any line search procedure that yields an α_k satisfying (5.42) will ensure that all directions p_k are descent directions for the function f.

THE POLAK–RIBIÈRE METHOD

There are many variants of the Fletcher–Reeves method that differ from each other mainly in the choice of the parameter β_k. The most important of these variants, proposed by Polak and Ribière, defines this parameter as follows:

$$\beta_{k+1}^{PR} = \frac{\nabla f_{k+1}^T (\nabla f_{k+1} - \nabla f_k)}{\|\nabla f_k\|^2}. \tag{5.43}$$

We refer to the algorithm in which (5.43) replaces (5.40a) as Algorithm PR-CG, and refer to Algorithm 5.4 as Algorithm FR-CG. They are identical when f is a strongly convex quadratic function and the line search is exact, since by (5.15) the gradients are mutually orthogonal, and so $\beta_{k+1}^{PR} = \beta_{k+1}^{FR}$. When applied to general nonlinear functions with inexact line searches, however, the behavior of the two algorithms differs markedly. Numerical experience indicates that Algorithm PR-CG tends to be the more robust and efficient of the two.

A surprising fact about Algorithm PR-CG is that the strong Wolfe conditions (5.42) do not guarantee that p_k is always a descent direction. If we define the β parameter as

$$\beta_{k+1}^+ = \max\{\beta_{k+1}^{PR}, 0\}, \tag{5.44}$$

giving rise to an algorithm we call Algorithm PR+, then a simple adaptation of the strong Wolfe conditions ensures that the descent property holds.

There are many other choices for β_{k+1} that coincide with the Fletcher–Reeves formula β_{k+1}^{FR} in the case where the objective is quadratic and the line search is exact. The Hestenes–Stiefel formula, which defines

$$\beta_{k+1}^{HS} = \frac{\nabla f_{k+1}^T (\nabla f_{k+1} - \nabla f_k)}{(\nabla f_{k+1} - \nabla f_k)^T p_k}, \tag{5.45}$$

gives rise to an algorithm that is similar to Algorithm PR-CG, both in terms of its theoretical convergence properties and in its practical performance. Formula (5.45) can be derived by demanding that consecutive search directions be conjugate with respect to the *average Hessian* over the line segment $[x_k, x_{k+1}]$, which is defined as

$$\bar{G}_k \equiv \int_0^1 [\nabla^2 f(x_k + \tau \alpha_k d_k)] d\tau.$$

Recalling from Taylor's theorem (2.5) that $\nabla f_{k+1} = \nabla f_k + \alpha_k \bar{G}_k p_k$, we see that for any direction of the form $p_{k+1} = -\nabla f_{k+1} + \beta_{k+1} p_k$, the condition $p_{k+1}^T \bar{G}_k p_k = 0$ requires β_{k+1} to be given by (5.45).

None of the other proposed definitions of β_k has proved to be significantly more efficient than the Polak–Ribière choice (5.43).

QUADRATIC TERMINATION AND RESTARTS

Implementations of nonlinear conjugate gradient methods usually preserve their close connections with the linear conjugate gradient method. Usually, a quadratic (or cubic) interpolation along the search direction p_k is incorporated into the line search procedure; see Chapter 3. This feature guarantees that when f is a strictly convex quadratic, the step length α_k is chosen to be the exact one-dimensional minimizer, so that the nonlinear conjugate gradient method reduces to the linear method, Algorithm 5.2.

Another modification that is often used in nonlinear conjugate gradient procedures is to *restart* the iteration at every n steps by setting $\beta_k = 0$ in (5.40a), that is, by taking a steepest descent step. Restarting serves to periodically refresh the algorithm, erasing old information that may not be beneficial. We can even prove a strong theoretical result about restarting: It leads to n-step quadratic convergence, that is,

$$\|x_{k+n} - x\| = O\left(\|x_k - x^*\|^2\right). \tag{5.46}$$

With a little thought, we can see that this result is not so surprising. Consider a function f that is strongly convex quadratic in a neighborhood of the solution, but is nonquadratic everywhere else. Assuming that the algorithm is converging to the solution in question, the iterates will eventually enter the quadratic region. At some point, the algorithm will be restarted in that region, and from that point onward, its behavior will simply be that of the linear conjugate gradient method, Algorithm 5.2. In particular, finite termination will occur within n steps of the restart. The restart is important, because the finite-termination property (and other appealing properties) of Algorithm 5.2 holds only when its initial search direction p_0 is equal to the negative gradient.

Even if the function f is not exactly quadratic in the region of a solution, Taylor's theorem (2.6) implies that it can still be approximated quite closely by a quadratic, provided that it is smooth. Therefore, while we would not expect termination in n steps after the restart, it is not surprising that substantial progress is made toward the solution, as indicated by the expression (5.46).

Though the result (5.46) is interesting from a theoretical viewpoint, it may not be relevant in a practical context, because nonlinear conjugate gradient methods can be recommended only for solving problems with large n. In such problems restarts may never occur, since an approximate solution is often located in fewer than n steps. Hence, nonlinear CG method are sometimes implemented without restarts, or else they include strategies for restarting that are based on considerations other than iteration counts. The most popular restart strategy makes use of the observation (5.15), which is that the gradients are mutually orthogonal when f is a quadratic function. A restart is performed whenever two consecutive gradients are far from orthogonal, as measured by the test

$$\frac{|\nabla f_k^T \nabla f_{k-1}|}{\|\nabla f_k\|^2} \geq \nu, \tag{5.47}$$

where a typical value for the parameter ν is 0.1.

Another modification to the restart strategy is to use a direction other than steepest descent as the restart direction. The Harwell subroutine VA14 [133], for instance, defines

p_{k+1} by using a three-term recurrence based on ∇f_{k+1}, p_k and a third direction that contains earlier information about the behavior of the objective function. An algorithm that takes this idea one step further is CONMIN, which is discussed in Chapter 9.

We could also think of formula (5.44) as a restarting strategy, because p_{k+1} will revert to the steepest descent direction whenever β_k^{PR} is negative. In contrast to (5.47), however, these restarts are rather infrequent because β_k^{PR} is positive most of the time.

NUMERICAL PERFORMANCE

Table 5.1 illustrates the performance of Algorithms FR-CG, PR-CG, and PR+ without restarts. For these tests, the parameters in the strong Wolfe conditions (5.42) were chosen to be $c_1 = 10^{-4}$ and $c_2 = 0.1$. The iterations were terminated when

$$\|\nabla f_k\|_\infty < 10^{-5}(1 + |f_k|),$$

or after 10,000 iterations (the latter is denoted by a $*$).

The final column, headed "mod," indicates the number of iterations of Algorithm PR+ for which the adjustment (5.44) was needed to ensure that $\beta_k^{PR} \geq 0$. An examination of the results of Algorithm FR-CG on problem GENROS shows that the method takes very short steps far from the solution that lead to tiny improvements in the objective function.

The Polak–Ribière algorithm, or its variation PR+, are not *always* more efficient than Algorithm FR-CG, and it has the slight disadvantage of requiring one more vector of storage. Nevertheless, we recommend that users choose Algorithm PR-CG or PR+ whenever possible.

BEHAVIOR OF THE FLETCHER–REEVES METHOD

We now investigate the Fletcher–Reeves algorithm, Algorithm 5.4, a little more closely, proving that it is globally convergent and explaining some of its observed inefficiencies.

The following result gives conditions on the line search that guarantee that all search directions are descent directions. It assumes that the level set $\mathcal{L} = \{x : f(x) \leq f(x_0)\}$ is

Table 5.1 Iterations and function/gradient evaluations required by three nonlinear conjugate gradient methods on a set of test problems.

Problem	n	Alg FR it/f-g	Alg PR it/f-g	Alg PR+ it/f-g	mod
CALCVAR3	200	2808/5617	2631/5263	2631/5263	0
GENROS	500	*	1068/2151	1067/2149	1
XPOWSING	1000	533/1102	212/473	97/229	3
TRIDIA1	1000	264/531	262/527	262/527	0
MSQRT1	1000	422/849	113/231	113/231	0
XPOWELL	1000	568/1175	212/473	97/229	3
TRIGON	1000	231/467	40/92	40/92	0

5.2. NONLINEAR CONJUGATE GRADIENT METHODS

bounded, and that f is twice continuously differentiable, so that we have from Lemma 3.1 that there exists a step length α_k that satisfies the strong Wolfe conditions.

Lemma 5.6.
Suppose that Algorithm 5.4 is implemented with a step length α_k that satisfies the strong Wolfe conditions (5.42) with $0 < c_2 < \frac{1}{2}$. Then the method generates descent directions p_k that satisfy the following inequalities:

$$-\frac{1}{1-c_2} \le \frac{\nabla f_k^T p_k}{\|\nabla f_k\|^2} \le \frac{2c_2 - 1}{1 - c_2}, \quad \text{for all } k = 0, 1, \ldots. \tag{5.48}$$

PROOF. Note first that the function $t(\xi) \stackrel{\text{def}}{=} (2\xi - 1)(1 - \xi)$ is monotonically increasing on the interval $[0, \frac{1}{2}]$ and that $t(0) = -1$ and $t(\frac{1}{2}) = 0$. Hence, because of $c_2 \in (0, \frac{1}{2})$, we have

$$-1 < \frac{2c_2 - 1}{1 - c_2} < 0. \tag{5.49}$$

The descent condition $\nabla f_k^T p_k < 0$ follows immediately once we establish (5.48).

The proof is by induction. For $k = 0$, the middle term in (5.48) is -1, so by using (5.49), we see that both inequalities in (5.48) are satisfied. Next, assume that (5.48) holds for some $k \ge 1$. From (5.40b) and (5.40a) we have

$$\frac{\nabla f_{k+1}^T p_{k+1}}{\|\nabla f_{k+1}\|^2} = -1 + \beta_{k+1} \frac{\nabla f_{k+1}^T p_k}{\|\nabla f_{k+1}\|^2} = -1 + \frac{\nabla f_{k+1}^T p_k}{\|\nabla f_k\|^2}. \tag{5.50}$$

By using the line search condition (5.42b), we have

$$|\nabla f_{k+1}^T p_k| \le -c_2 \nabla f_k^T p_k,$$

so by combining with (5.50), we obtain

$$-1 + c_2 \frac{\nabla f_k^T p_k}{\|\nabla f_k\|^2} \le \frac{\nabla f_{k+1}^T p_{k+1}}{\|\nabla f_{k+1}\|^2} \le -1 - c_2 \frac{\nabla f_k^T p_k}{\|\nabla f_k\|^2}.$$

Substituting for the term $\nabla f_k^T p_k / \|\nabla f_k\|^2$ from the left-hand-side of the induction hypothesis (5.48), we obtain

$$-1 - \frac{c_2}{1 - c_2} \le \frac{\nabla f_{k+1}^T p_{k+1}}{\|\nabla f_{k+1}\|^2} \le -1 + \frac{c_2}{1 - c_2},$$

which shows that (5.48) holds for $k + 1$ as well. □

This result used only the second strong Wolfe condition (5.42b); the first Wolfe condition (5.42a) will be needed in the next section to establish global convergence. The bounds on $\nabla f_k^T p_k$ in (5.48) impose a limit on how fast the norms of the steps $\|p_k\|$ can grow, and they will play a crucial role in the convergence analysis given below.

Lemma 5.6 can also be used to explain a weakness of the Fletcher–Reeves method. We will argue that if the method generates a bad direction and a tiny step, then the next direction and next step are also likely to be poor. As in Chapter 3, we let θ_k denote the angle between p_k and the steepest descent direction $-\nabla f_k$, defined by

$$\cos \theta_k = \frac{-\nabla f_k^T p_k}{\|\nabla f_k\| \|p_k\|}. \tag{5.51}$$

Suppose that p_k is a poor search direction, in the sense that it makes an angle of nearly 90° with $-\nabla f_k$, that is, $\cos \theta_k \approx 0$. By multiplying both sides of (5.48) by $\|\nabla f_k\|/\|p_k\|$ and using (5.51), we obtain

$$\chi_1 \frac{\|\nabla f_k\|}{\|p_k\|} \leq \cos \theta_k \leq \chi_2 \frac{\|\nabla f_k\|}{\|p_k\|}, \quad \text{for all } k = 0, 1, \ldots, \tag{5.52}$$

where χ_1 and χ_2 are two positive constants. From the right-hand inequality we can have $\cos \theta_k \approx 0$ if and only if

$$\|\nabla f_k\| \ll \|p_k\|.$$

Since p_k is almost orthogonal to the gradient, it is likely that the step from x_k to x_{k+1} is tiny, that is, $x_{k+1} \approx x_k$. If so, we have $\nabla f_{k+1} \approx \nabla f_k$, and therefore

$$\beta_{k+1}^{FR} \approx 1, \tag{5.53}$$

by the definition (5.40a). By using this approximation together with $\|\nabla f_{k+1}\| \approx \|\nabla f_k\| \ll \|p_k\|$ in (5.40b), we conclude that

$$p_{k+1} \approx p_k,$$

so the new search direction will improve little (if at all) on the previous one. It follows that if the condition $\cos \theta_k \approx 0$ holds at some iteration k and if the subsequent step is small, a long sequence of unproductive iterates will follow.

The Polak–Ribière method behaves quite differently in these circumstances. If, as in the previous paragraph, the search direction p_k satisfies $\cos \theta_k \approx 0$ for some k, and if the subsequent step is small, it follows by substituting $\nabla f_k \approx \nabla f_{k+1}$ into (5.43) that $\beta_{k+1}^{PR} \approx 0$. From the formula (5.40b), we find that the new search direction p_{k+1} will be close to the steepest descent direction $-\nabla f_{k+1}$, and $\cos \theta_{k+1}$ will be close to 1. Therefore, Algorithm

PR-CG essentially performs a restart after it encounters a bad direction. The same argument can be applied to Algorithms PR+ and HS-CG.

The inefficient behavior of the Fletcher–Reeves method predicted by the arguments given above can be observed in practice. For example, the paper [103] describes a problem with $n = 100$ in which $\cos\theta_k$ is of order 10^{-2} for hundreds of iterations, and the steps $\|x_k - x_{k-1}\|$ are of order 10^{-2}. Algorithm FR-CG requires thousands of iterations to solve this problem, while Algorithm PR-CG requires just 37 iterations. In this example, the Fletcher–Reeves method performs much better if it is periodically restarted along the steepest descent direction, since each restart terminates the cycle of bad steps. In general, Algorithm FR-CG should not be implemented without some kind of restart strategy.

GLOBAL CONVERGENCE

Unlike the linear conjugate gradient method, whose convergence properties are well understood and which is known to be optimal as described above, nonlinear conjugate gradient methods possess surprising, sometimes bizarre, convergence properties. The theory developed in the literature offers fascinating glimpses into their behavior, but our knowledge remains fragmentary. We now present a few of the main results known for the Fletcher–Reeves and Polak–Ribière methods using practical line searches.

For the purposes of this section, we make the following (nonrestrictive) assumptions on the objective function.

Assumption 5.1.

(i) *The level set $\mathcal{L} := \{x : f(x) \leq f(x_0)\}$ is bounded.*

(ii) *In some neighborhood $\mathcal{N}$ of $\mathcal{L}$, the objective function f is Lipschitz continuously differentiable, that is, there exists a constant $L > 0$ such that*

$$\|\nabla f(x) - \nabla f(\tilde{x})\| \leq L\|x - \tilde{x}\|, \quad \text{for all } x, \tilde{x} \in \mathcal{N}. \tag{5.54}$$

This assumption implies that there is a constant $\bar{\gamma}$ such that

$$\|\nabla f(x)\| \leq \bar{\gamma}, \text{ for all } x \in \mathcal{L}. \tag{5.55}$$

Our main analytical tool in this section is Zoutendijk's theorem—Theorem 3.2 in Chapter 3—which we restate here for convenience.

Theorem 5.7.

Suppose that Assumptions 5.1 hold. Consider any line search iteration of the form $x_{k+1} = x_k + \alpha_k p_k$, where p_k is a descent direction and α_k satisfies the Wolfe conditions (5.42). Then

$$\sum_{k=1}^{\infty} \cos^2\theta_k \, \|\nabla f_k\|^2 < \infty. \tag{5.56}$$

We can use this theorem to prove global convergence for algorithms that are periodically restarted by setting $\beta_k = 0$. If k_1, k_2, and so on denote the iterations on which restarts occur, we have from (5.56) that

$$\sum_{k=k_1,k_2,\ldots} \|\nabla f_k\|^2 < \infty. \tag{5.57}$$

If we allow no more than $\bar{n}$ iterations between restarts, the sequence $\{k_j\}_{j=1}^{\infty}$ will be infinite, and from (5.57) we have that $\lim_{j\to\infty} \|\nabla f_{k_j}\| = 0$. That is, a subsequence of gradients approaches zero, or equivalently,

$$\liminf_{k\to\infty} \|\nabla f_k\| = 0. \tag{5.58}$$

This result applies equally to restarted versions of all the algorithms discussed in this chapter.

It is more interesting, however, to study the global convergence of *unrestarted* conjugate gradient methods, because for large problems (say $n \geq 1000$) we expect to find a solution in many fewer than n iterations—the first point at which a regular restart would take place. Our study of large sequences of unrestarted conjugate gradient iterations reveals some surprising patterns in their behavior.

We can build on Lemma 5.6 and Theorem 5.7 to prove a global convergence result for the Fletcher–Reeves method. While we cannot show that the limit of the sequence of gradients $\{\nabla f_k\}$ is zero (as in the restarted method above), the following result shows that it is at least not bounded away from zero.

Theorem 5.8.

Suppose that Assumptions 5.1 hold, and that Algorithm 5.4 is implemented with a line search that satisfies the strong Wolfe conditions (5.42), with $0 < c_1 < c_2 < \frac{1}{2}$. Then

$$\liminf_{k\to\infty} \|\nabla f_k\| = 0. \tag{5.59}$$

PROOF. The proof is by contradiction. It assumes that the opposite of (5.59) holds, that is, there is a constant $\gamma > 0$ such that

$$\|\nabla f_k\| \geq \gamma, \quad \text{for all } k \text{ sufficiently large}, \tag{5.60}$$

and uses Lemma 5.6 and Theorem 5.7 to derive the contradiction.

From Lemma 5.6, we have that

$$\cos \theta_k \geq \frac{1}{1-c_2} \frac{\|\nabla f_k\|}{\|p_k\|}, \quad k = 1, 2, \ldots, \tag{5.61}$$

and by substituting this relation in Zoutendijk's condition (5.56), we obtain

$$\sum_{k=1}^{\infty} \frac{\|\nabla f_k\|^4}{\|p_k\|^2} < \infty. \tag{5.62}$$

By using (5.42b) and Lemma 5.6 again, we obtain that

$$|\nabla f_k^T p_{k-1}| \leq -c_2 \nabla f_{k-1}^T p_{k-1} \leq \frac{c_2}{1-c_2} \|\nabla f_{k-1}\|^2. \tag{5.63}$$

Thus, from (5.40b) and recalling the definition (5.40a) of β_k^{FR} we obtain

$$\|p_k\|^2 \leq \|\nabla f_k\|^2 + 2\beta_k^{\text{FR}} |\nabla f_k^T p_{k-1}| + (\beta_k^{\text{FR}})^2 \|p_{k-1}\|^2$$

$$\leq \|\nabla f_k\|^2 + \frac{2c_2}{1-c_2} \beta_k^{\text{FR}} \|\nabla f_{k-1}\|^2 + (\beta_k^{\text{FR}})^2 \|p_{k-1}\|^2$$

$$\leq \left(\frac{1+c_2}{1-c_2}\right) \|\nabla f_k\|^2 + (\beta_k^{\text{FR}})^2 \|p_{k-1}\|^2.$$

Applying this relation repeatedly, defining $c_3 \stackrel{\text{def}}{=} (1+c_2)/(1-c_2) \geq 1$, and using the definition (5.40a) of β_k^{FR}, we have

$$\|p_k\|^2 \leq c_3 \|\nabla f_k\|^2 + (\beta_k^{\text{FR}})^2 [\chi_3 \|\nabla f_{k-1}\|^2 + (\beta_{k-1}^{\text{FR}})^2 \|p_{k-2}\|^2]$$

$$= c_3 \|\nabla f_k\|^4 \left[\frac{1}{\|\nabla f_k\|^2} + \frac{1}{\|\nabla f_{k-1}\|^2}\right] + \frac{\|\nabla f_k\|^4}{\|\nabla f_{k-2}\|^4} \|p_{k-2}\|^2$$

$$\leq c_3 \|\nabla f_k\|^4 \sum_{j=1}^{k} \|\nabla f_j\|^{-2}, \tag{5.64}$$

where we used the facts that

$$(\beta_k^{\text{FR}})^2 (\beta_{k-1}^{\text{FR}})^2 \cdots (\beta_{k-i}^{\text{FR}})^2 = \frac{\|\nabla f_k\|^4}{\|\nabla f_{k-i-1}\|^4}$$

and $p_1 = -\nabla f_1$. By using the bounds (5.55) and (5.60) in (5.64), we obtain

$$\|p_k\|^2 \leq \frac{\chi_3 \bar{\gamma}^4}{\gamma^2} k, \tag{5.65}$$

which implies that

$$\sum_{k=1}^{\infty} \frac{1}{\|p_k\|^2} \geq \gamma_4 \sum_{k=1}^{\infty} \frac{1}{k}, \tag{5.66}$$

for some positive constant γ_4.

Suppose for contradiction that (5.60) holds. Then from (5.62), we have that

$$\sum_{k=1}^{\infty} \frac{1}{\|p_k\|^2} < \infty. \tag{5.67}$$

However, if we combine this inequality with (5.66), we obtain that $\sum_{k=1}^{\infty} 1/k < \infty$, which is not true. Hence, (5.60) does not hold, and the claim (5.59) is proved. □

Note that this global convergence result applies to a practical implementation of the Fletcher–Reeves method and is valid for general nonlinear objective functions. In this sense, it is more satisfactory than other convergence results that apply to specific types of objectives—for example, convex functions.

In general, if we can show that there is a constant $c_4 > 0$ such that

$$\cos \theta_k \geq c_4 \frac{\|\nabla f_k\|}{\|p_k\|}, \quad k = 1, 2, \ldots,$$

and another constant c_5 such that

$$\frac{\|\nabla f_k\|}{\|p_k\|} \geq c_5 > 0, \quad k = 1, 2, \ldots,$$

it follows from Theorem 5.7 that

$$\lim_{k \to \infty} \|\nabla f_k\| = 0.$$

This result can be established for the Polak–Ribière method under the assumption that f is strongly convex and that an exact line search is used.

For general (nonconvex) functions, is it not possible to prove a result like Theorem 5.8 for Algorithm PR-CG. This is unexpected, since the Polak–Ribière method performs better in practice than the Fletcher–Reeves method. In fact, the following surprising result shows that the Polak–Ribière method can cycle infinitely without approaching a solution point, even if an ideal line search is used. (By "ideal" we mean that line search returns a value α_k that is the first stationary point for the function $t(\alpha) = f(x_k + \alpha p_k)$.)

Theorem 5.9.

Consider the Polak–Ribière method method (5.43), with an ideal line search. There exists a twice continuously differentiable objective function $f : \mathbf{R}^3 \to \mathbf{R}$ and a starting point $x_0 \in \mathbf{R}^3$ such that the sequence of gradients $\{\|\nabla f_k\|\}$ is bounded away from zero.

5.2. NONLINEAR CONJUGATE GRADIENT METHODS

The proof of this result is given in [207], and is quite complex. It demonstrates the existence of the desired objective function without actually constructing this function explicitly. The result is interesting, since the step length assumed in the proof—the first stationary point—may be accepted by any of the practical line search algorithms currently in use.

The proof of Theorem 5.9 requires that some consecutive search directions become almost negatives of each other. In the case of ideal line searches, this can be achieved only if $\beta_k < 0$, so the analysis suggests a modification of the Polak–Ribière method in which we set

$$\beta_k^+ = \max\{\beta_k^{\text{PR}}, 0\}. \tag{5.68}$$

This method is exactly Algorithm PR+ discussed above. We mentioned earlier that a line search strategy based on a slight modification of the Wolfe conditions guarantees that all search directions generated by Algorithm PR+ are descent directions. Using these facts, it is possible to prove global convergence of Algorithm PR+ for general functions.

NOTES AND REFERENCES

The conjugate gradient method was developed in the 1950s by Hestenes and Stiefel [135] as an alternative to factorization methods for finding exact solutions of symmetric positive definite systems. It was not until some years later, in one of the most important developments in sparse linear algebra, that this method came to be viewed as an iterative method that could give good approximate solutions to systems in many fewer than n steps. Our presentation of the linear conjugate gradient method follows that of Luenberger [152].

Interestingly enough, the nonlinear conjugate gradient method of Fletcher and Reeves [88] was proposed after the linear conjugate gradient method had fallen out of favor, but several years before it was rediscovered as an iterative method. The Polak–Ribière method was proposed in [194], and the example showing that it may fail to converge on nonconvex problems is given by Powell [207]. Restart procedures are discussed in Powell [203].

Analysis due to Powell [200] provides further evidence of the inefficiency of the Fletcher–Reeves method using exact line searches. He shows that if the iterates enter a region in which the function is the two-dimensional quadratic

$$f(x) = \tfrac{1}{2} x^T x,$$

then the angle between the gradient ∇f_k and the search direction p_k stays constant. Therefore, if this angle is close to 90°, the method will converge very slowly. Indeed, since this angle can be arbitrarily close to 90°, the Fletcher–Reeves method can be slower than the steepest descent method. The Polak–Ribière method behaves quite differently in these circumstances: If a very small step is generated, the next search direction tends to the steepest descent direction, as argued above. This feature prevents a sequence of tiny steps.

The global convergence of nonlinear conjugate gradient methods is studied also in Al-Baali [3] and Gilbert and Nocedal [103].

Most of the theory on the *rate of convergence* of conjugate gradient methods assumes that the line search is exact. Crowder and Wolfe [61] show that the rate of convergence is linear, and show by constructing an example that Q-superlinear convergence is not achievable. Powell [200] studies the case in which the conjugate gradient method enters a region where the objective function is quadratic, and shows that either finite termination occurs or the rate of convergence is linear. Cohen [43] and Burmeister [33] prove n-step quadratic convergence for general objective functions, that is,

$$\|x_{k+n} - x^*\| = O(\|x_k - x^*\|^2).$$

Ritter [214] shows that in fact, the rate is *superquadratic*, that is,

$$\|x_{k+n} - x^*\| = o(\|x_k - x^*\|^2).$$

Powell [206] gives a slightly better result and performs numerical tests on small problems to measure the rate observed in practice. He also summarizes rate-of-convergence results for asymptotically exact line searches, such as those obtained by Baptist and Stoer [6] and Stoer [232].

Even faster rates of convergence can be established (see Schuller [225], Ritter [214]), under the assumption that the search directions are uniformly linearly independent, but this assumption is hard to verify and does not often occur in practice.

Nemirovsky and Yudin [180] devote some attention to the *global efficiency* of the Fletcher–Reeves and Polak–Ribière methods with exact line searches. For this purpose they define a measure of "laboriousness" and an "optimal bound" for it among a certain class of iterations. They show that on strongly convex problems not only do the Fletcher–Reeves and Polak–Ribière methods fail to attain the optimal bound, but they may also be slower than the steepest descent method. Subsequently, Nesterov [180] presented an algorithm that attains this optimal bound. It is related to PARTAN, the method of parallel tangents (see, for example, Luenberger [152]). We feel that this approach is unlikely to be effective in practice, but no conclusive investigation has been carried out, to the best of our knowledge.

Special line search strategies that ensure global convergence of the Polak–Ribière method have been proposed, but they are not without disadvantages.

The results in Table 5.1 are taken from Gilbert and Nocedal [103]. This paper also describes a line search that guarantees that Algorithm PR+ always generates descent directions, and proves global convergence.

✐ EXERCISES

✐ 5.1 Implement Algorithm 5.2 and use to it solve linear systems in which A is the Hilbert matrix, whose elements are $A_{i,j} = 1/(i+j-1)$. Set the right-hand-side to $b = (1, 1, \ldots, 1)^T$

and the initial point to $x_0 = 0$. Try dimensions $n = 5, 8, 12, 20$ and report the number of iterations required to reduce the residual below 10^{-6}.

✏ **5.2** Show that if the nonzero vectors $p_0, p_1, \ldots, p_l$ satisfy (5.4), where A is symmetric and positive definite, then these vectors are linearly independent. (This result implies that A has at most n conjugate directions.)

✏ **5.3** Verify (5.6).

✏ **5.4** Show that if $f(x)$ is a strictly convex quadratic, then the function $h(\sigma) \stackrel{\text{def}}{=} f(x_0 + \sigma_0 p_0 + \cdots + \sigma_{k-1} p_{k-1})$ also is a strictly convex quadratic in $\sigma = (\sigma_0, \ldots, \sigma_{k-1})^T$.

✏ **5.5** Using the form of the CG iteration prove directly that (5.16) and (5.17) hold for $k = 1$.

✏ **5.6** Show that (5.23d) is equivalent to (5.13d).

✏ **5.7** Let $\{\lambda_i, v_i\}\ i = 1, \ldots, n$ be the eigenpairs of A. Show that the eigenvalues and eigenvectors of $[I + P_k(A)A]^T A[I + P_k(A)A]$ are $\lambda_i [1 + \lambda_i P_k(\lambda_i)]^2$ and v_i, respectively.

✏ **5.8** Construct matrices with various eigenvalue distributions and apply the CG method to them. Then observe whether the behavior can be explained from Theorem 5.5.

✏ **5.9** Derive Algorithm 5.3 by applying the standard CG method in the variables $\hat{x}$ and then transforming back into the original variables.

✏ **5.10** Verify the modified conjugacy condition (5.39).

✏ **5.11** Show that when applied to a quadratic function, and with exact line searches, the Polak–Ribière formula given by (5.43) and the Hestenes–Stiefel formula given by (5.45) reduce to the Fletcher–Reeves formula (5.40a).

✏ **5.12** Prove that Lemma 5.6 holds for any choice of β_k satisfying $|\beta_k| \leq \beta_k^{\text{FR}}$.

Chapter 6

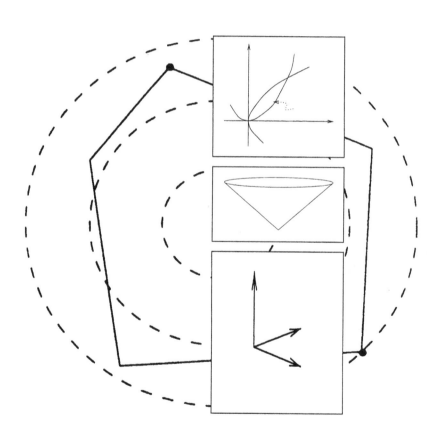

Practical Newton Methods

We saw in Chapter 3 that the pure Newton method with unit steps converges rapidly once it approaches a minimizer x^*. This simple algorithm is inadequate for general use, however, since it may fail to converge to a solution from remote starting points. Even if it does converge, its behavior may be erratic in regions where the function is not convex. Our first goal in designing a practical Newton-based algorithm is to make it robust and efficient in all cases.

Recall that the basic Newton step p_k^N is obtained by solving the symmetric $n \times n$ linear system

$$\nabla^2 f(x_k) p_k^N = -\nabla f(x_k). \tag{6.1}$$

To obtain global convergence we require the search direction p_k^N to be a descent direction, which will be true here if the Hessian $\nabla^2 f(x_k)$ is positive definite. However, if the Hessian is not positive definite, or is close to being singular, p_k^N may be an ascent direction or may be excessively long. In this chapter, we describe two strategies for ensuring that the step is always of good quality. In the first strategy we solve (6.1) using the conjugate gradient (CG) method (see Chapter 5), terminating if negative curvature is encountered. There are both line search and trust-region implementations of this strategy, which we call the *Newton–CG method*. The second approach consists in modifying the Hessian matrix $\nabla^2 f(x_k)$ before or during the solution of the system (6.1) so that it becomes sufficiently positive definite. We call this the *modified Newton method*.

Another concern in designing practical Newton methods is to keep the computational cost as small as possible. In the Newton–CG method, we accomplish this goal by terminating the CG iteration before an exact solution to (6.1) is found. This *inexact* Newton approach thus computes an approximation p_k to the pure Newton step p_k^N. When using a direct method to solve the Newton equations we can take advantage of any sparsity structure in the Hessian by using sparse Gaussian elimination techniques to factor $\nabla^2 f(x_k)$ and then using the factors to obtain an exact solution of (6.1).

The computation of the Hessian matrix $\nabla^2 f(x_k)$ usually represents a major task in the implementation of Newton's method. If the Hessian is not available in analytic form, we can use automatic differentiation techniques to compute it, or we can use finite differences to estimate it. A detailed treatment of these topics is deferred to Chapter 7. With all these ingredients—modification of the pure Newton iteration, exploitation of sparsity, and differentiation techniques—the Newton methods described in this chapter represent some of the most reliable and powerful methods for solving both small or large unconstrained optimization problems.

Since inexact Newton steps are useful in both line search and trust-region implementations, we discuss them first.

6.1 INEXACT NEWTON STEPS

We have noted that a drawback of the pure Newton method is the need to solve the equations (6.1) exactly. Techniques based on Gaussian elimination or another type of factorization of the coefficient matrix can be expensive when the number of variables is large. An accurate solution to (6.1) may not be warranted in any case, since the quadratic model used to derive the Newton equations may not provide a good prediction of the behavior of the function f, especially when the current iterate x_k is remote from the solution x^*. It is therefore appealing to apply an iterative method to (6.1), terminating the iterations at some approximate (inexact) solution of this system.

Most rules for terminating the iterative solver for (6.1) are based on the residual

$$r_k = \nabla^2 f(x_k) p_k + \nabla f(x_k), \tag{6.2}$$

where p_k is the inexact Newton step. Since the size of the residual changes if f is multiplied by a constant (i.e., r is not invariant to scaling of the objective function), we consider its size relative to that of the right-hand-side vector in (6.1), namely $\nabla f(x_k)$. We therefore terminate the iterative solver when

$$\|r_k\| \leq \eta_k \|\nabla f(x_k)\|, \tag{6.3}$$

where the sequence $\{\eta_k\}$ (with $0 < \eta_k < 1$ for all k) is called the *forcing sequence*.

We now study how the rate of convergence of inexact Newton methods based on (6.1)–(6.3) is affected by the choice of the forcing sequence. Our first result says that local convergence is obtained simply by ensuring that η_k is bounded away from 1.

Theorem 6.1.

Suppose that $\nabla f(x)$ is continuously differentiable in a neighborhood of a minimizer x^*, and assume that $\nabla^2 f(x^*)$ is positive definite. Consider the iteration $x_{k+1} = x_k + p_k$ where p_k satisfies (6.3), and assume that $\eta_k \leq \eta$ for some constant $\eta \in [0, 1)$. Then, if the starting point x_0 is sufficiently near x^*, the sequence $\{x_k\}$ converges to x^* linearly. That is, for all k sufficiently large, we have

$$\|x_{k+1} - x^*\| \leq c\|x_k - x^*\|,$$

for some constant $0 < c < 1$.

Note that the condition on the forcing sequence $\{\eta_k\}$ is not very restrictive, in the sense that if we relaxed it just a little, it would yield an algorithm that obviously does not converge. Specifically, if we allowed $\eta_k \geq 1$, the step $p_k = 0$ would satisfy (6.3) at every iteration, but the resulting method would set $x_k = x_0$ for all k and would not converge to the solution.

Rather than giving a rigorous proof of Theorem 6.1, we now present an informal derivation that contains the essence of the argument and motivates the results that follow.

Since the Hessian matrix $\nabla^2 f(x^*)$ is assumed to be positive definite, there is a positive constant L such that $\|\nabla^2 f(x_k)^{-1}\| \leq L$ for all x_k sufficiently close to x^*. We therefore have from (6.2) that the inexact Newton step satisfies

$$\|p_k\| \leq L(\|\nabla f(x_k)\| + \|r_k\|) \leq 2L\|\nabla f(x_k)\|,$$

where the second inequality follows from (6.3) and $\eta_k < 1$. By using this expression together with Taylor's theorem we obtain

$$\begin{aligned}
\nabla f(x_{k+1}) &= \nabla f(x_k) + \nabla^2 f(x_k) p_k + O\left(\|p_k\|^2\right) \\
&= \nabla f(x_k) - (\nabla f(x_k) - r_k) + O\left(L^2 \|\nabla f(x_k)\|^2\right) \\
&= r_k + O\left(\|\nabla f(x_k)\|^2\right).
\end{aligned} \quad (6.4)$$

By taking norms and recalling (6.3), we have that

$$\|\nabla f(x_{k+1})\| \leq \eta_k \|\nabla f(x_k)\| + O\left(\|\nabla f(x_k)\|^2\right). \quad (6.5)$$

Therefore, if x_k is close to x^*, we can expect $\|\nabla f(x)\|$ to decrease by a factor of approximately $\eta_k < 1$ at every iteration. More precisely, we have that

$$\limsup_{k\to\infty} \frac{\|\nabla f(x_{k+1})\|}{\|\nabla f(x_k)\|} \leq \eta < 1.$$

Relation (6.4) also suggests that if $r_k = o(\|\nabla f(x_k)\|)$, then the rate of convergence in the gradient will be superlinear, for in this case the limit satisfies

$$\limsup_{k\to\infty} \frac{\|\nabla f(x_{k+1})\|}{\|\nabla f(x_k)\|} = 0.$$

Similarly, if $r_k = O(\|\nabla f(x_k)\|^2)$, we have

$$\limsup_{k\to\infty} \frac{\|\nabla f(x_{k+1})\|}{\|\nabla f(x_k)\|^2} = c,$$

for some constant c, suggesting that the quadratic rate of convergence of the exact Newton method has been recaptured.

It is not difficult to show that the iterates $\{x_k\}$ converge to x^* at the same rate as the sequence of gradients $\{\nabla f(x_k)\}$ converges to zero.

Theorem 6.2.
Suppose that the conditions of Theorem 6.1 hold and assume that the iterates $\{x_k\}$ generated by the inexact Newton method converge to x^*. Then the rate of convergence is superlinear if $\eta_k \to 0$ and quadratic if $\eta_k = O(\|\nabla f(x_k)\|)$.

The proof makes use of Taylor's theorem and the relation (6.4). We leave the details as an exercise.

To obtain superlinear convergence we can set, for example, $\eta_k = \min(0.5, \sqrt{\|\nabla f(x_k)\|})$, while the choice $\eta_k = \min(0.5, \|\nabla f(x_k)\|)$ would yield quadratic convergence.

All the results presented in this section, which are proved by Dembo, Eisenstat, and Steihaug [66], are local in nature: They assume that the sequence $\{x_k\}$ eventually enters the near vicinity of the solution x^*. They also assume that the unit step length $\alpha_k = 1$ is taken, and hence that globalization strategies (line search, trust-region) do not get in the way of rapid convergence. In the next sections we show that inexact Newton strategies can, in fact, be incorporated in practical line search and trust-region implementations of Newton's method, yielding algorithms with good local and global convergence properties.

6.2 LINE SEARCH NEWTON METHODS

We now describe line search implementations of Newton's method that are practical for small and large problems. Each iteration has the form $x_{k+1} = x_k + \alpha_k p_k$, where α_k is the step length and p_k is either the pure Newton direction p_k^N or an approximation to it. We have dealt with the issue of selecting the step length α_k in Chapter 3. As stated there, α_k can be chosen to satisfy the Wolfe conditions (3.6) or the Goldstein conditions (3.11), or it can be obtained by an Armijo backtracking line search described in Section 3.1. We stress again that the line search should always try the unit step length $\alpha_k = 1$ first, so that this step length is used when acceptable. We now consider two different techniques for computing the search direction p_k.

LINE SEARCH NEWTON–CG METHOD

In the line search Newton-CG method, also known as the *truncated Newton method*, we compute the search direction by applying the CG method to the Newton equations (6.1), and attempt to satisfy a termination test of the form (6.3). Note, however, that the CG method is designed to solve positive definite systems, and that the Hessian could have negative eigenvalues away from the solution. Therefore, we terminate the CG iteration as soon as a direction of negative curvature is generated. This adaptation of the CG method ensures that the direction p_k is a descent direction, and that the fast convergence rate of the pure Newton method is preserved.

We discuss now the details of this inner CG iteration, which is a slight modification of Algorithm CG described in Chapter 5. The linear system to be solved is (6.1), and thus in the notation of Chapter 5, the coefficient matrix is $A = \nabla^2 f_k$ and the right-hand-side vector is $b = -\nabla f_k$. We will use superscripts to denote the iterates $\{x^{(i)}\}$ and search directions $\{p^{(i)}\}$ generated by the CG iteration, so as to distinguish them from the Newton iterates x_k and search directions p_k. We impose the following three requirements on the CG iteration for solving (6.1).

(a) The starting point for the CG iteration is $x^{(0)} = 0$.

(b) Negative curvature test. If a search direction $p^{(i)}$ generated by the CG iteration satisfies

$$(p^{(i)})^T A p^{(i)} \leq 0, \tag{6.6}$$

then we check whether this is the first CG iteration. If so ($i = 0$), we complete the first CG iteration, compute the new iterate $x^{(1)}$, and stop. If (6.6) holds and $i > 0$, then we stop the CG iteration immediately and return the most recently generated solution point, $x^{(i)}$.

(c) The Newton step p_k is defined as the final CG iterate $x^{(f)}$.

We could also choose $x^{(0)}$ to be the optimal solution of the previous linear system, which is the same as the most recently generated Newton–CG step. Both choices seem to perform equally well in this line search context.

If the condition (6.6) is not satisfied, then the ith CG iteration is exactly as described in Algorithm CG of Chapter 5. A vector $p^{(i)}$ satisfying (6.6) with a strict inequality is said to be a *direction of negative curvature* for A. Note that if negative curvature is encountered during the first CG iteration, then the Newton–CG direction p_k is the steepest descent direction $-\nabla f_k$. This is the reason for choosing the initial estimate in the CG iteration as $x^{(0)} = 0$. On the other hand, if the CG method performs more than one iteration, the Newton–CG direction will also be a descent direction; see Exercise 2. Preconditioning can be used in the CG iteration.

The Newton–CG method does not require explicit knowledge of the Hessian $\nabla^2 f(x_k)$. Rather, it requires only that we can supply matrix–vector products of the form $\nabla^2 f(x_k) p$ for any given vector p. This fact is useful for cases in which the user cannot easily supply code to calculate second derivatives, or where the Hessian requires too much storage. In these cases, the techniques of Chapter 7, which include automatic differentiation and finite differences, can be used to calculate the Hessian–vector products.

To illustrate the finite-differencing technique briefly, we note that we can use the approximation

$$\nabla^2 f(x_k) p \approx \frac{\nabla f(x_k + hp) - \nabla f(x_k)}{h}, \qquad (6.7)$$

for some small differencing interval h. It is easy to prove that the accuracy of this approximation is $O(h)$; appropriate choices of h are discussed in Chapter 7. The price we pay for bypassing the computation of the Hessian is one new gradient evaluation per CG iteration. An alternative to finite differencing is automatic differentiation, which can in principle be used to compute $\nabla^2 f(x_k) p$ exactly. (Again, details are given in Chapter 7.) Methods of this type are known as *Hessian-free* Newton methods.

We can now give a general description of the Newton–CG method. For concreteness, we choose the forcing sequence as $\eta_k = \min\left(0.5, \sqrt{\|\nabla^2 f_k\|}\right)$, so as to obtain a superlinear convergence rate, but other choices are possible.

Algorithm 6.1 (Line Search Newton–CG).
  Given initial point x_0;
    for $k = 0, 1, 2, \ldots$
        Compute a search direction p_k by applying the CG method to
            $\nabla^2 f(x_k) p = -\nabla f_k$, starting from $x^{(0)} = 0$. Terminate when
            $\|r_k\| \le \min(0.5, \sqrt{\|\nabla f_k\|}) \|\nabla f(x_k)\|$, or if negative curvature is
            encountered, as described in (b);
        Set $x_{k+1} = x_k + \alpha_k p_k$, where α_k satisfies the Wolfe, Goldstein, or
            Armijo backtracking conditions;
  end

This method is well suited for large problems, but it has a minor weakness, especially in the case where no preconditioning is used in the CG iteration. When the Hessian $\nabla^2 f_k$ is nearly singular, the Newton–CG direction can be excessively long, requiring many function evaluations in the line search. In addition, the reduction in the function may be very small in this case. To alleviate this difficulty we can try to normalize the Newton step, but good rules for doing so are difficult to determine. (They run the risk of undermining the rapid convergence of Newton's method in the case where the pure Newton step is well scaled.) It is preferable to introduce a threshold value in (6.6), but once again, good choices of this value are difficult to determine. The trust-region implementation of the Newton–CG method described in Section 6.4 deals more effectively with this problematic situation, and is therefore slightly preferable, in our opinion.

MODIFIED NEWTON'S METHOD

In many applications it is desirable to use a direct linear algebra technique, such as Gaussian elimination, to solve the Newton equations (6.1). If the Hessian is not positive definite, or is close to being singular, then we can modify this matrix before or during the solution process to ensure that the computed direction p_k satisfies a linear system identical to (6.1) except that the coefficient matrix is replaced with a positive definite approximation. The modified Hessian is obtained by adding either a positive diagonal matrix or a full matrix to the true Hessian $\nabla^2 f(x_k)$. Following is a general description of this method.

Algorithm 6.2 (Line Search Newton with Modification).
   given initial point x_0;
   for $k = 0, 1, 2, \ldots$
        Factorize the matrix $B_k = \nabla^2 f(x_k) + E_k$, where $E_k = 0$ if $\nabla^2 f(x_k)$
           is sufficiently positive definite; otherwise, E_k is chosen to
           ensure that B_k is sufficiently positive definite;
        Solve $B_k p_k = -\nabla f(x_k)$;
        Set $x_{k+1} = x_k + \alpha_k p_k$, where α_k satisfies the Wolfe, Goldstein, or
           Armijo backtracking conditions;
   end

The choice of Hessian modification E_k is crucial to the effectiveness of the method. Some approaches do not compute E_k explicitly, but rather introduce extra steps and tests into standard factorization procedures, modifying these procedures "on the fly" so that the computed factors are in fact the factors of a positive definite matrix. Strategies based on modifying a Cholesky factorization and on modifying a symmetric indefinite factorization of the Hessian are described in the next section.

We can establish fairly satisfactory global convergence results for Algorithm 6.2, provided that the strategy for choosing E_k (and hence B_k) satisfies the *bounded modified factorization* property. This property is that the matrices in the sequence $\{B_k\}$ have bounded condition number whenever the sequence of Hessians $\{\nabla^2 f(x_k)\}$ is bounded; that is,

$$\text{cond}(B_k) = \|B_k\| \, \|B_k^{-1}\| \leq C, \quad \text{for some } C > 0 \text{ and all } k = 0, 1, 2, \ldots. \tag{6.8}$$

If this property holds, global convergence of the modified line search Newton method follows directly from the results of Chapter 3, as we show in the following result.

Theorem 6.3.

Let f be twice continuously differentiable on an open set $\mathcal{D}$, and assume that the starting point x_0 of Algorithm 6.2 is such that the level set $L = \{x \in \mathcal{D} : f(x) \leq f(x_0)\}$ is compact. Then if the bounded modified factorization property holds, we have that

$$\lim_{k \to \infty} \nabla f(x_k) = 0.$$

PROOF. The line search ensures that all iterates x_k remain in the level set $\mathcal{D}$. Since $\nabla^2 f(x)$ is assumed to be continuous on $\mathcal{D}$, and since $\mathcal{D}$ is compact, we have that the sequence of Hessians $\{\nabla^2 f(x_k)\}$ is bounded, and therefore (6.8) holds. Since Algorithm 6.2 uses one of the line searches for which Zoutendijk's theorem (Theorem 3.2) holds, the result follows from (3.16). □

We now consider the convergence rate of Algorithm 6.2. Suppose that the sequence of iterates x_k converges to a point x^* where $\nabla^2 f(x^*)$ is sufficiently positive definite in the sense that the modification strategies described in the next section return the modification $E_k = 0$ for all sufficiently large k. By Theorem 3.5, we have that $\alpha_k = 1$ for all sufficiently large k, so that Algorithm 6.2 reduces to a pure Newton method, and the rate of convergence is quadratic.

For problems in which ∇f^* is close to singular, there is no guarantee that the modification E_k will eventually vanish, and the convergence rate may be only linear.

6.3 HESSIAN MODIFICATIONS

We now discuss procedures for modifying the Hessian matrices $\nabla^2 f(x_k)$ that implicitly or explicitly choose the modification E_k so that the matrix $B_k = \nabla^2 f(x_k) + E_k$ in Algorithm 6.2 is sufficiently positive definite. Besides requiring the modified matrix B_k to be well conditioned (so that Theorem 6.3 holds), we would like the modification to be as small as possible, so that the second-order information in the Hessian is preserved as far as possible. Naturally, we would also like the modified factorization to be computable at moderate cost.

To set the stage for the matrix factorization techniques that will be used in Algorithm 6.2 we will begin by describing an "ideal" strategy based on the eigenvalue decomposition of $\nabla^2 f(x_k)$.

EIGENVALUE MODIFICATION

Consider a problem in which, at the current iterate x_k, $\nabla f(x_k) = (1, -3, 2)$ and $\nabla^2 f(x_k) = \text{diag}(10, 3, -1)$, which is clearly indefinite. By the spectral decomposition theorem (see the Appendix) we can define $Q = I$ and $\Lambda = \text{diag}(\lambda_1, \lambda_2, \lambda_3)$, and write

$$\nabla^2 f(x_k) = Q \Lambda Q^T = \sum_{i=1}^{n} \lambda_i q_i q_i^T. \tag{6.9}$$

The pure Newton step—the solution of (6.1)—is $p_k^N = (-0.1, 1, 2)$, which is not a descent direction, since $\nabla f(x_k)^T p_k^N > 0$. One might suggest a modified strategy in which we replace $\nabla^2 f(x_k)$ by a positive definite approximation B_k, in which all negative eigenvalues in $\nabla^2 f(x_k)$ are replaced by a small positive number δ that is somewhat larger than machine accuracy u; say $\delta = \sqrt{u}$. Assuming that machine accuracy is 10^{-16}, the resulting matrix in our example is

$$B_k = \sum_{i=1}^{2} \lambda_i q_i q_i^T + \delta q_3 q_3^T = \text{diag}\left(10, 3, 10^{-8}\right), \tag{6.10}$$

which is numerically positive definite and whose curvature along the eigenvectors q_1 and q_2 has been preserved. Note, however, that the search direction based on this modified Hessian is

$$p_k = -B_k^{-1} \nabla f_k = -\sum_{i=1}^{2} \frac{1}{\lambda_i} q_i \left(q_i^T \nabla f_k\right) - \frac{1}{\delta} q_3 \left(q_3^T \nabla f(x_k)\right) \approx -\left(2 \times 10^8\right) q_3.$$

$$\tag{6.11}$$

For small δ, this step is nearly parallel to q_3 (with relatively small contributions from q_1 and q_2) and very long. Although f decreases along the direction p_k, its extreme length violates the spirit of Newton's method, which relies on a quadratic approximation of the objective function that is valid in a neighborhood of the current iterate x_k. It is therefore questionable that this search direction is effective.

A different type of search direction would be obtained by flipping the signs of the negative eigenvalues in (6.9), which amounts to setting $\delta = 1$ in our example. Again, there is no consensus as to whether this constitutes a desirable modification to Newton's method. Various other strategies can be considered: We could set the last term in (6.11) to zero, so that the search direction has no components along the negative curvature directions, or we could adapt the choice of δ to ensure that the length of the step is not excessive. (This last strategy has the flavor of trust-region methods.) We see that there is a great deal of freedom in devising modification strategies, and there is currently no agreement on which is the ideal strategy.

Setting the issue of the choice of δ aside for the moment, let us look more closely at the process of modifying a matrix so that it becomes positive definite. The modification (6.10) to the example matrix (6.9) can be shown to be optimal in the following sense. If A is a symmetric matrix with spectral decomposition $A = Q \Lambda Q^T$, then the correction matrix ΔA of *minimum Frobenius norm* that ensures that $\lambda_{\min}(A + \Delta A) \geq \delta$ is given by

$$\Delta A = Q \operatorname{diag}(\tau_i) Q^T, \quad \text{with} \quad \tau_i = \begin{cases} 0, & \lambda_i \geq \delta, \\ \delta - \lambda_i, & \lambda_i < \delta. \end{cases} \quad (6.12)$$

Here $\lambda_{\min}(A)$ denotes the smallest eigenvalue of A, and the Frobenius norm of a matrix is defined as $\|A\|_F^2 = \sum_{i,j=1}^n a_{ij}^2$ (see (A.40)). Note that ΔA is not diagonal in general, and that the modified matrix is given by

$$A + \Delta A = Q(\Lambda + \operatorname{diag}(\tau_i))Q^T.$$

By using a different norm we can obtain a *diagonal* modification. Suppose again that A is a symmetric matrix with spectral decomposition $A = Q \Lambda Q^T$. A correction matrix ΔA with minimum Euclidean norm that satisfies $\lambda_{\min}(A + \Delta A) \geq \delta$ is given by

$$\Delta A = \tau I, \quad \text{with} \quad \tau = \max(0, \delta - \lambda_{\min}(A)). \quad (6.13)$$

The modified matrix now has the form

$$A + \tau I, \quad (6.14)$$

which happens to have the same form as the matrix occurring in (unscaled) trust–region methods (see Chapter 4). All the eigenvalues of (6.14) have thus been shifted, and all are greater than δ.

These results suggest that both diagonal and nondiagonal modifications can be considered. Even though we have not answered the question of what constitutes a good modification, various practical diagonal and nondiagonal modifications have been proposed and implemented in software. They do not make use of the spectral decomposition of the Hessian, since this is generally too expensive to compute. Instead, they use Gaussian elimination, choosing the modifications indirectly and hoping that somehow they will produce good steps. Numerical experience indicates that the strategies described next often (but not always) produce good search directions.

ADDING A MULTIPLE OF THE IDENTITY

Perhaps the simplest idea is to find a scalar $\tau > 0$ such that $\nabla^2 f(x_k) + \tau I$ is sufficiently positive definite. From the previous discussion we know that τ must satisfy (6.13), but we normally don't have a good estimate of the smallest eigenvalue of the Hessian. A simple idea is to recall that the largest eigenvalue (in absolute value) of a matrix A is bounded by the Frobenius norm $\|A\|_F$. This suggests the following strategy; here a_{ii} denotes a diagonal element of A.

Algorithm 6.3 (Cholesky with Added Multiple of the Identity).
set $\beta \leftarrow \|A\|_F$;
if $\quad \min_i a_{ii} > 0$
$\quad\quad \tau_0 \leftarrow 0$
else
$\quad\quad \tau_0 \leftarrow \beta/2$;
end (if)
for $\quad k = 0, 1, 2, \ldots$
$\quad\quad$ Attempt to apply the incomplete Cholesky algorithm to obtain

$$LL^T = A + \tau_k I;$$

$\quad\quad$ if $\quad$ factorization is completed successfully
$\quad\quad\quad\quad$ **stop** and return L
$\quad\quad$ else
$\quad\quad\quad\quad \tau_{k+1} \leftarrow \max(2\tau_k, \beta/2)$;
$\quad\quad$ end (if)
end (for)

This strategy is quite simple and may be preferable to the modified factorization techniques described next, but it suffers from two drawbacks. The value of τ generated by this procedure may be unnecessarily large, which would bias the modified Newton direction too much toward the steepest descent direction. In addition, every value of τ_k requires a new factorization of $A + \tau_k I$, which can be quite expensive if several trial values are generated. (The symbolic factorization of A is performed only once, but the numerical factorization must be performed from scratch every time.)

MODIFIED CHOLESKY FACTORIZATION

A popular approach for modifying a Hessian matrix that is not positive definite is to perform a Cholesky factorization of $\nabla^2 f(x_k)$, but to increase the diagonal elements encountered during the factorization (where necessary) to ensure that they are sufficiently positive. This *modified Cholesky* approach is designed to accomplish two goals: It guarantees that the modified Cholesky factors exist and are bounded relative to the norm of the actual Hessian, and it does not modify the Hessian if it is sufficiently positive definite.

We begin our description of this approach by briefly reviewing the Cholesky factorization. Every symmetric and positive definite matrix A can be written as

$$A = LDL^T, \tag{6.15}$$

where L is a lower triangular matrix with unit diagonal elements and D is a diagonal matrix with positive elements on the diagonal. By equating the elements in (6.15), column by column, it is easy to derive formulas for computing L and D.

❏ EXAMPLE 6.1

Consider the case $n = 3$. The equation $A = LDL^T$ is given by

$$\begin{bmatrix} a_{11} & a_{21} & a_{31} \\ a_{21} & a_{22} & a_{32} \\ a_{31} & a_{32} & a_{33} \end{bmatrix} = \begin{bmatrix} 1 & 0 & 0 \\ l_{21} & 1 & 0 \\ l_{31} & l_{32} & 1 \end{bmatrix} \begin{bmatrix} d_1 & 0 & 0 \\ 0 & d_2 & 0 \\ 0 & 0 & d_3 \end{bmatrix} \begin{bmatrix} 1 & l_{21} & l_{31} \\ 0 & 1 & l_{32} \\ 0 & 0 & 1 \end{bmatrix}.$$

(The notation indicates that A is symmetric.) By equating the elements of the first column, we have

$$\begin{aligned} a_{11} &= d_1, \\ a_{21} &= d_1 l_{21} & \Rightarrow & \quad l_{21} = a_{21}/d_1, \\ a_{31} &= d_1 l_{31} & \Rightarrow & \quad l_{31} = a_{31}/d_1. \end{aligned}$$

Proceeding with the next two columns, we obtain

$$\begin{aligned} a_{22} &= d_1 l_{21}^2 + d_2 & \Rightarrow & \quad d_2 = a_{22} - d_1 l_{21}^2, \\ a_{32} &= d_1 l_{31} l_{21} + d_2 l_{32} & \Rightarrow & \quad l_{32} = (a_{32} - d_1 l_{31} l_{21})/d_2, \\ a_{33} &= d_1 l_{31}^2 + d_2 l_{32}^2 + d_3 & \Rightarrow & \quad d_3 = a_{33} - d_1 l_{31}^2 - d_2 l_{32}^2. \end{aligned}$$

❏

This procedure is generalized in the following algorithm.

Algorithm 6.4 (Cholesky Factorization, LDL^T Form).
    for $j = 1, 2, \ldots, n$
        $c_{jj} \leftarrow a_{jj} - \sum_{s=1}^{j-1} d_s l_{js}^2$;
        $d_j \leftarrow c_{jj}$;
        for $i = j+1, \ldots, n$
            $c_{ij} \leftarrow a_{ij} - \sum_{s=1}^{j-1} d_s l_{is} l_{js}$;

$$l_{ij} \leftarrow c_{ij}/d_j;$$
    end

end

One can show (see, for example, Golub and Van Loan [115, Section 4.2.3]) that the diagonal elements d_{jj} are all positive whenever A is positive definite. The scalars c_{ij} have been introduced only to facilitate the description of the modified factorization discussed below. We should note that Algorithm 6.4 differs a little from the standard form of the Cholesky factorization, which produces a lower triangular matrix M such that

$$A = MM^T. \tag{6.16}$$

In fact, we can make the identification $M = LD^{1/2}$ to relate M to the factors L and D computed in Algorithm 6.4. The technique for computing M appears as Algorithm A.2 in the Appendix.

If A is indefinite, the factorization $A = LDL^T$ may not exist. Even if it does exist, Algorithm 6.4 is numerically unstable when applied to such matrices, in the sense that the elements of L and D can become arbitrarily large. It follows that a strategy of computing the LDL^T factorization and then modifying the diagonal after the fact to force its elements to be positive may break down, or may result in a matrix that is drastically different from A.

Instead, we will modify the matrix A during the course of the factorization in such a way that all elements in D are sufficiently positive, and so that the elements of D and L are not too large. To control the quality of the modification, we choose two positive parameters δ and β and require that during the computation of the jth columns of L and D in Algorithm 6.4 (that is, for each j in the outer loop of the algorithm) the following bounds be satisfied:

$$d_j \geq \delta, \quad |m_{ij}| \leq \beta, \quad i = j+1, \ldots, n. \tag{6.17}$$

To satisfy these bounds we only need to change one step in Algorithm 6.4: The formula for computing the diagonal element d_j in Algorithm 6.4 is replaced by

$$d_j = \max\left(|c_{jj}|, \left(\frac{\theta_j}{\beta}\right)^2, \delta\right), \quad \text{with } \theta_j = \max_{j < i \leq n} |c_{ij}|. \tag{6.18}$$

To verify that (6.17) holds, we note from Algorithm 6.4 that $c_{ij} = l_{ij}d_j$, and therefore

$$|m_{ij}| = |l_{ij}\sqrt{d_j}| = \frac{|c_{ij}|}{\sqrt{d_j}} \leq \frac{|c_{ij}|\beta}{\theta_j} \leq \beta, \quad \text{for all } i > j.$$

We note that θ_j can be computed prior to d_j because the elements c_{ij} in the second **for** loop of Algorithm 6.4 do not involve d_j. In fact, this is the reason for introducing the quantities c_{ij} into the algorithm. This observation also suggests that the computations should be reordered so that the c_{ij} are computed before the diagonal element d_j. We use this procedure in Algorithm 6.5, the modified Cholesky factorization algorithm described below. To try to reduce the size of the modification, symmetric interchanges of rows and columns are introduced, so that at the jth step of the factorization, the jth row and column are those that yield the largest diagonal element. We also note that the computation of the elements c_{jj} can be carried out recursively after every column of the factorization is calculated.

Algorithm 6.5 (Modified Cholesky [104]).
given $\delta > 0, \beta > 0$:
for $k = 1, 2, \ldots, n$
$\quad c_{kk} = a_{kk};$ (initialize the diagonal elements)
end
Find index q such that $|c_{qq}| \geq |c_{ii}|, i = j, \ldots, n;$
Interchange row and column j and q;
for $j = 1, 2, \ldots, n$ (compute the jth column of L)
$\quad$ for $s = 1, 2, \ldots, j - 1$
$\quad\quad l_{js} \leftarrow c_{js}/d_s;$
$\quad$ end
$\quad$ for $i = j + 1, \ldots, n$
$\quad\quad c_{ij} \leftarrow a_{ij} - \sum_{s=1}^{j-1} l_{js} c_{is};$
$\quad$ end
$\quad \theta_j \leftarrow 0;$
$\quad$ if $j \leq n$
$\quad\quad \theta_j \leftarrow \max_{j < i \leq n} |c_{ij}|;$
$\quad$ end
$\quad d_j \leftarrow \max\{|c_{jj}|, (\theta_j/\beta)^2, \delta\};$
$\quad$ if $j < n$
$\quad\quad$ for $i = j + 1, \ldots, n$
$\quad\quad\quad c_{ii} \leftarrow c_{ii} - c_{ij}^2/d_j;$
$\quad\quad$ end
$\quad$ end
end.

The algorithm requires approximately $n^3/6$ arithmetic operations, which is roughly the same as the standard Cholesky factorization. However, the row and column interchanges require movement of data in the computer, and the cost of this operation may be significant on large problems. No additional storage is needed beyond the amount required to store A; the triangular factors L and D, as well as the intermediate scalars c_{ij}, can overwrite the elements of A.

Suppose that we use P to denote the permutation matrix associated with the row and column interchanges that occur during Algorithm 6.5. It is not difficult to see that the algorithm produces the Cholesky factorization of the permuted, modified matrix $PAP^T + E$, that is,

$$PAP^T + E = LDL^T = MM^T, \qquad (6.19)$$

where E is a nonnegative diagonal matrix that is zero if A is sufficiently positive definite. From an examination of the formulae for c_{jj} and d_j in Algorithm 6.5, it is clear that the diagonal entries of E are $e_j = d_j - c_{jj}$. It is also clear that incrementing c_{jj} by e_j in the factorization is equivalent to incrementing a_{jj} by e_j in the original data.

It remains only to specify the choice of the parameters δ and β in Algorithm 6.5. The constant δ is normally chosen to be close to machine accuracy $\mathbf{u}$; a typical choice is

$$\delta = \mathbf{u}\max(\gamma(A) + \xi(A), 1),$$

where $\gamma(A)$ and $\xi(A)$ are, respectively, the largest-magnitude diagonal and off-diagonal elements of the matrix A, that is,

$$\gamma = \max_{1 \le i \le n} |a_{ii}|, \qquad \xi = \max_{i \ne j} |a_{ij}|.$$

Gill, Murray, and Wright [104] suggest the following choice of β:

$$\beta = \max\left(\gamma(A), \frac{\xi(A)}{\sqrt{n^2 - 1}}, \mathbf{u}\right)^{1/2},$$

whose intent is to minimize the norm of the modification $\|E\|_\infty$.

❑ **EXAMPLE 6.2**

Consider the matrix

$$A = \begin{bmatrix} 4.0 & 2.0 & 1.0 \\ 2.0 & 6.0 & 3.0 \\ 1.0 & 3.0 & -0.004 \end{bmatrix},$$

whose eigenvalues are, to three digits, $-1.25, 2.87, 8.38$. Algorithm 6.5 gives the following Cholesky factor M and diagonal modification E:

$$M = \begin{bmatrix} 0.8165 & 1.8257 & 0 \\ 2.4495 & 0 & 0 \\ 1.2247 & -1.2 \times 10^{-16} & 1.2264 \end{bmatrix}, \qquad E = \mathrm{diag}(0, 0, 3.008).$$

The modified matrix is

$$A' = MM^T = \begin{bmatrix} 4.00 & 2.00 & 1.00 \\ 2.00 & 6.00 & 3.00 \\ 1.00 & 3.00 & 3.004 \end{bmatrix},$$

whose eigenvalues are 1.13, 3.00, 8.87, and whose condition number of 7.8 is quite moderate.

□

One can show (Moré and Sorensen [171]) that the matrices B_k obtained by applying Algorithm 6.5 to the exact Hessians $\nabla^2 f(x_k)$ have bounded condition numbers, that is, the bound (6.8) holds for some value of C.

GERSHGORIN MODIFICATION

We now give a brief outline of an alternative modified Cholesky algorithm that makes use of Gershgorin circle estimates to increase the diagonal elements as necessary. The first step of this strategy is to apply Algorithm 6.5 to the matrix A, terminating with the usual factorization (6.19). If the perturbation matrix E is zero, we are finished, since in this case A is already sufficiently positive definite. Otherwise, we compute two upper bounds b_1 and b_2 on the smallest eigenvalue $\lambda_{\min}(A)$ of A. The first estimate b_1 is obtained from the Gershgorin circle theorem; it guarantees that $A + b_1 I$ is strictly diagonally dominant. The second upper bound b_2 is simply

$$b_2 = \max_{1 \leq i \leq n} e_{ii}.$$

We now define

$$\mu = \min(b_1, b_2),$$

and conclude the algorithm by computing the factorization of $A + \mu I$, taking the Cholesky factor of this matrix as our modified Cholesky factor. The use of b_2 gives a much needed control on the process, since the estimate b_1 obtained from the Gershgorin circle theorem can be quite loose.

It is not known whether this alternative is preferable to Algorithm 6.5 alone. Neither strategy will modify a sufficiently positive definite matrix A, but it is difficult to quantify the meaning of the term "sufficient" in terms of the smallest eigenvalue $\lambda_{\min}(A)$. Both strategies may in fact modify a matrix A whose minimum eigenvalue is greater than the parameter δ of Algorithm 6.5.

MODIFIED SYMMETRIC INDEFINITE FACTORIZATION

Another strategy for modifying an indefinite Hessian is to use a procedure based on a symmetric indefinite factorization. Any symmetric matrix A, whether positive definite or not, can be written as

$$PAP^T = LBL^T, \tag{6.20}$$

where L is unit lower triangular, B is a block diagonal matrix with blocks of dimension 1 or 2, and P is a permutation matrix (see Golub and Van Loan [115, Section 4.4]). We mentioned earlier that attempting to compute the LDL^T factorization of an indefinite matrix (where D is a *diagonal* matrix) is inadvisable because even if the factors L and D are well-defined, they may contain entries that are larger than the original elements of A, thus amplifying rounding errors that arise during the computation. However, by using the block diagonal matrix B, which allows 2×2 blocks as well as 1×1 blocks on the diagonal, we can guarantee that the factorization (6.20) always exists and can be computed by a numerically stable process.

❑ EXAMPLE 6.3

The matrix

$$A = \begin{bmatrix} 0 & 1 & 2 & 3 \\ 1 & 2 & 2 & 2 \\ 2 & 2 & 3 & 3 \\ 3 & 2 & 3 & 4 \end{bmatrix}$$

can be written in the form (6.20) with $P = [e_1, e_4, e_3, e_2]$,

$$L = \begin{bmatrix} 1 & 0 & 0 & 0 \\ 0 & 1 & 0 & 0 \\ \frac{1}{9} & \frac{2}{3} & 1 & 0 \\ \frac{2}{9} & \frac{1}{3} & 0 & 1 \end{bmatrix}, \quad B = \begin{bmatrix} 0 & 3 & 0 & 0 \\ 3 & 4 & 0 & 0 \\ 0 & 0 & \frac{7}{9} & \frac{5}{9} \\ 0 & 0 & \frac{5}{9} & \frac{10}{9} \end{bmatrix}. \tag{6.21}$$

Note that both diagonal blocks in B are 2×2. ❑

The symmetric indefinite factorization allows us to determine the *inertia* of a matrix, that is, the number of positive, zero, and negative eigenvalues. One can show that the inertia of B equals the inertia of A. Moreover, the 2×2 blocks in B are always constructed to have one positive and one negative eigenvalue. Thus the number of positive eigenvalues in A equals the number of positive 1×1 blocks plus the number of 2×2 blocks.

The first step of the symmetric indefinite factorization proceeds as follows. We identify a submatrix E of A that is suitable to be used as a pivot block. The precise criteria that can be used to choose E are described below, but we note here that E is either a single diagonal element of A (giving rise to a 1×1 pivot block), or else it consists of two diagonal elements of A (say, a_{ii} and a_{jj}) together with the corresponding off-diagonal element (that is, a_{ji} and a_{ij}). In either case, E is chosen to be nonsingular. We then find a permutation matrix Π that makes E a leading principal submatrix of A, that is,

$$\Pi A \Pi^T = \begin{bmatrix} E & C^T \\ C & H \end{bmatrix}, \tag{6.22}$$

and then perform a block factorization on this rearranged matrix, using E as the pivot block, to obtain

$$\Pi A \Pi^T = \begin{bmatrix} I & 0 \\ CE^{-1} & I \end{bmatrix} \begin{bmatrix} E & 0 \\ 0 & H - CE^{-1}C^T \end{bmatrix} \begin{bmatrix} I & E^{-1}C^T \\ 0 & I \end{bmatrix}.$$

The next step of the factorization consists in applying exactly the same process to $H - CE^{-1}C^T$, known as the *remaining matrix* or the *Schur complement*, which has dimension either $(n-1) \times (n-1)$ or $(n-2) \times (n-2)$. The same procedure is applied recursively, until we terminate with the factorization (6.20).

The symmetric indefinite factorization requires approximately $n^3/3$ floating-point operations—the same as the cost of the Cholesky factorization of a positive definite matrix—but to this we must add the cost of identifying the pivot blocks E and performing the permutations Π, which can be considerable. There are various strategies for determining the pivot blocks, which have an important effect on both the cost of the factorization and its numerical properties. Ideally, our strategy for choosing E at each step of the factorization procedure should be inexpensive, should lead to at most modest growth in the elements of the remaining matrix at each step of the factorization, and should not lead to excessive fill-in (that is, L should not be too much more dense than A).

A well-known strategy, due to Bunch and Parlett [31], searches the whole working matrix and identifies the largest-magnitude diagonal and largest-magnitude off-diagonal elements, denoting their respective magnitudes by ξ_{dia} and ξ_{off}. If the diagonal element whose magnitude is ξ_{dia} is selected to be a 1×1 pivot block, the element growth in the remaining matrix is bounded by the ratio $\xi_{\text{dia}}/\xi_{\text{off}}$. If this growth rate is acceptable, we choose this diagonal element to be the pivot block. Otherwise, we select the off-diagonal element whose magnitude is ξ_{off} (a_{ij}, say), and choose E to be the 2×2 submatrix that includes this element, that is,

$$E = \begin{bmatrix} a_{ii} & a_{ij} \\ a_{ij} & a_{jj} \end{bmatrix}.$$

This pivoting strategy of Bunch and Parlett is numerically stable and guarantees to yield a matrix L whose maximum element is bounded by 2.781. Its drawback is that the evaluation of ξ_{dia} and ξ_{off} at each iteration requires many comparisons between floating-point numbers to be performed: $O(n^3)$ in total during the overall factorization. Since each comparison costs roughly the same as an arithmetic operation, this overhead is not insignificant, and the total run time may be roughly twice as long as that of Algorithm 6.5.

The more economical pivoting strategy of Bunch and Kaufman [30] searches at most two columns of the working matrix at each stage and requires just $O(n^2)$ comparisons in total. Its details are somewhat tricky, and we refer the interested reader to the original paper or to Golub and Van Loan [115, Section 4.4] for details. Unfortunately, this algorithm can give rise to arbitrarily large elements in the lower triangular factor L, making it unsuitable for use with a modified Cholesky strategy.

The bounded Bunch–Kaufman strategy is essentially a compromise between the Bunch–Parlett and Bunch–Kaufman strategies. It monitors the sizes of elements in L, accepting the (inexpensive) Bunch–Kaufman choice of pivot block when it yields only modest element growth, but searching further for an acceptable pivot when this growth is excessive. Its total cost is usually similar to that of Bunch–Kaufman, but in the worst case it can approach the cost of Bunch–Parlett.

So far, we have ignored the effect of the choice of pivot block E on the sparsity of the final L factor. This consideration is an important one when the matrix to be factored is large and sparse, since it greatly affects both the CPU time and the amount of storage required by the algorithm. Algorithms that modify the strategies above to take account of sparsity have been proposed by Duff et al. [76], Duff and Reid [74], and Fourer and Mehrotra [93].

As for the Cholesky factorization, the efficient and numerically stable indefinite symmetric factorization algorithms discussed above can be modified to ensure that the modified factors are the factors of a positive definite matrix. The strategy is first to compute the factorization (6.20), as well as the spectral decomposition $B = Q \Lambda Q^T$, which is very inexpensive to compute because B is block diagonal (Exercise 6). Then we construct a modification matrix F such that

$$L(B + F)L^T$$

is sufficiently positive definite. Motivated by the modified spectral decomposition (6.12) we will choose a parameter $\delta > 0$ and define F to be

$$F = Q \operatorname{diag}(\tau_i) Q^T, \quad \tau_i = \begin{cases} 0, & \lambda_i \geq \delta, \\ \delta - \lambda_i, & \lambda_i < \delta, \end{cases} \quad i = 1, 2, \ldots, n, \tag{6.23}$$

where λ_i are the eigenvalues of B. The matrix F is thus the modification of minimum Frobenius norm that ensures that all eigenvalues of the modified matrix $B + F$ are no less than δ. This strategy therefore modifies the factorization (6.20) as follows:

$$P(A + E)P^T = L(B + F)L^T, \qquad \text{where } E = P^T L F L^T P^T;$$

note that E will not be diagonal, in general. Hence, in contrast to the modified Cholesky approach, this modification strategy changes the entire matrix A and not just its diagonal. The aim of strategy (6.23) is that the modified matrix satisfies $\lambda_{\min}(A + E) \approx \delta$ whenever the original matrix A has $\lambda_{\min}(A) < \delta$. It is not clear, however, whether it always comes close to attaining this goal.

6.4 TRUST-REGION NEWTON METHODS

Unlike line search methods, trust-region methods do not require the Hessian of the quadratic model to be positive definite. Therefore, we can use the exact Hessian $B_k = \nabla^2 f(x_k)$ directly in this model and obtain steps p_k by solving

$$\min_{p \in \mathbb{R}^n} m_k(p) \stackrel{\text{def}}{=} f_k + \nabla f_k^T p + \tfrac{1}{2} p^T B_k p \qquad \text{s.t. } \|p\| \le \Delta_k. \qquad (6.24)$$

In Chapter 4 we described a variety of techniques for finding an approximate or accurate solution of this subproblem. Most of these techniques apply when B_k is *any* symmetric matrix, and we do not need to say much about the specific case in which $B_k = \nabla^2 f(x_k)$. We will, however, pay attention to the implementation of these trust-region Newton methods when the number of variables is large. Four techniques for solving (6.24) will be discussed: (i) the dogleg method; (ii) two-dimensional subspace minimization; (iii) accurate solution using iteration; (iv) the conjugate gradient (CG) method. For the CG-based method, we study the choice of norm defining the trust region, which can be viewed as a means of preconditioning the subproblem.

NEWTON–DOGLEG AND SUBSPACE-MINIMIZATION METHODS

If B_k is positive definite, the dogleg method described in Chapter 4 provides an approximate solution of (6.24) that is relatively inexpensive to compute and good enough to allow the method to be robust and rapidly convergent. However, since the Hessian matrix may not always be positive definite, the dogleg method is not directly applicable. To adapt it for the minimization of nonconvex functions, we can use one of the Hessian modifications described in Section 6.3 as B_k, in place of the true Hessian $\nabla^2 f(x_k)$, thus guaranteeing that we are working with a convex quadratic function in (6.24). This strategy for choosing B_k

and for finding an approximation Newton step allows us to implement the dogleg method exactly as in Chapter 4. That is, we choose the approximate solution of (6.24) to be the minimizer of the modified model function m_k in (6.24) along the dogleg path defined by

$$\tilde{p}(\tau) = \begin{cases} \tau p^{\text{U}}, & 0 \leq \tau \leq 1, \\ p^{\text{U}} + (\tau - 1)(p^{\text{B}} - p^{\text{U}}), & 1 \leq \tau \leq 2, \end{cases} \quad (6.25)$$

where p^{U} is defined as in (4.12) and p^{B} is the unconstrained minimizer of (6.24); see Figure 4.3.

Similarly, the two-dimensional subspace minimization algorithm of Chapter 4 can also be applied, when B_k is the exact Hessian or a modification that ensures its positive definiteness, to find an approximate solution of (6.24), as

$$\min_p m_k(p) = f_k + \nabla f_k^T p + \tfrac{1}{2} p^T B_k p \quad \text{s.t.} \quad \|p\| \leq \Delta_k, \ p \in \text{span}\{\nabla f_k, p^{\text{B}}\}. \quad (6.26)$$

The dogleg and two-dimensional subspace minimization approaches have the advantage that all the linear algebra computations can be performed with direct linear solvers, a feature that can be important in some applications. They are also globally convergent, as shown in Chapter 4. We hope, too, that the modified factorization strategy used to obtain B_k (and hence p^{B}) does not modify the exact Hessian $\nabla^2 f(x_k)$ when it is sufficiently positive definite, allowing the rapid local convergence that characterizes Newton's method to be observed.

The use of a modified factorization in the dogleg method is not completely satisfying from an intuitive viewpoint, however. A modified factorization perturbs $\nabla^2 f(x_k)$ in a somewhat random manner, giving more weight to certain directions than others, and the benefits of the trust-region approach may not be realized. In fact, the modification introduced during the factorization of the Hessian is redundant in some sense because the trust-region strategy introduces its own modification: The solution of the trust-region problem results in the factorization of the positive (semidefinite) matrix $\nabla^2 f(x_k) + \lambda I$, where λ depends on the size of the trust-region radius Δ_k.

We conclude that the dogleg method is most appropriate when the objective function is convex (that is, $\nabla^2 f(x_k)$ is always positive semidefinite). The techniques described next may be more suitable for the general case.

ACCURATE SOLUTION OF THE TRUST-REGION PROBLEM

We can also find a solution of (6.24) using Algorithm 4.4 described in Chapter 4. This algorithm requires the repeated factorization of matrices of the form $B_k + \lambda I$, for different values of λ. Practical experience indicates that between 1 and 3 such systems need to be solved, on average, at each iteration, so the cost of this approach may not be prohibitive. The resulting trust-region algorithm is quite robust—it can be expected to converge to a minimizer, not just to a stationary point. In some large-scale situations, however, the requirement of solving more than one linear system per iteration may become onerous.

TRUST-REGION NEWTON–CG METHOD

We can approximately solve the trust-region problem (6.24) by means of the conjugate gradient method (CG) with the termination tests proposed by Steihaug; see the CG–Steihaug algorithm, Algorithm 4.3 in Chapter 4. The step computation of this Newton–CG algorithm is obtained by setting $B_k = \nabla^2 f(x_k)$ at every iteration in Algorithm 4.3. This procedure amounts to applying the CG method to the system

$$B_k p_k = -\nabla f_k \qquad (6.27)$$

and stopping if (i) the size of the approximate solution exceeds the trust-region radius, (ii) the system (6.27) has been solved to a required accuracy, or (iii) if negative curvature is encountered. In the latter case we follow the direction of negative curvature to the boundary of the trust region $\|p\| \leq \Delta_k$. This is therefore the trust-region analogue of Algorithm 6.1.

Careful control of the accuracy in the inner CG iteration is crucial to keeping the cost of the Newton–CG method as low as possible. Near a well-behaved solution x^*, the trust-region constraint becomes inactive, and the Newton–CG method reduces to the inexact Newton method analyzed in Section 6.1. The results of that section, which relate the choice of forcing sequence $\{\eta_k\}$ to the rate of convergence, become relevant during the later stages of the Newton–CG method.

As discussed in the context of the line search Newton–CG method, it is not necessary to have explicit knowledge of the Hessian matrix, but we can compute products of the form $\nabla^2 f(x_k)v$ by automatic differentiation or using the finite difference approximation (6.7). Once again, the result is a *Hessian-free* method.

The trust-region Newton–CG method has a number of attractive computational and theoretical properties. First, it is globally convergent. Its first step along the direction $-\nabla f(x_k)$ identifies the Cauchy point for the subproblem (6.24) (see (4.5)), and any subsequent CG iterates only serve to improve the model value. Second, it requires no matrix factorizations, so we can exploit the sparsity structure of the Hessian $\nabla^2 f(x_k)$ without worrying about fill-in during a direct factorization. Moreover, the CG iteration—the most computationally intensive part of the algorithm—may be executed in parallel, since the key operation is a matrix–vector product. When the Hessian matrix is positive definite, the Newton–CG method approximates the pure Newton step more and more closely as the solution x^* is approached, so rapid convergence is also possible, as discussed above.

Two advantages, compared with the *line search* Newton–CG method, are that the lengths of the steps are controlled by the trust region and that directions of negative curvature are explored. Our computational experience shows that the latter is beneficial in practice, as it sometimes allows the iterates to move away from nonminimizing stationary points.

A limitation of the trust-region Newton–CG method is that it accepts *any* direction of negative curvature, even when this direction gives an insignificant reduction in the model. Consider, for example, the case where the subproblem (6.24) is

$$m(p) = 10^{-3} p_1 + 10^{-4} p_1^2 - p_2^2 \quad \text{subject to } \|p\| \leq 1,$$

where subscripts indicate elements of the vector p. The steepest descent direction at $p = 0$ is $(10^{-3}, 0)^T$, which is a direction of negative curvature for the model. Algorithm 4.3 would follow this direction to the boundary of the trust region, yielding a reduction in model function m of about 10^{-3}. A step along e_2—also a direction of negative curvature—would yield a much greater reduction of 1.

Several remedies have been proposed. We have seen in Chapter 4 that when the Hessian $\nabla^2 f(x_k)$ contains negative eigenvalues, the search direction should have a significant component along the eigenvector corresponding to the most negative eigenvalue of $\nabla^2 f(x_k)$. This feature would allow the algorithm to move away rapidly from stationary points that are not minimizers. A variation of the Newton–CG method that overcomes this drawback uses the *Lanczos method*, rather than CG, to solve the linear system (6.27). This approach does not terminate after encountering the first direction of negative curvature, but continues in search of a direction of sufficiently negative curvature; see [177], [121]. The additional robustness in the trust-region algorithm comes at the cost of a more expensive solution of the subproblem.

PRECONDITIONING THE NEWTON–CG METHOD

The unpreconditioned CG method can be inefficient when the Hessian is ill-conditioned, and may even fail to reach the desired accuracy. Hence, it is important to introduce *preconditioning* techniques into the CG iteration. Such techniques are based on finding a nonsingular matrix D such that the eigenvalues of $D^{-T} B_k D$ have a more favorable distribution, as discussed in Chapter 5.

The use of preconditioning in the Newton–CG method requires care because the preconditioned CG iteration no longer generates iterates of increasing ℓ_2 norm; this property holds only for the unpreconditioned CG iteration (see Theorem 4.2). Thus we cannot simply terminate the preconditioned CG method as soon as the iterates reach the boundary of the trust region $\|p\| \leq \Delta_k$, since later CG iterates could return to the trust region.

However, there exists a weighted ℓ_2 norm in which the magnitudes of the preconditioned CG iterates grow monotonically. This will allow us to develop an extension of Algorithm 4.3 of Chapter 4. (Not surprisingly, the weighting of the norm depends on the preconditioner D!) Consider the subproblem

$$\min_{p \in \mathbb{R}^n} m_k(p) \stackrel{\text{def}}{=} f_k + \nabla f_k^T p + \tfrac{1}{2} p^T B_k p \quad \text{s.t. } \|Dp\| \leq \Delta_k. \tag{6.28}$$

By making the change of variables $\hat{p} = Dp$ and defining

$$\hat{g}_k = D^{-T} \nabla f_k, \qquad \hat{B}_k = D^{-T} B_k D^{-1},$$

we can write (6.28) as

$$\min_{\hat{p} \in \mathbb{R}^n} f_k + \hat{g}_k^T \hat{p} + \tfrac{1}{2} \hat{p}^T \hat{B}_k \hat{p} \quad \text{s.t.} \ \|\hat{p}\| \le \Delta_k,$$

which has exactly the form of (6.24). We can now apply the CG algorithm without any modification to this subproblem. Algorithm 4.3 now monitors the norm $\|\hat{p}\| = \|Dp\|$, which is the quantity that grows monotonically during the CG iteration, and terminates when $\|Dp\|$ exceeds the bound Δ_k applied in (6.28).

Many preconditioners can be used within this framework; we discuss some of them in Chapter 5. Of particular interest is *incomplete Cholesky* factorization, which has proved to be useful in a wide range of optimization problems. The incomplete Cholesky factorization of a positive definite matrix B finds a lower triangular matrix L such that

$$B = LL^T - R,$$

where the amount of fill-in in L is restricted in some way. (For instance, it is constrained to have the same sparsity structure as the lower triangular part of B, or is allowed to have a number of nonzero entries similar to that in B.) The matrix R accounts for the "inexactness" in the approximate factorization. The situation is complicated a little further by the possible indefiniteness of the Hessian $\nabla^2 f(x_k)$; we must be able to handle this indefiniteness as well as maintain the sparsity. The following algorithm combines incomplete Cholesky and a form of modified Cholesky to define a preconditioner for Newton–CG methods.

Algorithm 6.6 (Inexact Modified Cholesky).
(scale B)
Compute $T = \text{diag}(\|Be_i\|)$, where e_i is the ith coordinate vector;
$\bar{B} \leftarrow T^{-1/2} B T^{-1/2}$; $\beta \leftarrow \|\bar{B}\|$;
(compute a shift to ensure positive definiteness)
if $\min_i b_{ii} > 0$
    $\alpha_0 \leftarrow 0$
else
    $\alpha_0 \leftarrow \beta/2$;
end
for $k = 0, 1, 2, \ldots$
Attempt to apply incomplete Cholesky algorithm to obtain

$$LL^T = \bar{B} + \alpha_k I;$$

>   **if** factorization is completed successfully
>   **stop** and return L
>   **else**
> $\alpha_{k+1} \leftarrow \max(2\alpha_k, \beta/2)$;
>   **end**
> **end**

We can then set the preconditioner to be $D = L$, where L is the lower triangular matrix output from Algorithm 6.6. A Newton–CG trust region method using this preconditioner is implemented in the forthcoming MINPACK-2 package under the name NMTR. The LANCELOT package also contains a Newton–CG method that makes use of slightly different preconditioning techniques.

LOCAL CONVERGENCE OF TRUST-REGION NEWTON METHODS

Since global convergence of trust-region methods that incorporate (possibly inexact) Newton steps is established above, we turn our attention now to local convergence. The key to attaining the fast rate of convergence associated with Newton's method is to show that the trust-region bound eventually does not interfere with the convergence. That is, when we reach the vicinity of a solution, the (approximate) solution of the subproblem is well inside the region defined by the trust-region constraint $\|p\| \leq \Delta_k$ and becomes closer and closer to the pure Newton step. Sequences of steps that satisfy the latter property are said to be *asymptotically exact*.

We first prove a general result that applies to any algorithm of the form of Algorithm 4.1 (see Chapter 4) that generates asymptotically exact steps whenever the true Newton step is well inside the trust region. It shows that the trust-region constraint eventually becomes inactive in algorithms with this property. The result assumes that the exact Hessian $B_k = \nabla^2 f(x_k)$ is used in (6.24) when x_k is close to a solution x^* that satisfies second-order conditions. Moreover, it assumes that the algorithm uses an approximate solution p_k of (6.24) that achieves at least the same decrease in the model function m_k as the Cauchy step. (All the methods discussed above satisfy this condition.)

Theorem 6.4.
Let f be twice Lipschitz continuously differentiable. Suppose the sequence $\{x_k\}$ converges to a point x^* that satisfies the second-order sufficient conditions (Theorem 2.4) and that for all k sufficiently large, the trust-region algorithm based on (6.24) with $B_k = \nabla^2 f(x_k)$ chooses steps p_k that achieve at least the same reduction as the Cauchy point (that is, $m_k(p_k) \leq m_k(p_k^C)$) and are asymptotically exact whenever $\|p_k^N\| \leq \frac{1}{2}\Delta_k$, that is,

$$\|p_k - p_k^N\| = o(\|p_k^N\|). \tag{6.29}$$

Then the trust-region bound Δ_k becomes inactive for all k sufficiently large.

PROOF. We show that $\|p_k^N\| \leq \frac{1}{2}\Delta_k$ and $\|p_k\| \leq \Delta_k$, for all sufficiently large k, so the near-optimal step p_k in (6.29) will eventually always be taken.

First, we seek a lower bound on the predicted reduction $m_k(0) - m_k(p_k)$ for all sufficiently large k. We assume that k is large enough that the $o(\|p_k^N\|)$ term in (6.29) is less than $\|p_k^N\|$. When $\|p_k^N\| \leq \frac{1}{2}\Delta_k$, we then have that $\|p_k\| \leq \|p_k^N\| + o(\|p_k^N\|) \leq 2\|p_k^N\|$, while if $\|p_k^N\| > \frac{1}{2}\Delta_k$, we have $\|p_k\| \leq \Delta_k < 2\|p_k^N\|$. In both cases, then, we have

$$\|p_k\| \leq 2\|p_k^N\| \leq 2\|\nabla^2 f(x_k)^{-1}\|\|\nabla f(x_k)\|,$$

and so $\|\nabla f(x_k)\| \geq \frac{1}{2}\|p_k\|/\|\nabla^2 f_k^{-1}\|$.

Because we assume that the step p_k achieves at least the same decrease as the Cauchy point, we have from the relation (4.34) that there is a constant $c_1 > 0$ such that

$$m_k(0) - m_k(p_k) \geq c_1\|\nabla f(x_k)\| \min\left(\Delta_k, \frac{\|\nabla f(x_k)\|}{\|B_k\|}\right)$$

$$\geq c_1 \frac{\|p_k\|}{2\|\nabla^2 f(x_k)^{-1}\|} \min\left(\|p_k\|, \frac{\|p_k\|}{2\|\nabla^2 f(x_k)\|\|\nabla^2 f(x_k)^{-1}\|}\right)$$

$$= c_1 \frac{\|p_k\|^2}{4\|\nabla^2 f(x_k)^{-1}\|^2\|\nabla^2 f(x_k)\|}.$$

Because $x_k \to x^*$, then by continuity of $\nabla^2 f(x)$ and positive definiteness of $\nabla^2 f(x^*)$, we have for k sufficiently large that

$$\frac{c_1}{4\|\nabla^2 f(x_k)^{-1}\|^2\|\nabla^2 f(x_k)\|} \geq \frac{c_1}{8\|\nabla^2 f(x^*)^{-1}\|^2\|\nabla^2 f(x^*)\|} \stackrel{\text{def}}{=} c_3,$$

and therefore

$$m_k(0) - m_k(p_k) \geq c_3\|p_k\|^2 \tag{6.30}$$

for all sufficiently large k. By Lipschitz continuity of $\nabla^2 f(x)$, we have by the argument leading to (4.41) that

$$|(f(x_k) - f(x_k + p_k)) - (m_k(0) - m_k(p_k))| \leq \frac{L}{2}\|p_k\|^3,$$

where $L > 0$ is the Lipschitz constant for $\nabla^2 f(\cdot)$ in the neighborhood of x^*. Hence, by definition of ρ_k (see (4.4)), we have for sufficiently large k that

$$|\rho_k - 1| \leq \frac{\|p_k\|^3(L/2)}{c_3\|p_k\|^2} = \frac{L}{2c_3}\|p_k\| \leq \frac{L}{2c_3}\Delta_k. \tag{6.31}$$

Now, the trust-region radius can be reduced only if $\rho_k < \frac{1}{4}$ (or some other fixed number less than 1), so it is clear from (6.31) that there is a threshold $\tilde{\Delta}$ such that Δ_k cannot be reduced further once it falls below $\tilde{\Delta}$. Hence, the sequence $\{\Delta_k\}$ is bounded away from zero. Since $x_k \to x^*$, we have $\|p_k^N\| \to 0$ and therefore $\|p_k\| \to 0$ from (6.29). Hence the trust-region bound is inactive for all k sufficiently large, and our proof is complete. $\square$

The conditions of Theorem 6.4 are satisfied trivially if $p_k = p_k^N$. In addition, all three algorithms discussed above also satisfy the assumptions of this theorem. We state this result formally as follows.

Lemma 6.5.
Suppose that $x_k \to x^*$, where x^* satisfies the second-order sufficient conditions of Theorem 2.4. Consider versions of Algorithm TR in which the inexact dogleg method (6.25) or the inexact two-dimensional subspace minimization method (6.26) with $B_k = \nabla^2 f(x_k)$ is used to obtain an approximate step p_k. Then for all sufficiently large k, the unconstrained minimum of the models (6.25) and (6.26) is simply p_k^N, so the conditions of Theorem 6.4 are satisfied.

When the termination criterion (6.3) is used with $\eta_k \to 0$ in the Newton–CG method (along with termination when the trust-region bound is exceeded or when negative curvature is detected), then this algorithm also satisfies the conditions of Theorem 6.4.

PROOF. For the case of the dogleg and two-dimensional subspace minimization methods, the exact step p_k^N is one of the candidates for p_k—it lies inside the trust region, along the dogleg path, and inside the two-dimensional subspace. Since in fact, p_k^N is the minimizer of m_k inside the trust region for k sufficiently large (since $B_k = \nabla^2 f(x_k)$ is positive definite), it is certainly the minimizer in these more restricted domains, so we have $p_k = p_k^N$.

For the Newton–CG method, the method does not terminate by finding a negative curvature direction, because $\nabla^2 f(x_k)$ is positive definite for all k sufficiently large. Also, since the norms of the CG iterates increase with iteration number, they do not exceed p_k^N, and hence stay inside the trust region. Hence, the CG iterations can terminate only because condition (6.3) is satisfied. Hence from (6.1), (6.2), and (6.3), we have that

$$\|p_k - p_k^N\| \leq \|\nabla^2 f(x_k)^{-1}\| \|\nabla^2 f(x_k) p_k - \nabla^2 f(x_k) p_k^N\|$$
$$= \|\nabla^2 f(x_k)^{-1}\| \|r_k\|$$
$$\leq \eta_k \|\nabla^2 f(x_k)^{-1}\| \|\nabla f(x_k)\|$$
$$\leq \eta_k \|\nabla^2 f(x_k)^{-1}\| \|\nabla^2 f(x_k)\| \|p_k^N\|.$$

Hence, the condition (6.29) is satisfied. $\square$

The nearly exact algorithm of Chapter 4—Algorithm 4.4—also satisfies the conditions of Theorem 6.4, since if the true Newton step p_k^N lies inside the trust region, the initial guesses

$\lambda = 0$ and $p_k = p_k^N$ will eventually satisfy any reasonable termination criteria for this method of determining the step.

Rapid convergence of all these algorithms now follows immediately from their eventual similarity to Newton's method. The methods that eventually take exact Newton steps $p_k = p_k^N$ converge quadratically. The asymptotic convergence rate of the Newton–CG algorithm with termination test (6.3) depends on the rate of convergence of the forcing sequence $\{\eta_k\}$ to zero, as described in Theorem 6.2.

NOTES AND REFERENCES

Newton methods in which the step is computed by an iterative algorithm have received much attention; see Sherman [229] and Ortega and Rheinboldt [185]. Our discussion of inexact Newton methods is based on Dembo, Eisenstat, and Steihaug [66].

For a more thorough treatment of the modified Cholesky factorization see Gill, Murray, and Wright [104] or Dennis and Schnabel [138]. The modified Cholesky factorization based on Gershgorin disk estimates is described in Schnabel and Eskow [223]. The modified indefinite factorization is from Cheng and Higham [41].

Another strategy for implementing a line search Newton method when the Hessian contains negative eigenvalues is to compute a direction of negative curvature and use it to define the search direction (see Moré and Sorensen [169] and Goldfarb [113]).

EXERCISES

6.1 Program a pure Newton iteration without line searches, where the search direction is computed by the CG method. Select stopping criteria such that the rate of convergence is linear, superlinear, and quadratic. Try your program on the following convex quartic function:

$$f(x) = \frac{1}{2}x^T x + 0.25\sigma(x^T A x)^2, \qquad (6.32)$$

where σ is a parameter and

$$A = \begin{bmatrix} 5 & 1 & 0 & 0.5 \\ 1 & 4 & 0.5 & 0 \\ 0 & 0.5 & 3 & 0 \\ 0.5 & 0 & 0 & 2 \end{bmatrix}.$$

This is a strictly convex function that allows us to control the deviation from a quadratic by means of the parameter σ. The starting point is

$$x_1 = (\cos 70°, \sin 70°, \cos 70°, \sin 70°)^T.$$

Try $\sigma = 1$ or larger values and observe the rate of convergence of the iteration.

✎ **6.2** Consider the line search Newton–CG method described in Section 6.2. Use the properties of the CG method described in Chapter 5 to prove that the search Newton direction p_k is always a descent direction for f.

✎ **6.3** Compute the eigenvalues of the 2 diagonal blocks of (6.21), and verify that each block has a positive and diagonal eigenvalue. Then compute the eigenvalues of A and verify that its inertia is the same as that of B.

✎ **6.4** Describe the effect that the modified Cholesky factorization of Algorithm 6.5 would have on the Hessian $\nabla^2 f(x_k) = \mathrm{diag}(-2, 12, 4)$.

✎ **6.5** Program a trust-region Newton–CG method in which the step is computed by the CG–Steihaug method (Algorithm 4.3), without preconditioning. Select the constants so that the rate of convergence is superlinear. Try it on (6.32). Then define an objective function that has negative curvature in a neighborhood of the starting point $x_0 = 0$, and observe the effect of the negative curvature steps.

✎ **6.6** Prove Theorem 6.2.

✎ **6.7** Consider a block diagonal matrix B with 1×1 and 2×2 blocks. Show that the eigenvalues and eigenvectors of B can be obtained by computing the spectral decomposition of each diagonal block separately.

Chapter 7

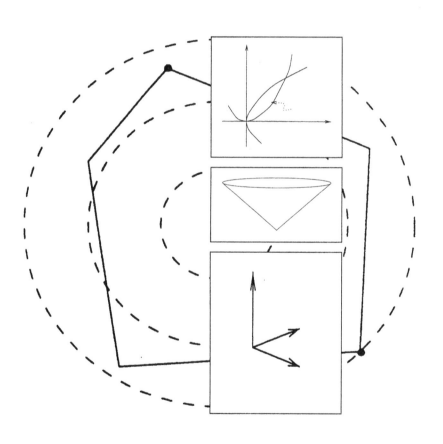

Calculating Derivatives

Most algorithms for nonlinear optimization and nonlinear equations require knowledge of derivatives. Sometimes the derivatives are easy to calculate by hand, and it is reasonable to expect the user to provide code to compute them. In other cases, the functions are too complicated, so we look for ways to calculate or approximate the derivatives automatically. A number of interesting approaches are available, of which the most important are probably the following.

Finite Differencing. This technique has its roots in Taylor's theorem (see Chapter 2). By observing the change in function values in response to small perturbations of the unknowns near a given point x, we can estimate the response to *infintesimal* perturbations, that is, the derivatives. For instance, the partial derivative of a smooth function $f : \mathbf{R}^n \to \mathbf{R}$ with respect to the ith variable x_i can be approximated by the central-difference formula

$$\frac{\partial f}{\partial x_i} \approx \frac{f(x + \epsilon e_i) - f(x - \epsilon e_i)}{2\epsilon},$$

where ϵ is a small positive scalar and e_i is the ith unit vector.

Automatic Differentiation. This technique takes the view that the computer code for evaluating the function can be broken down into a composition of elementary arithmetic operations, to which the chain rule (one of the basic rules of calculus) can be applied. Some software tools for automatic differentiation (such as ADIFOR [13]) produce new code that calculates both function and derivative values. Other tools (such as ADOL-C [126]) keep a record of the elementary computations that take place while the function evaluation code for a given point x is executing on the computer. This information is processed to produce the derivatives at the same point x.

Symbolic Differentiation. In this technique, the algebraic specification for the function f is manipulated by symbolic manipulation tools to produce new algebraic expressions for each component of the gradient. Commonly used symbolic manipulation tools can be found in the packages Mathematica [248], Maple [243], and Macsyma [153].

In this chapter we discuss the first two approaches: finite differencing and automatic differentiation.

The usefulness of derivatives is not restricted to *algorithms* for optimization. Modelers in areas such as design optimization and economics are often interested in performing postoptimal *sensitivity analysis*, in which they determine the sensitivity of the optimum to small perturbations in the parameter or constraint values. Derivatives are also important in other areas of numerical analysis such as nonlinear differential equations.

7.1 FINITE-DIFFERENCE DERIVATIVE APPROXIMATIONS

Finite differencing is an approach to the calculation of approximate derivatives whose motivation (like that of so many other algorithms in optimization) comes from Taylor's theorem. Many software packages perform automatic calculation of finite differences whenever the user is unable or unwilling to supply code to calculate exact derivatives. Although they yield only approximate values for the derivatives, the results are adequate in many situations.

By definition, derivatives are a measure of the sensitivity of the function to infinitesimal changes in the values of the variables. Our approach in this section is to make small *but finite* perturbations in the values of x and examine the resulting *differences* in the function values. By taking ratios of the function difference to variable difference, we obtain approximations to the derivatives.

APPROXIMATING THE GRADIENT

An approximation to the gradient vector $\nabla f(x)$ can be obtained by evaluating the function f at $(n+1)$ points and performing some elementary arithmetic. We describe this technique, along with a more accurate variant that requires additional function evaluations.

7.1. FINITE-DIFFERENCE DERIVATIVE APPROXIMATIONS

A popular formula for approximating the partial derivative $\partial f/\partial x_i$ at a given point x is the *forward-difference*, or *one-sided-difference*, approximation, defined as

$$\frac{\partial f}{\partial x_i}(x) \approx \frac{f(x + \epsilon e_i) - f(x)}{\epsilon}. \tag{7.1}$$

The gradient can be built up by simply applying this formula for $i = 1, 2, \ldots, n$. This process requires evaluation of f at the point x as well as the n perturbed points $x + \epsilon e_i$, $i = 1, 2, \ldots, n$: a total of $(n + 1)$ points.

The basis for the formula (7.1) is Taylor's theorem, which we stated as Theorem 2.1 in Chapter 2. When f is twice continuously differentiable, we have

$$f(x + p) = f(x) + \nabla f(x)^T p + \tfrac{1}{2} p^T \nabla^2 f(x + tp) p, \quad \text{some } t \in (0, 1) \tag{7.2}$$

(see (2.6)). If we choose L to be a bound on the size of $\|\nabla^2 f(\cdot)\|$ in the region of interest, it follows directly from this formula that the last term in this expression is bounded by $(L/2)\|p\|^2$, so that

$$\|f(x + p) - f(x) - \nabla f(x)^T p\| \leq (L/2)\|p\|^2. \tag{7.3}$$

We now choose the vector p to be ϵe_i, so that it represents a small change in the value of a single component of x (the ith component). For this p, we have that $\nabla f(x)^T p = \nabla f(x)^T e_i = \partial f/\partial x_i$, so by rearranging (7.3), we conclude that

$$\frac{\partial f}{\partial x_i}(x) = \frac{f(x + \epsilon e_i) - f(x)}{\epsilon} + \delta_\epsilon, \quad \text{where } |\delta_\epsilon| \leq (L/2)\epsilon. \tag{7.4}$$

We derive the forward-difference formula (7.1) by simply ignoring the error term δ_ϵ in this expression, which becomes smaller and smaller as ϵ approaches zero.

An important issue in implementing the formula (7.1) is the choice of the parameter ϵ. The error expression (7.4) suggests that we should choose ϵ as small as possible. Unfortunately, this expression ignores the roundoff errors that are introduced when the function f is evaluated on a real computer, in floating-point arithmetic. From our discussion in the Appendix (see (A.58) and (A.59)), we know that the quantity **u** known as *unit roundoff* is crucial: It is a bound on the relative error that is introduced whenever an arithmetic operation is performed on two floating-point numbers. (Typically, **u** is about 10^{-16} in double-precision arithmetic.) The effect of these errors on the final computed value of f depends on the way in which f is computed. It could come from an arithmetic formula, or from a differential equation solver, with or without refinement.

As a rough estimate, let us assume simply that the relative error in the computed f is bounded by **u**, so that the computed values of $f(x)$ and $f(x + \epsilon e_i)$ are related to the exact values in the following way:

$$|\text{comp}(f(x)) - f(x)| \leq \mathbf{u} L_f,$$
$$|\text{comp}(f(x + \epsilon e_i)) - f(x + \epsilon e_i)| \leq \mathbf{u} L_f,$$

where comp$(\cdot)$ denotes the computed value, and L_f is a bound on the value of $|f(\cdot)|$ in the region of interest. If we use these computed values of f in place of the exact values in (7.4) and (7.1), we obtain an error that is bounded by

$$(L/2)\epsilon + 2\mathbf{u} L_f/\epsilon. \tag{7.5}$$

Naturally, we would like to choose ϵ to make this error as small as possible; it is easy to see that the minimizing value is

$$\epsilon^2 = \frac{4 L_f \mathbf{u}}{L}.$$

If we assume that the problem is well scaled, then the ratio L_f/L (the ratio of function values to second derivative values) does not exceed a modest size. We can conclude that the following choice of ϵ is fairly close to optimal:

$$\epsilon = \sqrt{\mathbf{u}}. \tag{7.6}$$

(In fact, this value is used in many of the optimization software packages that use finite differencing as an option for estimating derivatives.) For this value of ϵ, we have from (7.5) that the total error in the forward-difference approximation is fairly close to $\sqrt{\mathbf{u}}$.

A more accurate approximation to the derivative can be obtained by using the *central-difference* formula, defined as

$$\frac{\partial f}{\partial x_i}(x) \approx \frac{f(x + \epsilon e_i) - f(x - \epsilon e_i)}{2\epsilon}. \tag{7.7}$$

As we show below, this approximation is more accurate than the forward-difference approximation (7.1). It is also about twice as expensive, since we need to evaluate f at the points $x + \epsilon e_i$, $i = 1, 2, \ldots, n$, and $x - \epsilon e_i$, $i = 1, 2, \ldots, n$: a total of $2n$ points.

The basis for the central difference approximation is again Taylor's theorem. When the second derivatives of f exist and are Lipschitz continuous, we have from (7.2) that

$$\begin{aligned} f(x + p) &= f(x) + \nabla f(x)^T p + \tfrac{1}{2} p^T \nabla^2 f(x + tp) p \quad \text{for some } t \in (0, 1) \\ &= f(x) + \nabla f(x)^T p + \tfrac{1}{2} p^T \nabla^2 f(x) p + O\left(\|p\|^3\right). \end{aligned} \tag{7.8}$$

7.1. FINITE-DIFFERENCE DERIVATIVE APPROXIMATIONS

By setting $p = \epsilon e_i$ and $p = -\epsilon e_i$, respectively, we obtain

$$f(x + \epsilon e_i) = f(x) + \epsilon \frac{\partial f}{\partial x_i} + \frac{1}{2}\epsilon^2 \frac{\partial^2 f}{\partial x_i^2} + O(\epsilon^3),$$

$$f(x - \epsilon e_i) = f(x) - \epsilon \frac{\partial f}{\partial x_i} + \frac{1}{2}\epsilon^2 \frac{\partial^2 f}{\partial x_i^2} + O(\epsilon^3).$$

(Note that the final error terms in these two expressions are generally not the same, but they are both bounded by some multiple of ϵ^3.) By subtracting the second equation from the first and dividing by 2ϵ, we obtain the expression

$$\frac{\partial f}{\partial x_i}(x) = \frac{f(x + \epsilon e_i) - f(x - \epsilon e_i)}{2\epsilon} + O(\epsilon^2).$$

We see from this expression that the error is $O(\epsilon^2)$, as compared to the $O(\epsilon)$ error in the forward-difference formula (7.1). However, when we take evaluation error in f into account, the accuracy that can be achieved in practice is less impressive; the same assumptions that were used to derive (7.6) lead to the estimate $\epsilon = \mathbf{u}^{2/3}$ (see the exercises). In some applications, however, the extra few digits of accuracy may improve the performance of the algorithm enough to make the extra expense worthwhile.

APPROXIMATING A SPARSE JACOBIAN

Consider now the case of a vector function $r : \mathbf{R}^n \to \mathbf{R}^m$, such as the residual vector that we consider in Chapter 10 or the system of nonlinear equations from Chapter 11. The matrix $J(x)$ of first derivatives for this function is shown in (7.34). The techniques described in the previous section can be used to evaluate the full Jacobian $J(x)$ one column at a time. When r is twice continuously differentiable, we can use Taylor's theorem to deduce that

$$\|r(x + p) - r(x) - J(x)p\| \leq (L/2)\|p\|^2, \tag{7.9}$$

where L is a Lipschitz constant for J in the region of interest. If we require an approximation to the Jacobian–vector product $J(x)p$ for a given vector p (as is the case with inexact Newton methods for nonlinear systems of equations; see Section 11.1), this expression immediately suggests choosing a small nonzero ϵ and setting

$$J(x)p \approx \frac{r(x + \epsilon p) - r(x)}{\epsilon}, \tag{7.10}$$

an approximation that is accurate to $O(\epsilon)$. (A two-sided approximation can be derived from the formula (7.7).)

If an approximation to the full Jacobian $J(x)$ is required, we can compute it a column at a time, analogously to (7.1), by setting set $p = \epsilon e_i$ in (7.9) to derive the following estimate of the ith column:

$$\frac{\partial r}{\partial x_i}(x) \approx \frac{r(x + \epsilon e_i) - r(x)}{\epsilon}. \tag{7.11}$$

A full Jacobian estimate can be obtained at a cost of $n + 1$ evaluations of the function r. When the Jacobian is sparse, however, we can often obtain the estimate at a much lower cost, sometimes just three or four evaluations of r. The key is to estimate a number of different columns of the Jacobian simultaneously, by judicious choices of the perturbation vector p in (7.9).

We illustrate the technique with a simple example. Consider the function $r : \mathbb{R}^n \to \mathbb{R}^n$ defined by

$$r(x) = \begin{bmatrix} 2(x_2^3 - x_1^2) \\ 3(x_2^3 - x_1^2) + 2(x_3^3 - x_2^2) \\ 3(x_3^3 - x_2^2) + 2(x_4^3 - x_3^2) \\ \vdots \\ 3(x_n^3 - x_{n-1}^2) \end{bmatrix}. \tag{7.12}$$

Each component of r depends on just two or three components of x, so that each row of the Jacobian contains just two or three nonzero elements. For the case of $n = 6$, the Jacobian has the following structure:

$$\begin{bmatrix} \times & \times & & & & \\ \times & \times & \times & & & \\ & \times & \times & \times & & \\ & & \times & \times & \times & \\ & & & \times & \times & \times \\ & & & & \times & \times \end{bmatrix}, \tag{7.13}$$

where each cross represents a nonzero element, and zeros are represented by blank space.

Staying for the moment with the case $n = 6$, suppose that we wish to compute a finite-difference approximation to the Jacobian. (Of course, it is easy to calculate this particular Jacobian by hand, but there are complicated functions with similar structure for which hand calculation is more difficult.) A perturbation $p = \epsilon e_1$ to the first component of x will affect only the first and second components of r. The remaining components will be unchanged, so that the right-hand-side of formula (7.11) will correctly evaluate to zero in the components 3, 4, 5, 6. It is wasteful, however, to reevaluate these components of r when we know in

7.1. FINITE-DIFFERENCE DERIVATIVE APPROXIMATIONS

advance that their values are not affected by the perturbation. Instead, we look for a way to modify the perturbation vector so that it does not have any further effect on components 1 and 2, but *does* produce a change in some of the components 3, 4, 5, 6, which we can then use as the basis of a finite-difference estimate for some *other* column of the Jacobian. It is not hard to see that the additional perturbation ϵe_4 has the desired property: It alters the 3rd, 4th, and 5th elements of r, but leaves the 1st and 2nd elements unchanged. The changes in r as a result of the perturbations ϵe_1 and ϵe_4 do not interfere with each other.

To express this discussion in mathematical terms, we set

$$p = \epsilon(e_1 + e_4),$$

and note that

$$r(x+p)_{1,2} = r(x + \epsilon(e_1 + e_4))_{1,2} = r(x + \epsilon e_1)_{1,2} \tag{7.14}$$

(where the notation $[\cdot]_{1,2}$ denotes the subvector consisting of the first and second elements), while

$$r(x+p)_{3,4,5} = r(x + \epsilon(e_1 + e_4))_{3,4,5} = r(x + \epsilon e_4)_{3,4,5}. \tag{7.15}$$

By substituting (7.14) into (7.9), we obtain

$$r(x+p)_{1,2} = r(x)_{1,2} + \epsilon[J(x)e_1]_{1,2} + O(\epsilon^2).$$

By rearranging this expression, we obtain the following difference formula for estimating the (1, 1) and (2, 1) elements of the Jacobian matrix:

$$\begin{bmatrix} \frac{\partial r_1}{\partial x_1}(x) \\ \frac{\partial r_2}{\partial x_1}(x) \end{bmatrix} = [J(x)e_1]_{1,2} \approx \frac{r(x+p)_{1,2} - r(x)_{1,2}}{\epsilon}. \tag{7.16}$$

A similar argument shows that the nonzero elements of the fourth column of the Hessian can be estimated by substituting (7.15) into (7.9); we obtain eventually that

$$\begin{bmatrix} \frac{\partial r_4}{\partial x_3}(x) \\ \frac{\partial r_4}{\partial x_4}(x) \\ \frac{\partial r_4}{\partial x_5}(x) \end{bmatrix} = [J(x)e_4]_{3,4,5} \approx \frac{r(x+p)_{3,4,5} - r(x)_{3,4,5}}{\epsilon}. \tag{7.17}$$

To summarize: We have been able to estimate *two* columns of the Jacobian $J(x)$ by evaluating the function r at the single extra point $x + \epsilon(e_1 + e_4)$.

We can approximate the remainder of $J(x)$ in an economical manner as well. Columns 2 and 5 can be approximated by choosing $p = \epsilon(e_2 + e_5)$, while we can use $p = \epsilon(e_3 + e_6)$ to approximate columns 3 and 6. In total, we need 3 extra gradient evaluations to estimate the entire matrix.

In fact, for *any* choice of n in (7.12) (no matter how large), three extra gradient evaluations are sufficient for the entire Hessian. The corresponding choices of perturbation vectors p are

$$p = \epsilon(e_1 + e_4 + e_7 + e_{10} + \cdots),$$
$$p = \epsilon(e_2 + e_5 + e_8 + e_{11} + \cdots),$$
$$p = \epsilon(e_3 + e_6 + e_9 + e_{12} + \cdots).$$

In the first of these vectors, the nonzero components are chosen so that no two of the columns 1, 4, 7, ... have a nonzero element in the same row. The same property holds for the other two vectors and, in fact, points the way to the criterion that we can apply to general problems to decide on a valid set of perturbation vectors.

Algorithms for choosing the perturbation vectors can be expressed conveniently in the language of graphs and graph coloring. For any function $r : \mathbf{R}^n \to \mathbf{R}^m$, we can construct a *column incidence graph G* with n nodes by drawing an arc between nodes i and k if there is some component of r that depends on both x_i and x_k. (In other words, the ith and kth columns of the Jacobian $J(x)$ each have a nonzero element in some row j, for some $j = 1, 2, \ldots, m$ and some value of x.) The intersection graph for the function defined in (7.12) (with $n = 6$) is shown in Figure 7.1. We now assign each node a "color" according to the following rule: Two nodes can have the same color if there is no arc that connects them. Finally, we choose one perturbation vector corresponding to each color: If nodes $i_1, i_2, \ldots, i_\ell$ have the same color, the corresponding p is $\epsilon(e_{i_1} + e_{i_2} + \cdots + e_{i_\ell})$.

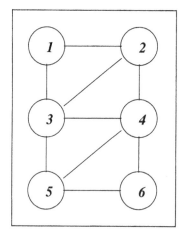

Figure 7.1
Column incidence graph for $r(x)$ defined in (7.12).

Usually, there are many ways to assign colors to the n nodes in the graph in a way that satisfies the required condition. The simplest way is just to assign each node a different color, but since that scheme produces n perturbation vectors, it is usually not the most efficient approach. It is generally very difficult to find the coloring scheme that uses the fewest possible colors, but there are simple algorithms that do a good job of finding a near-optimal coloring at low cost. Curtis, Powell, and Reid [62] and Coleman and Moré [48] provide descriptions of some methods and performance comparisons. Newsam and Ramsdell [182] show that by considering a more general class of perturbation vectors p, it is possible to evaluate the full Jacobian using no more than n_z evaluations of r (in addition to the evaluation at the point x), where n_z is the maximum number of nonzeros in each row of $J(x)$.

For some functions r with well-studied structures (those that arise from discretizations of differential operators, or those that give rise to banded Jacobians, as in the example above), optimal coloring schemes are known. For the tridiagonal Jacobian of (7.13) and its associated graph in Figure 7.1, the scheme with three colors is optimal.

APPROXIMATING THE HESSIAN

In some situations, the user may be able to provide a routine to calculate the gradient $\nabla f(x)$ but not the Hessian $\nabla^2 f(x)$. We can obtain the Hessian by applying the forward-difference or central-difference formula derived above to each element of the gradient vector in turn. When the second derivatives exist and are Lipschitz continuous, Taylor's theorem implies that

$$\nabla f(x + p) = \nabla f(x) + \nabla^2 f(x)p + O(\|p\|^2). \tag{7.18}$$

By substituting $p = \epsilon e_i$ in this expression and rearranging, we obtain

$$\nabla^2 f(x)e_i = \frac{\partial(\nabla f)}{\partial x_i}(x) \approx \frac{\nabla f(x + \epsilon e_i) - \nabla f(x)}{\epsilon}, \tag{7.19}$$

where the approximation error is $O(\epsilon)$. An approximation to the full Hessian can be obtained by setting $i = 1, 2, \ldots, n$ in turn, a process that requires a total of $n + 1$ evaluations of the gradient ∇f. Note that this column-at-a-time process does not necessarily lead to a symmetric result; however, we can recover symmetry by adding the approximation to its transpose and dividing the result by 2.

As in the case of sparse Jacobians discussed above, it usually is possible to approximate the entire Hessian using considerably fewer than n perturbation vectors. We discuss some approaches in the next section.

Some important algorithms—most notably the Newton–CG methods described in Chapter 6—do not require knowledge of the full Hessian. Instead, each iteration requires us to supply the Hessian–vector product $\nabla^2 f(x)p$, for a given vector p. An approximation to this product can be obtained by simply replacing p by ϵp in the expression (7.18), to obtain

$$\nabla^2 f(x) p \approx \frac{\nabla f(x + \epsilon p) - \nabla f(x)}{\epsilon} \qquad (7.20)$$

(see also (6.7)). The approximation error is again $O(\epsilon)$, and the cost of obtaining the approximation is a single gradient evaluation at the point $x + \epsilon p$. The formula (7.20) corresponds to the forward-difference approximation (7.1). A central difference formula (7.7) can be derived by evaluating $\nabla f(x - \epsilon p)$ as well.

For the case in which even gradients are not available, we can use Taylor's theorem once again to derive formulae for approximating the Hessian that use only function values. The main tool is the formula (7.8): By substituting the vectors $p = \epsilon e_i$, $p = \epsilon e_j$, and $p = \epsilon(e_i + e_j)$ into this formula and combining the results appropriately, we obtain

$$\frac{\partial^2 f}{\partial x_i \partial x_j}(x) = \frac{f(x + \epsilon e_i + \epsilon e_j) - f(x + \epsilon e_i) - f(x + \epsilon e_j) + f(x)}{\epsilon^2} + O(\epsilon). \qquad (7.21)$$

If we wished to approximate every element of the Hessian with this formula, then we would need to evaluate f at $x + \epsilon(e_i + e_j)$ for all possible i and j (a total of $n(n+1)/2$ points) as well as at the n points $x + \epsilon e_i$, $i = 1, 2, \ldots, n$. If the Hessian is sparse, we can, of course, reduce this operation count by skipping the evaluation whenever we know the element $\partial^2 f/\partial x_i \partial x_j$ to be zero.

APPROXIMATING A SPARSE HESSIAN

The Hessian $\nabla^2 f(x)$ can be thought of as the Jacobian of the vector function $\nabla f : \mathbf{R}^n \to \mathbf{R}^n$, so we could apply the sparse Jacobian estimation techniques described above to find an approximation, often at a cost of many fewer than $n + 1$ evaluations of the gradient vector ∇f. Such an approach, however, ignores the fact that the Hessian is symmetric. Because of symmetry, any estimate of the element $[\nabla^2 f(x)]_{i,j} = \partial^2 f(x)/\partial x_i \partial x_j$ is also an estimate of its symmetric counterpart $[\nabla^2 f(x)]_{j,i}$. Additional savings—sometimes very significant—can be had if we exploit the symmetry of the Hessian in choosing our perturbation vectors p.

We illustrate the point with the simple function $f : \mathbf{R}^n \to \mathbf{R}$ defined by

$$f(x) = x_1 \sum_{i=1}^{n} i^2 x_i^2. \qquad (7.22)$$

7.1. FINITE-DIFFERENCE DERIVATIVE APPROXIMATIONS

It is easy to show that the Hessian $\nabla^2 f$ has the "arrowhead" structure depicted below (for the case of $n = 6$):

$$\begin{bmatrix} \times & \times & \times & \times & \times & \times \\ \times & \times & & & & \\ \times & & \times & & & \\ \times & & & \times & & \\ \times & & & & \times & \\ \times & & & & & \times \end{bmatrix} \quad (7.23)$$

If we were to construct the intersection graph for the function ∇f (analogous to Figure 7.1), we would find that every node is connected to every other node, for the simple reason that row 1 has a nonzero in every column. According to the rule for coloring the graph, then, we would have to assign a different color to every node, which implies that we would need to evaluate ∇f at the $n + 1$ points x and $x + \epsilon e_i$ for $i = 1, 2, \ldots, n$.

We can construct a much more efficient scheme by taking the symmetry into account. Suppose we first use the perturbation vector $p = \epsilon e_1$ to estimate the first column of $\nabla^2 f(x)$. Because of symmetry, the same estimates apply to the first *row* of $\nabla^2 f$. From (7.23), we see that all that remains is to find the diagonal elements $\nabla^2 f(x)_{22}, \nabla^2 f(x)_{33}, \ldots, \nabla^2 f(x)_{66}$. The intersection graph for these remaining elements is completely disconnected, so we can assign them all the same color and choose the corresponding perturbation vector to be

$$p = \epsilon(e_2 + e_3 + \cdots + e_6) = \epsilon(0, 1, 1, 1, 1, 1)^T. \quad (7.24)$$

Note that the second component of ∇f is not affected by the perturbations in components 3, 4, 5, 6 of the unknown vector, while the third component of ∇f is not affected by perturbations in components 2, 4, 5, 6 of x, and so on. As in (7.14) and (7.15), we have for each component i that

$$\nabla f(x + p)_i = \nabla f(x + \epsilon(e_2 + e_3 + \cdots + e_6))_i = \nabla f(x + \epsilon e_i)_i.$$

By applying the forward-difference formula (7.1) to each of these individual components, we then obtain

$$\frac{\partial^2 f}{\partial x_i^2}(x) \approx \frac{\nabla f(x + \epsilon e_i)_i - \nabla f(x)_i}{\epsilon} = \frac{\nabla f(x + \epsilon p)_i - \nabla f(x)_i}{\epsilon}, \quad i = 2, 3, \ldots, 6.$$

By exploiting symmetry, we have been able to estimate the entire Hessian by evaluating ∇f only at x and two other points.

Again, graph-coloring techniques can be used as the basis of methods for choosing the perturbation vectors p economically. We use the *adjacency graph* in place of the intersection graph described earlier. The adjacency graph has n nodes, with arcs connecting nodes i and k whenever $i \neq k$ and $\partial^2 f(x)/(\partial x_i \partial x_k) \neq 0$ for some x. The requirements on the coloring scheme are a little more complicated than before, however. We require not only that connected nodes have different colors, but also that any path of length 3 through the graph contain at least three colors. In other words, if there exist nodes i_1, i_2, i_3, i_4 in the graph that are connected by arcs (i_1, i_2), (i_2, i_3), and (i_3, i_4), then at least three different colors must be used in coloring these four nodes. See Coleman and Moré [49] for an explanation of this rule and for algorithms to compute valid colorings. The perturbation vectors are constructed as before: Whenever the nodes $i_1, i_2, \ldots, i_\ell$ have the same color, we set the corresponding perturbation vector to be $p = \epsilon(e_{i_1} + e_{i_2} + \cdots + e_{i_\ell})$.

7.2 AUTOMATIC DIFFERENTIATION

Automatic differentiation is the generic name for techniques that use the computational representation of a function to produce analytic values for the derivatives. Some techniques produce code for the derivatives at a general point x by manipulating the function code directly. Other techniques keep a record of the computations made during the evaluation of the function at a specific point x and then review this information to produce a set of derivatives at x.

Automatic differentiation techniques are founded on the observation that any function, no matter how complicated, is evaluated by performing a sequence of simple elementary operations involving just one or two arguments at a time. Two-argument operations include addition, multiplication, division, and the power operation a^b. Examples of single-argument operations include the trigonometric, exponential, and logarithmic functions. Another common ingredient of the various automatic differentiation tools is their use of the *chain rule*. This is the well-known rule from elementary calculus that says that if h is a function of the vector $y \in \mathbf{R}^m$, which is in turn a function of the vector $x \in \mathbf{R}^n$, we can write the derivative of h with respect to x as follows:

$$\nabla_x h(y(x)) = \sum_{i=1}^{m} \frac{\partial h}{\partial y_i} \nabla y_i(x). \qquad (7.25)$$

See Appendix A for further details.

There are two basic modes of automatic differentiation: the *forward* and *reverse* modes. The difference between them can be illustrated by a simple example. We work through such an example below, and indicate how the techniques can be extended to general functions, including vector functions.

AN EXAMPLE

Consider the following function of 3 variables:

$$f(x) = (x_1 x_2 \sin x_3 + e^{x_1 x_2})/x_3. \tag{7.26}$$

The graph in Figure 7.2 shows how the evaluation of this function can be broken down into its elementary operations and also indicates the partial ordering associated with these operations. For instance, the multiplication $x_1 * x_2$ must take place prior to the exponentiation $e^{x_1 x_2}$, or else we would obtain the incorrect result $(e^{x_1}) x_2$. This graph introduces the *intermediate variables* $x_4, x_5, \ldots$ that contain the results of intermediate computations; they are distinguished from the *independent* variables x_1, x_2, x_3 that appear at the left of the graph. We can express the evaluation of f in arithmetic terms as follows:

$$\begin{aligned}
x_4 &= x_1 * x_2, \\
x_5 &= \sin x_3, \\
x_6 &= e^{x_4}, \\
x_7 &= x_4 * x_5, \\
x_8 &= x_6 + x_7, \\
x_9 &= x_8/x_3.
\end{aligned} \tag{7.27}$$

The final node x_9 in Figure 7.2 contains the function value $f(x)$. In the terminology of graph theory, node i is the *parent* of node j, and node j the *child* of node i, whenever there is a directed arc from i to j. Any node can be evaluated when the values of all its parents are known, so computation flows through the graph from left to right. Flow of computation in this direction is known as a *forward sweep*. It is important to emphasize

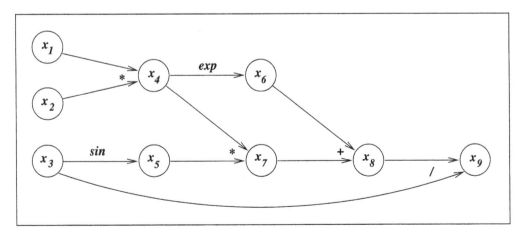

Figure 7.2 Computational graph for $f(x)$ defined in (7.26).

that software tools for automatic differentiation do not require the user to break down the code for evaluating the function into its elements, as in (7.27). Identification of intermediate quantities and construction of the computational graph is carried out, explicitly or implicitly, by the software tool itself.

We now describe how the forward mode can be used to obtain the derivatives of (7.26).

THE FORWARD MODE

In the forward mode of automatic differentiation, we evaluate and carry forward a directional derivative of each intermediate variable x_i in a given direction $p \in \mathbf{R}^m$, simultaneously with the evaluation of x_i itself. We introduce the following notation for the directional derivative for p associated with each variable:

$$D_p x_i \stackrel{\text{def}}{=} (\nabla x_i)^T p = \sum_{j=1}^{3} \frac{\partial x_i}{\partial x_j} p_j, \quad i = 1, 2, \ldots, 9. \tag{7.28}$$

Our goal is to evaluate $D_p x_9$, which is the same as the directional derivative $\nabla f(x)^T p$. We note immediately that initial values $D_p x_i$ for the independent variables x_i, $i = 1, 2, 3$, are simply the components p_1, p_2, p_3 of p. The direction p is referred to as the *seed vector*.

As soon as the value of x_i at any node is known, we can find the corresponding value of $D_p x_i$ from the chain rule. For instance, suppose we know the values of x_4, $D_p x_4$, x_5, and $D_p x_5$, and we are about to calculate x_7 in Figure 7.2. We have that $x_7 = x_4 x_5$; that is, x_7 is a function of the two variables x_4 and x_5, which in turn are functions of x_1, x_2, x_3. By applying the rule (7.25), we have that

$$\nabla x_7 = \frac{\partial x_7}{\partial x_4} \nabla x_4 + \frac{\partial x_7}{\partial x_5} \nabla x_5 = x_5 \nabla x_4 + x_4 \nabla x_5.$$

By taking the inner product of both sides of this expression with p and applying the definition (7.28), we obtain

$$D_p x_7 = \frac{\partial x_7}{\partial x_4} D_p x_4 + \frac{\partial x_7}{\partial x_5} D_p x_5 = x_5 D_p x_4 + x_4 D_p x_5. \tag{7.29}$$

The directional derivatives $D_p x_i$ are therefore evaluated side by side with the intermediate results x_i, and at the end of the process we obtain $D_p x_9 = D_p f = \nabla f(x)^T p$.

The principle of the forward mode is straightforward enough, but what of its practical implementation and computational requirements? First, we repeat that the user does *not* need to construct the computational graph, break the computation down into elementary operations as in (7.27), or identify intermediate variables. The automatic differentiation software should perform these tasks implicitly and automatically. Nor is it necessary to store the information x_i and $D_p x_i$ for *every* node of the computation graph at once (which is just as well, since this graph can be very large for complicated functions). Once all the children of any node have been evaluated, its associated values x_i and $D_p x_i$ are not needed further and may be overwritten in storage.

The key to practical implementation is the side-by-side evaluation of x_i and $D_p x_i$. The automatic differentiation software associates a scalar $D_p w$ with any scalar w that appears in the evaluation code. Whenever w is used in an arithmetic computation, the software performs an associated operation (based on the chain rule) with the gradient vector $D_p w$. For instance, if w is combined in a division operation with another value y to produce a new value z, that is,

$$z \leftarrow \frac{w}{y},$$

we use w, z, $D_p w$, and $D_p y$ to evaluate the directional derivative $D_p z$ as follows:

$$D_p z \leftarrow \frac{1}{y} D_p w - \frac{w}{y^2} D_p y. \tag{7.30}$$

To obtain the complete gradient vector, we can carry out this procedure simultaneously for the n seed vectors $p = e_1, e_2, \ldots, e_n$. By the definition (7.28), we see that $p = e_j$ implies that $D_p f = \partial f / \partial x_j$, $j = 1, 2, \ldots, n$. We note from the example (7.30) that the additional cost of evaluating f and ∇f (over the cost of evaluating f alone) may be significant. In this example, the single division operation on w and y needed to calculate z gives rise to approximately $2n$ multiplications and n additions in the computation of the gradient elements $D_{e_j} z$, $j = 1, 2, \ldots, n$. It is difficult to obtain an exact bound on the increase in computation, since the costs of retrieving and storing the data should also be taken into account. The storage requirements may also increase by a factor as large as n, since we now have to store n additional scalars $D_{e_j} x_i$, $j = 1, 2, \ldots, n$, alongside each intermediate variable x_i. It is usually possible to make savings by observing that many of these quantities are zero, particularly in the early stages of the computation (that is, toward the left of the computational graph), so sparse data structures can be used to store the vectors $D_{e_j} x_i$, $j = 1, 2, \ldots, n$ (see [16]).

The forward mode of automatic differentiation can be implemented by means of a precompiler, which transforms function evaluation code into extended code that evaluates the derivative vectors as well. An alternative approach is to use the operator-overloading facilities available in languages such as C++ to transparently extend the data structures and operations in the manner described above.

THE REVERSE MODE

The reverse mode of automatic differentiation does not perform function and gradient evaluations concurrently. Instead, after the evaluation of f is complete, it recovers the partial derivatives of f with respect to each variable x_i—independent and intermediate variables alike—by performing a *reverse sweep* of the computational graph. At the conclusion of this process, the gradient vector ∇f can be assembled from the partial derivatives $\partial f/\partial x_i$ with respect to the independent variables x_i, $i = 1, 2, \ldots, n$.

Instead of the gradient vectors $D_p x_i$ used in the forward mode, the reverse mode associates a scalar variable $\bar{x}_i$ with each node in the graph; information about the partial derivative $\partial f/\partial x_i$ is accumulated in $\bar{x}_i$ during the reverse sweep. The $\bar{x}_i$ are sometimes called the *adjoint variables*, and we initialize their values to zero, with the exception of the rightmost node in the graph (node N, say), for which we set $\bar{x}_N = 1$. This choice makes sense because x_N contains the final function value f, so we have $\partial f/\partial x_N = 1$.

The reverse sweep makes use of the following observation, which is again based on the chain rule (7.25): For any node i, the partial derivative $\partial f/\partial x_i$ can be built up from the partial derivatives $\partial f/\partial x_j$ corresponding to its child nodes j according to the following formula:

$$\frac{\partial f}{\partial x_i} = \sum_{j \text{ a child of } i} \frac{\partial f}{\partial x_j} \frac{\partial x_j}{\partial x_i}. \tag{7.31}$$

For each node i, we add the right-hand-side term in (7.31) to $\bar{x}_i$ as soon as it becomes known; that is, we perform the operation

$$\bar{x}_i \mathrel{+}= \frac{\partial f}{\partial x_j} \frac{\partial x_j}{\partial x_i}. \tag{7.32}$$

(In this expression and the ones below, we use the arithmetic notation of the programming language C, in which $x\mathrel{+}=a$ means $x \leftarrow x + a$.) Once contributions have been received from all the child nodes of i, we have $\bar{x}_i = \partial f/\partial x_i$, so we declare node i to be "finalized." At this point, node i is ready to contribute a term to the summation for each of its parent nodes according to the formula (7.31). The process continues in this fashion until all nodes are finalized. Note that the flow of computation in the graph is from children to parents. This is the opposite direction to the process of evaluating the function f, which proceeds from parents to children.

During the reverse sweep, we work with *numerical values*, not with formulae or computer code involving the variables x_i or the partial derivatives $\partial f/\partial x_i$. During the forward sweep—the evaluation of f—we not only calculate the values of each variable x_i, but we also calculate and store the numerical values of each partial derivative $\partial x_j/\partial x_i$. Each of these partial derivatives is associated with a particular arc of the computational graph. The numerical values of $\partial x_j/\partial x_i$ computed during the forward sweep are then used in the formula (7.32) during the reverse sweep.

We illustrate the reverse mode for the example function (7.26). In Figure 7.3 we fill in the graph of Figure 7.2 for a specific evaluation point $x = (1, 2, \pi/2)^T$, indicating the numerical values of the intermediate variables $x_4, x_5, \ldots, x_9$ associated with each node and the partial derivatives $\partial x_j/\partial x_i$ associated with each arc.

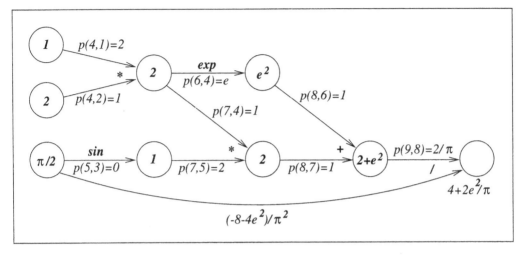

Figure 7.3 Computational graph for $f(x)$ defined in (7.26) showing numerical values of intermediate values and partial derivatives for the point $x = (1, 2, \pi/2)^T$. Notation: $p(j, i) = \partial x_j / \partial x_i$.

As mentioned above, we initialize the reverse sweep by setting all the adjoint variables $\bar{x}_i$ to zero, except for the rightmost node, for which we have $\bar{x}_9 = 1$. Since $f(x) = x_9$ and since node 9 has no children, we have $\bar{x}_9 = \partial f / \partial x_9$, and so we can immediately declare node 9 to be finalized.

Node 9 is the child of nodes 3 and 8, so we use formula (7.32) to update the values of $\bar{x}_3$ and $\bar{x}_8$ as follows:

$$\bar{x}_3 += \frac{\partial f}{\partial x_9} \frac{\partial x_9}{\partial x_3} = -\frac{2 + e^2}{(\pi/2)^2} = \frac{-8 - 4e^2}{\pi^2}, \tag{7.33a}$$

$$\bar{x}_8 += \frac{\partial f}{\partial x_9} \frac{\partial x_9}{\partial x_8} = \frac{1}{\pi/2} = \frac{2}{\pi}. \tag{7.33b}$$

Node 3 is not finalized after this operation; it still awaits a contribution from its other child, node 5. On the other hand, node 9 is the only child of node 8, so we can declare node 8 to be finalized with the value $\frac{\partial f}{\partial x_8} = 2/\pi$. We can now update the values of $\bar{x}_i$ at the two parent nodes of node 8 by applying the formula (7.32) once again; that is,

$$\bar{x}_6 += \frac{\partial f}{\partial x_8} \frac{\partial x_8}{\partial x_6} = \frac{2}{\pi};$$

$$\bar{x}_7 += \frac{\partial f}{\partial x_8} \frac{\partial x_8}{\partial x_7} = \frac{2}{\pi}.$$

At this point, nodes 6 and 7 are finalized, so we can use them to update nodes 4 and 5. At the end of this process, when all nodes are finalized, nodes 1, 2, and 3 contain

$$\begin{bmatrix} \bar{x}_1 \\ \bar{x}_2 \\ \bar{x}_3 \end{bmatrix} = \nabla f(x) = \begin{bmatrix} (4 + 4e^2)/\pi \\ (2 + 2e^2)/\pi \\ (-8 - 4e^2)/\pi^2 \end{bmatrix},$$

and the derivative computation is complete.

The main appeal of the reverse mode is that its computational complexity is low for the scalar functions $f : \mathbf{R}^n \to \mathbf{R}$ discussed here. The extra arithmetic associated with the gradient computation is at most four or five times the arithmetic needed to evaluate the function alone. Taking the division operation in (7.33) as an example, we see that two multiplications, a division, and an addition are required for (7.33a), while a division and an addition are required for (7.33b). This is about five times as much work as the single division involving these nodes that was performed during the forward sweep.

As we noted above, the forward mode may require up to n times more arithmetic to compute the gradient ∇f than to compute the function f alone, making it appear uncompetitive with the reverse mode. When we consider vector functions $r : \mathbf{R}^n \to \mathbf{R}^m$, the relative costs of the forward and reverse modes become more similar as m increases, as we describe in the next section.

An apparent drawback of the reverse mode is the need to store the entire computational graph, which is needed for the reverse sweep. In principle, storage of this graph is not too difficult to implement. Whenever an elementary operation is performed, we can form and store a new node containing the intermediate result, pointers to the (one or two) parent nodes, and the partial derivatives associated with these arcs. During the reverse sweep, the nodes can be read in the reverse order to that in which they were written, giving a particularly simple access pattern. The process of forming and writing the graph can be implemented as a straightforward extension to the elementary operations via operator overloading (as in ADOL-C [126]). The reverse sweep/gradient evaluation can be invoked as a simple function call.

Unfortunately, the computational graph may require a huge amount of storage. If each node can be stored in 20 bytes, then a function that requires one second of evaluation time on a 100 megaflop computer may produce a graph of up to 2 gigabytes in size. The storage requirements can be reduced, at the cost of some extra arithmetic, by performing partial forward and reverse sweeps on pieces of the computational graph, reevaluating portions of the graph as needed rather than storing the whole structure. Descriptions of this approach, sometimes known as *checkpointing*, can be found in Griewank [122] and Grimm, Pottier, and Rostaing-Schmidt [130], and an implementation can be found in the Odyssee software tool [219]. An implementation of checkpointing in the context of variational data assimilation can be found in Restrepo, Leaf, and Griewank [213].

VECTOR FUNCTIONS AND PARTIAL SEPARABILITY

So far, we have looked at automatic differentiation of general scalar-valued functions $f : \mathbf{R}^n \rightarrow \mathbf{R}$. In nonlinear least-squares problems (Chapter 10) and nonlinear equations (Chapter 11), we have to deal with vector functions $r : \mathbf{R}^n \rightarrow \mathbf{R}^m$ with m components r_j, $j = 1, 2, \ldots, m$. The rightmost column of the computational graph then consists of m nodes, none of which has any children, in place of the single node described above. The forward and reverse modes can be adapted in straightforward ways to find the Jacobian of r, that is, the $m \times n$ matrix $J(x)$ defined by

$$J(x) = \left[\frac{\partial r_j}{\partial x_i}\right]_{\substack{j=1,2,\ldots,m \\ i=1,2,\ldots,n}}. \tag{7.34}$$

Besides their applications to least-squares and nonlinear-equations problems, automatic differentiation of vector functions is a useful technique for dealing with partially separable functions. We recall that partial separability is commonly observed in large-scale optimization, and we saw in Chapter 9 that there exist efficient quasi-Newton procedures for the minimization of objective functions with this property. Since an automatic procedure for detecting the decomposition of a given function f into its partially separable representation was developed recently by Gay [98], it has become possible to exploit the efficiencies that accrue from this property without asking much information from the user.

In the simplest sense, a function f is partially separable if we can express it in the form

$$f(x) = \sum_{i=1}^{ne} f_i(x), \tag{7.35}$$

where each *element function* $f_i(\cdot)$ depends on just a few components of x. If we construct the vector function r from the partially separable components, that is,

$$r(x) = \begin{bmatrix} f_1(x) \\ f_2(x) \\ \vdots \\ f_{ne}(x) \end{bmatrix},$$

it follows from (7.35) that

$$\nabla f(x) = J(x)^T e, \tag{7.36}$$

where, as usual, $e = (1, 1, \ldots, 1)^T$. Because of the partial separability property, most columns of $J(x)$ contain just a few nonzeros. This structure makes it possible to calculate

$J(x)$ efficiently by applying graph-coloring techniques, as we discuss below. The gradient $\nabla f(x)$ can then be recovered from the formula (7.36).

In constrained optimization, it is often beneficial to evaluate the objective function f and the constraint functions $c_i, i \in \mathcal{I} \cup \mathcal{E}$, simultaneously. By doing so, we can take advantage of common expressions (which show up as shared intermediate nodes in the computation graph) and hence can reduce the total workload. In Figure 7.2, for instance, the intermediate variable x_4 is shared during the computation of x_6 and x_7. In this case, the vector function r can be defined as

$$r(x) = \begin{bmatrix} f(x) \\ [\,c_j(x)\,]_{j \in \mathcal{I} \cup \mathcal{E}} \end{bmatrix}.$$

CALCULATING JACOBIANS OF VECTOR FUNCTIONS

The forward mode is the same for vector functions as for scalar functions. Given a seed vector p, we continue to associate quantities $D_p x_i$ with the node that calculates each intermediate variable x_i. At each of the rightmost nodes (containing r_j, $j = 1, 2, \ldots, m$), this variable contains the quantity $D_p r_j = (\nabla r_j)^T p$, $j = 1, 2, \ldots, m$. By assembling these m quantities, we obtain $J(x)p$, the product of the Jacobian and our chosen vector p. As in the case of scalar functions ($m = 1$), we can evaluate the complete Jacobian by setting $p = e_1, e_2, \ldots, e_n$ and evaluating the n quantities $D_{e_j} x_i$ simultaneously. For sparse Jacobians, we can use the coloring techniques outlined above in the context of finite-difference methods to make more intelligent and economical choices of the seed vectors p. The factor of increase in cost of arithmetic, when compared to a single evaluation of r, is about equal to the number of seed vectors used.

The key to applying the reverse mode to a vector function $r(x)$ is to choose seed vectors $q \in \mathbb{R}^m$ and apply the reverse mode to the scalar functions $r(x)^T q$. The result of this process is the vector

$$\nabla [r(x)^T q] = \nabla \left[\sum_{j=1}^{m} q_j r_j(x) \right] = J(x)^T q.$$

Instead of the Jacobian–vector product that we obtain with the forward mode, the reverse mode yields a Jacobian-transpose–vector product. The technique can be implemented by seeding the variables $\bar{x}_i$ in the m dependent nodes that contain $r_1, r_2, \ldots, r_m$, with the components $q_1, q_2, \ldots, q_m$ of the vector q. At the end of the reverse sweep, the node for independent variables $x_1, x_2, \ldots, x_n$ will contain

$$\frac{d}{dx_i}[r(x)^T q], \quad i = 1, 2, \ldots, n,$$

which are simply the components of $J(x)^T q$.

As usual, we can obtain the full Jacobian by carrying out the process above for the m unit vectors $q = e_1, e_2, \ldots, e_m$. Alternatively, for sparse Jacobians, we can apply the usual coloring techniques to find a smaller number of seed vectors q—the only difference being that the graphs and coloring strategies are defined with reference to the transpose $J(x)^T$ rather than to $J(x)$ itself. The factor of increase in the number of arithmetic operations required, in comparison to an evaluation of r alone, is no more than 5 times the number of seed vectors. (The factor of 5 is the usual overhead from the reverse mode for a scalar function.) The space required for storage of the computational graph is no greater than in the scalar case. As before, we need only store the graph topology information together with the partial derivative associated with each arc.

The forward- and reverse-mode techniques can be combined to cumulatively reveal all the elements of $J(x)$. We can choose a set of seed vectors p for the forward mode to reveal some columns of J, then perform the reverse mode with another set of seed vectors q to reveal the rows that contain the remaining elements.

Finally, we note that for some algorithms, we do not need full knowledge of the Jacobian $J(x)$. For instance, iterative methods such as the inexact Newton method for nonlinear equations (see Section 11.1) require repeated calculation of $J(x)p$ for a succession of vectors p. Each such matrix–vector product can be computed using the forward mode by using a single forward sweep, at a similar cost to evaluation of the function alone.

CALCULATING HESSIANS: FORWARD MODE

So far, we have described how the forward and reverse modes can be applied to obtain first derivatives of scalar and vector functions. We now outline extensions of these techniques to the computation of the Hessian $\nabla^2 f$ of a scalar function f, and evaluation of the Hessian–vector product $\nabla^2 f(x)p$ for a given vector p.

Recall that the forward mode makes use of the quantities $D_p x_i$, each of which stores $(\nabla x_i)^T p$ for each node i in the computational graph and a given vector p. For a given *pair* of seed vectors p and q (both in $\mathbf{R}^n$) we now define another scalar quantity by

$$D_{pq} x_i = p^T (\nabla^2 x_i) q, \qquad (7.37)$$

for each node i in the computational graph. We can evaluate these quantities during the forward sweep through the graph, alongside the function values x_i and the first-derivative values $D_p x_i$. The initial values of D_{pq} at the independent variable nodes x_i, $i = 1, 2 \ldots, n$, will be 0, since the second derivatives of x_i are zero at each of these nodes. When the forward sweep is complete, the value of $D_{pq} x_i$ in the rightmost node of the graph will be $p^T \nabla^2 f(x) q$.

The formulae for transformation of the $D_{pq} x_i$ variables during the forward sweep can once again be derived from the chain rule. For instance, if x_i is obtained by adding the values at its two parent nodes, $x_i = x_j + x_k$, the corresponding accumulation operations on $D_p x_i$

and $D_{pq}x_i$ are as follows:

$$D_p x_i = D_p x_j + D_p x_k, \quad D_{pq} x_i = D_{pq} x_j + D_{pq} x_k. \tag{7.38}$$

The other binary operations $-, \times, /$ are handled similarly. If x_i is obtained by applying the unitary transformation L to x_j, we have

$$x_i = L(x_j), \tag{7.39a}$$
$$D_p x_i = L'(x_j)(D_p x_j), \tag{7.39b}$$
$$D_{pq} x_i = L''(x_j)(D_p x_j)(D_q x_j) + L'(x_j) D_{pq} x_j. \tag{7.39c}$$

We see in (7.39c) that computation of $D_{pq}x_i$ can rely on the first-derivative quantities $D_p x_i$ and $D_q x_i$, so both these quantities must be accumulated during the forward sweep as well.

We could compute a general dense Hessian by choosing the pairs (p, q) to be all possible pairs of unit vectors (e_j, e_k), for $j = 1, 2, \ldots, n$ and $k = 1, 2, \ldots, j$, a total of $n(n + 1)/2$ vector pairs. (Note that we need only evaluate the lower triangle of $\nabla^2 f(x)$, because of symmetry.) When we know the sparsity structure of $\nabla^2 f(x)$, we need evaluate $D_{e_j e_k} x_i$ only for the pairs (e_j, e_k) for which the (j, k) component of $\nabla^2 f(x)$ is possibly nonzero.

The total increase factor for the number of arithmetic operations, compared with the amount of arithmetic to evaluate f alone, is a small multiple of $1 + n + N_z(\nabla^2 f)$, where $N_z(\nabla^2 f)$ is the number of elements of $\nabla^2 f$ that we choose to evaluate. This number reflects the evaluation of the quantities x_i, $D_{e_j} x_i$ ($j = 1, 2, \ldots, n$), and $D_{e_j e_k} x_i$ for the $N_z(\nabla^2 f)$ vector pairs (e_j, e_k). The "small multiple" results from the fact that the update operations for $D_p x_i$ and $D_{pq} x_i$ may require a few times more operations than the update operation for x_i alone; see, for example, (7.39). One storage location per node of the graph is required for each of the $1 + n + N_z(\nabla^2 f)$ quantities that are accumulated, but recall that storage of node i can be overwritten once all its children have been evaluated.

When we do not need the complete Hessian, but only a matrix–vector product involving the Hessian (as in the Newton–CG algorithm of Chapter 6), the amount of arithmetic is, of course, smaller. Given a vector $q \in \mathbf{R}^n$, we use the techniques above to compute the first-derivative quantities $D_{e_1} x_i, \ldots D_{e_n} x_i$ and $D_q x_i$, as well as the second-derivative quantities $D_{e_1 q} x_i, \ldots, D_{e_n q} x_i$, during the forward sweep. The final node will contain the quantities

$$e_j^T \left(\nabla^2 f(x) \right) q = \left[\nabla^2 f(x) q \right]_j, \quad j = 1, 2, \ldots, n,$$

which are the components of the vector $\nabla^2 f(x) q$. Since $2n + 1$ quantities in addition to x_i are being accumulated during the forward sweep, the increase factor in the number of arithmetic operations increases by a small multiple of $2n$.

An alternative technique for evaluating sparse Hessians is based on the forward-mode propagation of first and second derivatives of *univariate* functions. To motivate this approach, note that the (i, j) element of the Hessian can be expressed as follows:

$$[\nabla^2 f(x)]_{ij} = e_i^T \nabla^2 f(x) e_j$$
$$= \frac{1}{2}\left[(e_i + e_j)^T \nabla^2 f(x)(e_i + e_j) - e_i^T \nabla^2 f(x) e_i - e_j^T \nabla^2 f(x) e_j\right]. \quad (7.40)$$

We can use this interpolation formula to evaluate $[\nabla^2 f(x)]_{ij}$, provided that the second derivatives $D_{pp} x_k$, for $p = e_i$, $p = e_j$, $p = e_i + e_j$, and all nodes x_k, have been evaluated during the forward sweep through the computational graph. In fact, we can evaluate all the nonzero elements of the Hessian, provided that we use the forward mode to evaluate $D_p x_k$ and $D_{pp} x_k$ for a selection of vectors p of the form $e_i + e_j$, where i and j are both indices in $\{1, 2, \ldots, n\}$, possibly with $i = j$.

One advantage of this approach is that it is no longer necessary to propagate "cross terms" of the form $D_{pq} x_k$ for $p \neq q$ (see, for example, (7.38) and (7.39c)). The propagation formulae therefore simplify somewhat. Each $D_{pp} x_k$ is a function of x_ℓ, $D_p x_\ell$, and $D_{pp} x_\ell$ for all parent nodes ℓ of node k.

Note, too, that if we define the univariate function ψ by

$$\psi(t) = f(x + tp), \quad (7.41)$$

then the values of $D_p f$ and $D_{pp} f$ (which emerge at the completion of the forward sweep) are simply the first two derivatives of ψ evaluated at $t = 0$; that is,

$$D_p f = p^T \nabla f(x) = \psi'(t)|_{t=0}, \quad D_{pp} f = p^T \nabla^2 f(x) p = \psi''(t)|_{t=0}.$$

Extension of this technique to third, fourth, and higher derivatives is possible. Interpolation formulae analogous to (7.40) can be used in conjunction with higher derivatives of the univariate functions ψ defined in (7.41), again for a suitably chosen set of vectors p, where each p is made up of a sum of unit vectors e_i. For details, see Bischof, Corliss, and Griewank [14].

CALCULATING HESSIANS: REVERSE MODE

We can also devise schemes based on the reverse mode for calculating Hessian–vector products $\nabla^2 f(x) q$, or the full Hessian $\nabla^2 f(x)$. A scheme for obtaining $\nabla^2 f(x) q$ proceeds as follows. We start by using the forward mode to evaluate both f and $\nabla f(x)^T q$, by accumulating the two variables x_i and $D_q x_i$ during the forward sweep in the manner described above. We then apply the reverse mode in the normal fashion to the computed function $\nabla f(x)^T q$. At the end of the reverse sweep, the nodes $i = 1, 2, \ldots, n$ of the computational graph that correspond to the independent variables will contain

$$\frac{\partial}{\partial x_i}(\nabla f(x)^T q) = [\nabla^2 f(x) q]_i, \quad i = 1, 2, \ldots, n.$$

The number of arithmetic operations required to obtain $\nabla^2 f(x)q$ by this procedure increases by only a modest factor, independent of n, over the evaluation of f alone. By the usual analysis for the forward mode, we see that the computation of f and $\nabla f(x)^T q$ jointly requires a small multiple of the operation count for f alone, while the reverse sweep introduces a further factor of at most 5. The total increase factor is approximately 12 over the evaluation of f alone. If the entire Hessian $\nabla^2 f(x)$ is required, we could apply the procedure just described with $q = e_1, e_2, \ldots, e_n$. This approach would introduce an additional factor of n into the operation count, leading to an increase of at most $12n$ over the cost of f alone.

Once again, when the Hessian is sparse with known structure, we may be able to use graph-coloring techniques to evaluate this entire matrix using many fewer than n seed vectors. The choices of q are similar to those used for finite-difference evaluation of the Hessian, described above. The increase in operation count over evaluating f alone is a multiple of up to $12 N_c(\nabla^2 f)$, where N_c is the number of seed vectors q used in calculating $\nabla^2 f$.

CURRENT LIMITATIONS

The current generation of automatic differentiation tools has proved its worth through successful application to some large and difficult design optimization problems. However, these tools can run into difficulties with some commonly used programming constructs and some implementations of computer arithmetic. As an example, if the evaluation of $f(x)$ depends on the solution of a partial differential equation (PDE), then the computed value of f may contain truncation error arising from the finite-difference or the finite-element technique that is used to solve the PDE numerically. That is, we have $\hat{f}(x) = f(x) + \tau(x)$, where $\hat{f}(\cdot)$ is the computed value of $f(\cdot)$ and $\tau(\cdot)$ is the truncation error. Though $|\tau(x)|$ is usually small, its derivative $\tau'(x)$ may not be, so the error in the computed derivative $\hat{f}'(x)$ is potentially large. (The finite-difference approximation techniques discussed in Section 7.1 experience the same difficulty.) Similar problems arise when the computer uses piecewise rational functions to approximate trigonometric functions.

Another source of potential difficulty is the presence of branching in the code to improve the speed or accuracy of function evaluation in certain domains. A pathological example is provided by the linear function $f(x) = x - 1$. If we used the following (perverse, but valid) piece of code to evaluate this function,

```
if (x = 1.0) then f = 0.0 else f = x - 1.0,
```

then by applying automatic differentiation to this procedure we would obtain the derivative value $f'(1) = 0$. For a discussion of such issues and an approach to dealing with them, see Griewank [123].

In conclusion, automatic differentiation should be regarded as a set of increasingly sophisticated techniques that enhances optimization algorithms, allowing them to be applied successfully to many practical problems involving complicated functions. It facilitates interpretations of the computed optimal solutions, allowing the modeler to extract more information from the results of the computation. Automatic differentiation should not be regarded as a panacea that absolves the user altogether from the responsibility of thinking about derivative calculations.

NOTES AND REFERENCES

The field of automatic differentiation has grown considerably in recent years, and a number of good software tools are now available. We mention in particular ODYSSEE [219], ADIFOR [13] and ADIC [17], and ADOL-C [126]. Current software tool development in automatic differentiation is moving away from the forward-reverse-mode dichotomy and toward "mixed modes" that combine the two approaches at different steps in the computation. For a summary of recent work in this area see Bischof and Haghighat [15].

A number of good survey papers and volumes on automatic differentiation are available. The volume edited by Griewank and Corliss [125] covers much of the state of the art up to 1991, while the update to this volume, edited by Berz, Bischof, Corliss, and Griewank [10], covers the developments of the following five years. (Particularly notable contributions in this volume are the paper of Corliss and Rall [57] and the extensive bibliography.) A survey of the applications of automatic differentiation in optimization is given by Griewank [124].

The technique for calculating the gradient of a partially separable function was described by Bischof et al. [12], whereas the computation of the Hessian matrix has been considered by several authors; see, for example, Gay [98].

The work of Coleman and Moré [49] on efficient estimation of Hessians was predated by Powell and Toint [210], who did not use the language of graph coloring but nevertheless devised highly effective schemes. Software for estimating sparse Hessians and Jacobians is described by Coleman, Garbow, and Moré [46, 47].

EXERCISES

7.1 Show that a suitable value for the perturbation ϵ in the central-difference formula is $\epsilon = \mathbf{u}^{1/3}$, and that the accuracy achievable by this formula when the values of f contain roundoff errors of size $\mathbf{u}$ is approximately $\mathbf{u}^{2/3}$. (Use similar assumptions to the ones used to derive the estimate (7.6) for the forward-difference formula.)

7.2 Derive a central-difference analogue of the Hessian–vector approximation formula (7.20).

7.3 Verify the formula (7.21) for approximating an element of the Hessian using only function values.

7.4 Verify that if the Hessian of a function f has nonzero diagonal elements, then its adjacency graph is a subgraph of the intersection graph for ∇f. In other words, show that any arc in the adjacency graph also belongs to the intersection graph.

7.5
Draw the adjacency graph for the function f defined by (7.22). Show that the coloring scheme in which node 1 has one color while nodes $2, 3, \ldots, n$ have another color is valid. Draw the intersection graph for ∇f.

7.6
Construct the adjacency graph for the function whose Hessian has the nonzero structure

$$\begin{bmatrix} \times & \times & \times & \times & & & \\ \times & \times & \times & & & \times & \\ \times & \times & \times & & & & \times \\ \times & & & & \times & & \\ & & \times & \times & & & \times \\ & & & \times & & \times & \end{bmatrix},$$

and find a valid coloring scheme with just three colors.

7.7
Trace the computations performed in the forward mode for the function $f(x)$ in (7.26), expressing the intermediate derivatives ∇x_i, $i = 6, \ldots, 12$, in terms of quantities available at their parent nodes and then in terms of the independent variables $x_1, \ldots, x_5$.

7.8
Formula (7.30) showed the gradient operations associated with scalar division. Derive similar formulae for the following operations:

$$\begin{aligned} (s, t) &\to s + t &&\text{addition;} \\ t &\to e^t &&\text{exponentiation;} \\ t &\to \tan(t) &&\text{tangent;} \\ (s, t) &\to s^t. \end{aligned}$$

7.9
By calculating the partial derivatives $\partial x_j / \partial x_i$ for the function (7.26) from the expressions (7.27), verify the numerical values for the arcs in Figure 7.3 for the evaluation point $x = (1, 2, \pi/2, -1, 3)^T$. Work through the remaining details of the reverse sweep process, indicating the order in which the nodes become finalized.

7.10
Using (7.33) as a guide, describe the reverse sweep operations corresponding to the following elementary operations in the forward sweep:

$$\begin{aligned} x_k &\leftarrow x_i x_j &&\text{multiplication;} \\ x_k &\leftarrow \cos(x_i) &&\text{cosine.} \end{aligned}$$

In each case, compare the arithmetic workload in the reverse sweep to the workload required for the forward sweep.

7.2. AUTOMATIC DIFFERENTIATION

✏ **7.11** Define formulae similar to (7.38) for accumulating the first derivatives $D_p x_i$ and the second derivatives $D_{pq} x_i$ when x_i is obtained from the following three binary operations: $x_i = x_j - x_k$, $x_i = x_j x_k$, and $x_i = x_j/x_k$.

✏ **7.12** By using the definitions (7.28) of $D_p x_i$ and (7.37) of $D_{pq} x_i$, verify the differentiation formulae (7.39) for the unitary operation $x_i = L(x_j)$.

✏ **7.13** Let $a \in \mathbf{R}^n$ be a fixed vector and define f as $f(x) = \frac{1}{2}\left(x^T x + \left(a^T x\right)^2\right)$. Count the number of operations needed to evaluate f, ∇f, $\nabla^2 f$, and the Hessian–vector product $\nabla^2 f(x) p$ for an arbitrary vector p.

CHAPTER 8

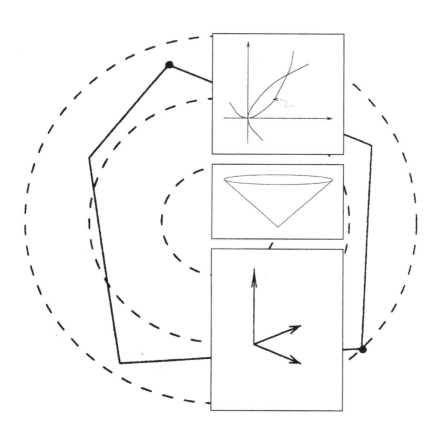

Quasi-Newton Methods

In the mid 1950s, W.C. Davidon, a physicist working at Argonne National Laboratory, was using the coordinate descent method (see Section 3.3) to perform a long optimization calculation. At that time computers were not very stable, and to Davidon's frustration, the computer system would always crash before the calculation was finished. So Davidon decided to find a way of accelerating the iteration. The algorithm he developed—the first quasi-Newton algorithm—turned out to be one of the most revolutionary ideas in nonlinear optimization. It was soon demonstrated by Fletcher and Powell that the new algorithm was much faster and more reliable than the other existing methods, and this dramatic advance transformed nonlinear optimization overnight. During the following twenty years, numerous variants were proposed and hundreds of papers were devoted to their study. An interesting historical irony is that Davidon's paper [64] was not accepted for publication; it remained as a technical report for more than thirty years until it appeared in the first issue of the *SIAM Journal on Optimization* in 1991 [65].

Quasi-Newton methods, like steepest descent, require only the gradient of the objective function to be supplied at each iterate. By measuring the changes in gradients, they construct a model of the objective function that is good enough to produce superlinear convergence. The improvement over steepest descent is dramatic, especially on difficult problems. Moreover, since second derivatives are not required, quasi-Newton methods are sometimes more efficient than Newton's method. Today, optimization software libraries contain a variety of quasi-Newton algorithms for solving unconstrained, constrained, and large-scale optimization problems. In this chapter we discuss quasi-Newton methods for small and medium-sized problems, and in Chapter 9 we consider their extension to the large-scale setting.

The development of automatic differentiation techniques has diminished the appeal of quasi-Newton methods, but only to a limited extent. Automatic differentiation eliminates the tedium of computing second derivatives by hand, as well as the risk of introducing errors in the calculation. More details are given in Chapter 7. Nevertheless, quasi-Newton methods remain competitive on many types of problems.

8.1 THE BFGS METHOD

The most popular quasi-Newton algorithm is the BFGS method, named for its discoverers Broyden, Fletcher, Goldfarb, and Shanno. In this section we derive this algorithm (and its close relative, the DFP algorithm) and describe its theoretical properties and practical implementation.

We begin the derivation by forming the following quadratic model of the objective function at the current iterate x_k:

$$m_k(p) = f_k + \nabla f_k^T p + \tfrac{1}{2} p^T B_k p. \tag{8.1}$$

Here B_k is an $n \times n$ symmetric positive definite matrix that will be revised or *updated* at every iteration. Note that the value and gradient of this model at $p = 0$ match f_k and ∇f_k, respectively. The minimizer p_k of this convex quadratic model, which we can write explicitly as

$$p_k = -B_k^{-1} \nabla f_k, \tag{8.2}$$

is used as the search direction, and the new iterate is

$$x_{k+1} = x_k + \alpha_k p_k, \tag{8.3}$$

where the step length α_k is chosen to satisfy the Wolfe conditions (3.6). This iteration is quite similar to the line search Newton method; the key difference is that the approximate Hessian B_k is used in place of the true Hessian.

Instead of computing B_k afresh at every iteration, Davidon proposed to update it in a simple manner to account for the curvature measured during the most recent step. Suppose that we have generated a new iterate x_{k+1} and wish to construct a new quadratic model, of the form

$$m_{k+1}(p) = f_{k+1} + \nabla f_{k+1}^T p + \tfrac{1}{2} p^T B_{k+1} p.$$

What requirements should we impose on B_{k+1}, based on the knowledge we have gained during the latest step? One reasonable requirement is that the gradient of m_{k+1} should match the gradient of the objective function f at the latest two iterates x_k and x_{k+1}. Since $\nabla m_{k+1}(0)$ is precisely ∇f_{k+1}, the second of these conditions is satisfied automatically. The first condition can be written mathematically as

$$\nabla m_{k+1}(-\alpha_k p_k) = \nabla f_{k+1} - \alpha_k B_{k+1} p_k = \nabla f_k.$$

By rearranging, we obtain

$$B_{k+1} \alpha_k p_k = \nabla f_{k+1} - \nabla f_k. \tag{8.4}$$

To simplify the notation it is useful to define the vectors

$$s_k = x_{k+1} - x_k, \quad y_k = \nabla f_{k+1} - \nabla f_k, \tag{8.5}$$

so that (8.4) becomes

$$B_{k+1} s_k = y_k. \tag{8.6}$$

We refer to this formula as the *secant equation*.

Given the displacement s_k and the change of gradients y_k, the secant equation requires that the symmetric positive definite matrix B_{k+1} map s_k into y_k. This will be possible only if s_k and y_k satisfy the *curvature condition*

$$s_k^T y_k > 0, \tag{8.7}$$

as is easily seen by premultiplying (8.6) by s_k^T. When f is strongly convex, the inequality (8.7) will be satisfied for any two points x_k and x_{k+1} (see the exercises). However, this condition will not always hold for nonconvex functions, and in this case we need to enforce (8.7) explicitly, by imposing restrictions on the line search procedure that chooses α. In fact, the condition (8.7) is guaranteed to hold if we impose the Wolfe (3.6) or strong Wolfe conditions (3.7) on the line search. To verify this claim, we note from (8.5) and (3.6b) that $\nabla f_{k+1}^T s_k \geq c_2 \nabla f_k^T s_k$, and therefore

$$y_k^T s_k \geq (c_2 - 1)\alpha_k \nabla f_k^T p_k. \tag{8.8}$$

Since $c_2 < 1$ and since p_k is a descent direction, the term on the right will be positive, and the curvature condition (8.7) holds.

196　Chapter 8.　Quasi-Newton Methods

When the curvature condition is satisfied, the secant equation (8.6) always has a solution B_{k+1}. In fact, it admits an infinite number of solutions, since there are $n(n+1)/2$ degrees of freedom in a symmetric matrix, and the secant equation represents only n conditions. The requirement of positive definiteness imposes n additional inequalities—all principal minors must be positive—but these conditions do not absorb the remaining degrees of freedom.

To determine B_{k+1} uniquely, then, we impose the additional condition that *among all symmetric matrices satisfying the secant equation, B_{k+1} is, in some sense, closest to the current matrix B_k*. In other words, we solve the problem

$$\min_B \|B - B_k\| \tag{8.9a}$$

$$\text{subject to} \quad B = B^T, \quad B s_k = y_k, \tag{8.9b}$$

where s_k and y_k satisfy (8.7) and B_k is symmetric and positive definite. Many matrix norms can be used in (8.9a), and each norm gives rise to a different quasi-Newton method. A norm that allows easy solution of the minimization problem (8.9), and that gives rise to a scale-invariant optimization method, is the weighted Frobenius norm

$$\|A\|_W \equiv \left\|W^{1/2} A W^{1/2}\right\|_F, \tag{8.10}$$

where $\|\cdot\|_F$ is defined by $\|C\|_F^2 = \sum_{i=1}^n \sum_{j=1}^n c_{ij}^2$. The weight W can be chosen as *any* matrix satisfying the relation $W y_k = s_k$. For concreteness, the reader can assume that $W = \bar{G}_k^{-1}$ where $\bar{G}_k$ is the *average Hessian* defined by

$$\bar{G}_k = \left[\int_0^1 \nabla^2 f(x_k + \tau \alpha_k p_k) d\tau\right]. \tag{8.11}$$

The property

$$y_k = \bar{G}_k \alpha_k p_k = \bar{G}_k s_k \tag{8.12}$$

follows from Taylor's theorem, Theorem 2.1. With this choice of weighting matrix W, the norm (8.10) is adimensional, which is a desirable property, since we do not wish the solution of (8.9) to depend on the units of the problem.

With this weighting matrix and this norm, the unique solution of (8.9) is

$$\text{(DFP)} \quad B_{k+1} = \left(I - \gamma_k y_k s_k^T\right) B_k \left(I - \gamma_k s_k y_k^T\right) + \gamma_k y_k y_k^T, \tag{8.13}$$

with

$$\gamma_k = \frac{1}{y_k^T s_k}.$$

This formula is called the DFP updating formula, since it is the one originally proposed by Davidon in 1959, and subsequently studied, implemented, and popularized by Fletcher and Powell.

The inverse of B_k, which we denote by

$$H_k = B_k^{-1},$$

is useful in the implementation of the method, since it allows the search direction (8.2) to be calculated by means of a simple matrix–vector multiplication. Using the Sherman–Morrison–Woodbury formula (A.56), we can derive the following expression for the update of the inverse Hessian approximation H_k that corresponds to the DFP update of B_k in (8.13):

$$\text{(DFP)} \quad H_{k+1} = H_k - \frac{H_k y_k y_k^T H_k}{y_k^T H_k y_k} + \frac{s_k s_k^T}{y_k^T s_k}. \tag{8.14}$$

Note that the last two terms in the right-hand-side of (8.14) are rank-one matrices, so that H_k undergoes a rank-two modification. It is easy to see that (8.13) is also a rank-two modification of B_k. This is the fundamental idea of quasi-Newton updating: Instead of recomputing the iteration matrices from scratch at every iteration, we apply a simple modification that combines the most recently observed information about the objective function with the existing knowledge embedded in our current Hessian approximation.

The DFP updating formula is quite effective, but it was soon superseded by the BFGS formula, which is presently considered to be the most effective of all quasi-Newton updating formulae. BFGS updating can be derived by making a simple change in the argument that led to (8.13). Instead of imposing conditions on the Hessian approximations B_k, we impose similar conditions on their inverses H_k. The updated approximation H_{k+1} must be symmetric and positive definite, and must satisfy the secant equation (8.6), now written as

$$H_{k+1} y_k = s_k.$$

The condition of closeness to H_k is now specified by the following analogue of (8.9):

$$\min_{H} \|H - H_k\| \tag{8.15a}$$

$$\text{subject to} \quad H = H^T, \quad H y_k = s_k. \tag{8.15b}$$

The norm is again the weighted Frobenius norm described above, where the weight matrix W is now any matrix satisfying $W s_k = y_k$. (For concreteness, we assume again that W is given by the average Hessian $\bar{G}_k$ defined in (8.11).) The unique solution H_{k+1} to (8.15) is given by

$$\text{(BFGS)} \quad H_{k+1} = (I - \rho_k s_k y_k^T) H_k (I - \rho_k y_k s_k^T) + \rho_k s_k s_k^T, \tag{8.16}$$

where

$$\rho_k = \frac{1}{y_k^T s_k}. \tag{8.17}$$

Just one issue has to be resolved before we can define a complete BFGS algorithm: How should we choose the initial approximation H_0? Unfortunately, there is no magic formula that works well in all cases. We can use specific information about the problem, for instance by setting it to the inverse of an approximate Hessian calculated by finite differences at x_0. Otherwise, we can simply set it to be the identity matrix, or a multiple of the identity matrix, where the multiple is chosen to reflect the scaling of the variables.

Algorithm 8.1 (BFGS Method).
 Given starting point x_0, convergence tolerance $\epsilon > 0$,
  inverse Hessian approximation H_0;
 $k \leftarrow 0$;
 while $\|\nabla f_k\| > \epsilon$;
  Compute search direction

$$p_k = -H_k \nabla f_k; \tag{8.18}$$

  Set $x_{k+1} = x_k + \alpha_k p_k$ where α_k is computed from a line search
   procedure to satisfy the Wolfe conditions (3.6);
  Define $s_k = x_{k+1} - x_k$ and $y_k = \nabla f_{k+1} - \nabla f_k$;
  Compute H_{k+1} by means of (8.16);
  $k \leftarrow k + 1$;
 end (while)

Each iteration can be performed at a cost of $O(n^2)$ arithmetic operations (plus the cost of function and gradient evaluations); there are no $O(n^3)$ operations such as linear system solves or matrix–matrix operations. The algorithm is robust, and its rate of convergence is superlinear, which is fast enough for most practical purposes. Even though Newton's method converges more rapidly (that is, quadratically), its cost per iteration is higher because it requires the solution of a linear system. A more important advantage for BFGS is, of course, that it does not require calculation of second derivatives.

We can derive a version of the BFGS algorithm that works with the Hessian approximation B_k rather than H_k. The update formula for B_k is obtained by simply applying the Sherman–Morrison–Woodbury formula (A.56) to (8.16) to obtain

$$\text{(BFGS)} \quad B_{k+1} = B_k - \frac{B_k s_k s_k^T B_k}{s_k^T B_k s_k} + \frac{y_k y_k^T}{y_k^T s_k}. \tag{8.19}$$

A naive implementation of this variant is not efficient for unconstrained minimization, because it requires the system $B_k p_k = -\nabla f_k$ to be solved for the step p_k, thereby increasing the cost of the step computation to $O(n^3)$. We discuss later, however, that less expensive implementations of this variant are possible by updating Cholesky factors of B_k.

PROPERTIES OF THE BFGS METHOD

It is usually easy to observe the superlinear rate of convergence of the BFGS method on practical problems. Below, we report the last few iterations of the steepest descent, BFGS, and an inexact Newton method on Rosenbrock's function (2.23). The table gives the value of $\|x_k - x^*\|$. The Wolfe conditions were imposed on the step length in all three methods. From the starting point $(-1.2, 1)$, the steepest descent method required 5264 iterations, whereas BFGS and Newton took only 34 and 21 iterations, respectively to reduce the gradient norm to 10^{-5}.

steep. desc.	BFGS	Newton
1.827e-04	1.70e-03	3.48e-02
1.826e-04	1.17e-03	1.44e-02
1.824e-04	1.34e-04	1.82e-04
1.823e-04	1.01e-06	1.17e-08

A few points in the derivation of the BFGS and DFP methods merit further discussion. Note that the minimization problem (8.15) that gives rise to the BFGS update formula does not explicitly require the updated Hessian approximation to be positive definite. It is easy to show, however, that H_{k+1} will be positive definite whenever H_k is positive definite, by using the following argument. First, note from (8.8) that $y_k^T s_k$ is positive, so that the updating formula (8.16), (8.17) is well-defined. For any nonzero vector z, we have

$$z^T H_{k+1} z = w^T H_k w + \rho_k (z^T s_k)^2 \geq 0,$$

where we have defined $w = z - \rho_k y_k (s_k^T z)$. The right hand side can be zero only if $s_k^T z = 0$, but in this case $w = z \neq 0$, which implies that the first term is greater than zero. Therefore, H_{k+1} is positive definite.

In order to obtain quasi-Newton updating formulae that are invariant to changes, in the variables, it is necessary that the objectives (8.9a) and (8.15a) be also invariant. The choice of the weighting matrices W used to define the norms in (8.9a) and (8.15a) ensures that this condition holds. Many other choices of the weighting matrix W are possible, each one of them giving a different update formula. However, despite intensive searches, no formula has been found that is significantly more effective than BFGS.

The BFGS method has many interesting properties when applied to quadratic functions. We will discuss these properties later on, in the more general context of the Broyden family of updating formulae, of which BFGS is a special case.

It is reasonable to ask whether there are situations in which the updating formula such as (8.16) can produce bad results. If at some iteration the matrix H_k becomes a very poor approximation, is there any hope of correcting it? If, for example, the inner product $y_k^T s_k$ is tiny (but positive), then it follows from (8.16)–(8.17) that H_{k+1} becomes huge. Is this behavior reasonable? A related question concerns the rounding errors that occur in finite-precision implementation of these methods. Can these errors grow to the point of erasing all useful information in the quasi-Newton approximate matrix?

These questions have been studied analytically and experimentally, and it is now known that the BFGS formula has very effective *self-correcting properties*. If the matrix H_k incorrectly estimates the curvature in the objective function, and if this bad estimate slows down the iteration, then the Hessian approximation will tend to correct itself within a few steps. It is also known that the DFP method is less effective in correcting bad Hessian approximations; this property is believed to be the reason for its poorer practical performance. The self-correcting properties of BFGS hold only when an adequate line search is performed. In particular, the Wolfe line search conditions ensure that the gradients are sampled at points that allow the model (8.1) to capture appropriate curvature information.

It is interesting to note that the DFP and BFGS updating formulae are *duals* of each other, in the sense that one can be obtained from the other by the interchanges $s \leftrightarrow y$, $B \leftrightarrow H$. This symmetry is not surprising, given the manner in which we derived these methods above.

IMPLEMENTATION

A few details and enhancements need to be added to Algorithm 8.1 to produce an efficient implementation. The line search, which should satisfy either the Wolfe conditions (3.6) or the strong Wolfe conditions (3.7), should always try the step length $\alpha_k = 1$ first, because this step length will eventually always be accepted (under certain conditions), thereby producing superlinear convergence of the overall algorithm. Computational observations strongly suggest that it is more economical, in terms of function evaluations, to perform a fairly inaccurate line search. The values $c_1 = 10^{-4}$ and $c_2 = 0.9$ are commonly used in (3.6).

As mentioned earlier, the initial matrix H_0 often is set to some multiple βI of the identity, but there is no good general strategy for choosing β. If β is "too large," so that the first step $p_0 = -\beta g_0$ is too long, many function evaluations may be required to find a suitable value for the step length α_0. Some software asks the user to prescribe a value δ for the norm of the first step, and then set $H_0 = \delta \|g_0\|^{-1} I$ to achieve this norm.

A heuristic that is often quite effective is to scale the starting matrix *after* the first step has been computed but before the first BFGS update is performed. We change the provisional value $H_0 = I$ by setting

$$H_0 \leftarrow \frac{y_k^T s_k}{y_k^T y_k} I, \qquad (8.20)$$

before applying the update (8.16), (8.17) to obtain H_1. This formula attempts to make the size of H_0 similar to that of $[\nabla^2 f(x_0)]^{-1}$, in the following sense. Assuming that the average Hessian defined in (8.11) is positive definite, there exists a square root $\bar{G}_k^{1/2}$ satisfying $\bar{G}_k = \bar{G}_k^{1/2} \bar{G}_k^{1/2}$ (see Exercise 5). Therefore, by defining $z_k = \bar{G}_k^{1/2} s_k$ and using the relation (8.12), we have

$$\frac{y_k^T s_k}{y_k^T y_k} = \frac{(\bar{G}_k^{1/2} s_k)^T \bar{G}_k^{1/2} s_k}{(\bar{G}_k^{1/2} s_k)^T \bar{G}_k \bar{G}_k^{1/2} s_k} = \frac{z_k^T z_k}{z_k^T \bar{G}_k z_k}. \tag{8.21}$$

The *reciprocal* of (8.21) is an approximation to one of the eigenvalues of $\bar{G}_k$, which in turn is close to an eigenvalue of $\nabla^2 f(x_k)$. Hence, the quotient (8.21) itself approximates an eigenvalue of $[\nabla^2 f(x_k)]^{-1}$. Other scaling factors can be used in (8.20), but the one presented here appears to be the most successful in practice.

In (8.19) we gave an update formula for a BFGS method that works with the Hessian approximation B_k instead of the the inverse Hessian approximation H_k. An efficient implementation of this approach does not store B_k explicitly, but rather the Cholesky factorization $L_k D_k L_k^T$ of this matrix. A formula that updates the factors L_k and D_k directly in $O(n^2)$ operations can be derived from (8.19). Since the linear system $B_k p_k = -\nabla f_k$ also can be solved in $O(n^2)$ operations (by performing triangular substitutions with L_k and L_k^T and a diagonal substitution with D_k), the total cost is quite similar to the variant described in Algorithm 8.1. A potential advantage of this alternative strategy is that it gives us the option of modifying diagonal elements in the D_k factor if they are not sufficiently large, to prevent instability when we divide by these elements during the calculation of p_k. However, computational experience suggests no real advantages for this variant, and we prefer the simpler strategy of Algorithm 8.1.

The performance of the BFGS method can degrade if the line search is not based on the Wolfe conditions. For example, some software implements an Armijo backtracking line search (see Section 3.1): The unit step length $\alpha_k = 1$ is tried first and is successively decreased until the sufficient decrease condition (3.6a) is satisfied. For this strategy, there is no guarantee that the curvature condition $y_k^T s_k > 0$ (8.7) will be satisfied by the chosen step, since a step length greater than 1 may be required to satisfy this condition. To cope with this shortcoming, some implementations simply *skip* the BFGS update by setting $H_{k+1} = H_k$ when $y_k^T s_k$ is negative or too close to zero. This approach is not recommended, because the updates may be skipped much too often to allow H_k to capture important curvature information for the objective function f. In Chapter 18 we discuss a *damped* BFGS update that is a more effective strategy for coping with the case where the curvature condition (8.7) is not satisfied.

8.2 THE SR1 METHOD

In the BFGS and DFP updating formulae, the updated matrix B_{k+1} (or H_{k+1}) differs from its predecessor B_k (or H_k) by a rank-2 matrix. In fact, as we now show, there is a simpler rank-1 update that maintains symmetry of the matrix and allows it to satisfy the secant equation. Unlike the rank-two update formulae, this *symmetric-rank-1*, or SR1, update does not guarantee that the updated matrix maintains positive definiteness. Good numerical results have been obtained with algorithms based on SR1, so we derive it here and investigate its properties.

The symmetric rank-1 update has the general form

$$B_{k+1} = B_k + \sigma v v^T,$$

where σ is either $+1$ or -1, and σ and v are chosen so that B_{k+1} satisfies the secant equation (8.6), that is, $y_k = B_{k+1} s_k$. By substituting into this equation, we obtain

$$y_k = B_k s_k + \left[\sigma v^T s_k\right] v. \tag{8.22}$$

Since the term in brackets is a scalar, we deduce that v must be a multiple of $y_k - B_k s_k$, that is, $v = \delta(y_k - B_k s_k)$ for some scalar δ. By substituting this form of v into (8.22), we obtain

$$(y_k - B_k s_k) = \sigma \delta^2 \left[s_k^T (y_k - B_k s_k)\right] (y_k - B_k s_k), \tag{8.23}$$

and it is clear that this equation is satisfied if (and only if) we choose the parameters δ and σ to be

$$\sigma = \text{sign}\left[s_k^T (y_k - B_k s_k)\right], \quad \delta = \pm \left|s_k^T (y_k - B_k s_k)\right|^{-1/2}.$$

Hence, we have shown that the only symmetric rank-1 updating formula that satisfies the secant equation is given by

$$\text{(SR1)} \quad B_{k+1} = B_k + \frac{(y_k - B_k s_k)(y_k - B_k s_k)^T}{(y_k - B_k s_k)^T s_k}. \tag{8.24}$$

By applying the Sherman–Morrison formula (A.55), we obtain the corresponding update formula for the inverse Hessian approximation H_k:

$$\text{(SR1)} \quad H_{k+1} = H_k + \frac{(s_k - H_k y_k)(s_k - H_k y_k)^T}{(s_k - H_k y_k)^T y_k}. \tag{8.25}$$

This derivation is so simple that the SR1 formula has been rediscovered a number of times.

It is easy to see that even if B_k is positive definite, B_{k+1} may not have this property; the same is, of course, true of H_k. This observation was considered a major drawback in the early days of nonlinear optimization when only line search iterations were used. However, with the advent of trust-region methods, the SR1 updating formula has proved to be quite useful, and its ability to generate indefinite Hessian approximations can actually be regarded as one of its chief advantages.

The main drawback of SR1 updating is that the denominator in (8.24) or (8.25) can vanish. In fact, even when the objective function is a convex quadratic, there may be steps on which there is no symmetric rank-1 update that satisfies the secant equation. It pays to reexamine the derivation above in the light of this observation.

By reasoning in terms of B_k (similar arguments can be applied to H_k), we see that there are three cases:

1. If $(y_k - B_k s_k)^T s_k \neq 0$, then the arguments above show that there is a unique rank-one updating formula satisfying the secant equation (8.6), and that it is given by (8.24).

2. If $y_k = B_k s_k$, then the only updating formula satisfying the secant equation is simply $B_{k+1} = B_k$.

3. If $y_k \neq B_k s_k$ and $(y_k - B_k s_k)^T s_k = 0$, then (8.23) shows that there is no symmetric rank-one updating formula satisfying the secant equation.

The last case clouds an otherwise simple and elegant derivation, and suggests that numerical instabilities and even breakdown of the method can occur. It suggests that rank-one updating does not provide enough freedom to develop a matrix with all the desired characteristics, and that a rank-two correction is required. This reasoning leads us back to the BFGS method, in which positive definiteness (and thus nonsingularity) of all Hessian approximations is guaranteed.

Nevertheless, we are interested in the SR1 formula for the following reasons.

(i) A simple safeguard seems to adequately prevent the breakdown of the method and the occurrence of numerical instabilities.

(ii) The matrices generated by the SR1 formula tend to be very good approximations of the Hessian matrix—often better than the BFGS approximations.

(iii) In quasi-Newton methods for constrained problems, or in methods for partially separable functions (see Chapters 18 and 9), it may not be possible to impose the curvature condition $y_k^T s_k > 0$, and thus BFGS updating is not recommended. Indeed, in these two settings, indefinite Hessian approximations are desirable insofar as they reflect indefiniteness in the true Hessian.

We now introduce a strategy to prevent the SR1 method from breaking down. It has been observed in practice that SR1 performs well simply by skipping the update if the denominator is small. More specifically, the update (8.24) is applied only if

$$\left| s_k^T (y_k - B_k s_k) \right| \geq r \|s_k\| \, \|y_k - B_k s_k\|, \tag{8.26}$$

where $r \in (0, 1)$ is a small number, say $r = 10^{-8}$. If (8.26) does not hold, we set $B_{k+1} = B_k$. Most implementations of the SR1 method use a skipping rule of this kind.

Why do we advocate skipping of updates for the SR1 method, when in the previous section we discouraged this strategy in the case of BFGS? The two cases are quite different. The condition $s_k^T(y_k - B_k s_k) \approx 0$ occurs infrequently, since it requires certain vectors to be aligned in a specific way. When it does occur, skipping the update appears to have no negative effects on the iteration. This is not surprising, since the skipping condition implies that $s_k^T \bar{G} s_k \approx s_k^T B_k s_k$, where $\bar{G}$ is the average Hessian over the last step—meaning that the curvature of B_k along s_k is already correct. In contrast, the curvature condition $s_k^T y_k \geq 0$ required for BFGS updating may easily fail if the line search does not impose the Wolfe conditions (e.g., if the step is not long enough), and therefore skipping the BFGS update can occur often and can degrade the quality of the Hessian approximation.

We now give a formal description of an SR1 method using a trust-region framework. We prefer it over a line search framework because it does not require us to modify the Hessian approximations to make them sufficiently positive definite.

Algorithm 8.2 (SR1 Trust-Region Method).

Given starting point x_0, initial Hessian approximation B_0,
    trust-region radius Δ_0, convergence tolerance $\epsilon > 0$,
    parameters $\eta \in (0, 10^{-3})$ and $r \in (0, 1)$;
$k \leftarrow 0$;
while $\|\nabla f_k\| > \epsilon$;
    Compute s_k by solving the subproblem

$$\min_s \nabla f_k^T s + \frac{1}{2} s^T B_k s \quad \text{subject to } \|s\| \leq \Delta_k. \quad (8.27)$$

    Compute

$$y_k = \nabla f(x_k + s_k) - \nabla f_k,$$
$$\textbf{ared} = f_k - f(x_k + s_k) \quad \text{(actual reduction)}$$
$$\textbf{pred} = -\left(\nabla f_k^T s_k + \frac{1}{2} s_k^T B_k s_k\right) \quad \text{(predicted reduction)};$$

    if ared/pred $> \eta$
        $x_{k+1} = x_k + s_k$
    else
        $x_{k+1} = x_k$;
    end (if)
    if ared/pred > 0.75

if $\|s_k\| \leq 0.8\Delta_k$
                        $\Delta_{k+1} = \Delta_k$
                else
                        $\Delta_{k+1} = 2\Delta_k$;
                end (if)
        elseif $0.1 \leq$ **ared**/**pred** ≤ 0.75
                $\Delta_{k+1} = \Delta_k$
        else
                $\Delta_{k+1} = 0.5\Delta_k$;
        end (if)
        if (8.26) holds
                Use (8.24) to compute B_{k+1} (even if $x_{k+1} = x_k$)
        else
                $B_{k+1} \leftarrow B_k$;
        end (if)
        $k \leftarrow k + 1$;
end (while)

This algorithm has the typical form of a trust region method (cf. Algorithm 4.1). For concreteness we have specified a particular strategy for updating the trust region radius, but other heuristics can be used instead.

To obtain a fast rate of convergence, it is important for the matrix B_k to be updated even along a failed direction s_k. The fact that the step was poor indicates that B_k is a poor approximation of the true Hessian in this direction. Unless the quality of the approximation is improved, steps along similar directions could be generated on later iterations, and repeated rejection of such steps could prevent superlinear convergence.

PROPERTIES OF SR1 UPDATING

One of the main advantages of SR1 updating is its ability to generate very good Hessian approximations. We demonstrate this property by first examining a quadratic function. For functions of this type, the choice of step length does not affect the update, so to examine the effect of the updates, we can assume for simplicity a uniform step length of 1, that is,

$$p_k = -H_k \nabla f_k, \qquad x_{k+1} = x_k + p_k. \tag{8.28}$$

It follows that $p_k = s_k$.

Theorem 8.1.
   Suppose that $f : \mathbf{R}^n \to \mathbf{R}$ is the strongly convex quadratic function $f(x) = b^T x + \frac{1}{2}x^T A x$, where A is symmetric positive definite. Then for any starting point x_0 and any symmetric starting matrix H_0, the iterates $\{x_k\}$ generated by the SR1 method (8.25), (8.28) converge to the

minimizer in at most n steps, provided that $(s_k - H_k y_k)^T y_k \neq 0$ for all k. Moreover, if n steps are performed, and if the search directions p_i are linearly independent, then $H_n = A^{-1}$.

PROOF. Because of our assumption $(s_k - H_k y_k)^T y_k \neq 0$, the SR1 update is always well-defined. We start by showing inductively that

$$H_k y_j = s_j \quad \text{for} \quad j = 0, \ldots, k-1. \tag{8.29}$$

In other words, we claim that the secant equation is satisfied not only along the most recent search direction, but along all previous directions.

By definition, the SR1 update satisfies the secant equation, so we have $H_1 y_0 = s_0$. Let us now assume that (8.29) holds for some value $k > 1$ and show that it holds also for $k+1$. From this assumption, we have from (8.29) that

$$(s_k - H_k y_k)^T y_j = s_k^T y_j - y_k^T (H_k y_j) = s_k^T y_j - y_k^T s_j = 0, \quad \text{for all } j < k, \tag{8.30}$$

where the last equality follows because $y_i = A s_i$ for the quadratic function we are considering here. By using (8.30) and the induction hypothesis (8.29) in (8.25), we have

$$H_{k+1} y_j = H_k y_j = s_j, \quad \text{for all } j < k.$$

Since $H_{k+1} y_k = s_k$ by the secant equation, we have shown that (8.29) holds when k is replaced by $k+1$. By induction, then, this relation holds for all k.

If the algorithm performs n steps and if these steps $\{s_j\}$ are linearly independent, we have

$$s_j = H_n y_j = H_n A s_j, \quad \text{for } j = 0, \ldots, n-1.$$

It follows that $H_n A = I$, that is, $H_n = A^{-1}$. Therefore, the step taken at x_n is the Newton step, and so the next iterate x_{n+1} will be the solution, and the algorithm terminates.

Consider now the case in which the steps become linearly dependent. Suppose that s_k is a linear combination of the previous steps, that is,

$$s_k = \xi_0 s_0 + \cdots + \xi_{k-1} s_{k-1}, \tag{8.31}$$

for some scalars ξ_i. From (8.31) and (8.29) we have that

$$\begin{aligned} H_k y_k &= H_k A s_k \\ &= \xi_0 H_k A s_0 + \cdots + \xi_{k-1} H_k A s_{k-1} \\ &= \xi_0 H_k y_0 + \cdots + \xi_{k-1} H_k y_{k-1} \\ &= \xi_0 s_0 + \cdots + \xi_{k-1} s_{k-1} \\ &= s_k. \end{aligned}$$

Since $y_k = \nabla f_{k+1} - \nabla f_k$ and since $s_k = p_k = -H_k \nabla f_k$ from (8.28), we have that

$$H_k(\nabla f_{k+1} - \nabla f_k) = -H_k \nabla f_k,$$

which, by the nonsingularity of H_k, implies that $\nabla f_{k+1} = 0$. Therefore, x_{k+1} is the solution point. □

The relation (8.29) shows that when f is quadratic, the secant equation is satisfied along all previous search directions, regardless of how the line search is performed. A result like this can be established for BFGS updating only under the restrictive assumption that the line search is exact, as we show in the next section.

For general nonlinear functions, the SR1 update continues to generate good Hessian approximations under certain conditions.

Theorem 8.2.
Suppose that f is twice continuously differentiable, and that its Hessian is bounded and Lipschitz continuous in a neighborhood of a point x^*. Let $\{x_k\}$ be any sequence of iterates such that $x_k \to x^*$ for some $x^* \in \mathbb{R}^n$. Suppose in addition that the inequality (8.26) holds for all k, for some $r \in (0, 1)$, and that the steps s_k are uniformly linearly independent. Then the matrices B_k generated by the SR1 updating formula satisfy

$$\lim_{k \to \infty} \|B_k - \nabla^2 f(x^*)\| = 0.$$

The term "uniformly linearly independent steps" means, roughly speaking, that the steps do not tend to fall in a subspace of dimension less than n. This assumption is usually, but not always, satisfied in practice. (See the Notes and Remarks at the end of this chapter.)

8.3 THE BROYDEN CLASS

So far, we have described the BFGS, DFP, and SR1 quasi-Newton updating formulae, but there are many others. Of particular interest is the *Broyden class*, a family of updates specified by the following general formula:

$$B_{k+1} = B_k - \frac{B_k s_k s_k^T B_k}{s_k^T B_k s_k} + \frac{y_k y_k^T}{y_k^T s_k} + \phi_k (s_k^T B_k s_k) v_k v_k^T, \tag{8.32}$$

where ϕ_k is a scalar parameter and

$$v_k = \left[\frac{y_k}{y_k^T s_k} - \frac{B_k s_k}{s_k^T B_k s_k} \right]. \tag{8.33}$$

The BFGS and DFP methods are members of the Broyden class—we recover BFGS by setting $\phi_k = 0$ and DFP by setting $\phi_k = 1$ in (8.32). We can therefore rewrite (8.32) as a "linear combination" of these two methods, that is,

$$B_{k+1} = (1 - \phi_k) B_{k+1}^{\text{BFGS}} + \phi_k B_{k+1}^{\text{DFP}}.$$

This relationship indicates that all members of the Broyden class satisfy the secant equation (8.6), since the BGFS and DFP matrices themselves satisfy this equation. Also, since BFGS and DFP updating preserve positive definiteness of the Hessian approximations when $s_k^T y_k > 0$, this relation implies that the same property will hold for the Broyden family if $0 \leq \phi_k \leq 1$.

Much attention has been given to the so-called *restricted Broyden class*, which is obtained by restricting ϕ_k to the interval $[0, 1]$. It enjoys the following property when applied to quadratic functions. Since the analysis is independent of the step length, we assume for simplicity that each iteration has the form

$$p_k = -B_k^{-1} \nabla f_k, \qquad x_{k+1} = x_k + p_k. \tag{8.34}$$

Theorem 8.3.

Suppose that $f : \mathbf{R}^n \to \mathbf{R}$ is the strongly convex quadratic function $f(x) = b^T x + \frac{1}{2} x^T A x$, where A is symmetric and positive definite. Let x_0 be any starting point for the iteration (8.34) and B_0 be any symmetric positive definite starting matrix, and suppose that the matrices B_k are updated by the Broyden formula (8.32) with $\phi_k \in [0, 1]$. Define $\lambda_1^k \leq \lambda_2^k \leq \cdots \leq \lambda_n^k$ to be the eigenvalues of the matrix

$$A^{\frac{1}{2}} B_k^{-1} A^{\frac{1}{2}}. \tag{8.35}$$

Then for all k, we have

$$\min\{\lambda_i^k, 1\} \leq \lambda_i^{k+1} \leq \max\{\lambda_i^k, 1\}, \quad i = 1, \ldots, n. \tag{8.36}$$

Moreover, the property (8.36) does not hold if the Broyden parameter ϕ_k is chosen outside the interval $[0, 1]$.

Let us discuss the significance of this result. If the eigenvalues λ_i^k of the matrix (8.35) are all 1, then the quasi-Newton approximation B_k is identical to the Hessian A of the quadratic objective function. This situation is the ideal one, so we should be hoping for these eigenvalues to be as close to 1 as possible. In fact, relation (8.36) tells us that the eigenvalues $\{\lambda_i^k\}$ converge monotonically (but not strictly monotonically) to 1. Suppose, for example, that at iteration k the smallest eigenvalue is $\lambda_1^k = 0.7$. Then (8.36) tells us that at the next iteration $\lambda_1^{k+1} \in [0.7, 1]$. We cannot be sure that this eigenvalue has actually gotten closer to 1, but it is reasonable to expect that it has. In contrast, the first eigenvalue can become smaller than 0.7 if we allow ϕ_k to be outside $[0, 1]$. Significantly, the result of Theorem 8.3 holds even if the linear searches are not exact.

Although Theorem 8.3 seems to suggest that the best update formulas belong to the restricted Broyden class, the situation is not at all clear. Some analysis and computational testing suggest that algorithms that allow ϕ_k to be negative (in a strictly controlled manner) may in fact be superior to the BFGS method. The SR1 formula is a case in point: It is a member of the Broyden class, obtained by setting

$$\phi_k = \frac{s_k^T y_k}{s_k^T y_k - s_k^T B_k s_k},$$

but it does not belong to the restricted Broyden class, because this value of ϕ_k may fall outside the interval $[0, 1]$.

We complete our discussion of the Broyden class by informally stating some of its main properties.

PROPERTIES OF THE BROYDEN CLASS

We have noted already that if B_k is positive definite, $y_k^T s_k > 0$, and $\phi_k \geq 0$, then B_{k+1} is also positive definite if a restricted Broyden class update, with $\phi_k \in [0, 1]$, is used. We would like to determine more precisely the range of values of ϕ_k that preserve positive definiteness.

The last term in (8.32) is a rank-one correction, which by the interlacing eigenvalue theorem (Theorem A.2) decreases the eigenvalues of the matrix when ϕ_k is negative. As we decrease ϕ_k, this matrix eventually becomes singular and then indefinite. A little computation shows that B_{k+1} is singular when ϕ_k has the value

$$\phi_k^c = \frac{1}{1 - \mu_k}, \tag{8.37}$$

where

$$\mu_k = \frac{(y_k^T B_k^{-1} y_k)(s_k^T B_k s_k)}{(y_k^T s_k)^2}. \tag{8.38}$$

By applying the Cauchy–Schwarz inequality to (8.38) we see that $\mu_k \geq 1$ and therefore $\phi_k^c \leq 0$. Hence, if the initial Hessian approximation B_0 is symmetric and positive definite, and if $s_k^T y_k > 0$ and $\phi_k > \phi_k^c$ for each k, then all the matrices B_k generated by Broyden's formula (8.32) remain symmetric and positive definite.

When the line search is exact, *all* methods in the Broyden class with $\phi_k \geq \phi_k^c$ generate the same sequence of iterates. This result applies to general nonlinear functions and is based on the observation that when all the line searches are exact, the directions generated by Broyden-class methods differ only in their lengths. The line searches identify the same minima along the chosen search direction, though the values of the line search parameter may differ because of the different scaling.

The Broyden class has several remarkable properties when applied with exact line searches to quadratic functions. We state some of these properties in the next theorem, whose proof is omitted.

Theorem 8.4.

Suppose that a method in the Broyden class is applied to a strongly convex quadratic function $f : \mathbb{R}^n \to \mathbb{R}$, where x_0 is the starting point and B_0 is any symmetric and positive definite matrix. Assume that α_k is the exact step length and that $\phi_k \geq \phi_k^c$ for all k. Then the following statements are true.

(i) The iterates converge to the solution in at most n iterations.

(ii) The secant equation is satisfied for all previous search directions, that is,

$$B_k s_j = y_j, \quad j = k-1, \ldots, 1.$$

(iii) If the starting matrix is $B_0 = I$, then the iterates are identical to those generated by the conjugate gradient method (see Chapter 5). In particular, the search directions are conjugate, that is,

$$s_i^T A s_j = 0 \quad \text{for } i \neq j,$$

where A is the Hessian of the quadratic function.

(iv) If n iterations are performed, we have $B_{n+1} = A$.

Note that parts (i), (ii), and (iv) of this result echo the statement and proof of Theorem 8.1, where similar results were derived for the SR1 update formula.

In fact, we can generalize Theorem 8.4 slightly: It continues to hold if the Hessian approximations remain nonsingular but not necessarily positive definite. (Hence, we could allow ϕ_k to be smaller than ϕ_k^c, provided that the chosen value did not produce a singular updated matrix.) We also can generalize point (iii) as follows: If the starting matrix B_0 is not the identity matrix, then the Broyden-class method is identical to the preconditioned conjugate gradient method that uses B_0 as preconditioner.

We conclude by commenting that results like Theorem 8.4 would appear to be mainly of theoretical interest, since the inexact line searches used in practical implementations of Broyden-class methods (and all other quasi-Newton methods) cause their performance to differ markedly. Nevertheless, this type of analysis guided most of the development of quasi-Newton methods.

8.4 CONVERGENCE ANALYSIS

In this section we present global and local convergence results for practical implementations of the BFGS and SR1 methods. We give many more details for BFGS because its analysis is more general and illuminating than that of SR1. The fact that the Hessian approximations evolve by means of updating formulas makes the analysis of quasi-Newton methods much more complex than that of steepest descent and Newton's method.

Although the BFGS and SR1 methods are known to be remarkably robust in practice, we will not be able to establish truly global convergence results for general nonlinear objective functions. That is, we cannot prove that the iterates of these quasi-Newton methods approach a stationary point of the problem from any starting point and any (suitable) initial Hessian approximation. In fact, it is not yet known if the algorithms enjoy such properties. In our analysis we will either assume that the objective function is convex or that the iterates satisfy certain properties. On the other hand, there are well known local, superlinear convergence results that are true under reasonable assumptions.

Throughout this section we use $\|\cdot\|$ to denote the Euclidean vector or matrix norm, and denote the Hessian matrix $\nabla^2 f(x)$ by $G(x)$.

GLOBAL CONVERGENCE OF THE BFGS METHOD

We study the global convergence of BFGS, with a practical line search, when applied to a smooth convex function from an arbitrary starting point x_0 and from any initial Hessian approximation B_0 that is symmetric and positive definite. We state our precise assumptions about the objective function formally:

Assumption 8.1.

(1) *The objective function f is twice continuously differentiable.*

(2) *The level set $\Omega = \{x \in \mathbf{R}^n : f(x) \leq f(x_0)\}$ is convex, and there exist positive constants m and M such that*

$$m\|z\|^2 \leq z^T G(x) z \leq M\|z\|^2 \tag{8.39}$$

for all $z \in \mathbf{R}^n$ and $x \in \Omega$.

The second part of this assumption implies that $G(x)$ is positive definite on Ω and that f has a unique minimizer x^* in Ω.

By using (8.12) and (8.39) we obtain

$$\frac{y_k^T s_k}{s_k^T s_k} = \frac{s_k^T \bar{G}_k s_k}{s_k^T s_k} \geq m, \tag{8.40}$$

where $\bar{G}_k$ is the average Hessian defined in (8.11). Assumption 8.1 implies that $\bar{G}_k$ is positive definite, so its square root is well-defined. Therefore, as in (8.21), we have by defining $z_k = \bar{G}_k^{1/2} s_k$ that

$$\frac{y_k^T y_k}{y_k^T s_k} = \frac{z_k^T \bar{G}_k z_k}{z_k^T z_k} \le M. \tag{8.41}$$

We are now ready to present the global convergence result for the BFGS method. As in Section 3.2, we could try to establish a bound on the condition number of the Hessian approximations B_k, but this approach does not seem to be possible. Instead, we will introduce two new tools in the analysis, the trace and determinant, to estimate the size of the largest and smallest eigenvalues of the Hessian approximations. The trace of a matrix (denoted by trace($\cdot$)) is the sum of its eigenvalues, while the determinant (denoted by det($\cdot$)) is the product of the eigenvalues; see the Appendix for a brief discussion of their properties.

Theorem 8.5.

Let B_0 be any symmetric positive definite initial matrix, and let x_0 be a starting point for which Assumption 8.1 is satisfied. Then the sequence $\{x_k\}$ generated by Algorithm 8.1 converges to the minimizer x^* of f.

PROOF. Let us define

$$m_k = \frac{y_k^T s_k}{s_k^T s_k}, \qquad M_k = \frac{y_k^T y_k}{y_k^T s_k}, \tag{8.42}$$

and note from (8.40) and (8.41) that

$$m_k \ge m, \qquad M_k \le M. \tag{8.43}$$

By computing the trace of the BFGS approximation (8.19), we obtain that

$$\text{trace}(B_{k+1}) = \text{trace}(B_k) - \frac{\|B_k s_k\|^2}{s_k^T B_k s_k} + \frac{\|y_k\|^2}{y_k^T s_k} \tag{8.44}$$

(see Exercise 10). We can also show (Exercise 9) that

$$\det(B_{k+1}) = \det(B_k) \frac{y_k^T s_k}{s_k^T B_k s_k}. \tag{8.45}$$

Let us also define

$$\cos \theta_k = \frac{s_k^T B_k s_k}{\|s_k\| \|B_k s_k\|}, \qquad q_k = \frac{s_k^T B_k s_k}{s_k^T s_k}, \tag{8.46}$$

so that θ_k is the angle between s_k and $B_k s_k$. We then obtain that

$$\frac{\|B_k s_k\|^2}{s_k^T B_k s_k} = \frac{\|B_k s_k\|^2 \|s_k\|^2}{(s_k^T B_k s_k)^2} \frac{s_k^T B_k s_k}{\|s_k\|^2} = \frac{q_k}{\cos^2 \theta_k}. \tag{8.47}$$

In addition, we have from (8.42) that

$$\det(B_{k+1}) = \det(B_k) \frac{y_k^T s_k}{s_k^T s_k} \frac{s_k^T s_k}{s_k^T B_k s_k} = \det(B_k) \frac{m_k}{q_k}. \tag{8.48}$$

We now combine the trace and determinant by introducing the following function of a positive definite matrix B:

$$\psi(B) = \text{trace}(B) - \ln(\det(B)), \tag{8.49}$$

where $\ln(\cdot)$ denotes the natural logarithm. It is not difficult to show that $\psi(B) > 0$; see Exercise 8. By using (8.42) and (8.44)–(8.49), we have that

$$\psi(B_{k+1}) = \psi(B_k) + M_k - \frac{q_k}{\cos^2 \theta_k} - \ln(\det(B_k)) - \ln m_k + \ln q_k$$
$$= \psi(B_k) + (M_k - \ln m_k - 1)$$
$$+ \left[1 - \frac{q_k}{\cos^2 \theta_k} + \ln \frac{q_k}{\cos^2 \theta_k} \right] + \ln \cos^2 \theta_k. \tag{8.50}$$

Now, since the function $h(t) = 1 - t + \ln t \leq 0$ is nonpositive for all $t > 0$ (see Exercise 7), the term inside the square brackets is nonpositive, and thus from (8.43) and (8.50) we have

$$0 < \psi(B_{k+1}) \leq \psi(B_1) + ck + \sum_{j=1}^{k} \ln \cos^2 \theta_j, \tag{8.51}$$

where we can assume the constant $c = M - \ln m - 1$ to be positive, without loss of generality.

We now relate these expressions to the results given in Section 3.2. Note from the form $s_k = -\alpha_k B_k^{-1} \nabla f_k$ of the quasi-Newton iteration that $\cos \theta_k$ defined by (8.46) is the angle between the steepest descent direction and the search direction, which plays a crucial role in the global convergence theory of Chapter 3. From (3.21), (3.22) we know that the sequence $\|\nabla f_k\|$ generated by the line search algorithm is bounded away from zero only if $\cos \theta_j \to 0$.

Let us then proceed by contradiction and assume that $\cos \theta_j \to 0$. Then there exists $k_1 > 0$ such that for all $j > k_1$, we have

$$\ln \cos^2 \theta_j < -2c,$$

where c is the constant defined above. Using this inequality in (8.51) we find the following relations to be true for all $k > k_1$:

$$0 < \psi(B_1) + ck + \sum_{j=1}^{k_1} \ln \cos^2 \theta_j + \sum_{j=k_1+1}^{k} (-2c)$$

$$= \psi(B_1) + \sum_{j=1}^{k_1} \ln \cos^2 \theta_j + 2ck_1 - ck.$$

However, the right-hand-side is negative for large k, giving a contradiction. Therefore, there exists a subsequence of indices $\{j_k\}$ such that $\{\cos \theta_{j_k}\} \geq \delta > 0$. By Zoutendijk's result (3.14) this limit implies that $\liminf \|\nabla f_k\| \to 0$. Since the problem is strongly convex, the latter limit is enough to prove that $x_k \to x^*$. $\square$

Theorem 8.5 has been generalized to the entire restricted Broyden class, *except for the DFP method*. In other words, Theorem 8.5 can be shown to hold for all $\phi_k \in [0, 1)$ in (8.32), but the argument seems to break down as ϕ_k approaches 1 because some of the self-correcting properties of the update are weakened considerably.

An extension of the analysis just given shows that the rate of convergence of the iterates is linear. In particular, we can show that the sequence $\|x_k - x^*\|$ converges to zero rapidly enough that

$$\sum_{k=1}^{\infty} \|x_k - x^*\| < \infty. \qquad (8.52)$$

We will not prove this claim, but rather establish that if (8.52) holds, then the rate of convergence is actually superlinear.

SUPERLINEAR CONVERGENCE OF BFGS

The analysis of this section makes use of the Dennis and Moré characterization (3.31) of superlinear convergence. It applies to general nonlinear—not just convex—objective functions. For the results that follow we need to make an additional assumption.

Assumption 8.2.

The Hessian matrix G is Lipschitz continuous at x^*, that is,

$$\|G(x) - G(x^*)\| \leq L\|x - x^*\|,$$

for all x near x^*, where L is a positive constant.

8.4. CONVERGENCE ANALYSIS

We start by introducing the quantities

$$\tilde{s}_k = G_*^{1/2} s_k, \qquad \tilde{y}_k = G_*^{-1/2} y_k, \qquad \tilde{B}_k = G_*^{-1/2} B_k G_*^{-1/2},$$

where $G_* = G(x^*)$ and x^* is a minimizer of f. As in (8.46), we define

$$\cos \tilde{\theta}_k = \frac{\tilde{s}_k^T \tilde{B}_k \tilde{s}_k}{\|\tilde{s}_k\| \|\tilde{B}_k \tilde{s}_k\|}, \qquad \tilde{q}_k = \frac{\tilde{s}_k^T \tilde{B}_k \tilde{s}_k}{\|\tilde{s}_k\|^2},$$

while we echo (8.42) and (8.43) in defining

$$\tilde{M}_k = \frac{\|\tilde{y}_k\|^2}{\tilde{y}_k^T \tilde{s}_k}, \qquad \tilde{m}_k = \frac{\tilde{y}_k^T \tilde{s}_k}{\tilde{s}_k^T \tilde{s}_k}.$$

By pre- and postmultiplying the BFGS update formula (8.19) by $G_*^{-1/2}$ and grouping terms appropriately, we obtain

$$\tilde{B}_{k+1} = \tilde{B}_k - \frac{\tilde{B}_k \tilde{s}_k \tilde{s}_k^T \tilde{B}_k}{\tilde{s}_k^T \tilde{B}_k \tilde{s}_k} + \frac{\tilde{y}_k \tilde{y}_k^T}{\tilde{y}_k^T \tilde{s}_k}.$$

Since this expression has precisely the same form as the BFGS formula (8.19), it follows from the argument leading to (8.50) that

$$\psi(\tilde{B}_{k+1}) = \psi(\tilde{B}_k) + (\tilde{M}_k - \ln \tilde{m}_k - 1) \\ + \left[1 - \frac{\tilde{q}_k}{\cos^2 \tilde{\theta}_k} + \ln \frac{\tilde{q}_k}{\cos^2 \tilde{\theta}_k} \right] \\ + \ln \cos^2 \tilde{\theta}_k. \qquad (8.53)$$

Recalling (8.12), we have that

$$y_k - G_* s_k = (\bar{G}_k - G_*) s_k,$$

and thus

$$\tilde{y}_k - \tilde{s}_k = G_*^{-1/2} (\bar{G}_k - G_*) G_*^{-1/2} \tilde{s}_k.$$

By Assumption 8.2, and recalling the definition (8.11), we have

$$\|\tilde{y}_k - \tilde{s}_k\| \le \|G_*^{-1/2}\|^2 \|\tilde{s}_k\| \|\bar{G}_k - G_*\| \le \|G_*^{-1/2}\|^2 \|\tilde{s}_k\| L \epsilon_k,$$

216 CHAPTER 8. QUASI-NEWTON METHODS

where ϵ_k is defined by

$$\epsilon_k = \max\{\|x_{k+1} - x^*\|, \|x_k - x^*\|\}.$$

We have thus shown that

$$\frac{\|\tilde{y}_k - \tilde{s}_k\|}{\|\tilde{s}_k\|} \leq \bar{c}\epsilon_k, \tag{8.54}$$

for some positive constant $\bar{c}$. This inequality and (8.52) play an important role in superlinear convergence, as we now show.

Theorem 8.6.

Suppose that f is twice continuously differentiable and that the iterates generated by the BFGS algorithm converge to a minimizer x^* at which Assumption 8.2 holds. Suppose also that (8.52) holds. Then x_k converges to x^* at a superlinear rate.

PROOF. From (8.54), we have from the triangle inequality (A.36a) that

$$\|\tilde{y}_k\| - \|\tilde{s}_k\| \leq \bar{c}\epsilon_k \|\tilde{s}_k\|, \qquad \|\tilde{s}_k\| - \|\tilde{y}_k\| \leq \bar{c}\epsilon_k \|\tilde{s}_k\|,$$

so that

$$(1 - \bar{c}\epsilon_k)\|\tilde{s}_k\| \leq \|\tilde{y}_k\| \leq (1 + \bar{c}\epsilon_k)\|\tilde{s}_k\|. \tag{8.55}$$

By squaring (8.54) and using (8.55), we obtain

$$(1 - \bar{c}\epsilon_k)^2 \|\tilde{s}_k\|^2 - 2\tilde{y}_k^T \tilde{s}_k + \|\tilde{s}_k\|^2 \leq \|\tilde{y}_k\|^2 - 2\tilde{y}_k^T \tilde{s}_k + \|\tilde{s}_k\|^2 \leq \bar{c}^2 \epsilon_k^2 \|\tilde{s}_k\|^2,$$

and therefore

$$2\tilde{y}_k^T \tilde{s}_k \geq (1 - 2\bar{c}\epsilon_k + \bar{c}^2\epsilon_k^2 + 1 - \bar{c}^2\epsilon_k^2)\|\tilde{s}_k\|^2 = 2(1 - \bar{c}\epsilon_k)\|\tilde{s}_k\|^2.$$

It follows from the definition of $\tilde{m}_k$ that

$$\tilde{m}_k = \frac{\tilde{y}_k^T \tilde{s}_k}{\|\tilde{s}_k\|^2} \geq 1 - \bar{c}\epsilon_k. \tag{8.56}$$

By combining (8.55) and (8.56), we obtain also that

$$\tilde{M}_k = \frac{\|\tilde{y}_k\|^2}{\tilde{y}_k^T \tilde{s}_k} \leq \frac{1 + \bar{c}\epsilon_k}{1 - \bar{c}\epsilon_k}. \tag{8.57}$$

Since $x_k \to x^*$, we have that $\epsilon_k \to 0$, and thus by (8.57) there exists a positive constant $c > \bar{c}$ such that the following inequalities hold for all sufficiently large k:

$$\tilde{M}_k \leq 1 + \frac{2\bar{c}}{1 - \bar{c}\epsilon_k}\epsilon_k \leq 1 + c\epsilon_k. \tag{8.58}$$

We again make use of the nonpositiveness of the function $h(t) = 1 - t + \ln t$. Therefore, we have

$$\frac{-x}{1-x} - \ln(1-x) = h\left(\frac{1}{1-x}\right) \leq 0.$$

Now, for k large enough we can assume that $\bar{c}\epsilon_k < \frac{1}{2}$, and therefore

$$\ln(1 - \bar{c}\epsilon_k) \geq \frac{-\bar{c}\epsilon_k}{1 - \bar{c}\epsilon_k} \geq -2\bar{c}\epsilon_k.$$

This relation and (8.56) imply that for sufficiently large k, we have

$$\ln \tilde{m}_k \geq \ln(1 - \bar{c}\epsilon_k) \geq -2\bar{c}\epsilon_k > -2c\epsilon_k. \tag{8.59}$$

We can now deduce from (8.53), (8.58), and (8.59) that

$$0 < \psi(\tilde{B}_{k+1}) \leq \psi(\tilde{B}_k) + 3c\epsilon_k + \ln \cos^2 \tilde{\theta}_k + \left[1 - \frac{\tilde{q}_k}{\cos^2 \tilde{\theta}_k} + \ln \frac{\tilde{q}_k}{\cos^2 \tilde{\theta}_k}\right]. \tag{8.60}$$

By summing this expression and making use of (8.52) we have that

$$\sum_{j=0}^{\infty} \left(\ln \frac{1}{\cos^2 \tilde{\theta}_j} - \left[1 - \frac{\tilde{q}_j}{\cos^2 \tilde{\theta}_j} + \ln \frac{\tilde{q}_j}{\cos^2 \tilde{\theta}_j}\right]\right) \leq \psi(\tilde{B}_0) + 3c \sum_{j=0}^{\infty} \epsilon_j < +\infty.$$

Since the term in the square brackets is nonpositive, and since $\ln\left(1/\cos^2 \tilde{\theta}_j\right) \geq 0$ for all j, we obtain the two limits

$$\lim_{j \to \infty} \ln \frac{1}{\cos^2 \tilde{\theta}_j} = 0, \qquad \lim_{j \to \infty}\left(1 - \frac{\tilde{q}_j}{\cos^2 \tilde{\theta}_j} + \ln \frac{\tilde{q}_j}{\cos^2 \tilde{\theta}_j}\right) = 0,$$

which imply that

$$\lim_{j \to \infty} \cos \tilde{\theta}_j = 1, \qquad \lim_{j \to \infty} \tilde{q}_j = 1. \tag{8.61}$$

The essence of the result has now been proven; we need only to interpret these limits in terms of the Dennis–Moré characterization of superlinear convergence.

Recalling (8.47), we have

$$\frac{\|G_*^{-1/2}(B_k - G_*)s_k\|^2}{\|G_*^{1/2}s_k\|^2} = \frac{\|(\tilde{B}_k - I)\tilde{s}_k\|^2}{\|\tilde{s}_k\|^2}$$

$$= \frac{\|\tilde{B}_k \tilde{s}_k\|^2 - 2\tilde{s}_k^T \tilde{B}_k \tilde{s}_k + \tilde{s}_k^T \tilde{s}_k}{\tilde{s}_k^T \tilde{s}_k}$$

$$= \frac{\tilde{q}_k^2}{\cos\tilde{\theta}_k^2} - 2\tilde{q}_k + 1.$$

Since by (8.61) the right-hand-side converges to 0, we conclude that

$$\lim_{k \to \infty} \frac{\|(B_k - G_*)s_k\|}{\|s_k\|} = 0.$$

The limit (3.31) and Theorem 3.5 imply that the unit step length $\alpha_k = 1$ will satisfy the Wolfe conditions near the solution, and hence that the rate of convergence is superlinear. □

CONVERGENCE ANALYSIS OF THE SR1 METHOD

The convergence properties of the SR1 method are not as well understood as those of the BFGS method. No global results like Theorem 8.5 or local superlinear results like Theorem 8.6 have been established, except the results for quadratic functions discussed earlier. There is, however, an interesting result for the trust-region SR1 algorithm, Algorithm 8.2. It states that when the objective function has a unique stationary point and the condition (8.26) holds at every step (so that the SR1 update is never skipped) and the Hessian approximations B_k are bounded above, then the iterates converge to x^* at an $(n+1)$-step superlinear rate. The result does not require exact solution of the trust-region subproblem (8.27).

We state the result formally as follows.

Theorem 8.7.

Suppose that the iterates x_k are generated by Algorithm 8.2. Suppose also that the following conditions hold:

(c1) *The sequence of iterates does not terminate, but remains in a closed, bounded, convex set D, on which the function f is twice continuously differentiable, and in which f has a unique stationary point x^*;*

(c2) *the Hessian $\nabla^2 f(x^*)$ is positive definite, and $\nabla^2 f(x)$ is Lipschitz continuous in a neighborhood of x^*;*

(c3) the sequence of matrices $\{B_k\}$ is bounded in norm;

(c4) condition (8.26) holds at every iteration, where r is some constant in $(0, 1)$.

Then $\lim_{k \to \infty} x_k = x^*$, and we have that

$$\lim_{k \to \infty} \frac{\|x_{k+n+1} - x^*\|}{\|x_k - x^*\|} = 0.$$

Note that the BFGS method does not require the boundedness assumption (c3) to hold. As we have mentioned already, the SR1 update does not necessarily maintain positive definiteness of the Hessian approximations B_k. In practice, B_k may be indefinite at any iteration, which means that the trust region bound may continue to be active for arbitrarily large k. Interestingly, however, it can be shown that the SR1 Hessian approximations tend to be positive definite most of the time. The precise result is that

$$\lim_{k \to \infty} \frac{\text{number of indices } j = 1, 2, \ldots, k \text{ for which } B_j \text{ is positive semidefinite}}{k} = 1,$$

under the assumptions of Theorem 8.7. This result holds regardless of whether the initial Hessian approximation is positive definite or not.

NOTES AND REFERENCES

For a comprehensive treatment of quasi-Newton methods see Dennis and Schnabel [69], Dennis and Moré [68], and Fletcher [83]. A formula for updating the Cholesky factors of the BFGS matrices is given in Dennis and Schnabel [69].

Several safeguards and modifications of the SR1 method have been proposed, but the condition (8.26) is favored in the light of the analysis of Conn, Gould, and Toint [52]. Computational experiments by Conn, Gould, and Toint [51, 54] and Khalfan, Byrd, and Schnabel [143], using both line search and trust-region approaches, indicate that the SR1 method appears to be competitive with the BFGS method. The proof of Theorem 8.7 is given in Byrd, Khalfan, and Schnabel [35].

A study of the convergence of BFGS matrices for nonlinear problems can be found in Ge and Powell [100] and Boggs and Tolle [22]; however, the results are not as satisfactory as for SR1 updating.

The global convergence of the BFGS method was established by Powell [201]. This result was extended to the restricted Broyden class, except for DFP, by Byrd, Nocedal, and Yuan [38]. For a discussion of the self-correcting properties of quasi-Newton methods see Nocedal [184]. Most of the early analysis of quasi-Newton methods was based on the *bounded deterioration* principle. This is a tool for the local analysis that quantifies the worst-case behavior of quasi-Newton updating. Assuming that the starting point is sufficiently close to the solution x^* and that the initial Hessian approximation is sufficiently close to $\nabla^2 f(x^*)$, one can use the bounded deterioration bounds to prove that the iteration cannot stray away from the solution. This property can then be used to show that the quality of the quasi-Newton approximations is good enough to yield superlinear convergence. For details, see Dennis and Moré [68] or Dennis and Schnabel [69].

EXERCISES

✎ 8.1

(a) Show that if f is strongly convex, then (8.7) holds for any vectors x_k and x_{k+1}.

(b) Give an example of a function of one variable satisfying $g(0) = -1$ and $g(1) = -\frac{1}{4}$ and show that (8.7) does not hold in this case.

✎ 8.2 Show that the second strong Wolfe condition (3.7b) implies the curvature condition (8.7).

✎ 8.3 Verify that (8.19) and (8.16) are inverses of each other.

✎ 8.4 Use the Sherman–Morrison formula (A.55) to show that (8.24) is the inverse of (8.25).

✎ 8.5 Prove the statements (ii) and (iii) given in the paragraph following (8.25).

✎ 8.6 The square root of a matrix A is a matrix $A^{1/2}$ such that $A^{1/2} A^{1/2} = A$. Show that any symmetric positive definite matrix A has a square root, and that this square root is itself symmetric and positive definite. (Hint: Use the factorization $A = U D U^T$ (A.46), where U is orthogonal and D is diagonal with positive diagonal elements.)

✎ 8.7 Define $h(t) = 1 - t + \ln t$, and note that $h'(t) = -1 + 1/t$, $h''(t) = -1/t^2 < 0$, $h(1) = 0$, and $h'(1) = 0$. Show that $h(t) \le 0$ for all $t > 0$.

✎ 8.8 Denote the eigenvalues of the positive definite matrix B by $\lambda_1, \lambda_2, \ldots, \lambda_n$, where $0 < \lambda_1 \le \lambda_2 \le \cdots \le \lambda_n$. Show that the ψ function defined in (8.49) can be written as

$$\psi(B) = \sum_{i=1}^{n} (\lambda_i - \ln \lambda_i).$$

Use this form to show that $\psi(B) > 0$.

✎ 8.9 The object of this exercise is to prove (8.43).

(a) First show that $\det(I + xy^T) = 1 + y^T x$, where x and y are n-vectors. Hint: Assuming that $x \ne 0$, we can find vectors $w_1, w_2, \ldots, w_{n-1}$ such that the matrix Q defined by

$$Q = [x, w_1, \ldots, w_{n-1}]$$

is nonsingular and $x = Qe_1$, where $e_1 = (1, 0, \ldots, 0)^T$. If we define

$$y^T Q = (z_1, \ldots, z_n),$$

then

$$z_1 = y^T Q e_1 = y^T Q(Q^{-1}x) = y^T x,$$

and

$$\det(I + xy^T) = \det(Q^{-1}(I + xy^T)Q) = \det(I + e_1 y^T Q).$$

(b) Use a similar technique to prove that

$$\det(I + xy^T + uv^T) = (1 + y^T x)(1 + v^T u) - (x^T v)(y^T u).$$

(c) Now use this relation to establish (8.45).

✏ **8.10**
Use the properties of the trace of a symmetric matrix and the formula (8.19) to prove (8.44).

✏ **8.11** Show that if f satisfies Assumption 8.1 and if the sequence of gradients satisfies $\liminf \|\nabla f_k\| = 0$, then the whole sequence of iterates x converges to the solution x^*.

CHAPTER 9

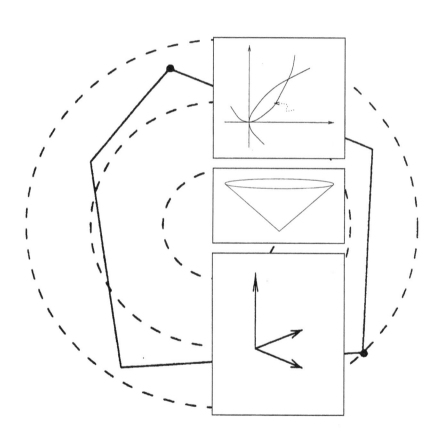

Large-Scale Quasi-Newton and Partially Separable Optimization

The quasi-Newton methods of Chapter 8 are not directly applicable to large optimization problems because their approximations to the Hessian or its inverse are usually dense. The storage and computational requirements grow in proportion to n^2, and become excessive for large n. We can, however, modify and extend quasi-Newton methods in several ways to make them suitable for large problems.

The first such approach—limited-memory quasi-Newton methods—modifies the techniques described in Chapter 8 to obtain Hessian approximations that can be stored compactly in just a few vectors of length n, where n is the number of unknowns in the problem. These methods are fairly robust, inexpensive, and easy to implement, but they do not converge rapidly. A second approach is to define quasi-Newton approximations that preserve sparsity, for example by mimicking the sparsity pattern of the true Hessian. We mention but do not dwell on these sparse quasi-Newton methods, since they have not proved to be particularly effective.

A third approach is based on the observation that many (perhaps most) large-scale objective functions possess a structural property known as *partial separability*. Effective Newton and quasi-Newton methods that exploit this property have been developed. Such methods usually converge rapidly and are robust, but they require detailed information about the objective function, which can be difficult to obtain in some applications.

9.1 LIMITED-MEMORY BFGS

Limited-memory quasi-Newton methods are useful for solving large problems whose Hessian matrices cannot be computed at a reasonable cost or are too dense to be manipulated easily. These methods maintain simple and compact approximations of Hessian matrices: Instead of storing fully dense $n \times n$ approximations, they save just a few vectors of length n that represent the approximations implicitly. Despite these modest storage requirements, they often yield an acceptable (albeit linear) rate of convergence. Various limited-memory methods have been proposed; we will focus mainly on an algorithm known as L-BFGS, which as its name suggests, is based on the BFGS updating formula. The main idea of this method is to use curvature information from only the most recent iterations to construct the Hessian approximation. Curvature information from earlier iterations, which is less likely to be relevant to the actual behavior of the Hessian at the current iteration, is discarded in the interests of saving storage.

We begin our description of the L-BFGS method by recalling its parent, the BFGS method, which was described in Algorithm 8.1. Each step of the BFGS method has the form

$$x_{k+1} = x_k - \alpha_k H_k \nabla f_k, \qquad k = 0, 1, 2, \ldots, \tag{9.1}$$

where α_k is the step length, and H_k is updated at every iteration by means of the formula

$$H_{k+1} = V_k^T H_k V_k + \rho_k s_k s_k^T \tag{9.2}$$

(see (8.16)), where

$$\rho_k = \frac{1}{y_k^T s_k}, \qquad V_k = I - \rho_k y_k s_k^T, \tag{9.3}$$

and

$$s_k = x_{k+1} - x_k, \qquad y_k = \nabla f_{k+1} - \nabla f_k. \tag{9.4}$$

We say that the matrix H_{k+1} is obtained by updating H_k using the pair $\{s_k, y_k\}$.

The inverse Hessian approximation H_k will generally be dense, so that the cost of storing and manipulating it is prohibitive when the number of variables is large. To circumvent this problem, we store a *modified* version of H_k implicitly, by storing a certain number (say m) of the vector pairs $\{s_i, y_i\}$ that are used in the formulae (9.2)–(9.4). The product $H_k \nabla f_k$ can be obtained by performing a sequence of inner products and vector summations involving

∇f_k and the pairs $\{s_i, y_i\}$. After the new iterate is computed, the oldest vector pair in the set of pairs $\{s_i, y_i\}$ is deleted and replaced by the new pair $\{s_k, y_k\}$ obtained from the current step (9.4). In this way, the set of vector pairs includes curvature information from the m most recent iterations. Practical experience has shown that modest values of m (between 3 and 20, say) often produce satisfactory results. Apart from the modified matrix updating strategy and a modified technique for handling the initial Hessian approximation (described below), the implementation of L-BFGS is identical to that of the standard BFGS method given in Algorithm 8.1. In particular, the same line search strategy can be used.

We now describe the updating process in a little more detail. At iteration k, the current iterate is x_k and the set of vector pairs contains $\{s_i, y_i\}$ for $i = k - m, \ldots, k - 1$. We first choose some initial Hessian approximation H_k^0 (in contrast to the standard BFGS iteration, this initial approximation is allowed to vary from iteration to iteration), and find by repeated application of the formula (9.2) that the L-BFGS approximation H_k satisfies the following formula:

$$
\begin{aligned}
H_k = & \left(V_{k-1}^T \cdots V_{k-m}^T\right) H_k^0 \left(V_{k-m} \cdots V_{k-1}\right) \\
& + \rho_{k-m} \left(V_{k-1}^T \cdots V_{k-m+1}^T\right) s_{k-m} s_{k-m}^T \left(V_{k-m+1} \cdots V_{k-1}\right) \\
& + \rho_{k-m+1} \left(V_{k-1}^T \cdots V_{k-m+2}^T\right) s_{k-m+1} s_{k-m+1}^T \left(V_{k-m+2} \cdots V_{k-1}\right) \\
& + \cdots \\
& + \rho_{k-1} s_{k-1} s_{k-1}^T.
\end{aligned}
\tag{9.5}
$$

From this expression we can derive a recursive procedure to compute the product $H_k \nabla f_k$ efficiently.

Algorithm 9.1 (L-BFGS two-loop recursion).
   $q \leftarrow \nabla f_k$;
   for $i = k - 1, k - 2, \ldots, k - m$
      $\alpha_i \leftarrow \rho_i s_i^T q$;
      $q \leftarrow q - \alpha_i y_i$;
   end (for)
   $r \leftarrow H_k^0 q$;
   for $i = k - m, k - m + 1, \ldots, k - 1$
      $\beta \leftarrow \rho_i y_i^T r$;
      $r \leftarrow r + s_i (\alpha_i - \beta)$
   end (for)
   stop with result $H_k \nabla f_k = r$.

Without considering the multiplication $H_k^0 q$, the two-loop recursion scheme requires $4mn$ multiplications; if H_k^0 is diagonal, then n additional multiplications are needed. Apart from being inexpensive, this recursion has the advantage that the multiplication by the initial matrix H_k^0 is isolated from the rest of the computations, allowing this matrix to be chosen freely and to vary between iterations. We may even use an implicit choice of H_k^0 by defining some initial approximation B_k^0 to the Hessian (not its inverse), and obtaining r by solving the system $B_k^0 r = q$.

A method for choosing H_k^0 that has proved to be effective in practice is to set $H_k^0 = \gamma_k I$, where

$$\gamma_k = \frac{s_{k-1}^T y_{k-1}}{y_{k-1}^T y_{k-1}}. \tag{9.6}$$

As discussed in Chapter 8, γ_k is the scaling factor that attempts to estimate the size of the true Hessian matrix along the most recent search direction (see (8.21)). This choice helps to ensure that the search direction p_k is well-scaled, and as a result the step length $\alpha_k = 1$ is accepted in most iterations. As discussed in Chapter 8, it is important that the line search be based on the Wolfe conditions (3.6) or strong Wolfe conditions (3.7), so that BFGS updating is stable.

The limited-memory BFGS algorithm can be stated formally as follows.

Algorithm 9.2 (L-BFGS).
Choose starting point x_0, integer $m > 0$;
$k \leftarrow 0$;
repeat
    Choose H_k^0 (for example, by using (9.6));
    Compute $p_k \leftarrow -H_k \nabla f_k$ from Algorithm 9.1;
    Compute $x_{k+1} \leftarrow x_k + \alpha_k p_k$, where α_k is chosen to
        satisfy the Wolfe conditions;
    if $k > m$
        Discard the vector pair $\{s_{k-m}, y_{k-m}\}$ from storage;
    Compute and save $s_k \leftarrow x_{k+1} - x_k$, $y_k = \nabla f_{k+1} - \nabla f_k$;
    $k \leftarrow k + 1$;
until convergence.

During its first $m - 1$ iterations, Algorithm 9.2 is equivalent to the BFGS algorithm of Chapter 8 if the initial matrix H_0 is the same in both methods, and if L-BFGS chooses $H_k^0 = H_0$ at each iteration. In fact, we could reimplement the standard BFGS method by setting m to some large value in Algorithm 9.2 (larger than the number of iterations required to find the solution). However, as m approaches n (specifically, $m > n/2$), this approach would be more costly in terms of computer time and storage than the approach of Algorithm BFGS.

The strategy of keeping the m most recent correction pairs $\{s_i, y_i\}$ works well in practice, and no other strategy has yet proved to be consistently better. Other criteria may be considered, however—for example, one in which we maintain the set of correction pairs for which the matrix formed by the s_i components has the best conditioning, thus tending to avoid sets of vector pairs in which some of the s_i's are nearly collinear.

Table 9.1 Performance of Algorithm 9.2.

Problem	n	L-BFGS $m=3$		L-BFGS $m=5$		L-BFGS $m=17$		L-BFGS $m=29$	
		nfg	time	nfg	time	nfg	time	nfg	time
DIXMAANL	1500	146	16.5	134	17.4	120	28.2	125	44.4
EIGENALS	110	821	21.5	569	15.7	363	16.2	168	12.5
FREUROTH	1000	>999	—	>999	—	69	8.1	38	6.3
TRIDIA	1000	876	46.6	611	41.4	531	84.6	462	127.1

Table 9.1 presents a few results illustrating the behavior of Algorithm 9.2 for various levels of memory m. It gives the number of function and gradient evaluations (nfg) and the total CPU time. The test problems are taken from the CUTE collection [25], the number of variables is indicated by n, and the termination criterion $\|\nabla f_k\| \leq 10^{-5}$ is used. The table shows that the algorithm tends to be less robust when m is small. As the amount of storage increases, the number of function evaluations tends to decrease, but since the cost of each iteration increases with the amount of storage, the best CPU time is often obtained for small values of m. Clearly, the optimal choice of m is problem-dependent.

Algorithm 9.2 is often the approach of choice for large problems in which the true Hessian is not sparse, because some rival algorithms become inefficient. In particular, a Newton method, in which the true Hessian is computed and factorized, is not practical in such circumstances. The L-BFGS approach may even outperform Hessian-free Newton methods such as Algorithm Newton–CG discussed in Chapter 6, in which Hessian–vector products are calculated by finite differences or automatic differentiation. Computational experience to date also indicates that L-BFGS is more rapid and robust than nonlinear conjugate gradient methods.

When the Hessian is dense but the objective function has partially separable structure (see Section 9.4), the methods that exploit this structure normally outperform L-BFGS by a wide margin in terms of function evaluations. In terms of computing time, however, L-BFGS can be more efficient due to the low cost of its iteration.

The main weaknesses of the L-BFGS method are that it often converges slowly, which usually leads to a relatively large number of function evaluations, and that it is inefficient on highly ill-conditioned problems—specifically, on problems where the Hessian matrix contains a wide distribution of eigenvalues.

RELATIONSHIP WITH CONJUGATE GRADIENT METHODS

Limited-memory methods evolved gradually as an attempt to improve nonlinear conjugate gradient methods, and early implementations resembled conjugate gradient methods more than quasi-Newton methods. The relationship between the two classes is the basis of a *memoryless BFGS iteration*, which we now outline.

We start by considering the Hestenes–Stiefel form of the nonlinear conjugate gradient method (5.45). Recalling that $s_k = \alpha_k p_k$, we have that the search direction for this method is given by

$$p_{k+1} = -\nabla f_{k+1} + \frac{\nabla f_{k+1}^T y_k}{y_k^T p_k} p_k = -\left(I - \frac{s_k y_k^T}{y_k^T s_k}\right) \nabla f_{k+1} \equiv -\hat{H}_{k+1} \nabla f_{k+1}. \qquad (9.7)$$

This formula resembles a quasi-Newton iteration, but the matrix $\hat{H}_{k+1}$ is neither symmetric nor positive definite. We could symmetrize it as $\hat{H}_{k+1}^T \hat{H}_{k+1}$, but this matrix does not satisfy the secant equation $\hat{H}_{k+1} y_k = s_k$, and is, in any case, singular. An iteration matrix that is symmetric, positive definite, and satisfies the secant equation is given by

$$H_{k+1} = \left(I - \frac{s_k y_k^T}{y_k^T s_k}\right)\left(I - \frac{y_k s_k^T}{y_k^T s_k}\right) + \frac{s_k s_k^T}{y_k^T s_k}. \qquad (9.8)$$

Interestingly enough, this matrix is exactly the one obtained by applying a single BFGS update (9.2) to the identity matrix. By using the notation of (9.3) and (9.4), we can rewrite (9.8) as

$$H_{k+1} = V_k^T V_k + \rho_k s_k s_k^T.$$

Hence, an algorithm whose search direction is given by $p_{k+1} = -H_{k+1} \nabla f_{k+1}$, with H_{k+1} defined by (9.8), can be thought of as a "memoryless" BFGS method in which the previous Hessian approximation is always reset to the identity matrix before updating, and where only the most recent correction pair (s_k, y_k) is kept at every iteration. Alternatively, we can view the method as a variant of Algorithm 9.2 in which $m = 1$ and $H_k^0 = I$ at each iteration. In this sense, L-BFGS is a natural extension of the memoryless method that uses extra memory to store additional curvature information.

This discussion suggests that for any quasi-Newton method (in particular, for every formula in the Broyden class) there is a memoryless counterpart in which the current Hessian approximation is always reset to the identity in the update formula. The form and storage requirements of these methods are similar to those of the nonlinear conjugate gradient methods discussed in Chapter 5. One connection can be seen if we consider the memoryless BFGS formula (9.8) in conjunction with an exact line search, for which $\nabla f_{k+1}^T p_k = 0$ for all k. We then obtain

$$p_{k+1} = -H_{k+1} \nabla f_{k+1} = -\nabla f_{k+1} + \frac{\nabla f_{k+1}^T y_k}{y_k^T p_k} p_k, \qquad (9.9)$$

which is none other than the Hestenes–Stiefel conjugate gradient method. Moreover, it is easy to verify that when $\nabla f_{k+1}^T p_k = 0$, the Hestenes–Stiefel formula reduces to the Polak–Ribière formula (5.43). Even though the assumption of exact line searches is unrealistic, it is intriguing that the BFGS formula, which is considered to be the most effective quasi-Newton update, is related in this way to the Polak–Ribiére and Hestenes–Stiefel methods, which are the most efficient nonlinear conjugate gradient methods.

We summarize the storage requirements of various conjugate gradient, memoryless quasi-Newton, and limited-memory methods in the following table.

Method	Storage
Fletcher–Reeves	$3n$
Polak–Ribière	$4n$
Harwell VA14	$6n$
CONMIN	$7n$
L-BFGS	$2mn + 4n$

The program CONMIN [228] is an extension of the memoryless BFGS method described above. It outperforms nonlinear conjugate gradient methods because of a few important enhancements, such as automatic restarts along a carefully chosen direction. Restarts are also used in the Harwell routine VA14, which implements an extension of the Polak–Ribière method. The efficiency of these methods, in terms of function evaluations, and their robustness tends to increase with storage as we proceed down the table.

9.2 GENERAL LIMITED-MEMORY UPDATING

Limited-memory quasi-Newton approximations are useful in a variety of optimization methods. L-BFGS, Algorithm 9.2, is a line search method for unconstrained optimization that (implicitly) updates an approximation H_k to the inverse of the Hessian matrix. Trust-region methods, on the other hand, require an approximation B_k to the Hessian matrix, not to its inverse. We would also like to develop limited-memory methods based on the SR1 formula, which is an attractive alternative to BFGS; see Chapter 8. In this section we consider limited-memory updating in a general setting and show that by representing quasi-Newton matrices in a compact (or outer product) form, we can derive efficient implementations of *all* popular quasi-Newton update formulae, and their inverses. These compact representations will also be useful in designing limited-memory methods for constrained optimization, where approximations to the Hessian or reduced Hessian of the Lagrangian are needed; see Chapter 18.

We will only consider limited-memory methods (like L-BFGS) that continuously refresh the correction pairs by removing and adding information at each stage. A different approach proceeds by saving correction pairs until the available storage is exhausted, and then discarding all correction pairs (except perhaps one) and starting the process anew. Computational experience suggests that this second approach is less effective in practice.

Throughout this chapter we let B_k denote an approximation to a Hessian matrix and H_k the approximation to the inverse. In particular, we always have that $B_k^{-1} = H_k$.

COMPACT REPRESENTATION OF BFGS UPDATING

We now describe an approach to limited-memory updating that is based on representing quasi-Newton matrices in outer-product form. We illustrate it for the case of a BFGS approximation B_k to the Hessian.

Theorem 9.1.
Let B_0 be symmetric and positive definite and assume that the k vector pairs $\{s_i, y_i\}_{i=0}^{k-1}$ satisfy $s_i^T y_i > 0$. Let B_k be obtained by applying k BFGS updates with these vector pairs to B_0, using the formula (8.19). We then have that

$$B_k = B_0 - \begin{bmatrix} B_0 S_k & Y_k \end{bmatrix} \begin{bmatrix} S_k^T B_0 S_k & L_k \\ L_k^T & -D_k \end{bmatrix}^{-1} \begin{bmatrix} S_k^T B_0 \\ Y_k^T \end{bmatrix}, \qquad (9.10)$$

where S_k and Y_k are the $n \times k$ matrices defined by

$$S_k = [s_0, \ldots, s_{k-1}], \qquad Y_k = [y_0, \ldots, y_{k-1}], \qquad (9.11)$$

while L_k and D_k are the $k \times k$ matrices

$$(L_k)_{i,j} = \begin{cases} s_{i-1}^T y_{j-1} & \text{if } i > j, \\ 0 & \text{otherwise,} \end{cases} \qquad (9.12)$$

$$D_k = \mathrm{diag}\left[s_0^T y_0, \ldots, s_{k-1}^T y_{k-1}\right]. \qquad (9.13)$$

This result can be proved by induction. We should note that the conditions $s_i^T y_i > 0$ $i = 0, \ldots, k-1$ ensure that the middle matrix in (9.10) is nonsingular, so that this expression is well-defined. The utility of this representation becomes apparent when we consider limited-memory updating.

As in the L-BFGS algorithm, we will keep the m most recent correction pairs $\{s_i, y_i\}$ and refresh this set at every iteration by removing the oldest pair and adding a newly generated pair. During the first m iterations, the update procedure described in Theorem 9.1 can be used without modification, except that usually we make the specific choice $B_k^0 = \delta_k I$ for the basic matrix, where $\delta_k = 1/\gamma_k$ and γ_k is defined by (9.6).

At subsequent iterations $k > m$, the update procedure needs to be modified slightly to reflect the changing nature of the set of vector pairs $\{s_i, y_i\}$ for $i = k-m, \ldots, k-1$. Defining the $n \times m$ matrices S_k and Y_k by

$$S_k = [s_{k-m}, \ldots, s_{k-1}], \qquad Y_k = [y_{k-m}, \ldots, y_{k-1}], \qquad (9.14)$$

we find that the matrix B_k resulting from m updates to the basic matrix $B_0^{(k)} = \delta_k I$ is given by

$$B_k = \delta_k I - \begin{bmatrix} \delta_k S_k & Y_k \end{bmatrix} \begin{bmatrix} \delta_k S_k^T S_k & L_k \\ L_k^T & -D_k \end{bmatrix}^{-1} \begin{bmatrix} \delta_k S_k^T \\ Y_k^T \end{bmatrix}, \qquad (9.15)$$

where L_k and D_k are now the $m \times m$ matrices defined by

$$(L_k)_{i,j} = \begin{cases} (s_{k-m-1+i})^T (y_{k-m-1+j}) & \text{if } i > j, \\ 0 & \text{otherwise}, \end{cases}$$

$$D_k = \text{diag} \left[s_{k-m}^T y_{k-m}, \ldots, s_{k-1}^T y_{k-1} \right].$$

After the new iterate x_{k+1} is generated, we obtain S_{k+1} by deleting s_{k-m} from S_k and adding the new displacement s_k, and we update Y_{k+1} in a similar fashion. The new matrices L_{k+1} and D_{k+1} are obtained in an analogous way.

Since the middle matrix in (9.15) is small—of dimension $2m$—its factorization requires a negligible amount of computation. The key idea behind the compact representation (9.15) is that the corrections to the basic matrix can be expressed as an outer product of two long and narrow matrices—$[\delta_k S_k \ Y_k]$ and its transpose—with an intervening multiplication by a small $2m \times 2m$ matrix. See Figure 9.1 for a graphical illustration.

The limited-memory updating procedure of B_k requires approximately $2mn + O(m^3)$ operations, and matrix–vector products of the form $B_k v$ can be performed at a cost of $(4m + 1)n + O(m^2)$ multiplications. These operation counts indicate that updating and manipulating the direct limited-memory BFGS matrix B_k is quite economical when m is small.

This approximation B_k can be used in a trust-region method for unconstrained optimization or, more significantly, in methods for bound-constrained and general-constrained optimization. The program L-BFGS-B [259] makes extensive use of compact limited-memory approximations to solve large nonlinear optimization problems with bound constraints. In this situation, projections of B_k into subspaces defined by the constraint gra-

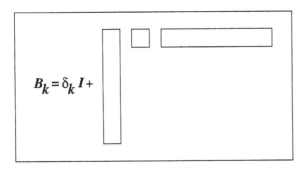

Figure 9.1
Compact (or outer product) representation of B_k in (9.15).

dients must be calculated repeatedly. Many algorithms for general-constrained optimization (for instance the SQP methods of Chapter 18) can make use of the limited-memory matrix B_k just described to approximate the Hessian (or reduced Hessian) of the Lagrangian.

We can derive a formula, similar to (9.10), that provides a compact representation of the inverse BFGS approximation H_k; see [37] for details. An implementation of the L-BFGS algorithm based on this expression requires essentially the same amount of computation as the algorithm described in the previous section. However, when the compact form is used, the product $H_k \nabla f_k$ can be performed more rapidly on advanced computer systems with hierarchical memory, since it can be implemented with BLAS-2 matrix–vector operations.

SR1 MATRICES

Compact representations can also be derived for matrices generated by the symmetric rank-one (SR1) formula. We recall that the SR1 update (8.24) is well-defined only if the denominator $(y_k - B_k s_k)^T s_k$ is nonzero. We can assume that this condition always holds, since if not, we can simply skip the update. The following result uses this assumption to derive the form of the approximation B_k.

Theorem 9.2.

Suppose that k updates are applied to the symmetric matrix B_0 using the vector pairs $\{s_i, y_i\}_{i=0}^{k-1}$ and the SR1 formula (8.24), and assume that each update is well-defined in the sense that $(y_i - B_i s_i)^T s_i \neq 0, i = 0, \ldots, k-1$. The resulting matrix B_k can be expressed as

$$B_k = B_0 + (Y_k - B_0 S_k)(D_k + L_k + L_k^T - S_k^T B_0 S_k)^{-1}(Y_k - B_0 S_k)^T, \qquad (9.16)$$

where $S_k, Y_k, D_k,$ and L_k are as defined in (9.11), (9.12), and (9.13).

The conditions of the theorem imply that the matrix $(D_k + L_k + L_k^T - S_k^T B_0 S_k)$ is nonsingular. Since the SR1 method is self-dual, the inverse formula H_k can be obtained simply by replacing $B, s,$ and y by $H, y,$ and s, respectively. Limited-memory SR1 methods can be derived in the same way as for the BFGS method. We replace B_0 with the basic matrix B_k^0 at the kth iteration, and we redefine S_k and Y_k to contain the m most recent corrections, as in (9.14).

UNROLLING THE UPDATE

The reader may wonder whether limited-memory updating can be implemented in simpler ways. In fact, as we show here, the most obvious implementation of limited-memory BFGS updating is considerably more expensive than the approach based on compact representations discussed in the previous section.

The direct BFGS formula (8.19) can be written as

$$B_{k+1} = B_k - a_k a_k^T + b_k b_k^T, \qquad (9.17)$$

where the vectors a_k and b_k are defined by

$$a_k = \frac{B_k s_k}{(s_k^T B_k s_k)^{\frac{1}{2}}}, \qquad b_k = \frac{y_k}{(y_k^T s_k)^{\frac{1}{2}}}. \tag{9.18}$$

We could continue to save the vector pairs $\{s_i, y_i\}$ but use the formula (9.17) to compute matrix–vector products. A limited-memory BFGS method that uses this approach would proceed by defining the basic matrix B_k^0 at each iteration and then updating according to the formula

$$B_k = B_k^0 + \sum_{i=k-m}^{k-1} \left[b_i b_i^T - a_i a_i^T \right]. \tag{9.19}$$

The vector pairs $\{a_i, b_i\}$, $i = k - m, \ldots, k - 1$, would then be recovered from the stored vector pairs $\{s_i, y_i\}$, $i = k - m, \ldots, k - 1$, by the following procedure:

Procedure 9.3 (Unrolling the BFGS formula).
   for $i = k - m, \ldots, k - 1$
      $b_i \leftarrow y_i / (y_i^T s_i)^{1/2}$;
      $a_i \leftarrow B_k^0 s_i + \sum_{j=k-m}^{i-1} \left[(b_j^T s_i) b_j - (a_j^T s_i) a_j \right]$;
      $a_i \leftarrow a_i / (s_i^T a_i)^{1/2}$;
   end (for)

Note that the vectors a_i must be recomputed at each iteration, since they all depend on the vector pair $\{s_{k-m}, y_{k-m}\}$, which is removed at the end of iteration k. On the other hand, the vectors b_i and the inner products $b_j^T s_i$ can be saved from the previous iteration, so only the new values b_{k-1} and $b_j^T s_{k-1}$ need to be computed at the current iteration.

By taking all these computations into account, and assuming that $B_k^0 = I$, we find that approximately $\frac{3}{2} m^2 n$ operations are needed to determine the limited-memory matrix. The actual computation of the inner product $B_m v$ (for arbitrary $v \in \mathbb{R}^n$) requires $4mn$ multiplications. Overall, therefore, this approach is less efficient than the one based on the compact matrix representation described above. Indeed, while the product $B_k v$ costs the same in both cases, updating the representation of the limited-memory matrix using the compact form requires only $2mn$ multiplications, compared to $\frac{3}{2} m^2 n$ multiplications needed when the BFGS formula is unrolled.

9.3 SPARSE QUASI-NEWTON UPDATES

We now discuss a quasi-Newton approach to large-scale problems that has intuitive appeal: We demand that the quasi-Newton approximations B_k have the same (or similar) sparsity pattern as the true Hessian. This would reduce the storage requirements of the algorithm and perhaps give rise to more accurate Hessian approximations.

Suppose that we know which components of the Hessian may be nonzero at some point in the domain of interest. That is, we know the contents of the set Ω defined by

$$\Omega \stackrel{\text{def}}{=} \{(i,j) \mid [\nabla^2 f(x)]_{ij} \neq 0 \text{ for some } x \text{ in the domain of } f\}.$$

Suppose also that the current Hessian approximation B_k mirrors the nonzero structure of the exact Hessian, i.e., $(B_k)_{ij} = 0$ for $(i,j) \notin \Omega$. In updating B_k to B_{k+1}, then, we could try to find the matrix B_{k+1} that satisfies the secant condition, has the same sparsity pattern, and is as close as possible to B_k. Specifically, we define B_{k+1} to be the solution of the following quadratic program:

$$\min_B \|B - B_k\|_F^2 = \sum_{(i,j) \in \Omega} [B_{ij} - (B_k)_{ij}]^2, \tag{9.20a}$$

$$\text{subject to } B s_k = y_k, \ B = B^T, \text{ and } B_{ij} = 0 \text{ for } (i,j) \notin \Omega. \tag{9.20b}$$

It can be shown that the solution B_{k+1} of this problem can be obtained by solving an $n \times n$ linear system whose sparsity pattern is Ω, the same as the sparsity of the true Hessian. Once B_{k+1} has been computed we can use it, within a trust region method, to obtain the new iterate x_{k+1}. We note that B_{k+1} is not guaranteed to be positive definite.

We omit further details of this approach because it has several major drawbacks: The updating process does not possess scale invariance under linear transformations of the variables, and, more significantly, its practical performance has been disappointing. A standard implementation of this sparse quasi-Newton method typically requires at least as many function evaluations as L-BFGS, but its cost per iteration is higher. The fundamental weakness of this approach is that (9.20a) is an inadequate model and can produce poor Hessian approximations.

An alternative approach is to relax the secant equation, making sure that it is approximately satisfied along the last few steps rather than requiring it to hold strictly on the latest step. To do so, we define S_k and Y_k by (9.14) so that they contain the m most recent difference pairs. We can then define the new Hessian approximation B_{k+1} to be the solution of

$$\min_B \|BS_k - Y_k\|_F^2$$

$$\text{subject to } B = B^T \text{ and } B_{ij} = 0 \text{ for } (i,j) \notin \Omega.$$

This convex optimization problem has a solution, but it is not easy to compute. Moreover, this approach can produce singular or poorly conditioned Hessian approximations. Even though it frequently outperforms methods based on (9.20a), its performance on large problems has not been impressive.

A much more promising approach to exploiting structure in large-scale optimization is outlined in the next section.

9.4 PARTIALLY SEPARABLE FUNCTIONS

In a *separable* unconstrained optimization problem, the objective function can be decomposed into a sum of simpler functions that can be optimized independently. For example, if we have

$$f(x) = f_1(x_1, x_3) + f_2(x_2, x_4, x_6) + f_3(x_5),$$

we can find the optimal value of x by minimizing each function f_i, $i = 1, 2, 3$, independently, since no variable appears in more than one function. The cost of performing n lower-dimensional optimizations is much less in general than the cost of optimizing an n-dimensional function.

In many large problems the objective function $f : \mathbf{R}^n \to \mathbf{R}$ is not separable, but it can still be written as the sum of simpler functions, known as *element functions*. Each element function has the property that it is unaffected when we move along a large number of linearly independent directions. If this property holds, we say that f is *partially separable*. We will see that all functions whose Hessians $\nabla^2 f$ are sparse are partially separable, but so are many functions whose Hessian is not sparse. Partial separability allows for economical problem representation, efficient automatic differentiation, and effective quasi-Newton updating.

The simplest form of partial separability arises when the objective function can be written as

$$f(x) = \sum_{i=1}^{ne} f_i(x), \tag{9.21}$$

where each of the element functions f_i depends on only a few components of x. It follows that the gradients ∇f_i and Hessians $\nabla^2 f_i$ of each element function contain just a few nonzeros. By differentiating (9.21), we obtain

$$\nabla f(x) = \sum_{i=1}^{ne} \nabla f_i(x), \qquad \nabla^2 f(x) = \sum_{i=1}^{ne} \nabla^2 f_i(x),$$

and it is natural to ask whether it is more effective to maintain quasi-Newton approximations to each of the element Hessians $\nabla^2 f_i(x)$ separately, rather than approximating the entire Hessian $\nabla^2 f(x)$. We will show that the answer is affirmative, provided that the quasi-Newton approximation fully exploits the structure of each element Hessian.

A SIMPLE EXAMPLE

Consider the objective function

$$f(x) = (x_1 - x_3^2)^2 + (x_2 - x_4^2)^2 + (x_3 - x_2^2)^2 + (x_4 - x_1^2)^2 \quad (9.22)$$
$$\equiv f_1(x) + f_2(x) + f_3(x) + f_4(x),$$

where the element functions f_i have been defined in the obvious way. The Hessians of these element functions are 4×4 sparse, singular matrices with 4 nonzero entries.

Let us focus on f_1; all other element functions have exactly the same form. Even though f_1 is formally a function of all components of x, it depends only on x_1 and x_3, that we call the *element variables* for f_1. We assemble the element variables into a vector that we call $x_{[1]}$, that is,

$$x_{[1]} = \begin{bmatrix} x_1 \\ x_3 \end{bmatrix},$$

and note that

$$x_{[1]} = U_1 x \quad \text{with} \quad U_1 = \begin{bmatrix} 1 & 0 & 0 & 0 \\ 0 & 0 & 1 & 0 \end{bmatrix}.$$

If we define the function ϕ_1 by

$$\phi_1(z_1, z_2) = (z_1 - z_2^2)^2,$$

then we can write $f_1(x) = \phi_1(U_1 x)$. By applying the chain rule to this representation, we obtain

$$\nabla f_1(x) = U_1^T \nabla \phi_1(U_1 x), \qquad \nabla^2 f_1(x) = U_1^T \nabla^2 \phi_1(U_1 x) U_1. \quad (9.23)$$

In our case, we have

$$\nabla^2 \phi(U_1 x) = \begin{bmatrix} 2 & -4x_3 \\ -4x_3 & 12x_3^2 \end{bmatrix}, \qquad \nabla^2 f_1(x) = \begin{bmatrix} 2 & 0 & -4x_3 & 0 \\ 0 & 0 & 0 & 0 \\ -4x_3 & 0 & 12x_3^2 & 0 \\ 0 & 0 & 0 & 0 \end{bmatrix}.$$

It is easy to see that U_1 performs a "compactification" of the variable vector; it allows us to map the derivative information for the low-dimensional function ϕ_1 into the derivative information for the element function f_1.

Now comes the key idea: Instead of maintaining a quasi-Newton approximation to $\nabla^2 f_1$, we maintain a 2×2 quasi-Newton approximation $B_{[1]}$ of $\nabla^2 \phi_1$ and use the relation (9.23) to transform it into a quasi-Newton approximation to $\nabla^2 f_1$. To update $B_{[1]}$ after a typical step from x to x^+, we record the information

$$s_{[1]} = x_{[1]}^+ - x_{[1]}, \qquad y_{[1]} = \nabla \phi_1(x_{[1]}^+) - \nabla \phi_1(x_{[1]}), \qquad (9.24)$$

and use BFGS or SR1 updating to obtain the new approximation $B_{[1]}^+$. We therefore update small, dense quasi-Newton approximations with the property

$$B_{[1]} \approx \nabla^2 \phi_1(U_1 x) = \nabla^2 \phi_1(x_{[1]}). \qquad (9.25)$$

To obtain an approximation of the element Hessian $\nabla^2 f_1$, we use the transformation suggested by the relationship (9.23); that is,

$$\nabla^2 f_1(x) \approx U_1^T B_{[1]} U_1.$$

This operation has the effect of mapping the elements of $B_{[1]}$ to the correct positions in the full $n \times n$ Hessian approximation.

The previous discussion concerned only the first element function f_1. We do the same for all other element functions, maintaining a quasi-Newton approximation $B_{[i]}$ for each one of them. To obtain a complete approximation to the full Hessian $\nabla^2 f$, we simply sum the element Hessian approximations as follows:

$$B = \sum_{i=1}^{ne} U_i^T B_{[i]} U_i.$$

We argue later that this strategy will produce a very good quasi-Newton approximation, much better than one that would be obtained if we ignored the partially separable structure of the objective function f.

Before discussing these quasi-Newton methods further, we examine the concept of partial separability in detail. We first describe how to determine whether a function is partially separable and how to obtain its best representation. Next, we explain the concept of an invariant subspace, and we explore the relationship between sparsity and partial separability.

INTERNAL VARIABLES

In the simple example (9.22), the best strategy for decomposing the objective function (into the four element functions that we chose) was obvious. In many applications, however, there are many alternative ways to choose the partially separable decomposition, and it is important to the efficiency of quasi-Newton methods that we select the finest decomposition.

We illustrate this issue with the following example, which arises in the solution of the minimum surface area problem (see Exercise 13).

The objective function is given by (9.21), where each of the ne element functions f_i has the form

$$f_i(x) = \frac{1}{q^2}\left[1 + \frac{q^2}{2}[(x_j - x_{j+q+1})^2 + (x_{j+1} - x_{j+q})^2]\right]^{\frac{1}{2}}. \tag{9.26}$$

Here q is a parameter that determines the size of the discretization, $ne = n = q^2$, and the integer j, which determines which components of x influence f_i, is a function of i and q. (Details of the formulation are given in Exercise 13.)

Each element function f_i depends on only four components of x. The gradient of f_i (with respect to the full vector x) contains at most four nonzeros and has the form

$$\nabla f_i(x) = \frac{1}{2q^2} f_i^{-1}(x) \begin{bmatrix} 0 \\ \vdots \\ x_j - x_{j+q+1} \\ x_{j+1} - x_{j+q} \\ \vdots \\ -x_{j+1} + x_{j+q} \\ -x_j + x_{j+q+1} \\ \vdots \\ 0 \end{bmatrix}.$$

Note that two of these four nonzero components are negatives of the other two. The Hessian of the element function f_i has the sparsity pattern

$$\begin{bmatrix} & & & & & & \\ & * & * & & * & * & \\ & * & * & & * & * & \\ & & & & & & \\ & * & * & & * & * & \\ & * & * & & * & * & \\ & & & & & & \end{bmatrix}, \tag{9.27}$$

and a close examination shows that some of the nonzero entries differ only in sign. In fact, only three different magnitudes are represented by the 16 nonzero elements, and the 4 × 4 matrix that can be assembled from these nonzero elements happens to be singular.

The obvious way to define the element variables $x_{[i]}$ for f_i is simply to take the four components of x that appear in (9.26). However, the fact that repeated information is contained in the derivatives suggests that there is a more economical representation that avoids these redundancies. The key observation is that f_i is *invariant* in the subspace

$$\mathcal{N}_i = \{w \in \mathbf{R}^n \mid w_j = w_{j+q+1} \text{ and } w_{j+1} = w_{j+q}\}, \tag{9.28}$$

which means that for any x and for any $w \in \mathcal{N}_i$ we have that $f_i(x+w) = f_i(x)$. In other words, any move along a direction in $\mathcal{N}_i$ (a subspace of $\mathbf{R}^n$ whose dimension is $n-2$) does not change the value of f_i, so it is not useful to try to gather curvature information about f_i along $\mathcal{N}_i$. Put another way, we can say that there is an orthogonal basis for $\mathbf{R}^n$ for which f_i changes along just two of the vectors in this basis. It therefore makes sense to look for a compact representation of $\nabla^2 f_i$ that is based on a 2 × 2 matrix.

We can find such a representation almost immediately by examining the definition (9.28). We define the *internal variables* u_j and u_{j+1} by

$$u_j = x_j - x_{j+q+1}, \qquad u_{j+1} = x_{j+1} - x_{j+q}, \tag{9.29}$$

and the *internal function* ϕ_i by

$$\phi_i(u_j, u_{j+1}) = \frac{1}{q^2}\left[1 + \frac{q^2}{2}(u_j^2 + u_{j+1}^2)\right]^{\frac{1}{2}}. \tag{9.30}$$

In fact, this representation of f_i is minimal, since it allows for just two distinct nonzero values in ∇f_i and three distinct nonzero values in the Hessian $\nabla^2 f_i$, which is exactly what we observed earlier. Hence, we call this representation the *finest* partially separable representation of f.

Formally, we can represent (9.29) in matrix form,

$$u_{[i]} = U_i x, \tag{9.31}$$

where the internal variable vector $u_{[i]}$ has two components and U_i is a 2 × n matrix that has zero elements except for

$$U_{1,j} = 1, \quad U_{1,j+q+1} = -1, \quad U_{2,j+1} = 1, \quad U_{2,j+q} = -1.$$

Note that the null space of U_i is exactly the invariant subspace $\mathcal{N}_i$. The full objective function for the minimum surface problem can now be written as

$$f(x) = \sum_{i=1}^{ne} \phi_i(U_i x). \qquad (9.32)$$

The gradient and Hessian of f are given by

$$\nabla f(x) = \sum_{i=1}^{ne} U_i^T \nabla \phi_i(U_i x), \qquad \nabla^2 f(x) = \sum_{i=1}^{ne} U_i^T \nabla^2 \phi_i(U_i x) U_i, \qquad (9.33)$$

which clearly exhibit the underlying structure. All the information is now contained in the transformations U_i and in the internal gradient and Hessians—$\nabla \phi_i(\cdot)$ and $\nabla^2 \phi_i(\cdot)$—which contain just two and four nonzeros, respectively. We will see that effective algorithms result when we assemble quasi-Newton approximations of $\nabla^2 f$ from quasi-Newton approximations of the internal Hessians $\nabla^2 \phi_i$.

Returning to the example (9.22), we see that the internal variables for f_1 should be defined as $u_{[1]} = x_{[1]}$, and that the internal function is given by the function ϕ described there.

9.5 INVARIANT SUBSPACES AND PARTIAL SEPARABILITY

We now generalize the previous examples and formally describe the key concepts that underlie partial separability.

Definition 9.1 (Invariant Subspace).
    The invariant subspace $\mathcal{N}_i$ of a function $f(x) : \mathbf{R}^n \to \mathbf{R}$ is the largest subspace in $\mathbf{R}^n$ such that for all $w \in \mathcal{N}_i$, we have $f(x + w) = f(x)$ whenever x and $x + w$ are in the domain of f.

We have seen one example of an invariant subspace in (9.28). Another simple example can be derived from the element function f_i defined by

$$f_i(x_1, \ldots, x_n) = x_{50}^2 \qquad (n > 50), \qquad (9.34)$$

for which $\mathcal{N}_i$ is the set of all vectors $w \in \mathbf{R}^n$ for which $w_{50} = 0$. This subspace has dimension $n - 1$, so that all the nontrivial behavior of the function is contained in the one-dimensional subspace orthogonal to $\mathcal{N}_i$. In this case, the obvious compact representation of f_i is the simple function $\phi_i(z) = z^2$, where the compactifying matrix U_i is the $1 \times n$ matrix defined by

$$U_i = \begin{bmatrix} 0 & \cdots & 0 & 1 & 0 & \cdots & 0 \end{bmatrix},$$

9.5. INVARIANT SUBSPACES AND PARTIAL SEPARABILITY

where the nonzero element occurs in the 50th position.

It is interesting to compare (9.34) with the element function f_i defined by

$$f_i(x_1, \ldots, x_n) = (x_1 + \ldots + x_n)^2. \tag{9.35}$$

At first glance, the two functions appear to be quite different. The first one depends on only one variable, and as a result has a very sparse gradient and Hessian, while (9.35) depends on all variables, and its gradient and Hessian are completely dense. However, the invariant subspace $\mathcal{N}_i$ of (9.35) is

$$\mathcal{N}_i = \{w \in \mathbf{R}^n \mid e^T w = 0\}, \qquad \text{where } e = (1, 1, \ldots, 1)^T. \tag{9.36}$$

Again, the dimension of this subspace is $n - 1$, so that as in (9.34), all the nontrivial behavior is confined to a one-dimensional subspace. We can derive a compact representation of the function in (9.35) by defining

$$U_i = \begin{bmatrix} 1 & 1 & \cdots & 1 \end{bmatrix}, \qquad \phi_i(z) = z^2.$$

Note that the compact representation function $\phi_i(z)$ is the same for both (9.34) and (9.35); the two functions differ only in the makeup of the matrix U_i.

We now use the concept of an invariant subspace to define partial separability.

Definition 9.2 (Partial Separability).

A function f is said to be partially separable *if it is the sum of element functions, $f(x) = \sum_{i=1}^{ne} f_i(x)$, where each f_i has a large invariant subspace. In other words, f can be written in the form (9.32), where the matrices U_i (whose null spaces coincide with the invariant subspaces $\mathcal{N}_i$) have dimension $n_i \times n$, with $n_i \ll n$.*

This definition is a little vague in that the term "large invariant subspace" is not precisely defined. However, such vagueness is also present in the definition of sparsity, for instance, which depends not just on the proportion of nonzeros in a matrix but also on its structure, size, and the underlying application. For practical purposes, it makes sense to exploit partial separability only if the dimensions of all the invariant subspaces are close to n.

Note that in all our examples above, the matrices U_i whose null spaces spanned each $\mathcal{N}_i$ could have been chosen in a number of different ways; a basis for a subspace is not unique, after all. In all cases, however, a definition of U_i was fairly clear from the analytic definition of f_i. The situation may be different when the function is defined by a complicated computer program. In this case we would prefer to circumvent the task of analyzing the function in order to identify the matrix U_i. An automatic procedure for detection of partial separability has been proposed by Gay [98] and has been implemented in the AMPL modeling language [92] in conjunction with its automatic differentiation software. We show in Chapter 7 that the decomposition of partially separable functions also improves the efficiency of automatic differentiation.

SPARSITY VS. PARTIAL SEPARABILITY

We have already seen in (9.35) that functions can be partially separable even if their Hessians are not sparse. The converse is also true: Functions with sparse Hessians are always partially separable. We prove this result in the following theorem. In this sense, the concept of partial separability is more general than sparsity, and it yields more information from the optimization viewpoint.

Theorem 9.3.

Every twice continuously differentiable function with a sparse Hessian is partially separable. More specifically, suppose that $f : D \to \mathbb{R}^n$ is twice continuously differentiable in an open subset D of $\mathbb{R}^n$, and that

$$\frac{\partial^2}{\partial x_i \partial x_j} f(x_1, \ldots, x_n) = 0$$

for all $x \in D$. Then

$$f(x_1, \ldots, x_n) = f(x_1, \ldots, x_{j-1}, 0, x_{j+1}, \ldots, x_n)$$
$$+ f(x_1, \ldots, x_{i-1}, 0, x_{i+1}, \ldots, x_n)$$
$$- f(x_1, \ldots, x_{i-1}, 0, x_{i+1}, \ldots, x_{j-1}, 0, x_{j+1}, \ldots, x_n)$$

for all $(x_1, \ldots, x_n)$.

PROOF. Without loss of generality, consider the case $n = 2$ so that $(x_1, \ldots, x_n) = (x, y)$. The assumption and the integral mean value theorem imply that for all (x, y),

$$\begin{aligned}
0 &= \int_0^x \int_0^z \frac{\partial^2 f(\xi, \zeta)}{\partial x \partial y} \, d\xi \, d\zeta \\
&= \int_0^x \left[\frac{\partial f(\xi, \zeta)}{\partial x} \, d\xi \right]_0^y \\
&= \int_0^x \frac{\partial f(\xi, y)}{\partial x} \, d\xi - \int_0^x \frac{\partial f(\xi, 0)}{\partial x} \, d\xi \\
&= [f(\xi, y)]_0^x - [f(\xi, 0)]_0^x \\
&= f(x, y) - f(0, y) - f(x, 0) + f(0, 0).
\end{aligned}$$
□

Theorem 9.3 states that even if only one of the partials vanishes, then f is the sum of element functions, each of which has a nontrivial invariant subspace.

GROUP PARTIAL SEPARABILITY

Partial separability is an important concept, but it is not quite as general as we would like. Consider a nonlinear least-squares problem of the form

$$f(x) = \sum_{k=1}^{l}(f_k(x) + f_{k+1}(x) + c_k)^2, \tag{9.37}$$

where the functions f_j are partially separable and each c_k is a constant. The definition of partial separability given above forces us either to expand the quadratic function and end up with products of the f_k, or to regard the whole kth term in the summation as the kth element function. A more intuitive approach would avoid this somewhat artificial expansion and recognize that each term contains two element functions, grouped together and squared. By extending slightly the concept of partial separability, we can make better use of this type of structure.

We say that a function $f : \mathbf{R}^n \to \mathbf{R}$ is *group partially separable* if it can be written in the form

$$f(x) = \sum_{k=1}^{l} \psi_k(h_k(x)) \tag{9.38}$$

where each h_k is a partially separable function from $\mathbf{R}^n$ to $\mathbf{R}$, and each ψ_k is a scalar twice continuously differentiable function defined on the range of h_k. We refer to ψ_k as the *group function*. Group partial separability is a useful concept in many important areas of optimization, such as nonlinear least-squares problems and penalty and merit functions arising in constrained optimization.

Compact representations of the derivatives of f can be built up from the representations of the h_k functions (and their derivatives), together with the values of ψ_k, ψ_k', and ψ_k'', for each k. By using the chain rule, we obtain

$$\nabla[\psi_k(h_k(x))] = \psi_k'(h_k(x))\nabla h_k(x),$$
$$\nabla^2[\psi_k(h_k(x))] = \psi_k''(h_k(x))\nabla h_k(x)\nabla h_k(x)^T + \psi_k'(h_k(x))\nabla^2 h_k(x).$$

These formulae, together with the compact representations of the quantities ∇h_k and $\nabla^2 h_k$ derived in (9.33), can be used to represent the derivatives of the function f of (9.38) efficiently. Hence, all the ideas we have discussed so far in the context of partially separable functions generalize easily to group partially separable objectives.

EXAMPLE 9.1

Consider the function

$$\tfrac{1}{2}\left((x_5 - x_{25})x_7 + x_{11}\right)^2.$$

The group function is $\psi(z) = \tfrac{1}{2}z^2$. The partially separable function $h(x) = (x_5-x_{25})x_7+x_{11}$ that serves as the argument of $\psi(\cdot)$ can be partitioned into two element functions defined by $f_1 = (x_5 - x_{25})x_7$ and $f_2 = x_{11}$. Note that the function f_1 has three element variables, but these can be transformed into the two internal variables $u_1 = x_5 - x_{25}$ and $u_2 = x_7$.

9.6 ALGORITHMS FOR PARTIALLY SEPARABLE FUNCTIONS

The concept of partial separability has most to offer when the number of variables is large. Savings can be obtained by representing the Hessian matrix economically in a Newton method or by developing special quasi-Newton updating techniques.

EXPLOITING PARTIAL SEPARABILITY IN NEWTON'S METHOD

We consider first an inexact (truncated) Newton method in which the search direction p_k is an approximate solution of the Newton equations

$$\nabla^2 f(x_k) p = -\nabla f(x_k). \tag{9.40}$$

We solve this system by the conjugate gradient method. To obtain a complete algorithm, we embed this step computation into either a line search or a trust-region framework, as discussed in Chapter 6.

Suppose that we know how to express the objective function f in partially separable form, that is, we have identified the number of element functions ne, the internal variables u_i, and the transformations U_i for each element function. Recalling that the conjugate gradient method requires matrix–vector products to be formed with the Hessian $\nabla^2 f(x)$, we see from the representation (9.33) that such products can be obtained by means of matrix–vector operations involving the matrices U_i and $\nabla^2 \phi_i$, for $i = 1, 2, \ldots, ne$. By using this structure, we often find that the Hessian–vector product can be computed more economically than if we formed $\nabla^2 f(x)$ and performed the multiplication explicitly. The representation (9.33) can therefore yield faster computations as well as a reduction in storage.

As an example, consider the minimum surface problem where the element functions are of the form (9.26). In this function, there are $ne = n$ Hessians $\nabla^2 \phi_i$, each of which is a 2×2 symmetric matrix that can be stored in three floating-point numbers. The matrices U_i all have similar structure; each contains four nonzeros, two 1's and two -1's, in matrix locations that are determined by the index i. It is not necessary to store these matrices explicitly at all, since all we need to know to perform matrix–vector products with them is the value of i. Hence, the Hessian can be stored in approximately $3n$ floating-point memory locations. By contrast, it can be shown that storage of the lower triangle of the full Hessian $\nabla^2 f(x)$ in the standard format requires about $5n$ locations. Hence, in this example, the partially separable representation has yielded a 40% savings in storage. In other applications, the savings can be even more significant.

We should note, however, that the storage reduction that we achieve by the use of internal variables comes at a price: The mapping performed by the transformations U_i requires memory access and computation. The efficiency loss due to these operations depends on the problem structure and on the computer architecture, but sometimes it can be significant. Consequently, it may not be worth our while to seek the finest possible partially separable representation for a given problem, but to settle instead for an easily identified representation that captures most of the structure, or even to ignore the partial separability altogether. For example, in the minimum surface problem, we could represent each element function as a function of 4 variables, and not as a function of two variables, as in (9.30).

Another implementation of Newton's method that is suitable for partially separable functions is based on solving (9.40) by using a direct method instead of conjugate gradients. The multifrontal factorization technique described by Duff and Reid [74], which was developed initially in connection with finite element systems, is well suited for the Hessians of partially separable functions. In this approach, partial assembly of the Hessian matrix is used in conjunction with dense factorization of submatrices to obtain an efficient factorization. The LANCELOT software package [53] includes an option to allow the user to choose between conjugate gradients and the multifrontal approach in its implementation of Newton's method.

In conclusion, the effectiveness of Newton's method for partially separable functions—as measured by total computing time—varies according to the problem structure and computer architecture. In some cases it may be profitable to exploit partial separability and in other cases it may not. The situation is less ambiguous in the case of quasi-Newton methods, where exploitation of the partially separable structure often yields a dramatic reduction in the number of function and gradient evaluations.

QUASI-NEWTON METHODS FOR PARTIALLY SEPARABLE FUNCTIONS

As indicated in (9.25), (9.24), we can construct quasi-Newton Hessian approximations for partially separable functions by storing and updating approximations $B_{[i]}$ to each internal function $\nabla^2 \phi_i$ and assembling these into an approximation B to the full Hessian $\nabla^2 f(x)$ using the formula

$$B = \sum_{i=1}^{ne} U_i^T B_{[i]} U_i. \tag{9.41}$$

We may use this approximate Hessian in a trust-region algorithm, obtaining an approximate solution p_k of the system

$$B_k p_k = -\nabla f(x_k). \tag{9.42}$$

As in the case of Newton's method, we need not assemble B_k explicitly, but rather use the conjugate gradient approach to solve (9.42), computing matrix–vector products of the form $B_k v$ by performing operations with the matrices U_i and $B_{[i]}$.

As for any quasi-Newton method, we update the approximations $B_{[i]}$ by requiring them to satisfy the secant equation for each element function, namely,

$$B_{[i]} s_{[i]} = y_{[i]}, \tag{9.43}$$

where

$$s_{[i]} = u_{[i]}^+ - u_{[i]} \tag{9.44}$$

is the change in the internal variables corresponding to the ith element function, and

$$y_{[i]} = \nabla \phi_i(u_{[i]}^+) - \nabla \phi(u_{[i]}) \tag{9.45}$$

is the corresponding change in gradients. Here u^+ and u indicate the vectors of internal variables at the current and previous iterations, respectively. Thus the formulae (9.44) and (9.45) are obvious generalizations of (9.24) to the case of internal variables.

We return again to the minimum surface problem (9.26) to illustrate the usefulness of this element-by-element updating technique. In this case, the functions ϕ_i depend on only two internal variables, so that each Hessian approximation $B_{[i]}$ is a 2×2 matrix. After just a few iterations, we will have sampled enough directions $s_{[i]}$ to make each $B_{[i]}$ an accurate approximation to $\nabla^2 \phi_i$. Hence the full quasi-Newton approximation (9.41) will tend to be a very good approximation to $\nabla^2 f(x)$.

By contrast, a quasi-Newton method that ignores the partially separable structure of the objective function will attempt to estimate the total average curvature—the sum of the individual curvatures of the element functions—by constructing a dense $n \times n$ matrix. When the number of variables n is large, many iterations will be required before this quasi-Newton approximation is of good quality. Hence an algorithm of this type (e.g., standard BFGS or L-BFGS) will require many more iterations than a method based on the partially separable approximate Hessian.

It is not always possible to use the BFGS formula in conjunction with the formulae (9.44) and (9.45) to update the partial Hessian $B_{[i]}$, because there is no guarantee that the curvature condition $s_{[i]}^T y_{[i]} > 0$ will be satisfied. (Recall from Chapter 8 that this curvature condition is needed to ensure that the BFGS approximate Hessian remains positive definite.) That is, even though the full Hessian $\nabla^2 f(x)$ is at least positive semidefinite at the solution x^*, some of the individual Hessians $\nabla^2 \phi_i(\cdot)$ may be indefinite. One way to overcome this obstacle is to apply the SR1 update to each of the element Hessians, using simple precautions to ensure that it is well-defined. SR1 updating has proved to be very effective in the LANCELOT package [53], which is designed to take full advantage of partial separability. The performance of the SR1-based quasi-Newton method is often comparable to that of Newton's method, provided that we find the *finest* partially separable decomposition of the problem; otherwise, there can be a substantial loss of efficiency. For the problems listed in Table 9.1, the Newton and quasi-Newton versions of LANCELOT required a similar number of function evaluations and similar computing time.

NOTES AND REFERENCES

For further discussion on the L-BFGS method see Nocedal [183], Liu and Nocedal [151], and Gilbert and Lemaréchal [102]. The last paper also discusses various ways in which the scaling parameter can be chosen.

Algorithm 9.1, the two-loop L-BFGS recursion, constitutes an economical procedure for computing the product $H_k \nabla f_k$. It is based, however, on the specific form of the BFGS update formula (9.2), and recursions of this type have not yet been developed (and may not exist) for other members of the Broyden class (for instance, the SR1 and DFP methods).

Limited-memory methods using roughly half of the storage employed by Algorithm 9.2 are discussed by Siegel [230], Deuflhard et al. [71], and Gill and Leonard [106]. It is not yet known, however, whether such approaches outperform L-BFGS. A limited-memory method that combines cycles of quasi-Newton and CG iterations and that is often competitive with L-BFGS is described by Buckley and Le Nir [28]. Our discussion of compact representations of limited-memory matrices is based on Byrd, Nocedal, and Schnabel [37].

Limited-memory methods for solving systems of nonlinear equations can also be developed; see Chapter 11. In particular, the paper [37] shows that compact representations of Broyden matrices for these problems can be derived.

Sparse quasi-Newton updates have been studied by Toint [237, 238] and Fletcher et al. [84, 85], among others.

The concept of partial separability was introduced by Griewank and Toint [129, 127]. For an extensive treatment of the subject see Conn, Gould, and Toint [53]. Theorem 9.3 was proved by Griewank and Toint [128].

If f is partially separable, a general affine transformation will not in general preserve the partially separable structure. The quasi-Newton method for partially separable functions described here is not invariant to affine transformations of the variables, but this is not a drawback because it is invariant to transformations that preserve separability.

Exercises

⬥ **9.1** Code Algorithm 9.2 and test it on the extended Rosenbrock function

$$f(x) = \sum_{i=1}^{n/2} \left[\alpha(x_{2i} - x_{2i-1}^2)^2 + (1 - x_{2i-1})^2\right],$$

where α is a parameter that you can vary (e.g., 1, 100). The solution is $x^* = (1, 1, \ldots, 1)$, $f^* = 0$.

Choose the starting point as $(-1, -1, \ldots, -1)^T$. Observe the behavior of your program for various values of the memory parameter m.

⬥ **9.2** Show that the matrix $\hat{H}_{k+1}$ in (9.7) is singular.

⬥ **9.3** Derive the formula (9.9) under the assumption that line searches are exact.

⬥ **9.4** Describe how to take advantage of parallelism when computing the matrix–vector product $B_k v$, when B_k is given by (9.10). Compare this with the level of parallelism that can be achieved in the two-loop recursion for computing $H_k \nabla f_k$.

⬥ **9.5** Consider limited-memory SR1 updating based on (9.16). Explain how the storage can be cut in half if the basic matrix B_k^0 is kept fixed for all k. (Hint: Consider the matrix $Q_k = [q_0, \ldots, q_{k-1}] = Y_k - B_0 S_k$.)

⬥ **9.6** Compute the elements in the Hessian (9.27) and verify that if differences in the signs are ignored, only three of the elements are different.

⬥ **9.7** Write the function defined by

$$f(x) = x_2 x_3 e^{x_1 + x_3 - x_4} + (x_2 x_3)^2 + (x_3 - x_4)$$

in the form (9.32). In particular, give the definition of each of the compactifying transformations U_i.

⬥ **9.8** Show that the dimension of (9.28) is $n - 2$ and that the dimension of (9.36) is $n - 1$.

⬥ **9.9** For the following two functions, find the invariant subspace, and define the internal variables. Make sure that the number of internal variables is $n - \dim(\mathcal{N})$.

1. $f(x) = (x_9 + x_3 - 2x_7) \exp(x_9 - x_7)$.
2. $f(x) = \left(\sum_{i=1}^{100} i x_i\right)^2 + \left(\sum_{i=1}^{100} i^2 x_{n-1}\right)^2$.

⬥ **9.10** Give an example of a partially separable function that is strictly convex but that contains an element function f_i that is concave.

◇ **9.11** Does the approximation B obtained by the partially separable quasi-Newton updating (9.41), (9.45) satisfy the secant equation $Bs = y$?

◇ **9.12** (Griewank and Toint [128]) Let t be a nonsingular affine transformation $t(x) = Ax + b$. Show that f and $f \circ t$ have the same separability structure if the linear part of t is block-diagonal, where the ith and jth variables belong to the same block only if the ith and jth components of e are identical for all $e \in E$ (see the proof of Theorem 9.3 for a definition of e and E). In other words, t is only allowed to mix variables that occur either together or not at all in the element functions f_i.

◇ **9.13** The minimum surface problem is a classical application of the calculus of variations and can be found in many textbooks. We wish to find the surface of minimum area, defined on the unit square, that interpolates a prescribed continuous function on the boundary of the square. In the standard discretization of this problem, the unknowns are the values of the sought-after function $z(x, y)$ on a $q \times q$ rectangular mesh of points over the unit square.

More specifically, we divide each edge of the square into q intervals of equal length, yielding $(q+1)^2$ grid points. We label the grid points as

$$x_{(i-1)(q+1)+1}, \ldots, x_{i(q+1)} \quad \text{for } i = 1, 2, q+1,$$

so that each value of i generates a line. With each point we associate a variable z_i that represents the height of the surface at this point. For the $4q$ grid points on the boundary of the unit square, the values of these variables are determined by the given function. The optimization problem is to determine the other $(q+1)^2 - 4q$ variables z_i so that the total surface area is minimized.

A typical subsquare in this partition looks as follows:

$$\begin{array}{|cc|} \hline x_{j+q+1} & x_{j+q+2} \\ & \\ & \\ x_j & x_{j+1} \\ \hline \end{array}$$

We denote this square by A_j and note that its area is q^2. The desired function is $z(x, y)$, and we wish to compute its surface over A_j. It is shown in calculus books that the area of the surface is given by

$$f_j(x) \equiv \int\int_{(x,y) \in A_j} \sqrt{1 + \left(\frac{\partial z}{\partial x}\right)^2 + \left(\frac{\partial z}{\partial y}\right)^2} \, dx \, dy.$$

Approximate the derivatives by finite differences and show that f_j has the form (9.26).

CHAPTER 10

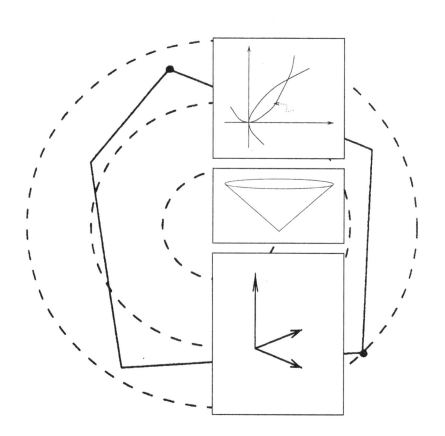

Nonlinear Least-Squares Problems

In least-squares problems, the objective function f has the following special form:

$$f(x) = \tfrac{1}{2} \sum_{j=1}^{m} r_j^2(x), \tag{10.1}$$

where each r_j is a smooth function from $\mathbf{R}^n$ to $\mathbf{R}$. We refer to each r_j as a *residual*, and we assume throughout this chapter that $m \geq n$.

These problems have been a fruitful area of study for over 30 years, mainly because of their applicability to many practical problems. Almost anyone who formulates a parametrized model for a chemical, physical, financial, or economic application uses a function of the form (10.1) to measure the discrepancy between the model and the output of the system at various observation points. By minimizing this function, they select values for the parameters that best match the model to the data. Researchers in numerical analysis and optimization have been able to devise efficient, robust minimization algorithms by exploiting the special structure in f and its derivatives.

To see why the special form of f often makes least-squares problems easier to solve than general unconstrained minimization problems, we first assemble the individual components

r_j from (10.1) into a *residual vector* $r : \mathbf{R}^n \to \mathbf{R}^m$ defined by

$$r(x) = (r_1(x), r_2(x), \ldots, r_m(x))^T. \tag{10.2}$$

Using this notation, we can rewrite f as $f(x) = \frac{1}{2}\|r(x)\|_2^2$. The derivatives of $f(x)$ can be expressed in terms of the Jacobian of r, which is the $m \times n$ matrix of first partial derivatives defined by

$$J(x) = \left[\frac{\partial r_j}{\partial x_i}\right]_{\substack{j=1,2,\ldots,m \\ i=1,2,\ldots,n}}. \tag{10.3}$$

We have

$$\nabla f(x) = \sum_{j=1}^{m} r_j(x) \nabla r_j(x) = J(x)^T r(x), \tag{10.4}$$

$$\nabla^2 f(x) = \sum_{j=1}^{m} \nabla r_j(x) \nabla r_j(x)^T + \sum_{j=1}^{m} r_j(x) \nabla^2 r_j(x)$$

$$= J(x)^T J(x) + \sum_{j=1}^{m} r_j(x) \nabla^2 r_j(x). \tag{10.5}$$

In many applications, it is possible to calculate the first partial derivatives that make up the Jacobian matrix $J(x)$ explicitly. These could be used to calculate the gradient $\nabla f(x)$ as written in formula (10.4). However, the distinctive feature of least-squares problems is that by knowing the Jacobian we can compute the first part of the Hessian $\nabla^2 f(x)$ for free. Moreover, this term $J(x)^T J(x)$ is often more important than the second summation term in (10.5), either because of near-linearity of the model near the solution (that is, $\nabla^2 r_j$ small) or because of small residuals (that is, r_j small). Most algorithms for nonlinear least-squares exploit these structural properties of the Hessian.

The major algorithms for minimizing (10.1) fit nicely into the line search and trust-region frameworks we have already described, with a few simplifications. They are based on the Newton and quasi-Newton approaches described in earlier sections, with modifications that exploit the particular structure of f.

In some large-scale applications, it is more practical to compute the gradient $\nabla f(x)$ directly using computational differentiation techniques (see Chapter 7) than to compute the first partial derivatives (10.3) separately and forming $J(x)^T r(x)$ from (10.4). Since we do not have access to $J(x)$ for these problems, we cannot exploit the special structure associated with the least-squares objective (10.1), and therefore the algorithms of this chapter do not apply. Instead, we can apply the algorithms for general large-scale unconstrained optimization problems described in Chapters 6 and 9. Many large-scale problems do, however, allow ready computation of the first partial derivatives and can therefore make use of the algorithms in this chapter.

10.1 BACKGROUND

MODELING, REGRESSION, STATISTICS

We start with an example of process modeling and show why least-squares functions are often used to fit this model to a set of observed data.

❑ EXAMPLE 10.1

We would like to study the effect of a certain medication on a patient. We draw blood samples at certain times after the patient takes a dose, and measure the concentration of the medication in each sample, tabulating the time t_j and concentration y_j for each sample.

We wish to construct a model that indicates the concentration of the medication as a function of time, choosing the parameters of this model so that its predictions agree as closely as possible with the observations we made by drawing the blood samples. Based on our previous experience with projects of this type, we choose the following model function:

$$\phi(x;t) = x_1 + tx_2 + e^{-x_3 t}. \qquad (10.6)$$

Here, x_1, x_2, x_3, and t are real numbers; the *ordinate* t indicates time, while the x_i's are the *parameters* of the model. The predicted concentration at time t is given by $\phi(x; t)$. The differences between model predictions and observed values are combined in the following least-squares function:

$$\tfrac{1}{2}\sum_{j=1}^{m}[y_j - \phi(x;t_j)]^2. \qquad (10.7)$$

This function has precisely the form (10.1) if we define $r_j(x) = y_j - \phi(x;t_j)$. Graphically, each term in (10.7) represents the square of the vertical distance between the curve $\phi(x;t)$ (plotted as a function of t) and the point (t_j, y_j); see Figure 10.1. We choose the minimizer x^* of the least-squares problem as the best estimate of the parameters, and use $\phi(x^*; t)$ to estimate the concentration remaining in the patient's bloodstream at any time t. ❑

This model is an example of what statisticians call a *fixed-regressor model*. It assumes that the times t_j at which the blood samples are drawn are known to high accuracy, while

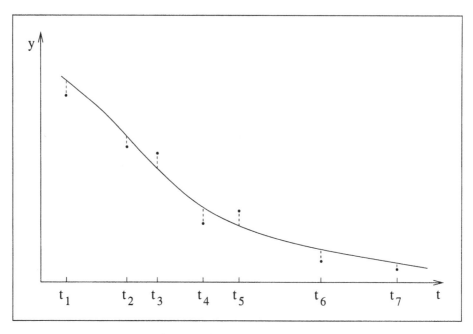

Figure 10.1 Deviation between the model (10.7) (smooth curve) and the observed measurements are indicated by the vertical bars.

the observations y_j may contain more or less random errors due to the limitations of the equipment (or the lab technician!)

In general data-fitting problems of the type just described, the ordinate t in the model $\phi(x;t)$ could be a vector instead of a scalar. (In the example above, for instance, t could contain one dimension for each of a large number of patients in a study of this particular medication.)

The sum-of-squares function (10.7) is not the only way of measuring the discrepancy between the model and the observations. Instead, we could use a sum of sixth-power terms

$$\sum_{j=1}^{m}[y_j - \phi(x;t_j)]^6,$$

or a sum of absolute values

$$\sum_{j=1}^{m}|y_j - \phi(x;t_j)|, \qquad (10.8)$$

or even some more complicated function. In fact, there are statistical motivations for choosing one fitting criterion over the alternatives. The least-squares criterion is solidly grounded in statistics, as we show below, though other criteria such as (10.8) are appropriate in some circumstances.

Changing the notation slightly, let the discrepancies between model and observation in a general data-fitting problem be denoted by ϵ_j, that is,

$$\epsilon_j = y_j - \phi(x; t_j).$$

It is often reasonable to assume that the ϵ_j's are independent and identically distributed with a certain variance σ^2 and probability density function $g_\sigma(\cdot)$. (This assumption will often be true, for instance, when the model accurately reflects the actual process, and when the errors made in obtaining the measurements y_i do not contain a "systematic" component.) Under this assumption, the likelihood of a particular set of observations y_j, $j = 1, 2, \ldots, m$, given that the actual parameter vector is x, is given by the function

$$p(y; x, \sigma) = \prod_{j=1}^{m} g_\sigma(\epsilon_j) = \prod_{j=1}^{m} g_\sigma(y_j - \phi(x; t_j)). \tag{10.9}$$

Since we *know* the value of the observations $y_1, y_2, \ldots, y_m$, the "most likely" value of x is obtained by maximizing $p(y; x, \sigma)$ with respect to x. The resulting value of x is called the *maximum likelihood estimate* of the parameters.

When we assume that the discrepancies follow a *normal* distribution, we have

$$g_\sigma(\epsilon) = \frac{1}{\sqrt{2\pi\sigma^2}} \exp\left(-\frac{\epsilon^2}{2\sigma^2}\right).$$

Substitution in (10.9) yields

$$p(y; x, \sigma) = (2\pi\sigma^2)^{-m/2} \exp\left(-\frac{1}{2} \sum_{j=1}^{m} \frac{[y_i - \phi(x; t_i)]^2}{\sigma^2}\right).$$

For any fixed value of the variance σ^2, it is obvious that p is maximized when the sum of squares (10.7) is minimized. To summarize: When the discrepancies are assumed to be independent, identically distributed with a normal distribution function, the maximum likelihood estimate is obtained by minimizing the sum of squares.

The assumptions on ϵ_j in the previous paragraph are common, but they do not describe the only situation for which the minimizer of the sum of squares makes statistical sense. Seber and Wild [227] describe many instances in which minimization of functions like (10.7), or generalizations of this function such as

$$r^T V r, \text{ where } V \in \mathbb{R}^{m \times m} \text{ is symmetric and } r_j(x) = y_j - \phi(x; t_j),$$

is the crucial step in obtaining estimates of the parameters from observed data.

Before discussing algorithms for nonlinear least-squares problems, we consider first the case in which the function $r(x)$ is *linear*. Linear least-squares problems have a wide range of applications of their own, and they also arise as subproblems in algorithms for the nonlinear problem. The algorithms that we discuss in the following section are therefore an essential component of most algorithms for nonlinear least-squares.

LINEAR LEAST-SQUARES PROBLEMS

In the special case in which each function r_i is linear, the Jacobian J is constant, and we can write

$$f(x) = \tfrac{1}{2}\|Jx + r\|_2^2, \qquad (10.10)$$

where $r = r(0)$. We also have

$$\nabla f(x) = J^T(Jx + r), \qquad \nabla^2 f(x) = J^T J.$$

(Note that the second term in $\nabla^2 f(x)$ (see (10.5)) disappears, because $\nabla^2 r_i = 0$ for all i.) The function $f(x)$ in (10.10) is convex—a property that does not necessarily hold for the nonlinear problem (10.1). By setting $\nabla f(x^*) = 0$, we see that any solution x^* of (10.10) must satisfy

$$J^T J x^* = -J^T r. \qquad (10.11)$$

The equations (10.11) are the *normal equations* for (10.10).

The linear problem (10.10) is interesting in its own right, since many models $\phi(x; t)$ are linear. (Our example (10.6) would be linear but for the existence of the x_3 term.) Algorithms for linear least-squares problems fall under the aegis of numerical linear algebra rather than optimization, and there is a great deal of literature on the topic. We give additional details in the Notes and References.

We outline briefly three major algorithms for the unconstrained linear least-squares problem. We assume for the present that $m \geq n$ and that J has full column rank.

The first and most obvious algorithm is simply to form and solve the system (10.11) by the following three-step procedure:

- compute the coefficient matrix $J^T J$ and the right-hand-side $-J^T r$;
- compute the Cholesky factorization of the symmetric matrix $J^T J$;
- perform two triangular substitutions with the Cholesky factors to recover the solution x^*.

The Cholesky factorization

$$J^T J = \bar{R}^T \bar{R} \qquad (10.12)$$

(where $\bar{R}$ is $n \times n$ upper triangular) is guaranteed to exist when $m \geq n$ and J has rank m (see the exercises). This method is frequently used in practice and is often effective, but it has one significant disadvantage, namely, that the condition number of $J^T J$ is the square of the condition number of J. Since the relative error in the computed solution of any problem is usually proportional to the condition number, the solution obtained by the Cholesky-based method may be much less accurate than solutions obtained from algorithms that work directly with the formulation (10.10). When J is ill conditioned, the Cholesky factorization process may even break down, since roundoff errors may cause small negative elements to appear on the diagonal.

A second approach is based on a QR factorization of the matrix J. Since the Euclidean norm of any vector is not affected by orthogonal transformations, we have

$$\|Jx + r\|_2 = \|Q^T(Jx + r)\|_2 \tag{10.13}$$

for any $m \times m$ orthogonal matrix Q. Suppose we perform a QR factorization with column pivoting on the matrix J (see (A.54)) to obtain

$$J\Pi = Q \begin{bmatrix} R \\ 0 \end{bmatrix} = \begin{bmatrix} Q_1 & Q_2 \end{bmatrix} \begin{bmatrix} R \\ 0 \end{bmatrix} = Q_1 R, \tag{10.14}$$

where

Π is an $n \times n$ permutation matrix (hence, orthogonal);
Q is $m \times m$ orthogonal;
Q_1 is the first n columns of Q, while Q_2 contains the last $m - n$ columns;
R is $n \times n$ upper triangular.

By combining (10.13) and (10.14), we obtain

$$\begin{aligned} \|Jx + r\|_2^2 &= \left\| \begin{bmatrix} Q_1^T \\ Q_2^T \end{bmatrix} (J \Pi \Pi^T x + r) \right\|_2^2 \\ &= \left\| \begin{bmatrix} R \\ 0 \end{bmatrix} (\Pi^T x) + \begin{bmatrix} Q_1^T r \\ Q_2^T r \end{bmatrix} \right\|_2^2 \\ &= \|R(\Pi^T x) + Q_1^T r\|_2^2 + \|Q_2^T r\|_2^2. \end{aligned} \tag{10.15}$$

No choice of x has any effect on the second term of this last expression, but we can minimize $\|Jx + r\|$ by driving the first term to zero, that is, by setting

$$x^* = -\Pi R^{-1} Q_1^T r.$$

(In practice, we perform a triangular substitution to solve $Rz = -Q_1^T r$, then permute the components of z to get $x^* = \Pi z$.)

This QR-based approach does not degrade the conditioning of the problem unnecessarily. The relative error in the final computed solution x^* is usually proportional to the condition number of J, not its square, and this method is usually reliable. Some situations, however, call for greater robustness or more information about the sensitivity of the solution to perturbations in the data (J or r). A third approach, based on the singular-value decomposition (SVD) of J, can be used in these circumstances. Recall from (A.45) that the SVD of J is given by

$$J = U \begin{bmatrix} S \\ 0 \end{bmatrix} V^T = \begin{bmatrix} U_1 & U_2 \end{bmatrix} \begin{bmatrix} S \\ 0 \end{bmatrix} V^T = U_1 S V^T, \qquad (10.16)$$

where

U is $m \times m$ orthogonal;
U_1 contains the first n columns of U, U_2 the last $m - n$ columns;
V is $n \times n$ orthogonal;
S is $n \times n$ diagonal, with diagonal elements $\sigma_1 \geq \sigma_2 \geq \cdots \geq \sigma_n > 0$.

(Note that $J^T J = V S^2 V^T$, so that the columns of V are eigenvectors of J^T with eigenvalues σ_j^2, $j = 1, 2, \ldots, n$.) By following the same logic that led to (10.15), we obtain

$$\|Jx + r\|_2^2 = \left\| \begin{bmatrix} S \\ 0 \end{bmatrix} (V^T x) + \begin{bmatrix} U_1^T \\ U_2^T \end{bmatrix} r \right\|_2^2$$
$$= \|S(V^T x) + U_1^T r\|_2^2 + \|U_2^T r\|_2^2. \qquad (10.17)$$

Again, the optimum is found by choosing x to make the first term equal to zero; that is,

$$x^* = V S^{-1} U_1^T r.$$

If we denote the ith columns of U and V by $u_i \in \mathbf{R}^m$ and $v_i \in \mathbf{R}^n$, respectively, we can write

$$x^* = \sum_{i=1}^n \frac{u_i^T r}{\sigma_i} v_i. \qquad (10.18)$$

This formula yields useful information about the sensitivity of x^*. When σ_i is small, x^* is particularly sensitive to perturbations in r that affect $u_i^T r$, and also to perturbations in J that affect this same quantity. Information such as this is particularly useful when J is nearly rank-deficient, that is, when $\sigma_n/\sigma_1 \ll 1$. This information is not generally available when the QR algorithm is used, and it is sometimes worth the extra cost of the SVD algorithm to obtain it.

All three approaches above have their place. The Cholesky-based algorithm is particularly useful when $m \gg n$ and it is practical to store $J^T J$ but not J itself. However, it must be modified when J is rank-deficient or ill conditioned to allow pivoting of the diagonal elements of $J^T J$. In the QR approach, ill conditioning usually (but not always) causes the elements in the lower right-hand corner of R to be much smaller than the other elements in this matrix. In this situation, the strategy above can be modified to produce a solution to a nearby problem in which J is slightly perturbed. The SVD approach is the most robust and reliable one for ill-conditioned problems. When J is actually rank-deficient, some of the singular values σ_i are exactly zero, and any vector x^* of the form

$$x^* = \sum_{\sigma_i \neq 0} \frac{u_i^T r}{\sigma_i} v_i + \sum_{\sigma_i = 0} \tau_i v_i \tag{10.19}$$

(for arbitrary coefficients τ_i) is a minimizer of (10.17). Frequently, the solution with smallest norm is most desirable, and we obtain this by setting each $\tau_i = 0$ in (10.19) (see the exercises). When J has full rank but is ill conditioned, the last few singular values $\sigma_n, \sigma_{n-1}, \ldots$ are small. The coefficients $u_i^T r / \sigma_i$ in (10.19) are then particularly sensitive to perturbations in $u_i^T r$ when σ_i is small, so a "robust" approximate solution is sometimes obtained by omitting these terms from the summation.

10.2 ALGORITHMS FOR NONLINEAR LEAST-SQUARES PROBLEMS

THE GAUSS–NEWTON METHOD

We now describe methods for minimizing the nonlinear objective function (10.1) that exploit the structure in the gradient ∇f (10.4) and Hessian $\nabla^2 f$ (10.5). The simplest of these methods—the Gauss–Newton method—can be viewed as a modification of Newton's method with line search. Instead of generating the search direction p_k by solving the standard Newton equations $\nabla^2 f(x_k) p = -\nabla f(x_k)$, we exclude the second-order term from $\nabla^2 f(x_k)$ and obtain p_k^{GN} by solving

$$J_k^T J_k p_k^{GN} = -J_k^T r_k. \tag{10.20}$$

This simple modification gives a surprising number of advantages over the plain Newton's method. First, our use of the approximation

$$\nabla^2 f_k \approx J_k^T J_k \tag{10.21}$$

saves us the trouble of computing the individual Hessians $\nabla^2 r_i$, $i = 1, 2, \ldots, m$, of the residuals, which are needed in the second term in (10.5). In fact, if we calculated the Jacobian

$J(x_k)$ in the course of evaluating the gradient $\nabla f_k = J_k^T f_k$, the approximation (10.21) is essentially free, and the savings in function and derivative evaluation time can be very significant. Second, there are many interesting situations in which the first term $J^T J$ in (10.5) is much more significant than the second term, so that the Gauss–Newton method gives performance quite similar to that of Newton's method even when the second term $\sum_{j=1}^m r_j \nabla^2 r_j$ is omitted. A sufficient condition for the first term in (10.5) to dominate the second term is that the size of each second-order term (that is, $|r_j(x)| \|\nabla^2 r_j(x)\|$) be significantly smaller than the eigenvalues of $J^T J$. This happens, for instance, when the residuals r_j are small (the so-called *small-residual case*), or when each r_j is nearly a linear function (so that $\|\nabla^2 r_j\|$ is small). In practice, many least-squares problems have small residuals at the solution, and rapid local convergence of Gauss–Newton often is observed on these problems.

A third advantage of Gauss–Newton is that whenever $J(x_k)$ has full rank and the gradient ∇f is nonzero, the direction p_k^{GN} is a descent direction for $f(\cdot)$, and therefore a suitable direction for a line search. From (10.4) and (10.20) we have

$$(p_k^{\text{GN}})^T \nabla f_k = (p_k^{\text{GN}})^T J_k^T r_k = -(p_k^{\text{GN}})^T J_k^T J_k p_k^{\text{GN}} = -\|J_k p_k^{\text{GN}}\|_2^2 \le 0.$$

The final inequality is strict unless we have $J_k p_k^{\text{GN}} = 0$, which by (10.20) is equivalent to $J_k^T r_k = \nabla f_k = 0$. Finally, the fourth advantage of this method arises from the similarity between the Gauss–Newton equations (10.20) and the normal equations (10.11) for the linear least-squares problem. This connection tells us that p_k^{GN} is the solution of the linear least-squares problem

$$\min_p \|J_k p + f_k\|_2^2. \tag{10.22}$$

Hence, we can find the search direction by applying the algorithms of the previous section to the subproblem (10.22). If the QR or SVD algorithms are used, there is no need to calculate the Hessian approximation $J_k^T J_k$ explicitly.

The subproblem (10.22) suggests another motivation for the Gauss–Newton step. Rather than forming a quadratic model of the function $f(x)$, we are forming a linear model of the vector function $r(x+p) \approx r(x) + J(x)p$. The step p is obtained by substituting this linear model into the expression $f(x) = \frac{1}{2}\|r(x)\|_2^2$ and minimizing over p.

As mentioned above, we usually perform a line search in the Gauss–Newton direction, finding a step length α_k that satisfies the Wolfe conditions (3.6). The theory of Chapter 3 can then be applied to ensure global convergence of this method. Assume for the moment that the Jacobians $J(x)$ have their singular values uniformly bounded away from zero in the region of interest; that is, there is a constant $\gamma > 0$ such that

$$\|J(x)z\|_2 \ge \gamma \|z\|_2 \tag{10.23}$$

10.2. ALGORITHMS FOR NONLINEAR LEAST-SQUARES PROBLEMS

for all x in a neighborhood $\mathcal{N}$ of the level set

$$\mathcal{L} = \{x \mid f(x) \le f(x_0)\}, \qquad (10.24)$$

where x_0 is the starting point. This assumption is the key to proving the following result, which is a consequence of Theorem 3.2.

Theorem 10.1.
 Suppose that each residual function r_j is Lipschitz continuously differentiable in a neighborhood $\mathcal{N}$ of the level set (10.24), and that the Jacobians $J(x)$ satisfy the uniform full-rank condition (10.23). Then if the iterates x_k are generated by the Gauss–Newton method with step lengths α_k that satisfy (3.6), we have

$$\lim_{k \to \infty} J_k^T r_k = 0.$$

PROOF. First, we check that the angle θ_k between the search direction p_k^{GN} and the negative gradient $-\nabla f_k$ is uniformly bounded away from $\pi/2$. Note that Lipschitz continuity of each $r_j(x)$ implies continuity of $J(x)$; hence there is a constant $\beta > 0$ such that $\|J(x)^T\|_2 = \|J(x)\|_2 \le \beta$ for all $x \in \mathcal{L}$. From (3.12), we have for $x = x_k \in \mathcal{L}$ and $p^{\text{GN}} = p_k^{\text{GN}}$ that

$$\cos \theta_k = -\frac{(\nabla f)^T p^{\text{GN}}}{\|p^{\text{GN}}\| \|\nabla f\|}$$

$$= -\frac{r^T J p^{\text{GN}}}{\|p^{\text{GN}}\| \|J^T r\|} = \frac{\|J p^{\text{GN}}\|^2}{\|p^{\text{GN}}\| \|J^T J p^{\text{GN}}\|} \ge \frac{\gamma^2 \|p^{\text{GN}}\|^2}{\beta^2 \|p^{\text{GN}}\|^2} = \frac{\gamma^2}{\beta^2} > 0.$$

Lipschitz continuity of ∇r_j also implies Lipschitz continuity of $\nabla f(\cdot)$ over the neighborhood $\mathcal{N}$, so we can apply Theorem 3.2 to deduce that $\nabla f(x_k) \to 0$, giving the result. □

If $J(x_k)$ is rank-deficient for some k (that is, a condition like (10.23) is not satisfied), the coefficient matrix in (10.20) is singular. The system (10.20) still has a solution, however, because of the equivalence between this linear system and the minimization problem (10.22). In fact, there are infinitely many solutions for p_k^{GN} in this case; each of them has the form shown in (10.19). However, there is no longer an assurance that $\cos \theta_k$ is uniformly bounded away from zero, so we cannot prove a result like Theorem 10.1 in general.

The speed of convergence of Gauss–Newton near a solution x^* depends on how much the leading term $J^T J$ dominates the second-order term in the Hessian (10.5). Suppose that x_k is close to x^* and that an assumption like (10.23) is satisfied. Then, applying an argument like the Newton's method analysis (3.34), (3.35), (3.36) in Chapter 3, we have

$$x_k + p_k^{\text{GN}} - x^* = x_k - x^* - [J^T J(x_k)]^{-1} \nabla f(x_k)$$
$$= [J^T J(x_k)]^{-1} \left[J^T J(x_k)(x_k - x^*) + \nabla f(x^*) - \nabla f(x_k) \right],$$

where $J^T J(x)$ is shorthand notation for $J(x)^T J(x)$. Using $H(x)$ to denote the second-order term in (10.5), we can show that

$$\nabla f(x_k) - \nabla f(x^*) = \int_0^1 J^T J(x^* + t(x_k - x^*))(x_k - x^*)\,dt$$
$$+ \int_0^1 H(x^* + t(x_k - x^*))(x_k - x^*)\,dt.$$

The same argument as in (3.35), (3.36) shows that

$$\|x_k + p_k^{\text{GN}} - x^*\|$$
$$\leq \int_0^1 \|[J^T J(x_k)]^{-1} H(x^* + t(x_k - x^*))\| \|x_k - x^*\|\,dt + O(\|x_k - x^*\|^2)$$
$$\approx \|[J^T J(x^*)]^{-1} H(x^*)\| \|x_k - x^*\| + O(\|x_k - x^*\|^2). \quad (10.25)$$

Therefore, we cannot expect a unit step of Gauss–Newton to move us closer to the solution unless $\|[J^T J(x^*)]^{-1} H(x^*)\| < 1$. In the small residual case discussed above, we often have $\|[J^T J(x^*)]^{-1} H(x^*)\| \ll 1$, so (10.25) implies that the convergence of x_k to x^* is rapid. When $H(x^*) = 0$, the convergence is quadratic.

THE LEVENBERG–MARQUARDT METHOD

Recall that the Gauss–Newton method is like Newton's method with line search, except that we use the convenient and often effective approximation (10.21) for the Hessian. The Levenberg–Marquardt method can be derived by replacing the line search strategy with a trust-region strategy. The use of a trust region avoids one of the weaknesses of Gauss–Newton, namely, its behavior when the Jacobian $J(x)$ is rank-deficient, or nearly so. The second-order Hessian component in (10.5) is still ignored, however, so the local convergence properties of the two methods are similar.

The Levenberg–Marquardt method can be described and analyzed using the trust-region framework of Chapter 4. (In fact, the Levenberg–Marquardt method is generally considered to be the progenitor of the trust-region approach for general unconstrained optimization discussed in Chapter 4.) For a spherical trust region, the subproblem to be solved at each iteration is

$$\min_p \tfrac{1}{2} \|J_k p + r_k\|_2^2, \qquad \text{subject to } \|p\| \leq \Delta_k, \quad (10.26)$$

where $\Delta_k > 0$ is the trust-region radius. In effect, we are choosing the model function $m_k(\cdot)$ in (4.3) to be

$$m_k(p) = \tfrac{1}{2} \|r_k\|^2 + p^T J_k^T r_k + \tfrac{1}{2} p^T J_k^T J_k p. \quad (10.27)$$

10.2. Algorithms for Nonlinear Least-Squares Problems

The following convergence result can be obtained as a direct consequence of Theorem 4.8.

Theorem 10.2.

Let $\eta \in \left(0, \frac{1}{4}\right)$ in Algorithm 4.1 of Chapter 4, and suppose that the functions $r_i(\cdot)$ are twice continuously differentiable in a neighborhood of the level set $\mathcal{L} = \{x \mid f(x) \leq f(x_0)\}$, and that for each k, the approximate solution p_k of (10.26) satisfies the inequality

$$m_k(0) - m_k(p_k) \geq c_1 \|J_k^T r_k\| \min\left(\Delta_k, \frac{\|J_k^T r_k\|}{\|J_k^T J_k\|}\right), \tag{10.28}$$

for some constant $c_1 > 0$. We then have that

$$\lim_{k \to \infty} \nabla f_k = \lim_{k \to \infty} J_k^T r_k = 0.$$

PROOF. The smoothness assumption on $r_i(\cdot)$ implies that $\|J^T J\|$ and $\|\nabla^2 f\|$ are bounded over a neighborhood of $\mathcal{L}$, so the assumptions of Theorem 4.8 are satisfied, and the result follows. □

As in Chapter 4, there is no need to calculate the right-hand-side in the inequality (10.28) or to check it explicitly. Instead, we can simply require the decrease given by our approximate solution p_k of (10.26) to at least match the decrease given by the Cauchy point, which can be calculated inexpensively in the same way as in Chapter 4.

Let us drop the iteration counter k during the rest of this section and concern ourselves with the subproblem (10.26). It happens that we can characterize the solution of (10.26) in the following way: When the solution p^{GN} of the Gauss–Newton equations (10.20) lies strictly inside the trust region (that is, $\|p^{\text{GN}}\| < \Delta$), then this step p^{GN} also solves the subproblem (10.26). Otherwise, there is a $\lambda > 0$ such that the solution $p = p^{\text{LM}}$ of (10.26) satisfies $\|p\| = \Delta$ and

$$\left(J^T J + \lambda I\right) p = -J^T r. \tag{10.29}$$

This claim is verified in the following lemma, which is a straightforward consequence of Theorem 4.3 from Chapter 4.

Lemma 10.3.

The vector p^{LM} is a solution of the trust-region subproblem

$$\min_p \|Jp + r\|_2^2, \quad \text{subject to } \|p\| \leq \Delta,$$

for some $\Delta > 0$ if and only if there is a scalar $\lambda \geq 0$ such that

$$(J^T J + \lambda I) p^{\text{LM}} = -J^T r, \tag{10.30a}$$

$$\lambda(\Delta - \|p^{\text{LM}}\|) = 0. \tag{10.30b}$$

PROOF. In Theorem 4.3, the semidefiniteness condition (4.19c) is satisfied automatically, since $J^T J$ is positive semidefinite and $\lambda \geq 0$. The two conditions (10.30a) and (10.30b) follow from (4.19a) and (4.19b), respectively. □

Note that the equations (10.29) are just the normal equations for the linear least-squares problem

$$\min_p \frac{1}{2} \left\| \begin{bmatrix} J \\ \sqrt{\lambda} I \end{bmatrix} p + \begin{bmatrix} r \\ 0 \end{bmatrix} \right\|^2. \tag{10.31}$$

Just as in the Gauss–Newton case, the equivalence between (10.29) and (10.31) gives us a way of solving the subproblem without computing the matrix–matrix product $J^T J$ and its Cholesky factorization.

The original description of the Levenberg–Marquardt algorithm [150, 160] did not make the connection with the trust-region concept. Rather, it adjusted the value of λ in (10.29) directly, increasing or decreasing it by a certain factor according to whether or not the previous trial step was effective in decreasing $f(\cdot)$. (The heuristics for adjusting λ were similar to those used for adjusting the trust-region radius Δ_k in Algorithm 4.1.) Similar convergence results to Theorem 10.2 can be proved for algorithms that use this approach (see, for instance, Osborne [186]), independently of trust-region analysis. The connection with trust regions was firmly established in a paper by Moré [166]. The trust-region viewpoint has proved interesting from both a theoretical and practical perspective, since it has given rise to efficient, robust software, as we note below.

IMPLEMENTATION OF THE LEVENBERG–MARQUARDT METHOD

To find a value of λ that approximately matches the given Δ in Lemma 10.3, we can use the root-finding algorithm described in Chapter 4. It is easy to safeguard this procedure: The Cholesky factor R is guaranteed to exist whenever the current estimate $\lambda^{(\ell)}$ is positive, since the approximate Hessian $B = J^T J$ is already positive semidefinite. Because of the special structure of B, we do not need to compute the Cholesky factorization naively as in Chapter 4. Rather, it is easy to show that the following QR factorization of the coefficient matrix in (10.31),

$$\begin{bmatrix} R_\lambda \\ 0 \end{bmatrix} = Q_\lambda^T \begin{bmatrix} J \\ \sqrt{\lambda} I \end{bmatrix} \tag{10.32}$$

(Q_λ orthogonal, R_λ upper triangular), yields an R_λ that satisfies $R_\lambda^T R_\lambda = (J^T J + \lambda I)$.

We can save a little computer time in the calculation of the factorization (10.32) by using a combination of Householder and Givens transformations. Suppose we use Householder

10.2. Algorithms for Nonlinear Least-Squares Problems

transformations to calculate the QR factorization of J alone as

$$J = Q \begin{bmatrix} R \\ 0 \end{bmatrix}. \tag{10.33}$$

We then have

$$\begin{bmatrix} R \\ 0 \\ \sqrt{\lambda}I \end{bmatrix} = \begin{bmatrix} Q^T & \\ & I \end{bmatrix} \begin{bmatrix} J \\ \sqrt{\lambda}I \end{bmatrix}. \tag{10.34}$$

The leftmost matrix in this formula is upper triangular except for the n nonzero terms of the matrix λI. These can be eliminated by a sequence of $n(n+1)/2$ Givens rotations, in which the diagonal elements of the upper triangular part are used to eliminate the nonzeros of λI and the fill-in terms that arise in the process. The first few steps of this process are as follows:

rotate row n of R with row n of $\sqrt{\lambda}I$, to eliminate the (n, n) element of $\sqrt{\lambda}I$;

rotate row $n-1$ of R with row $n-1$ of $\sqrt{\lambda}I$ to eliminate the $(n-1, n-1)$ element of the latter matrix. This step introduces fill-in in position $(n-1, n)$ of $\sqrt{\lambda}I$, which is eliminated by rotating row n of R with row $n-1$ of $\sqrt{\lambda}I$, to eliminate the fill-in element at position $(n-1, n)$;

rotate row $n-2$ of R with row $n-2$ of $\sqrt{\lambda}I$, to eliminate the $(n-2)$ diagonal in the latter matrix. This step introduces fill-in in the $(n-2, n-1)$ and $(n-2, n)$ positions, which we eliminate by ...

and so on. If we gather all the Givens rotations into a matrix $\bar{Q}_\lambda$, we obtain from (10.34) that

$$\bar{Q}_\lambda^T \begin{bmatrix} R \\ 0 \\ \sqrt{\lambda}I \end{bmatrix} = \begin{bmatrix} R_\lambda \\ 0 \\ 0 \end{bmatrix},$$

and hence (10.32) holds with

$$Q_\lambda = \begin{bmatrix} Q & \\ & I \end{bmatrix} \bar{Q}_\lambda.$$

The advantage of this combined approach is that when the value of λ is changed in the root-finding algorithm, we need only recalculate $\bar{Q}_\lambda$ and not the Householder part of the factorization (10.34). This feature can save a lot of computation in the case of $m \gg n$, since just $O(n^3)$ operations are required to recalculate $\bar{Q}_\lambda$ and R_λ for each value of λ, after the initial cost of $O(mn^2)$ operations needed to calculate Q in (10.33).

Least-squares problems are often poorly scaled. Some of the variables could have values of about 10^4, while other variables could be of order 10^{-6}. If such wide variations are ignored, the algorithms above may encounter numerical difficulties or produce solutions of poor quality. One way to reduce the effects of poor scaling is to use an *ellipsoidal* trust region in place of the spherical trust region defined above. The step is confined to an ellipse in which the lengths of the principal axes are related to the typical values of the corresponding variables. Analytically, the trust-region subproblem becomes

$$\min_p \tfrac{1}{2}\|J_k p + r_k\|_2^2, \qquad \text{subject to } \|D_k p\| \le \Delta_k, \tag{10.35}$$

where D_k is a diagonal matrix with positive diagonal entries. Instead of (10.29), the solution of (10.35) satisfies an equation of the form

$$\left(J_k^T J_k + \lambda D_k^2\right) p_k^{\text{LM}} = -J_k^T r_k, \tag{10.36}$$

and, equivalently, solves the linear least-squares problem

$$\min_p \left\| \begin{bmatrix} J_k \\ \sqrt{\lambda} D_k \end{bmatrix} p + \begin{bmatrix} r_k \\ 0 \end{bmatrix} \right\|. \tag{10.37}$$

The diagonals of the scaling matrix D_k can change from iteration to iteration, as we gather information about the typical range of values for each component of x. If the variation in these elements is kept within certain bounds, then the convergence theory for the spherical case continues to hold. Moreover, the technique described above for calculating R_λ needs no modification. Seber and Wild [227] suggest choosing the diagonals of D_k^2 to match those of $J_k^T J_k$, to make the algorithm invariant under diagonal scaling of the components of x. This is analogous to the technique of scaling by diagonal elements of the Hessian, which was described in Chapter 4 in the context of trust-region algorithms for unconstrained optimization.

The local convergence behavior of Levenberg–Marquardt is similar to the Gauss–Newton method. Near a solution with small residuals, the model function in (10.26) will give an accurate picture of the behavior of f. The trust region will eventually become inactive, and the algorithm will take unit Gauss–Newton steps, giving the rapid linear convergence rate that we derived in our analysis of the Gauss–Newton method.

LARGE-RESIDUAL PROBLEMS

In large-residual problems, the quadratic model in (10.26) is an inadequate representation of the function f, because the second-order part of the Hessian $\nabla^2 f(x)$ is too significant to be ignored. Some statisticians downplay the importance of algorithms for such problems. They argue that if the residuals are large at the solution, then the model is not up to the task of matching the observations. Instead of solving the least-squares problem, they would prefer to construct a new and better model. Often, however, large residuals are caused by "outliers" in the observations due to human error: An operator may read a dial incorrectly, or a geologist may assign a particular seismic reading to the wrong earthquake. In these cases, we may still want to solve the least-squares problem because the solution may be useful in identifying the outliers. These can then be removed from the data set or deemphasized by assigning them a lower weight.

Performance of the Gauss–Newton and Levenberg–Marquardt algorithms is usually poor in the large-residual case. Asymptotic convergence is only linear—slower than the superlinear convergence rate attained by algorithms for general unconstrained problems, such as Newton or quasi-Newton. Newton's method with trust region or line search is an option, particularly if the individual Hessians $\nabla^2 r_j(\cdot)$ are easy to calculate. Quasi-Newton methods will also converge at a faster asymptotic rate than Gauss–Newton or Levenberg–Marquardt in the large-residual case. However, the behavior of both Newton and quasi-Newton on early iterations (before the iterations reach a neighborhood of the solution) may be inferior to Gauss–Newton and Levenberg–Marquardt, and in the case of Newton's method, we have to pay the additional cost of computing second derivatives.

Of course, we often do not know beforehand whether a problem will turn out to have small or large residuals at the solution. It seems reasonable, therefore, to consider *hybrid algorithms*, which would behave like Gauss–Newton or Levenberg–Marquardt if the residuals turn out to be small (and hence take advantage of the cost savings associated with these methods) but switch to Newton or quasi-Newton steps if the residuals at the solution appear to be large.

There are a couple of ways to construct hybrid algorithms. One approach, due to Fletcher and Xu (see Fletcher [83]), maintains a sequence of positive definite Hessian approximations B_k. If the Gauss–Newton step from x_k reduces the function f by a certain fixed amount (say, a factor of 5), then this step is taken and B_k is overwritten by $J_k^T J_k$. Otherwise, a direction is computed using B_k, and the new point x_{k+1} is obtained by doing a line search. In either case, a BFGS-like update is applied to B_k to obtain a new approximation B_{k+1}. In the zero-residual case, the method eventually always takes Gauss–Newton steps (giving quadratic convergence), while it eventually reduces to BFGS in the nonzero-residual case (giving superlinear convergence). Numerical results in Fletcher [83, Tables 6.1.2, 6.1.3] show good results for this approach on small-, large-, and zero-residual problems.

A second way to combine Gauss–Newton and quasi-Newton ideas is to maintain approximations to just the second-order part of the Hessian. That is, we maintain a sequence of matrices S_k that approximate the summation term $\sum_{j=1}^{m} r_j(x_k) \nabla^2 r_j(x_k)$ in (10.5), and then use the overall Hessian approximation

$$B_k = J_k^T J_k + S_k$$

in a trust-region or line search model for calculating the step p_k. Updates to S_k are devised so that the approximate Hessian B_k, or its constituent parts, mimics the behavior of the corresponding exact quantities over the step just taken. Here is another instance of the *secant equation*, which also arose in the context of unconstrained minimization (8.6) and nonlinear equations (11.26). In the present case, there are a number of different ways to define the secant equation and to specify the other conditions needed for a complete update formula for S_k. We describe the algorithm of Dennis, Gay, and Welsch [67], which is probably the best-known algorithm in this class because of its implementation in the well-known NL2SOL package.

Dennis, Gay, and Welsch motivate their form of the secant equation in the following way. Ideally, S_{k+1} should be a close approximation to the exact second-order term at $x = x_{k+1}$; that is,

$$S_{k+1} \approx \sum_{j=1}^{m} r_j(x_{k+1}) \nabla^2 r_j(x_{k+1}).$$

Since we do not want to calculate the individual Hessians $\nabla^2 r_j$ in this formula, we could replace each of them with an approximation $(B_j)_{k+1}$ and impose the condition that $(B_j)_{k+1}$ should mimic the behavior of its exact counterpart $\nabla^2 r_j$ over the step just taken; that is,

$$(B_j)_{k+1}(x_{k+1} - x_k) = \nabla r_j(x_{k+1}) - \nabla r_j(x_k)$$
$$= (\text{row } j \text{ of } J(x_{k+1}))^T - (\text{row } j \text{ of } J(x_k))^T.$$

This condition leads to a secant equation on S_{k+1}, namely,

$$S_{k+1}(x_{k+1} - x_k) = \sum_{j=1}^{m} r_j(x_{k+1})(B_j)_{k+1}(x_{k+1} - x_k)$$
$$= \sum_{j=1}^{m} r_j(x_{k+1}) \left[(\text{row } j \text{ of } J(x_{k+1}))^T - (\text{row } j \text{ of } J(x_k))^T \right]$$
$$= J_{k+1}^T r_{k+1} - J_k^T r_k.$$

As usual, this condition does not completely specify the new approximation S_{k+1}. Dennis, Gay, and Welsch add requirements that S_{k+1} be symmetric and that the difference $S_{k+1} - S_k$ from the previous estimate S_k be minimized in a certain sense, and derive the following update formula:

$$S_{k+1} = S_k + \frac{(y^\sharp - S_k s) y^T + y (y^\sharp - S_k s)^T}{y^T s} - \frac{(y^\sharp - S_k s)^T s}{(y^T s)^2} y y^T, \qquad (10.38)$$

where

$$s = x_{k+1} - x_k,$$

$$y = J_{k+1}^T r_{k+1} - J_k^T r_k,$$
$$y^\sharp = J_{k+1}^T r_{k+1} - J_k^T r_{k+1}.$$

Note that (10.38) is a slight variant on the DFP update for unconstrained minimization. It would be identical if $y^\sharp$ and y were the same.

Dennis, Gay, and Welsch use their approximate Hessian $J_k^T J_k + S_k$ in conjunction with a trust-region strategy, but a few more features are needed to enhance its performance. One deficiency of its basic update strategy for S_k is that this matrix is not guaranteed to vanish as the iterates approach a zero-residual solution, so it can interfere with superlinear convergence. This problem is avoided by scaling S_k prior to its update; we replace S_k by $\tau_k S_k$ on the right-hand-side of (10.38), where

$$\tau_k = \min\left(1, \frac{|s^T y^\sharp|}{|s^T S_k s|}\right).$$

A final modification in the overall algorithm is that the S_k term is omitted from the Hessian approximation when the resulting Gauss–Newton model produces a sufficiently good step.

LARGE-SCALE PROBLEMS

The problem (10.1) can be classified as large if the number of parameters n and number of residuals m are both large. If n is small while m is large, then the algorithms described in earlier sections can be applied directly, possibly with some modifications to account for peculiarities of the problem. If, for instance, m is of order 10^6 while n is about 100, the Jacobian $J(x)$ may be too large to store conveniently. However, it is easy to calculate the matrix $J^T J$ and gradient vector $J^T r$ by evaluating r_j and ∇r_j successively for $j = 1, 2, \ldots, m$ and performing the accumulations

$$J^T J = \sum_{i=1}^m (\nabla r_j)(\nabla r_j)^T, \quad J^T r = \sum_{i=1}^m r_j (\nabla r_j).$$

The Gauss–Newton and Levenberg–Marquardt steps can then be computed by solving the systems (10.20) and (10.29) of normal equations directly. Note that there is no need to recalculate $J^T J$ in (10.29) for each new value of λ in the Levenberg–Marquardt step; we need only add the new value of λI to the previously computed $J^T J$ and then recompute the factorization.

When n and m are *both* large, however, and the Jacobian $J(x)$ is sparse, the cost of computing steps exactly via the formulae (10.20) and (10.29) may become quite expensive relative to the cost of function and gradient evaluations. In this case, we can design *inexact* variants of the Gauss–Newton and Levenberg–Marquardt algorithms that are analogous to the inexact Newton algorithms discussed in Chapters 6. We simply replace the

Hessian $\nabla^2 f(x_k)$ in these methods by its approximation $J_k^T J_k$. The positive semidefiniteness of this approximation simplifies the resulting algorithms in several places. Naturally, we cannot expect the superlinear convergence results to hold, except in the case in which the approximation is exact at the solution, that is, $\nabla^2 f(x^*) = J(x^*)^T J(x^*)$.

We mention two other approaches for large-scale problems that take advantage of the special structure of least squares. Toint [239] combines the Gauss–Newton and Dennis–Gay–Welsch approaches described earlier with the partially separable ideas discussed in Chapter 9. A compact representation of each second-order term $\nabla^2 r_j$, $j = 1, 2, \ldots, m$, of the Hessian is maintained, and at each iteration a choice is made as to whether this approximation should be used in the step calculation or simply ignored. (In the latter case, the Gauss–Newton step will be calculated.) The step equations are solved by a conjugate gradient procedure.

The inexact Levenberg–Marquardt algorithm described by Wright and Holt [257] takes steps $\bar{p}$ that are inexact solutions of the system (10.29). Analogously to (6.2) and (6.3) in Chapter 6, they satisfy

$$\left\| \left(J^T J + \lambda I \right) \bar{p} + J^T r \right\| \leq \eta \| J^T r \|, \qquad \text{for some } \eta \in (0, 1). \tag{10.39}$$

Instead of using a trust-region radius to dictate the choice of λ, however, this method reverts to the original Levenberg–Marquardt strategy of manipulating λ directly. Specifically, λ is chosen sufficiently large at each iteration to force a "sufficient decrease" in the sum of squares at the new point $x + \bar{p}$, according to the following criterion:

$$\frac{\|r(x)\|^2 - \|r(x+\bar{p})\|^2}{\|r(x)\|^2 - \|r(x) + J(x)\bar{p}\|^2 - \lambda^2 \|\bar{p}\|^2} \geq \gamma_1, \tag{10.40}$$

for some chosen constant $\gamma_1 \in (0, 1)$. By choosing λ to be not too much larger than the smallest value for which this bound in satisfied, we are able to prove a global convergence result.

Rather than recomputing approximate solutions of the system (10.29) for a number of different λ values in turn, until one such value yields a step $\bar{p}$ for which (10.39) holds, the algorithm in [257] applies Algorithm LSQR of Paige and Saunders [188] to the least-squares formulation (10.31), solving this system simultaneously for a number of different values of λ. This approach takes advantage of the fact that each additional value of λ entails just a small marginal cost for LSQR: to be precise, storage of just two additional vectors and some additional vector arithmetic (but no extra matrix–vector multiplications). We can therefore identify a value of λ and corresponding step $\bar{p}$ that satisfy the conditions of the global convergence result at a cost not too much higher than that of the approximate solution of a single linear least-squares problem of the form (10.31) by the LSQR algorithm.

10.3 ORTHOGONAL DISTANCE REGRESSION

In Example 10.1 we assumed that no errors were made in noting the time at which the blood samples were drawn, so that the differences between the model $\phi(x; t_j)$ and the observation y_j were due to inadequacy in the model or in errors in making the measurement of y_j. We assumed that any errors in the ordinates—the times t_j—are tiny by comparison with the errors in the observations. This assumption often is reasonable, but there are cases where the answer can be seriously distorted if we fail to take possible errors in the ordinates into account. Models that take these errors into account are known in the statistics literature as *errors-in-variables models* [227, Chapter 10], and the resulting optimization problems are referred to as *total least squares* in the case of a linear model (see Golub and Van Loan [115, Chapter 5]) or as *orthogonal distance regression* in the nonlinear case (see Boggs, Byrd, and Schnabel [20]).

We formulate this problem mathematically by introducing perturbations δ_j for the ordinates t_j, as well as perturbations ϵ_j for y_j, and seeking the values of these $2m$ perturbations that minimize the discrepancy between the model and the observations, as measured by a weighted least-squares objective function. To be precise, we relate the quantities t_j, y_j, δ_j, and ϵ_j by

$$y_j = \phi(x; t_j + \delta_j) + \epsilon_j, \qquad j = 1, 2, \ldots, m, \tag{10.41}$$

and define the minimization problem as

$$\min_{x, \delta_j, \epsilon_j} \tfrac{1}{2} \sum_{j=1}^{m} w_j^2 \epsilon_j^2 + d_j^2 \delta_j^2, \qquad \text{subject to (10.41)}. \tag{10.42}$$

The quantities w_i and d_i are weights, selected either by the modeler or by some automatic estimate of the relative significance of the error terms.

It is easy to see how the term "orthogonal distance regression" originates when we graph this problem; see Figure 10.2. If all the weights w_i and d_i are equal, then each term in the summation (10.42) is simply the shortest distance between the point (t_j, y_j) and the curve $\phi(x; t)$ (plotted as a function of t). The shortest path between the point and the curve will be normal (orthogonal) to the curve at the point of intersection.

It is easy to use the constraints (10.41) to eliminate the variables ϵ_j from (10.42) and obtain the unconstrained least-squares problem

$$\min_{x, \delta_j} F(x, \delta) = \tfrac{1}{2} \sum_{j=1}^{m} w_i^2 [y_j - \phi(x; t_j + \delta_j)]^2 + d_j^2 \delta_j^2 = \tfrac{1}{2} \sum_{j=1}^{2m} r_j(x, \delta), \tag{10.43}$$

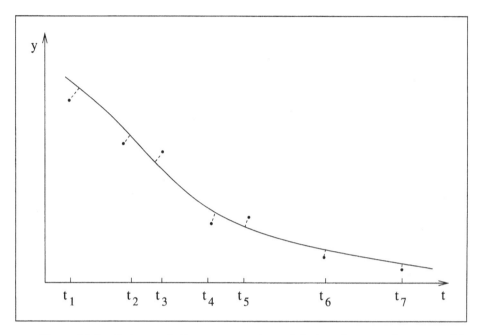

Figure 10.2 Orthogonal distance regression minimizes the sum of squares of the distance from each point to the curve.

where we have defined

$$r_j(x, \delta) = \begin{cases} w_j[\phi(x; t_j + \delta_j) - y_j], & j = 1, 2, \ldots, m, \\ d_{j-m}\delta_{j-m}, & j = m+1, \ldots, 2m. \end{cases} \quad (10.44)$$

Note that (10.43) is now a standard least-squares problem with $2m$ terms and $m + n$ unknowns, which we can solve by using any of the techniques in this chapter. A naive implementation of this strategy may, however, be quite expensive, since the number of parameters ($2n$) and the number of observations ($m + n$) may both be much larger than for the original problem.

Fortunately, the Jacobian matrix for (10.43) has a special structure that can be exploited in implementing methods such as Gauss–Newton or Levenberg–Marquardt. Many of its components are zero; for instance, we have

$$\frac{\partial r_j}{\partial \delta_i} = \frac{\partial [\phi(t_j + \delta_j; x) - y_j]}{\partial \delta_i} = 0, \quad i, j = 1, 2, \ldots, m, \ i \neq j,$$

whereas

$$\frac{\partial r_j}{\partial x_i} = 0, \quad j = m+1, \ldots, 2n, \ i = 1, 2, \ldots, n.$$

10.3. ORTHOGONAL DISTANCE REGRESSION

Additionally, we have for $j = 1, 2, \ldots, m$ and $i = 1, 2, \ldots, m$ that

$$\frac{\partial r_{m+j}}{\partial \delta_i} = \begin{cases} d_j & \text{if } i = j, \\ 0 & \text{otherwise.} \end{cases}$$

Hence, we can partition the Jacobian of the residual function r defined by (10.44) into blocks and write

$$J(x, \delta) = \begin{bmatrix} \hat{J} & V \\ 0 & D \end{bmatrix}, \tag{10.45}$$

where V and D are $m \times m$ diagonal matrices and $\hat{J}$ is the $m \times n$ matrix of partial derivatives of the functions $w_j \phi(t_j + \delta_j; x)$ with respect to x. Boggs, Byrd, and Schnabel apply the Levenberg–Marquardt algorithm to (10.43) and note that block elimination can be used to solve the subproblems (10.29), (10.31) efficiently. Given the partitioning (10.45), we can partition the step vector p and the residual vector r accordingly as

$$p = \begin{bmatrix} p_x \\ p_\delta \end{bmatrix}, \quad r = \begin{bmatrix} \hat{r}_1 \\ \hat{r}_2 \end{bmatrix},$$

and write the normal equations (10.29) in the partitioned form

$$\begin{bmatrix} \hat{J}^T \hat{J} + \lambda I & \hat{J}^T V \\ V \hat{J} & V^2 + D^2 + \lambda I \end{bmatrix} \begin{bmatrix} p_x \\ p_\delta \end{bmatrix} = -\begin{bmatrix} \hat{J}^T \hat{r}_1 \\ V \hat{r}_1 + D \hat{r}_2 \end{bmatrix}. \tag{10.46}$$

Since the lower right submatrix $V^2 + D^2 + \lambda I$ is diagonal, it is easy to eliminate p_δ from this system and obtain a smaller $n \times n$ system to be solved for p_x alone. The total cost of finding a step is only marginally greater than for the $m \times n$ problem arising from the standard least-squares model.

NOTES AND REFERENCES

Interesting examples of large-scale linear least-squares problems arise in structural engineering [192, 18] and in numerical geodesy [191]. Algorithms for linear least squares are discussed comprehensively by Lawson and Hanson [148], who include detailed error analyses of the different algorithms and software listings. They consider not just the basic problem (10.10) but also the situation in which there are bounds (for example, $x \geq 0$) or linear constraints (for example, $Ax \geq b$) on the variables. Golub and Van Loan [115, Chapter 5] survey the state of the art, including discussion of the suitability of the different approaches (e.g., normal equations vs. QR factorization) for different problem types. The recent book of Björck [19] gives a comprehensive survey of the whole field.

Problems in which m is of order 10^6 while n is less than 100 occur in a Laue crystallography application; see Ren and Moffatt [212].

Software for nonlinear least squares is fairly prevalent because of the high demand for it. An overview of available programs with details on how to obtain them is given in the NEOS Guide on the World-Wide Web (see the Preface) and by Moré and Wright [173, Chapter 3]. Seber and Wild [227, Chapter 15] describe some of the important practical issues in selecting software for statistical applications. For completeness, we mention a few of the more popular programs here.

The pure Gauss–Newton method is apparently not available in production software because it lacks robustness. However, as we discuss above, many algorithms do take Gauss–Newton steps when these steps are effective in reducing the objective function; consider, for instance, the NL2SOL program, which implements the method of Dennis, Gay, and Welsch [67]. The Levenberg–Marquardt algorithm is also available in a high-quality implementation in the MINPACK package. The orthogonal distance regression algorithm is implemented by ODRPACK [21]. All these routines give the user the option of either supplying Jacobians explicitly or else allowing the code to compute them by finite differencing. (In the latter case, the user need only write code to compute the residual vector $r(x)$; see Chapter 7.) All routines mentioned are freely available.

Major numerical software libraries such as NAG and Harwell [133] also contain robust nonlinear least-squares implementations.

EXERCISES

10.1 When J is an $m \times n$ matrix with $m \geq n$, show that J has full column rank if and only if $J^T J$ is nonsingular.

10.2 Show that $\bar{R}$ in (10.12) and R in (10.14) are identical if $\Pi = I$ and if both have nonnegative diagonal elements.

10.3 Prove that when J is rank-deficient, the minimizer of (10.10) with smallest Euclidean norm is obtained by setting each $\tau_i = 0$ in (10.19).

10.4 Suppose that each residual function r_j and its gradient are Lipschitz continuous with Lipschitz constant L, that is,

$$\|r_j(x_1) - r_j(x_2)\| \leq L\|x_1 - x_2\|, \quad \|\nabla r_j(x_1) - \nabla r_j(x_2)\| \leq L\|x_1 - x_2\|$$

for all $j = 1, 2, \ldots, m$ and all $x_1, x_2 \in \mathcal{D}$, where $\mathcal{D}$ is a compact subset of $\mathbf{R}^n$. Find Lipschitz constants for the Jacobian $J(x)$ (10.3) and the gradient $\nabla f(x)$ (10.4) over $\mathcal{D}$.

10.5 For the Gauss–Newton step p^{GN}, use the singular-value decomposition (10.16) of J_k and the formula (10.19) to express $(\nabla f_k)^T p_k^{\text{GN}}$ in terms of r_k, u_i, v_i, and σ_i. Express

$\|p_k^{GN}\|$ and $\|\nabla f_k\|$, and therefore $\cos\theta_k$ defined by (3.12), in these same terms, and explain why there is no guarantee that $\cos\theta_k > 0$ when J_k is rank-deficient.

🖉 **10.6** Express the solution p of (10.29) in terms of the singular-value decomposition of $J(x)$ and the scalar λ. Express its squared-norm $\|p\|^2$ in these same terms, and show that

$$\lim_{\lambda\to 0} p = \sum_{\sigma_i \neq 0} \frac{u_i^T r}{\sigma_i} v_i.$$

🖉 **10.7** Eliminate p_δ from (10.46) to obtain an $n \times n$ linear system in p_x alone. This system happens to be the normal equations for a certain (somewhat messy) linear least-squares problem in p_x. What is this problem?

CHAPTER 11

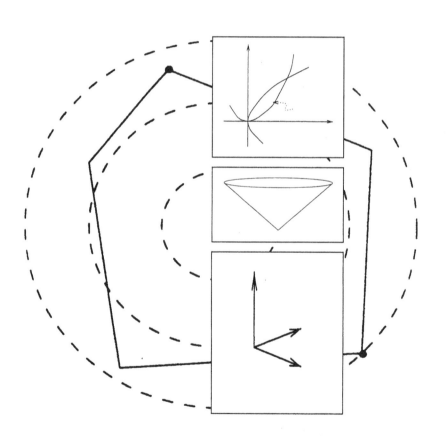

Nonlinear Equations

In many applications we do not need to optimize an objective function explicitly, but rather to find values of the variables in a model that satisfy a number of given relationships. When these relationships take the form of n equalities—the same number of equality conditions as variables in the model—the problem is one of solving a system of *nonlinear equations*. We write this problem mathematically as

$$r(x) = 0, \qquad (11.1)$$

where $r : \mathbf{R}^n \to \mathbf{R}^n$ is a vector function, that is,

$$r(x) = \begin{bmatrix} r_1(x) \\ r_2(x) \\ \vdots \\ r_n(x) \end{bmatrix},$$

where each $r_i : \mathbf{R}^n \to \mathbf{R}, i = 1, 2, \ldots, n$, is smooth. A vector x^* for which (11.1) is satisfied is called a *solution* or *root* of the nonlinear equations. A simple example is the system

$$r(x) = \begin{bmatrix} x_2^2 - 1 \\ \sin x_1 - x_2 \end{bmatrix} = 0,$$

which is a system of $n = 2$ equations with infinitely many solutions, two of which are $x^* = (3\pi/2, -1)^T$ and $x^* = (\pi/2, 1)^T$. In general, the system (11.1) may have no solutions, a unique solution, or many solutions.

The techniques for solving nonlinear equations overlap in their motivation, analysis, and implementation with optimization techniques discussed in earlier chapters. In both optimization and nonlinear equations, Newton's method lies at the heart of many important algorithms. Features such as line searches, trust regions, and inexact solution of the linear algebra subproblems at each iteration are important in both areas, as are other issues such as derivative evaluation and global convergence.

Because some important algorithms for nonlinear equations proceed by minimizing a sum of squares of the equations, that is,

$$\min_x \sum_{i=1}^{n} r_i^2(x),$$

there are particularly close connections with the nonlinear least-squares problem discussed in Chapter 10. The differences are that in nonlinear equations, the number of equations *equals* the number of variables (instead of exceeding the number of variables as in Chapter 10), and that we expect all equations to be satisfied at the solution, rather than just minimizing the sum of squares. This point is important because the nonlinear equations may represent physical or economic constraints such as conservation laws or consistency principles, which must hold exactly in order for the solution to be meaningful.

Many applications require us to solve a sequence of closely related nonlinear systems, as in the following example.

❏ **EXAMPLE 11.1**   (RHEINBOLDT; SEE [168])

An interesting problem in control is to analyze the stability of an aircraft in response to the commands of the pilot. The following is a simplified model based on force-balance equations, in which gravity terms have been neglected.

The equilibrium equations for a particular aircraft are given by a system of 5 equations in 8 unknowns of the form

$$F(x) \equiv Ax + \phi(x) = 0, \qquad (11.2)$$

where $F : \mathbf{R}^8 \to \mathbf{R}^5$, the matrix A is given by

$$A = \begin{bmatrix} -3.933 & 0.107 & 0.126 & 0 & -9.99 & 0 & -45.83 & -7.64 \\ 0 & -0.987 & 0 & -22.95 & 0 & -28.37 & 0 & 0 \\ 0.002 & 0 & -0.235 & 0 & 5.67 & 0 & -0.921 & -6.51 \\ 0 & 1.0 & 0 & -1.0 & 0 & -0.168 & 0 & 0 \\ 0 & 0 & -1.0 & 0 & -0.196 & 0 & -0.0071 & 0 \end{bmatrix},$$

and the nonlinear part is defined by

$$\phi(x) = \begin{bmatrix} -0.727 x_2 x_3 + 8.39 x_3 x_4 - 684.4 x_4 x_5 + 63.5 x_4 x_2 \\ 0.949 x_1 x_3 + 0.173 x_1 x_5 \\ -0.716 x_1 x_2 - 1.578 x_1 x_4 + 1.132 x_4 x_2 \\ -x_1 x_5 \\ x_1 x_4 \end{bmatrix}.$$

The first three variables x_1, x_2, x_3, represent the rates of roll, pitch, and yaw, respectively, while x_4 is the incremental angle of attack and x_5 the sideslip angle. The last three variables x_6, x_7, x_8 are the controls; they represent the deflections of the elevator, aileron, and rudder, respectively.

For a given choice of the control variables x_6, x_7, x_8 we obtain a system of 5 equations and 5 unknowns. If we wish to study the behavior of the aircraft as the controls are changed, we need to solve a system of nonlinear equations with unknowns $x_1, \ldots, x_5$ for each setting of the controls.

❐

Despite the many similarities between nonlinear equations and unconstrained and least-squares optimization algorithms, there are also some important differences. To obtain quadratic convergence in optimization we require second derivatives of the objective function, whereas knowledge of the first derivatives is sufficient in nonlinear equations. Quasi-Newton methods are perhaps less useful in nonlinear equations than in optimization. In unconstrained optimization, the objective function is the natural choice of merit function that gauges progress towards the solution, but in nonlinear equations various merit functions can be used, all of which have some drawbacks. Line search and trust-region techniques play an equally important role in optimization, but one can argue that trust-region algorithms have certain theoretical advantages in solving nonlinear equations.

Some of the difficulties that arise in trying to solve nonlinear equation problems can be illustrated by a simple scalar example ($n = 1$). Suppose we have

$$r(x) = \sin(5x) - x, \tag{11.3}$$

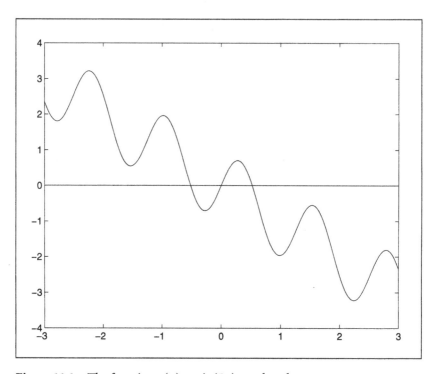

Figure 11.1 The function $r(x) = \sin(5x) - x$ has three roots.

as plotted in Figure 11.1. From this figure we see that there are three solutions of the problem $r(x) = 0$, also known as *roots of r*, located at zero and approximately ±0.519148. This situation of multiple solutions is similar to optimization problems where, for example, a function may have more than one local minimum. It is not *quite* the same, however: In the case of optimization, one of the local minima may have a lower function value than the others (making it a "better" solution), while in nonlinear equations all solutions are equally good from a mathematical viewpoint. (If the modeler decides that the solution found by the algorithm makes no sense on physical grounds, their model may need to be reformulated.)

In this chapter we start by outlining algorithms related to Newton's method and examining their local convergence properties. Besides Newton's method itself, these include Broyden's quasi-Newton method, inexact Newton methods, and tensor methods. We then address global convergence, which is the issue of trying to force convergence to a solution from a remote starting point. Finally, we discuss a class of methods in which an "easy" problem—one to which the solution is well known—is gradually transformed into the problem $F(x) = 0$. In these so-called continuation (or homotopy) methods, we track the solution as the problem changes, with the aim of finishing up at a solution of $F(x) = 0$.

Throughout this chapter we make the assumption that the vector function r is continuously differentiable in the region $\mathcal{D}$ containing the values of x we are interested in. In other words, the Jacobian $J(x) = \nabla r(x)$ exists and is continuous. For some local convergence results, we will use the stronger assumption of Lipschitz continuity of the Jacobian, that is, existence of a constant $\beta_L > 0$ such that

$$\|J(x_0) - J(x_1)\| \leq \beta_L \|x_0 - x_1\|, \tag{11.4}$$

for all $x_0, x_1 \in \mathcal{D}$. We say that x^* satisfying $r(x^*) = 0$ is a *degenerate solution* if $J(x^*)$ is singular, and a *nondegenerate solution* otherwise.

11.1 LOCAL ALGORITHMS

NEWTON'S METHOD FOR NONLINEAR EQUATIONS

Recall from Theorem 2.1 that Newton's method for minimizing $f : \mathbf{R}^n \to \mathbf{R}$ forms a quadratic model function by taking the first three terms of the Taylor series approximation of f around the current iterate x_k. Unless trust-region constraints or step lengths interfere, the Newton step is the vector that minimizes this model. In the case of nonlinear equations, Newton's method is derived in a similar way, by taking a Taylor series approximation to r around the current iterate x_k. The relevant variant of Taylor's theorem is as follows.

Theorem 11.1.
Suppose that $r : \mathbf{R}^n \to \mathbf{R}^n$ is continuously differentiable in some convex open set $\mathcal{D}$ and that x and $x + p$ are vectors in $\mathcal{D}$. Then we have that

$$r(x + p) = r(x) + \int_0^1 J(x + tp)p \, dt. \tag{11.5}$$

We can define a linear model $M_k(p)$ of $r(x_k + p)$ by approximating the second term on the right-hand-side of (11.5) by $J(x)p$, and writing

$$M_k(p) \stackrel{\text{def}}{=} r(x_k) + J(x_k)p. \tag{11.6}$$

The difference between $M_k(p)$ and $r(x_k + p)$ is

$$r(x_k + p) = M_k(p) + \int_0^1 [J(x_k + tp) - J(x_k)]p \, dt.$$

We can quantify this difference by noting that J is continuous, so that

$$\|J(x + tp) - J(x)\| \to 0 \text{ as } p \to 0, \text{ for all } t \in [0, 1],$$

and therefore

$$\left\| \int_0^1 [J(x+tp) - J(x)] p \, dt \right\| \leq \int_0^1 \|J(x+tp) - J(x)\| \, \|p\| \, dt = o(\|p\|). \quad (11.7)$$

When we make the stronger assumption that J is Lipschitz continuous (11.4), we have the stronger estimate

$$\left\| \int_0^1 [J(x+tp) - J(x)] p \, dt \right\| = O(\|p\|^2). \quad (11.8)$$

Newton's method, in its pure form, chooses the step p_k to be the vector for which $M_k(p_k) = 0$, that is, $p_k = -J(x_k)^{-1} r(x_k)$. We define it formally as follows.

Algorithm 11.1 (Newton's Method for Nonlinear Equations).
Choose x_0;
for $k = 0, 1, 2, \ldots$
    Calculate a solution p_k to the Newton equations

$$J(x_k) p_k = -r(x_k); \quad (11.9)$$

    $x_{k+1} \leftarrow x_k + p_k$;
end (for)

We use a linear model to derive the Newton step—rather than a quadratic model as in unconstrained optimization—because the linear model normally has a solution and yields an algorithm with rapid convergence properties. In fact, Newton's method for unconstrained optimization (see (2.14)) can be derived by applying Algorithm 11.1 to the nonlinear equations $\nabla f(x) = 0$. We see also in Chapter 18 that sequential quadratic programming for equality-constrained optimization can be derived by applying Algorithm 11.1 to the nonlinear equations formed by the first-order optimality conditions (18.3) for this problem. Another connection is with the Gauss–Newton method for nonlinear least squares; the formula (11.9) is equivalent to (10.20) in the usual case in which $J(x_k)$ is nonsingular.

When the iterate x_k is close to a nondegenerate root x^*, Newton's method has local superlinear convergence when the Jacobian J is a continuous function of x, and local quadratic convergence when J is Lipschitz continuous. (We outline these convergence properties below.) Potential shortcomings of the method include the following.

- When the starting point is remote from a solution, Algorithm 11.1 can behave erratically. When $J(x_k)$ is singular, the Newton step may not even be defined.

- First-derivative information (required for the Jacobian matrix J) may be difficult to obtain.

- It may be too expensive to find and calculate the Newton step p_k exactly when n is large.

- The root x^* in question may be degenerate, that is, $J(x^*)$ may be singular.

An example of a degenerate problem is the scalar function $r(x) = x^2$, which has a single degenerate root at $x^* = 0$. Algorithm 11.1, when started from any nonzero x_0, generates the sequence of iterates

$$x_k = \frac{1}{2^k} x_0,$$

which converges to the solution 0, but at only a linear rate.

As we show later in this chapter, Newton's method can be modified and enhanced in various ways to get around most of these problems. The variants we describe form the basis of much of the available software for solving nonlinear equations.

We summarize the local convergence properties of Algorithm 11.1 in the following theorem.

Theorem 11.2.

Suppose that r is continuously differentiable in a convex open set $\mathcal{D} \subset \mathbf{R}^n$. Let $x^* \in \mathcal{D}$ be a nondegenerate solution of $r(x) = 0$, and let $\{x_k\}$ be the sequence of iterates generated by Algorithm 11.1. Then when $x_k \in \mathcal{D}$ is sufficiently close to x^*, we have

$$x_{k+1} - x^* = o(\|x_k - x^*\|), \tag{11.10}$$

indicating local Q-superlinear convergence. When r is Lipschitz continuously differentiable near x^*, we have for all x_k sufficiently close to x^* that

$$x_{k+1} - x^* = O(\|x_k - x^*\|^2), \tag{11.11}$$

indicating local Q-quadratic convergence.

PROOF. Since $r(x^*) = 0$, an estimate similar to (11.7) yields that

$$\begin{aligned} r(x_k) &\\ &= r(x_k) - r(x^*) \\ &= J(x_k)(x_k - x^*) + \int_0^1 \left[J(x_k + t(x^* - x_k)) - J(x_k) \right] (x_k - x^*) \, dt \\ &= J(x_k)(x_k - x^*) + o(\|x_k - x^*\|). \end{aligned} \tag{11.12}$$

Since $J(x^*)$ is nonsingular, there is a radius $\delta > 0$ and a positive constant β^* such that for all x in the ball $\mathcal{B}(x^*, \delta)$ defined by

$$\mathcal{B}(x^*, \delta) = \{x \mid \|x - x^*\| \leq \delta\}, \tag{11.13}$$

we have that

$$\|J(x)^{-1}\| \leq \beta^* \quad \text{and} \quad x \in \mathcal{D}. \tag{11.14}$$

Assuming that $x_k \in \mathcal{B}(x^*, \delta)$, and recalling the definition (11.9), we multiply both sides of (11.12) by $J(x_k)^{-1}$ to obtain

$$-p_k = (x_k - x^*) + \|J(x_k)^{-1}\| o(\|x_k - x^*\|),$$
$$\Rightarrow x_k + p_k - x^* = o(\|x_k - x^*\|),$$
$$\Rightarrow x_{k+1} - x^* = o(\|x_k - x^*\|), \tag{11.15}$$

which yields (11.10).

When the Lipschitz continuity assumption (11.4) is satisfied, we can write

$$r(x_k) = r(x_k) - r(x^*) = J(x_k)(x_k - x^*) + w(x_k, x^*), \tag{11.16}$$

where the remainder term $w(x_k, x^*)$ is defined by

$$w(x_k, x^*) = \int_0^1 \left[J(x_k + t(x^* - x_k)) - J(x_k) \right] (x_k - x^*) \, dt.$$

We can use the same argument that led to the bound (11.8) to estimate $w(x_k, x^*)$ as follows:

$$\|w(x_k, x^*)\| = O(\|x_k - x^*\|^2). \tag{11.17}$$

By multiplying (11.16) by $J(x_k)^{-1}$ as above and using (11.9), we obtain

$$-p_k + (x_k - x^*) = J(x_k)^{-1} w(x_k, x^*),$$

so the estimate (11.11) follows by taking norms of both sides and using the bound (11.17). □

INEXACT NEWTON METHODS

Instead of solving (11.9) exactly, inexact Newton methods use search directions p_k that satisfy the condition

$$\|r_k + J_k p_k\| \leq \eta_k \|r_k\|, \quad \text{for some } \eta_k \in [0, \eta], \tag{11.18}$$

where $\eta \in [0, 1)$ is a constant. We refer to η_k as the *forcing parameter*. Different methods make different choices of the sequence $\{\eta_k\}$, and they use different algorithms for finding the approximate solutions p_k. The general framework for this class of methods can be stated as follows.

Framework 11.2 (Inexact Newton for Nonlinear Equations).
  Given η;
  Choose x_0;
  for $k = 0, 1, 2, \ldots$
    Choose forcing parameter $\eta_k \in [0, \eta]$;
    Find a vector p_k that satisfies (11.18);
    $x_{k+1} \leftarrow x_k + p_k$;
  end (for)

The convergence theory for these methods depends only on the condition (11.18) and not on the particular technique used to calculate p_k. The most important methods in this class, however, make use of iterative techniques for solving linear systems of the form $Jp = -r$, such as GMRES (Saad and Schultz [221], Walker [242]) or other Krylov-space methods. Like the conjugate-gradient algorithm of Chapter 5 (which is not directly applicable here, since the coefficient matrix J is not symmetric positive definite), these methods typically require us to perform a matrix–vector multiplication of the form Jd for some d at each iteration, and to store a number of work vectors of length n. GMRES itself requires an additional vector to be stored at each iteration, and so must be restarted periodically (often after every 10 or 20 iterations) to keep memory requirements at a reasonable level.

The matrix–vector products Jd can be computed without explicit knowledge of the Jacobian J. A finite difference approximation to Jd that requires one evaluation of $r(\cdot)$ is given by the formula (7.10). Calculation of Jd exactly (at least, to within the limits of finite-precision arithmetic) can be performed by using the forward mode of automatic differentiation, at a cost of at most a small multiple of an evaluation of $r(\cdot)$. Details of this procedure are given in Section 7.2.

We do not discuss the iterative methods for sparse linear systems here, but refer the interested reader to Kelley [141] for comprehensive descriptions and implementations of the most interesting techniques. We prove a local convergence theorem for the method, similar to Theorem 11.2.

Theorem 11.3.
  Suppose that r is continuously differentiable in a convex open set $\mathcal{D} \subset \mathbf{R}^n$. Let $x^* \in \mathcal{D}$ be a nondegenerate solution of $r(x) = 0$, and let $\{x_k\}$ be the sequence of iterates generated by the Framework 11.2. Then when $x_k \in \mathcal{D}$ is sufficiently close to x^*, the following are true:

  (i) *If η in (11.18) is sufficiently small, then the convergence of $\{x_k\}$ to x^* is Q-linear.*

  (ii) *If $\eta_k \to 0$, the convergence is Q-superlinear.*

(iii) *If, in addition, $J(\cdot)$ is Lipschitz continuous on D and $\eta_k = O(\|r_k\|)$, then the convergence is Q-quadratic.*

PROOF. We first rewrite (11.18) as

$$J(x_k)p_k = -r(x_k) + v_k, \quad \text{where } \|v_k\| \leq \eta_k \|r(x_k)\|. \tag{11.19}$$

Since x^* is a nondegenerate root, we have as in (11.14) that there is a radius $\delta > 0$ such that $\|J(x)^{-1}\| \leq \beta^*$ for some constant β^* and all $x \in \mathcal{B}(x^*, \delta)$. By multiplying both sides of (11.19) by $J(x_k)^{-1}$ and rearranging, we find that

$$p_k + J(x_k)^{-1}r(x_k) = J(x_k)^{-1}v_k = \beta^*\eta_k \|r(x_k)\|. \tag{11.20}$$

As in (11.12), we have that

$$r(x) = J(x)(x - x^*) + w(x), \tag{11.21}$$

where $\rho(x) \stackrel{\text{def}}{=} \|w(x)\|/\|x - x^*\| \to 0$ as $x \to x^*$. By reducing δ if necessary, we have from this expression that the following bound holds for all $x \in \mathcal{B}(x^*, \delta)$:

$$\|r(x)\| \leq 2\|J(x^*)\| \|x - x^*\| + o(\|x - x^*\|) \leq 4\|J(x^*)\| \|x - x^*\|. \tag{11.22}$$

We now set $x = x_k$ in (11.21), and substitute into (11.22) to obtain

$$\|x_k + p_k - x^*\| \leq 4\|J(x^*)\|\beta^*\eta_k \|x_k - x^*\| + \|J(x_k)^{-1}\| \|w(x_k)\|$$
$$\leq (4\|J(x^*)\|\beta^*\eta_k + \beta^*\rho(x_k)) \|x_k - x^*\|. \tag{11.23}$$

By choosing x_k close enough to x^* such that $\rho(x_k) \leq 1/(4\beta^*)$, and choosing $\eta = 1/(8\|J(x^*)\|\beta^*)$, we have that the term in parentheses in (11.23) is at most $\frac{1}{2}$. Hence, since $x_{k+1} = x_k + p_k$, this formula indicates Q-linear convergence of $\{x_k\}$ to x^*, proving part (i).

Part (ii) follows immediately from the fact that the term in brackets in (11.23) goes to zero as $x_k \to x^*$. For part (iii), we combine the techniques above with the logic of the second part of the proof of Theorem 11.2. Details are left as an exercise. □

BROYDEN'S METHOD

Secant methods, also known as quasi-Newton methods, do not require calculation of the Jacobian $J(x)$. Instead, they construct their own approximation to this matrix, updating it at each iteration so that it mimics the behavior of the true Jacobian J over the step just taken. The approximate Jacobian is then used in place of $J(x_k)$ in a formula like (11.9) to compute the new search direction.

11.1. LOCAL ALGORITHMS

To formalize these ideas, let the Jacobian approximation at iteration k be denoted by B_k. Assuming that B_k is nonsingular, we define the step to the next iterate by

$$p_k = -B_k^{-1} r(x_k), \qquad x_{k+1} = x_k + p_k. \qquad (11.24)$$

Let s_k and y_k denote the differences between successive iterates and successive functions, respectively:

$$s_k = x_{k+1} - x_k, \qquad y_k = r(x_{k+1}) - r(x_k).$$

(Note that $p_k = s_k$, but we change the notation because we will soon consider replacing p_k by αp_k for some positive step length $\alpha > 0$.) From Theorem 11.1, we have that s_k and y_k are related by the expression

$$y_k = \int_0^1 J(x_k + t s_k) s_k \, dt \approx J(x_{k+1}) s_k + o(\|s_k\|). \qquad (11.25)$$

We require the updated Jacobian approximation B_{k+1} to satisfy the following equation, which is known as the *secant equation*,

$$y_k = B_{k+1} s_k, \qquad (11.26)$$

which ensures that B_{k+1} and $J(x_{k+1})$ have similar behavior along the direction s_k. (Note the similarity with the secant equation (8.6) in quasi-Newton methods for unconstrained optimization; the motivation is the same in both cases.) The secant equation does not say anything about how B_{k+1} should behave along directions orthogonal to s_k. In fact, we can view (11.26) as a system of n linear equations in n^2 unknowns, where the unknowns are the components of B_{k+1}, so for $n > 1$ the equation (11.26) does not determine all the components of B_{k+1} uniquely. (The scalar case of $n = 1$ gives rise to the scalar secant method; see (A.32).)

The most successful practical algorithm is *Broyden's method*, for which the update formula is

$$B_{k+1} = B_k + \frac{(y_k - B_k s_k) s_k^T}{s_k^T s_k}. \qquad (11.27)$$

The Broyden update makes the smallest possible change to the Jacobian (as measured by the Euclidean norm $\|B_k - B_{k+1}\|$) that is consistent with (11.26), as we show in the following Lemma.

Lemma 11.4 (Dennis and Schnabel [69, Lemma 8.1.1]).
Among all matrices B satisfying $B s_k = y_k$, the matrix B_{k+1} defined by (11.27) minimizes the difference $\|B - B_k\|$.

PROOF. Let B be any matrix that satisfies $Bs_k = y_k$. By the properties of the Euclidean norm (see (A.41)) and the fact that $\|ss^T/s^Ts\| = 1$ for any vector s (see the exercises), we have

$$\|B_{k+1} - B_k\| = \left\|\frac{(y_k - B_k s_k)s_k^T}{s_k^T s_k}\right\|$$

$$= \left\|\frac{(B - B_k)s_k s_k^T}{s_k^T s_k}\right\| \leq \|B - B_k\| \left\|\frac{s_k s_k^T}{s_k^T s_k}\right\| = \|B - B_k\|.$$

Hence, we have that

$$B_{k+1} \in \arg\min_{B\,:\,y_k = Bs_k} \|B - B_k\|,$$

and the result is proved. □

We summarize the algorithm as follows.

Algorithm 11.3 (Broyden).
  Choose x_0 and B_0;
  for $k = 0, 1, 2, \ldots$
    Calculate a solution p_k to the linear equations

$$B_k p_k = -r(x_k); \tag{11.28}$$

   Choose α_k by performing a line search along p_k;
   $x_{k+1} \leftarrow x_k + \alpha_k p_k$;
   $s_k \leftarrow x_{k+1} - x_k$;
   $y_k \leftarrow r(x_{k+1}) - r(x_k)$;
   Obtain B_{k+1} from the formula (11.27);
  end (for)

Under certain assumptions, Broyden's method converges *superlinearly*, that is,

$$\|x_{k+1} - x^*\| = o(\|x_k - x^*\|). \tag{11.29}$$

This local convergence rate is fast enough for most practical purposes, though not as fast as the Q-quadratic convergence of Newton's method.

We illustrate the difference between the convergence rates of Newton's and Broyden's method with a small example. The function $r : \mathbf{R}^2 \to \mathbf{R}^2$ defined by

$$r(x) = \begin{bmatrix} (x_1 + 3)(x_2^3 - 7) + 18 \\ \sin(x_2 e^{x_1} - 1) \end{bmatrix} \tag{11.30}$$

Table 11.1 Convergence of Iterates in Broyden and Newton Methods

Iteration k	$\|x_k - x^*\|_2$	
	Broyden	Newton
0	0.64×10^0	0.64×10^0
1	0.62×10^{-1}	0.62×10^{-1}
2	0.52×10^{-3}	0.21×10^{-3}
3	0.25×10^{-3}	0.18×10^{-7}
4	0.43×10^{-4}	0.12×10^{-15}
5	0.14×10^{-6}	
6	0.57×10^{-9}	
7	0.18×10^{-11}	
8	0.87×10^{-15}	

has a nondegenerate root at $x^* = (0, 1)^T$. We start both methods from the point $x_0 = (-0.5, 1.4)^T$, and use the exact Jacobian $J(x_0)$ at this point as the initial Jacobian approximation B_0. Results are shown in Table 11.1.

Newton's method clearly exhibits Q-quadratic convergence, which is characterized by doubling of the exponent of the error at each iteration. Broyden's method takes twice as many iterations as Newton's, and reduces the error at a rate that accelerates slightly towards the end. Table 11.2 shows that the ratio of successive errors $\|x_{k+1} - x^*\|/\|x_k - x^*\|$ in Broyden's method does indeed approach zero, despite some dithering on iterations 5 and 6.

The function norms $\|r(x_k)\|$ approach zero at a similar rate to the iteration errors $\|x_k - x^*\|$. As in (11.16), we have that

$$r(x_k) = r(x_k) - r(x^*) \approx J(x^*)(x_k - x^*),$$

so by nonsingularity of $J(x^*)$, the norms of $r(x_k)$ and $(x_k - x^*)$ are bounded above and below by multiples of each other. For our example problem (11.30), convergence of the sequence of function norms in the two methods is shown in Table 11.3.

The convergence analysis of Broyden's method is more complicated than that of Newton's method. We state the following result without proof.

Theorem 11.5.

Suppose the assumptions of Theorem 11.2 hold. Then there are positive constants ϵ and δ such that if the starting point x_0 and the starting approximate Jacobian B_0 satisfy

$$\|x_0 - x^*\| \leq \delta, \quad \|B_0 - J(x^*)\| \leq \epsilon, \tag{11.31}$$

the sequence $\{x_k\}$ generated by Broyden's method (11.24), (11.27) is well-defined and converges Q-superlinearly to x^*.

The second condition in (11.31)—that the initial Jacobian approximation B_0 must be close to the true Jacobian at the solution $J(x^*)$—is difficult to guarantee in practice. In

Table 11.2 Ratio of Successive Errors in Broyden's Method

Iteration k	$\|x_{k+1} - x^*\|_2 / \|x_k - x^*\|_2$
0	0.097
1	0.0085
2	0.47
3	0.17
4	0.0033
5	0.0041
6	0.0030
7	0.00049

Table 11.3 Convergence of Function Norms in Broyden and Newton Methods

	$\|r(x_k)\|_2$	
Iteration k	Broyden	Newton
0	0.74×10^1	0.74×10^1
1	0.59×10^0	0.59×10^0
2	0.20×10^{-2}	0.23×10^{-2}
3	0.21×10^{-2}	0.16×10^{-6}
4	0.37×10^{-3}	0.22×10^{-15}
5	0.12×10^{-5}	
6	0.49×10^{-8}	
7	0.15×10^{-10}	
8	0.11×10^{-18}	

contradistinction to the case of unconstrained minimization, a good choice of B_0 can be critical to the performance of the algorithm. Some implementations of Broyden's method recommend choosing B_0 to be $J(x_0)$, or some finite difference approximation to this matrix.

The Broyden matrix B_k will be dense in general, even if the true Jacobian J is sparse. Therefore, when n is large, an implementation of Broyden's method that stores B_k as a full $n \times n$ matrix may be inefficient. Instead, we can use limited-memory methods in which B_k is stored implicitly in the form of a number of vectors of length n, while the system (11.28) is solved by a technique based on application of the Sherman–Morrison–Woodbury formula (A.56). These methods are similar to the ones described in Chapter 9 for large-scale unconstrained optimization.

TENSOR METHODS

In tensor methods, the linear model $M_k(p)$ used by Newton's method is augmented with an extra term that aims to capture some of the nonlinear, higher-order, behavior of r. By doing so, it achieves more rapid and reliable convergence to degenerate roots—in particular, to roots x^* for which the Jacobian $J(x^*)$ has rank $n-1$ or $n-2$.

11.1. LOCAL ALGORITHMS

We give a broad outline of the method here, and refer to the paper of Schnabel and Frank [224] for details.

We use $\hat{M}_k(p)$ to denote the model function on which tensor methods are based; this function has the form

$$\hat{M}_k(p) = r(x_k) + J(x_k)p + \tfrac{1}{2}T_k pp, \tag{11.32}$$

where T_k is a tensor defined by n^3 elements $(T_k)_{ijl}$ whose action on a pair of arbitrary vectors u and v in $\mathbf{R}^n$ is defined by

$$(T_k uv)_i = \sum_{j=1}^{n}\sum_{l=1}^{n}(T_k)_{ijl}u_j v_l.$$

If we followed the reasoning behind Newton's method, we could consider building T_k from the *second* derivatives of r at the point x_k, that is,

$$(T_k)_{ijl} = [\nabla^2 r_i(x_k)]_{jl}.$$

For instance, in the example (11.30), we have that

$$(T(x)uv)_1 = u^T \nabla^2 r_1(x) v = u^T \begin{bmatrix} 0 & 3x_2^2 \\ 3x_2^2 & 6x_2(x_1+3) \end{bmatrix} v$$
$$= 3x_2^2(u_1 v_2 + u_2 v_1) + 6x_2(x_1+3)u_2 v_2.$$

However, use of the exact second derivatives is not practical in most instances. If we were to store this information explicitly, about $n^3/2$ memory locations would be needed: about n times the requirements of Newton's method. Moreover, there may be no vector p for which $\hat{M}_k(p) = 0$, so the step may not even be defined.

Instead, the approach described in [224] defines T_k in a way that requires little additional storage, but which gives $\hat{M}_k$ some potentially appealing properties. Specifically, T_k is chosen so that $\hat{M}_k(p)$ interpolates the function $r(x_k + p)$ at some previous iterates visited by the algorithm. That is, we require that

$$\hat{M}_k(x_{k-j} - x_k) = r(x_{k-j}), \quad \text{for } j = 1, 2, \ldots, q, \tag{11.33}$$

for some integer $q > 0$. By substituting from (11.32), we see that T_k must satisfy the condition

$$\tfrac{1}{2}T_k s_{jk} s_{jk} = r(x_{k-j}) - r(x_k) - J(x_k)s_{jk},$$

where

$$s_{jk} \stackrel{\text{def}}{=} x_{k-j} - x_k \text{ and } j = 1, 2, \ldots, q.$$

In [224] it is shown that this condition can be ensured by choosing T_k so that its action on arbitrary vectors u and v is

$$T_k u v = \sum_{j=1}^{q} a_j (s_{jk}^T u)(s_{jk}^T v),$$

where a_j, $j = 1, 2, \ldots, q$, are vectors of length n. The number of interpolating points q is typically chosen to be quite modest, usually less than $\sqrt{n}$. This T_k can be stored in $2nq$ locations, which contain the vectors a_j and s_{jk} for $j = 1, 2, \ldots, q$. Note the connection between this idea and Broyden's method, which also chooses information in the model (albeit in the *first-order* part of the model) to interpolate the function value at the previous iterate.

This technique can be refined in various ways. The points of interpolation can be chosen to make the collection of directions s_{jk} more linearly independent. There may still not be a vector p for which $\hat{M}_k(p) = 0$, but we can instead take the step to be the vector that minimizes $\|\hat{M}_k(p)\|_2^2$, which can be found by using a specialized least-squares technique. There is no assurance that the step obtained in this way is a descent direction for the merit function $\frac{1}{2}\|r(x)\|^2$ (which is discussed in the next section), and in this case it can be replaced by the standard Newton direction $-J_k^{-1} r_k$.

Numerical testing by Schnabel and Frank [224] shows that the method generally requires fewer iterations on standard test problems, while on degenerate problems it outperforms Newton's method (which is known to behave poorly in these cases).

11.2 PRACTICAL METHODS

MERIT FUNCTIONS

As mentioned above, neither Newton's method (11.9) nor Broyden's method (11.24), (11.27) with unit step lengths can be guaranteed to converge to a solution of $r(x) = 0$ unless they are started close to that solution. Sometimes, components of the unknown or function vector or the Jacobian will blow up. Another, more exotic, kind of behavior is *cycling*, where the iterates move between distinct regions of the parameter space without approaching a root. An example is the scalar function

$$r(x) = -x^5 + x^3 + 4x,$$

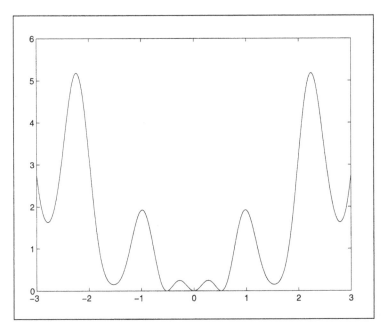

Figure 11.2 Plot of $\frac{1}{2}[\sin(5x) - x]^2$, showing its many local minima.

which has five nondegenerate roots. When started from the point $x_0 = 1$, Newton's method produces a sequence of iterates that oscillates between 1 and -1 (see the exercises) without converging to any of the roots.

The Newton and Broyden methods can, however, be made more robust by using line search and trust-region techniques similar to those described in Chapters 3 and 4. Before doing so, however, we need to define a *merit function*, which is a scalar-valued function of x whose value indicates whether a new candidate iterate is better or worse than the current iterate, in the sense of making progress toward a root of r. In unconstrained optimization, the objective function f is itself a natural merit function; algorithms for minimizing f typically require a decrease in f at each iteration. In nonlinear equations, the merit function is obtained by combining the n components of the vector r in some way.

The most widely used merit function is the sum of squares, defined by

$$f(x) = \tfrac{1}{2}\|r(x)\|^2 = \tfrac{1}{2}\sum_{i=1}^{n} r_i^2(x). \tag{11.34}$$

(The factor $\frac{1}{2}$ is introduced for convenience.) Any root x^* of r obviously has $f(x^*) = 0$, and since $f(x) \geq 0$ for all x, each root is at least a local minimizer of f. The converse is not true—local minimizers of f are not necessarily roots of r—but the merit function (11.34) has been used successfully in many applications and is implemented in a number of software packages.

The merit function for the example (11.3) is plotted in Figure 11.2. It shows three local minima corresponding to the three roots, but there are many other local minima (for example, those at around ± 1.53053). Local minima like these that are *not* roots of f satisfy an interesting property. Since

$$\nabla f(x^*) = J(x^*)^T r(x^*) = 0, \qquad (11.35)$$

we can have $r(x^*) \neq 0$ only if $J(x^*)$ is singular.

Since local minima for the sum-of-squares merit function may be points of attraction for the algorithms described in this section, global convergence results for the algorithms discussed here are less satisfactory than for similar algorithms applied to unconstrained optimization.

Other merit functions have also been proposed. One such is the ℓ_1 norm merit function defined by

$$f_1(x) = \|r(x)\|_1 = \sum_{i=1}^{m} |r_i(x)|.$$

There is a connection here with ℓ_1 merit functions for nonlinear programming, which contain a sum of absolute values of the constraint violations; see Chapter 15.

LINE SEARCH METHODS

We can obtain algorithms with global convergence properties by applying the line search approach of Chapter 3 to the sum-of-squares merit function $f(x) = \frac{1}{2}\|r(x)\|^2$. From any point x_k, the search direction p_k must be a descent direction for r; that is,

$$\cos \theta_k = \frac{-p_k^T \nabla f(x_k)}{\|p_k\| \|\nabla f(x_k)\|} > 0. \qquad (11.36)$$

Step lengths α_k are chosen by one of the procedures of Chapter 3, and the iterates are defined by the formula

$$x_{k+1} = x_k + \alpha_k p_k, \qquad k = 0, 1, 2, \ldots. \qquad (11.37)$$

The line search procedures in Section 3.1 can be used to calculate the step α_k. For the case of line searches that choose α_k that satisfy the Wolfe conditions (3.6), we have the following convergence result, which follows directly from Theorem 3.2.

Theorem 11.6.

Suppose that $J(x)$ is Lipschitz continuous in a neighborhood $\mathcal{D}$ of the level set $\mathcal{L} = \{x : f(x) \leq f(x_0)\}$. Suppose that a line search algorithm (11.37) is applied to r, where the search directions p_k satisfy (11.36) while the step lengths α_k satisfy the Wolfe conditions (3.6). Then for $\cos\theta_k$ defined as in (11.36), we have that the Zoutendijk condition holds, that is,

$$\sum_{k\geq 0} \cos^2\theta_k \|J_k^T r_k\|^2 < \infty.$$

PROOF. Since $J(x)$ is Lipschitz continuous in $\mathcal{D}$ (with constant β_L defined in (11.4)), we can define a constant β_R by

$$\beta_R \stackrel{\text{def}}{=} \max\left(\sup_{x\in\mathcal{D}} f(x), \sup_{x\in\mathcal{D}} \|J(x)\|\right). \tag{11.38}$$

Then for any y and z in $\mathcal{D}$, we have

$$\begin{aligned}
\|\nabla f(y) - \nabla f(z)\| &= \|J(y)^T r(y) - J(z)^T r(z)\| \\
&\leq \|J(y) - J(z)\|\|r(y)\| + \|J(z)\|\|r(y) - r(z)\| \\
&\leq (\beta_L \beta_R + \beta_R^2)\|y - z\|,
\end{aligned}$$

showing that ∇r is Lipschitz continuous on $\mathcal{D}$. Since in addition, $f(x)$ is bounded below by 0 on $\mathcal{D}$, the conditions of Theorem 3.2 are satisfied. The result follows from $\nabla f(x_k) = J_k^T r_k$. □

Provided that we ensure that

$$\cos\theta_k \geq \delta, \quad \text{for some } \delta \in (0,1) \text{ and all } k \text{ sufficiently large}, \tag{11.39}$$

Theorem 11.6 guarantees that $J_k^T r_k \to 0$. Moreover, if we know that $\|J(x)^{-1}\|$ is bounded on the set $\mathcal{D}$ defined in the theorem, then we must have $r_k \to 0$, so that the iterates x_k must approach a limit point x^* that solves the nonlinear equations problem $r(x) = 0$.

We now investigate the values of $\cos\theta_k$ for the directions generated by the Newton and inexact Newton methods, and describe how the Newton direction can be modified to ensure that the lower bound (11.39) is satisfied.

When it is well-defined, the Newton step (11.9) is a descent direction for $f(\cdot)$ whenever $r_k \neq 0$, since

$$p_k^T \nabla f(x_k) = -p_k^T J_k^T r_k = -\|r_k\|^2 < 0.$$

From (11.36), we have

$$\cos\theta_k = -\frac{p_k^T \nabla f(x_k)}{\|p_k\|\|\nabla f(x_k)\|} = \frac{\|r_k\|^2}{\|J_k^{-1} r_k\|\|J_k^T r_k\|} \geq \frac{1}{\|J_k^T\|\|J_k^{-1}\|} = \frac{1}{\kappa(J_k)}. \tag{11.40}$$

If the condition number $\kappa(J_k)$ is uniformly bounded for all k, then $\cos\theta_k$ is bounded below in the style of (11.39). When $\kappa(J_k)$ is large, however, this lower bound is close to zero, and use of the Newton direction may cause poor performance of the algorithm.

The importance of the condition (11.39) is illustrated by an example of Powell [196]. In this example, a line search Newton approach with exact line searches converges to a point that is not a solution and not even a stationary point for the merit function (11.34). Powell's function is

$$r(x) = \begin{bmatrix} x_1 \\ \dfrac{10x_1}{x_1+0.1} + 2x_2^2 \end{bmatrix}, \qquad (11.41)$$

with unique solution $x^* = 0$, whose Jacobian is

$$J(x) = \begin{bmatrix} 1 & 0 \\ \dfrac{1}{(x_1+0.1)^2} & 4x_2 \end{bmatrix}.$$

Note that the Jacobian is singular at all x for which $x_2 = 0$, and that for points of this type, we have

$$\nabla f(x) = \begin{bmatrix} x_1 + \dfrac{10x_1}{(x_1+0.1)^3} \\ 0 \end{bmatrix},$$

so that the gradient points in the direction of the positive x_1 axis whenever $x_1 > 0$. Powell shows, however, that the Newton step generated from an iterate that is close to (but not quite on) the x_1 axis tends to be parallel to the x_2 axis, making it nearly orthogonal to the gradient $\nabla f(x)$. That is, $\cos\theta_k$ for the Newton direction may be arbitrarily close to zero. Powell uses the starting point $(3, 1)^T$ and shows that convergence to the point $(1.8016, 0)^T$ (to four digits of accuracy) is attained in four iterations. However, this point is not a solution of (11.41)—in fact, it is easy to see that a step from this point in the direction $-x_1$ will produce a decrease in both components of r.

To ensure that (11.39) holds, we may have to modify the Newton direction. One possibility is to add some multiple $\tau_k I$ of the identity to $J_k^T J_k$, and define the step p_k to be

$$p_k = -(J_k^T J_k + \tau_k I)^{-1} J_k^T r_k. \qquad (11.42)$$

For each iterate k, we choose τ_k such that the condition (11.39) is satisfied, for some given value of $\delta \in (0, 1)$. Note that we can always choose τ_k large enough to ensure that this condition holds, since p_k approaches a multiple of $-J_k^T r_k$ as $\tau_k \uparrow \infty$. Note that instead of forming $J_k^T J_k$ explicitly and then performing trial Cholesky factorizations of matrices of the form $(J_k^T J_k + \tau I)$, we can use the technique (10.32), illustrated earlier for the least-squares case, which uses the fact that the Cholesky factor of $(J_k^T J_k + \tau I)$ is identical to R^T, where R is the upper triangular factor from the QR factorization of the matrix

$$\begin{bmatrix} J_k \\ \tau^{1/2} I \end{bmatrix}. \tag{11.43}$$

A combination of Householder and Givens transformations can be used, as for (10.32), and the savings noted in the discussion following (10.32) continue to hold if we need to perform this calculation for several candidate values of τ_k.

When p_k is an inexact Newton direction—that is, one that satisfies the condition (11.18)—it is not hard to derive a lower bound for $\cos \theta_k$ in the style of (11.39). Discarding the subscripts k, we have by squaring (11.18) that

$$\|r + Jp\|^2 \leq \eta_k^2 \|r\|^2 \Rightarrow 2p^T J^T r + \|r\|^2 + \|Jp\|^2 \leq \eta^2 \|r\|^2$$
$$\Rightarrow p^T \nabla r = p^T J^T r \leq [(\eta^2 - 1)/2] \|r\|^2.$$

Meanwhile,

$$\|p\| \leq \|J^{-1}\| [\|r + Jp\| + \|r\|] \leq \|J^{-1}\|(\eta + 1)\|r\|,$$

and

$$\|\nabla r\| = \|J^T r\| \leq \|J\| \|r\|.$$

By combining these estimates, we obtain

$$\cos \theta_k = -\frac{p^T \nabla r}{\|p\| \|\nabla r\|} \geq \frac{1 - \eta^2}{2\|J\| \|J^{-1}\|(1 + \eta)} \geq \frac{1 - \eta}{2\kappa(J)},$$

and we conclude that a bound of the form (11.39) is satisfied, provided that $\kappa(J)$ is bounded over $\mathcal{D}$. In other words, the inexact Newton method satisfies the bound (11.39) whenever Newton's method does (though for a different value of the constant δ), so by allowing inexactness we do not compromise global convergence behavior.

A line search method that includes the enhancements mentioned above and that uses the Newton direction (11.9) and unit step length $\alpha_k = 1$ whenever these are acceptable can be stated formally as follows.

Algorithm 11.4 (Line Search Newton for Nonlinear Equations).
Given $\delta \in (0, 1)$ and c_1, c_2 with $0 < c_2 < c_1 < \frac{1}{2}$;
Choose x_0;
for $k = 0, 1, 2, \ldots$
  if Newton step (11.9) satisfies (11.39)

                Set p_k to the Newton step;
        else
                Obtain p_k from (11.42), choosing τ_k to ensure that (11.39) holds;
        end (if)
        if $\alpha = 1$ satisfies the Wolfe conditions (3.6)
                Set $\alpha_k = 1$;
        else
                Perform a line search to find $\alpha_k > 0$ that satisfies (3.6);
        end (if)
        $x_{k+1} \leftarrow x_k + \alpha_k p_k$;
end (for)

Rapid local convergence of line search methods such as Algorithm 11.4 follows from a result similar to Theorem 6.2. For generality, we state this result so that it applies to *any* algorithm that eventually uses the Newton search direction (11.9) and that accepts the unit step $\alpha_k = 1$ whenever this step satisfies the Wolfe conditions.

Theorem 11.7.

Suppose that a line search algorithm that uses Newton search directions p_k from (11.9) yields a sequence $\{x_k\}$ that converges to x^*, where $r(x^*) = 0$ and $J(x^*)$ is nonsingular. Suppose also that there is an open neighborhood $\mathcal{D}$ of x^* such that the components r_i, $i = 1, 2, \ldots, n$, are twice differentiable, with $\|\nabla^2 r_i(x)\|$, $i = 1, 2, \ldots, n$, bounded for $x \in \mathcal{D}$. If the unit step length $\alpha_k = 1$ is accepted whenever it satisfies the Wolfe conditions (3.6), with $c_1 < \frac{1}{2}$, then the convergence is Q-quadratic; that is, $\|x_{k+1} - x^*\| = O(\|x_k - x^*\|^2)$.

We omit the proof of this result, which is similar in spirit to the corresponding result for trust-region methods, Theorem 11.10. The body of the proof shows that the step length $\alpha_k = 1$ is eventually always acceptable. Hence the method eventually reduces to a pure Newton method, and the rapid convergence rate follows from Theorem 11.2.

TRUST-REGION METHODS

The most widely used trust-region methods for nonlinear equations simply apply Algorithm 4.1 from Chapter 4 to the merit function $f(x) = \frac{1}{2}\|r(x)\|_2^2$, using $B_k = J(x_k)^T J(x_k)$ as the approximate Hessian in the model function $m_k(p)$. Global convergence properties follow directly from Theorems 4.7 and 4.8. Rapid local convergence for an algorithm that computes exact solutions of the trust-region subproblems can be proved under an assumption of Lipschitz continuity of the Jacobian $J(x)$. For background on the motivation and analysis of the trust-region approach, see Chapter 4.

For the least-squares merit function $f(x) = \frac{1}{2}\|r(x)\|_2^2$, the model function $m_k(p)$ is defined as

$$m_k(p) = \tfrac{1}{2}\|r_k + J_k p\|_2^2 = f_k + p^T J_k^T r_k + \tfrac{1}{2} p^T J_k^T J_k p_k;$$

cf. the model function (4.1) for general objective functions. The step p_k is generated by finding an approximate solution of the subproblem

$$\min_p m_k(p), \quad \text{subject to } \|p\| \le \Delta_k, \tag{11.44}$$

where Δ_k is the radius of the trust region. Given our merit function $f(x) = \|r(x)\|^2$ and model function m_k, the ratio ρ_k of actual to predicted reduction (see (4.4)) is defined as

$$\rho_k = \frac{\|r(x_k)\|^2 - \|r(x_k + p_k)\|^2}{\|r(x_k)\|^2 - \|r(x_k) + J(x_k)p_k\|^2}. \tag{11.45}$$

We can state the trust-region framework that results from this model as follows

Algorithm 11.5 (Trust Region for Nonlinear Equations).
Given $\bar{\Delta} > 0$, $\Delta_0 \in (0, \bar{\Delta})$, and $\eta \in [0, \frac{1}{4})$:
for $k = 0, 1, 2, \ldots$
    Calculate p_k as an (approximate) solution of (11.44);
    Evaluate ρ_k from (11.45);
    if $\rho_k < \frac{1}{4}$
        $\Delta_{k+1} = \frac{1}{4}\|p_k\|$
    else
        if $\rho_k > \frac{3}{4}$ and $\|p_k\| = \Delta_k$
            $\Delta_{k+1} = \min(2\Delta_k, \bar{\Delta})$
        else
            $\Delta_{k+1} = \Delta_k$;
        end (if)
    end (if)
    if $\rho_k > \eta$
        $x_{k+1} = x_k + p_k$
    else
        $x_{k+1} = x_k$;
    end (if)
end (for).

The dogleg method is a special case of the trust-region algorithm, Algorithm 4.1, that constructs an approximate solution to (11.44) based on the Cauchy point p_k^C and the unconstrained minimizer p_k^J. The Cauchy point is

$$p_k^C = -\tau_k(\Delta_k/\|J_k^T r_k\|)J_k^T r_k, \tag{11.46}$$

where

$$\tau_k = \min\left\{1, \|J_k^T r_k\|^3/(\Delta_k r_k^T J_k(J_k^T J_k)J_k^T r_k)\right\}; \tag{11.47}$$

By comparing with the general definition (4.7), (4.8) we see that it is not necessary to consider the case of an indefinite Hessian approximation in $m_k(p)$, since the model Hessian $J_k^T J_k$ that we use is positive semidefinite. The unconstrained minimizer of $m_k(p)$ is unique when J_k has full rank. In this case, we denote it by p_k^J and write

$$p_k^J = -(J_k^T J_k)^{-1}(J_k^T r_k) = -J_k^{-1} r_k.$$

The selection of p_k proceeds as follows.

Procedure 11.6 (Dogleg).
    Calculate p_k^C;
    if $\|p_k^C\| = \Delta_k$
        $p_k \leftarrow p_k^C$;
    else
        Calculate p_k^J;
        $p_k \leftarrow p_k^C + \tau(p_k^J - p_k^C)$, where τ is the largest value in $[0, 1]$
        such that $\|p_k\| \leq \Delta_k$
    end (if).

Lemma 4.1 shows that when J_k is nonsingular, the vector p_k chosen above is the minimizer of m_k along the piecewise linear path that leads from the origin to the Cauchy point and then to the unconstrained minimizer p_k^J. Hence, the reduction in model function at least matches the reduction obtained by the Cauchy point, which can be estimated by specializing the bound (4.34) to the least-squares case by writing

$$m_k(0) - m_k(p_k) \geq c_1 \|J_k^T r_k\| \min\left(\Delta_k, \frac{\|J_k^T r_k\|}{\|J_k^T J_k\|}\right), \qquad (11.48)$$

where c_1 is some positive constant.

From Theorem 4.3, we know that the exact solution of (11.44) has the form

$$p_k = -(J_k^T J_k + \lambda_k I)^{-1} J_k^T r_k, \qquad (11.49)$$

for some $\lambda_k \geq 0$, and that $\lambda_k = 0$ if the unconstrained solution p_k^J satisfies $\|p_k^J\| \leq \Delta_k$. (Note that (11.49) is identical to the formula (10.30a) from Chapter 10. In fact, the Levenberg–Marquardt approach for nonlinear equations is in a sense a special case of the same algorithm for nonlinear least-squares problems.) The Levenberg–Marquardt algorithm uses the techniques of Section 4.2 to search for the value of λ_k that satisfies (11.49). The procedure described in the "exact" trust-region algorithm, Algorithm 4.4, is based on Cholesky factorizations, but as in Chapter 10, we can replace these by specialized QR algorithms to compute the factorization (10.34). Even if the exact λ_k corresponding to the solution of (11.44) is not found, the p_k calculated from (11.49) will still yield global convergence if it satisfies the condition (11.48) for some value of c_1, together with

$$\|p_k\| \leq \gamma \Delta_k, \quad \text{for some constant } \gamma \geq 1. \tag{11.50}$$

The dogleg method has the advantage over methods that search for the exact solution of (11.44) that just one linear system needs to be solved per iteration. As in Chapter 4, there is a tradeoff to be made between the amount of effort to spend on each iteration and the total number of function and derivative evaluations required.

We can also consider alternative trust-region approaches that are based on different merit functions and different definitions of the trust region. An algorithm based on the ℓ_1 merit function with an ℓ_∞-norm trust region gives rise to subproblems of the form

$$\min_p \|J_k p + r_k\|_1 \quad \text{subject to } \|p\|_\infty \leq \Delta, \tag{11.51}$$

which can be formulated and solved using linear programming techniques. This approach is closely related to the Sℓ_1QP approach for nonlinear programming discussed in Section 18.8, for which the subproblem is (18.51).

Global convergence results of Algorithm 11.5 when the steps p_k satisfy (11.48) and (11.50) are given in the following two theorems. They can be proved by referring directly to Theorems 4.7 and 4.8. The first result is for the case of $\eta = 0$, in which the algorithm accepts all steps that produce a decrease in the merit function f_k.

Theorem 11.8.

Let $\eta = 0$ in Algorithm 11.5. Suppose that $J(x)$ is continuous in a neighborhood $\mathcal{D}$ of the level set $\mathcal{L} = \{x : f(x) \leq f(x_0)\}$ and that $\|J(x)\|$ is bounded above on $\mathcal{L}$. Suppose in addition that all approximate solutions of (11.44) satisfy the bounds (11.48) and (11.50). We then have that

$$\liminf_{k \to \infty} \|J_k^T r_k\| = 0.$$

The second result requires a strictly positive choice of η and Lipschitz continuity of J, and produces a correspondingly stronger result: convergence of the sequence $\{J_k^T r_k\}$ to zero.

Theorem 11.9.

Let $\eta \in \left(0, \frac{1}{4}\right)$ in Algorithm 11.5. Suppose that $J(x)$ is Lipschitz continuous in a neighborhood $\mathcal{D}$ of the level set $\mathcal{L} = \{x : f(x) \leq f(x_0)\}$ and that $\|J(x)\|$ is bounded above on $\mathcal{L}$. Suppose in addition that all approximate solutions of (11.44) satisfy the bounds (11.48) and (11.50). We then have that

$$\lim_{k \to \infty} \|J_k^T r_k\| = 0.$$

We turn now to local convergence of the trust-region algorithm for the case in which the subproblem (11.44) is solved exactly. We assume that the sequence $\{x_k\}$ converges to a nondegenerate solution x^* of the nonlinear equations $r(x) = 0$. The significance of this result is that the algorithmic enhancements needed for global convergence do not, in well-designed algorithms, interfere with the fast local convergence properties described in Section 11.1.

Theorem 11.10.

Suppose that the sequence $\{x_k\}$ generated by Algorithm 11.5 converges to a nondegenerate solution x^* of the problem $r(x) = 0$. Suppose also that $J(x)$ is Lipschitz continuous in an open neighborhood $\mathcal{D}$ of x^* and that the trust-region subproblem (11.44) is solved exactly for all sufficiently large k. Then the sequence $\{x_k\}$ converges quadratically to x^*.

PROOF. We prove this result by showing that there is an index K such that the trust-region radius is not reduced further after iteration K; that is, $\Delta_k \geq \Delta_K$ for all $k \geq K$. We then show that the algorithm eventually takes the pure Newton step at every iteration, so that quadratic convergence follows from Theorem 11.2.

Let p_k denote the exact solution of (11.44). Note first that p_k will simply be the unconstrained Newton step $-J_k^{-1} r_k$ whenever this step satisfies the trust-region bound. Otherwise, we have $\|J_k^{-1} r_k\| > \Delta_k$, while the solution p_k satisfies $\|p_k\| \leq \Delta_k$. In either case, we have

$$\|p_k\| \leq \|J_k^{-1} r_k\|. \tag{11.52}$$

We consider the ratio ρ_k of actual to predicted reduction defined by (11.45). We have directly from the definition that

$$|1 - \rho_k| \leq \frac{\left| \|r_k + J_k p_k\|^2 - \|r(x_k + p_k)\|^2 \right|}{\|r(x_k)\|^2 - \|r(x_k) + J(x_k) p_k\|^2}. \tag{11.53}$$

From Theorem 11.1, we have for the second term in the numerator that

$$\|r(x_k + p_k)\|^2 = \|[r(x_k) + J(x_k) p_k] + w(x_k, p_k)\|^2,$$

where

$$w(x_k, p_k) \stackrel{\text{def}}{=} \int_0^1 [J(x_k + t p_k) - J(x_k)] p_k \, dt.$$

Because of Lipschitz continuity of J (with Lipschitz constant β_L as in (11.4)), we have

$$\|w(x_k, p_k)\| \leq \int_0^1 \|J(x_k + t p_k) - J(x_k)\| \|p_k\| \, dt$$

11.2. PRACTICAL METHODS

$$\leq \int_0^1 \beta_L \|p_k\|^2 \, dt = (\beta_L/2)\|p_k\|^2,$$

so that using the fact that $\|r_k + J_k p_k\| \leq \|r_k\| = f(x_k)^{1/2}$ (since p_k is the solution of (11.44)), we can bound the numerator as follows:

$$\begin{aligned}
\left|\|r_k + J_k p_k\|^2 - \|r(x_k + p_k)\|^2\right| &\leq 2\|r_k + J_k p_k\|\|w(x_k, p_k)\| + \|w(x_k, p_k)\|^2 \\
&\leq f(x_k)^{1/2} \beta_L \|p_k\|^2 + (\beta_L/2)^2 \|p_k\|^4 \\
&\leq \epsilon(x_k) \|p_k\|^2,
\end{aligned} \quad (11.54)$$

where we define

$$\epsilon(x_k) = f(x_k)^{1/2} \beta_L + (\beta_L/2)^2 \|p_k\|^2.$$

Since by assumption, we have $x_k \to x^*$, it follows that $f(x_k) \to 0$ and $\|r_k\| \to 0$. Because x^* is a nondegenerate root, we have as in (11.14) that $\|J(x_k)^{-1}\| \leq \beta^*$ for all k sufficiently large, so from (11.52), we have

$$\|p_k\| \leq \|J_k^{-1} r_k\| \leq \beta^* \|r_k\| \to 0. \quad (11.55)$$

Therefore, we have that

$$\lim_{k \to \infty} \epsilon(x_k) \to 0. \quad (11.56)$$

Turning now to the denominator of (11.53), we define $\bar{p}_k$ to be a step of the same length as the solution p_k in the Newton direction $-J_k^{-1} r_k$, that is,

$$\bar{p}_k = -\frac{\|p_k\|}{\|J_k^{-1} r_k\|} J_k^{-1} r_k.$$

Since $\bar{p}_k$ is feasible for (11.44), and since p_k is optimal for this subproblem, we have

$$\begin{aligned}
\|r_k\|^2 - \|r_k + J_k p_k\|^2 &\geq \|r_k\|^2 - \left\| r_k - \frac{\|p_k\|}{\|J_k^{-1} r_k\|} r_k \right\|^2 \\
&= 2 \frac{\|p_k\|}{\|J_k^{-1} r_k\|} \|r_k\|^2 - \frac{\|p_k\|^2}{\|J_k^{-1} r_k\|^2} \|r_k\|^2 \\
&\geq \frac{\|p_k\|}{\|J_k^{-1} r_k\|} \|r_k\|^2,
\end{aligned}$$

where for the last inequality we have used (11.52). By using (11.55) again, we have from this bound that

$$\|r_k\|^2 - \|r_k + J_k p_k\|^2 \geq \frac{\|p_k\|}{\|J_k^{-1} r_k\|}\|r_k\|^2 \geq \frac{1}{\beta^*}\|p_k\|\|r_k\|. \quad (11.57)$$

By substituting (11.54) and (11.57) into (11.53), and then applying (11.55) again, we have

$$|1 - \rho_k| \leq \frac{\beta^* \epsilon(x_k)\|p_k\|^2}{\|p_k\|\|r_k\|} \leq (\beta^*)^2 \epsilon(x_k) \to 0. \quad (11.58)$$

Therefore, for all k sufficiently large, we have $\rho_k > \frac{1}{4}$, and so the trust region radius Δ_k will not be increased beyond this point. As claimed, there is an index K such that

$$\Delta_k \geq \Delta_K, \quad \text{for all } k \geq K.$$

Since $\|J_k^{-1} r_k\| \leq \beta^* \|r_k\| \to 0$, the Newton step $-J_k^{-1} r_k$ will eventually be smaller than Δ_K (and hence Δ_k), so it will eventually always be accepted as the solution of (11.44). The result now follows from Theorem 11.2. □

We can replace the assumption that $x_k \to x^*$ with an assumption that the nondegenerate solution x^* is just one of the limit points of the sequence. (In fact, this condition implies that $x_k \to x^*$; see the exercises.)

11.3 CONTINUATION/HOMOTOPY METHODS

MOTIVATION

We mentioned above that Newton-based methods all suffer from one shortcoming: Unless $J(x)$ is nonsingular in the region of interest—a condition that often cannot be guaranteed—they are in danger of converging to a local minimum of the merit function rather than to a solution of the nonlinear system. Continuation methods, which we outline in this section, aim explicitly for a solution of $r(x) = 0$ and are more likely to converge to such a solution in difficult cases. Their underlying motivation is simple to describe: Rather than dealing with the original problem $r(x) = 0$ directly, we set up an "easy" system of equations for which the solution is obvious. We then gradually transform the "easy" system of equations into the original system $r(x)$, and follow the solution as it moves from the solution of the easy problem to the solution of the original problem.

To be specific, we define the *homotopy map* $H(x, \lambda)$ as follows:

$$H(x, \lambda) = \lambda r(x) + (1 - \lambda)(x - a), \quad (11.59)$$

11.3. CONTINUATION/HOMOTOPY METHODS

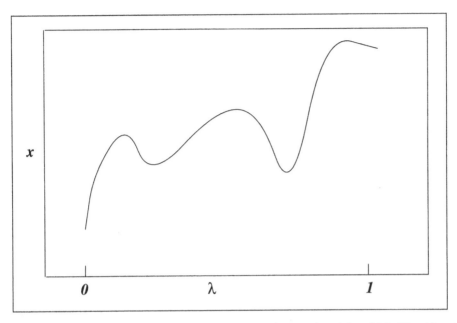

Figure 11.3 Plot of a zero path: the trajectory of points (x, λ) for which $H(x, \lambda) = 0$.

where λ is a scalar parameter and $a \in \mathbf{R}^n$ is a fixed vector. When $\lambda = 0$, (11.59) defines the artificial, easy problem $H(x, 0) = x - a$, whose solution is obviously $x = a$. When $\lambda = 1$, we have $H(x, 1) = r(x)$, the original system of equations.

To solve $r(x) = 0$, consider the following algorithm: First, set $\lambda = 0$ in (11.59) and set $x = a$. Then, increase λ from 0 to 1 in small increments, and for each value of λ, calculate the solution of the system $H(x, \lambda) = 0$. The final value of x corresponding to $\lambda = 1$ will solve the original problem $r(x) = 0$.

This naive approach sounds plausible, and Figure 11.3 illustrates a situation in which it would be successful. In this figure, there is a unique solution x of the system $H(x, \lambda) = 0$ for each value of λ in the range $[0, 1]$. The trajectory of points (x, λ) for which $H(x, \lambda) = 0$ is called the *zero path*.

Unfortunately, however, the approach often fails, as illustrated in Figure 11.4. Here, the algorithm follows the lower branch of the curve from $\lambda = 0$ to $\lambda = \lambda_T$, but it then loses the trail unless it is lucky enough to jump to the top branch of the path. The value λ_T is known as a *turning point*, since at this point we can follow the path smoothly only if we no longer insist on increasing λ at every step, but rather allow it to decrease where necessary. In fact, practical continuation methods work by doing exactly as Figure 11.4 suggests, that is, they follow the zero path explicitly, even if this means allowing λ to decrease from time to time, and even to roam outside the interval $[0, 1]$.

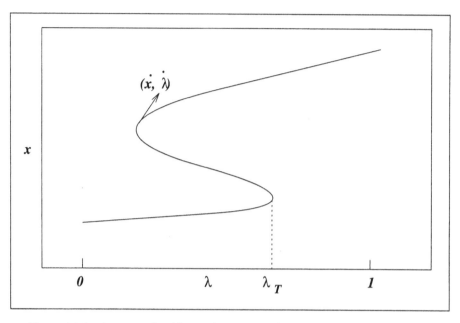

Figure 11.4 A zero path with turning points. The path joins $(a, 0)$ to $(x^*, 1)$, but it cannot be followed by merely increasing λ monotonically from 0 to 1.

PRACTICAL CONTINUATION METHODS

In one practical technique, we model the zero path by allowing both x and λ to be functions of an independent variable s that represents arc length along the path. That is, $(x(s), \lambda(s))$ is the point that we arrive at by traveling a distance s along the path from the initial point $(x(0), \lambda(0)) = (a, 0)$. Because we have that

$$H(x(s), \lambda(s)) = 0, \quad \text{for all } s \geq 0,$$

we can take the total derivative of this expression with respect to s to obtain

$$\frac{\partial}{\partial x} H(x, \lambda) \dot{x} + \frac{\partial}{\partial \lambda} H(x, \lambda) \dot{\lambda} = 0, \quad \text{where } (\dot{x}, \dot{\lambda}) = \left(\frac{dx}{ds}, \frac{d\lambda}{ds} \right). \quad (11.60)$$

The vector $(\dot{x}(s), \dot{\lambda}(s))$ is the tangent vector to the zero path, as we illustrate in Figure 11.4. From (11.60), we see that it lies in the null space of the $n \times (n+1)$ matrix

$$\left[\frac{\partial}{\partial x} H(x, \lambda) \quad \frac{\partial}{\partial \lambda} H(x, \lambda) \right]. \quad (11.61)$$

When this matrix has full rank, its null space has dimension 1, so to complete the definition of $(\dot{x}, \dot{\lambda})$ in this case, we need to assign it a length and direction. The length is fixed by imposing the normalization condition

$$\|\dot{x}(s)\|^2 + |\dot{\lambda}(s)|^2 = 1, \quad \text{for all } s, \tag{11.62}$$

which ensures that s is the true arc length along the path from $(0, a)$ to $(x(s), \lambda(s))$. We need to choose the sign to ensure that we keep moving forward along the zero path. A heuristic that works well is to choose the sign so that the tangent vector $(\dot{x}, \dot{\lambda})$ at the current value of s makes an angle of less than $\pi/2$ with the tangent point at the previous value of s.

We can outline the complete procedure for computing $(\dot{x}, \dot{\lambda})$ as follows:

Procedure 11.7 (Tangent Vector Calculation).
Compute a vector in the null space of (11.61) by performing a QR factorization with column pivoting,

$$Q^T \left[\frac{\partial}{\partial x} H(x, \lambda) \quad \frac{\partial}{\partial \lambda} H(x, \lambda) \right] \Pi = \left[\begin{array}{cc} R & w \end{array} \right],$$

where Q is $n \times n$ orthogonal, R is $n \times n$ upper triangular, Π is an $(n+1) \times (n+1)$ permutation matrix, and $w \in \mathbb{R}^n$.
Set

$$v = \Pi \begin{bmatrix} R^{-1} w \\ -1 \end{bmatrix};$$

Set $(\dot{x}, \dot{\lambda}) = \pm v / \|v\|_2$, where the sign is chosen to satisfy the angle criterion mentioned above.

Details of the QR factorization procedure are given in the Appendix.

Since we can obtain the tangent at any given point (x, λ) and since we know the initial point $(x(0), \lambda(0)) = (a, 0)$, we can trace the zero path by calling a standard initial-value first-order ordinary differential equation solver, terminating the algorithm when it finds a value of s for which $\lambda(s) = 1$.

A second approach for following the zero path is quite similar to the one just described, except that it takes an algebraic viewpoint instead of a differential-equations viewpoint. Given a current point (x, λ), we compute the tangent vector $(\dot{x}, \dot{\lambda})$ as above, and take a small step (of length ϵ, say) along this direction to produce a "predictor" point (x^P, λ^P); that is,

$$(x^P, \lambda^P) = (x, \lambda) + \epsilon(\dot{x}, \dot{\lambda}).$$

Usually, this new point will not lie exactly on the zero path, so we apply some "corrector" iterations to bring it back to the path, thereby identifying a new iterate (x^+, λ^+) that satisfies

CHAPTER 11. NONLINEAR EQUATIONS

$H(x^+, \lambda^+) = 0$. (This process is illustrated in Figure 11.5.) During the corrections, we choose a component of the predictor step (x^P, λ^P)—one of the components that has been changing most rapidly during the past few steps—and hold this component fixed during the correction process. If the index of this component is i, and if we use a pure Newton corrector process (often adequate, since (x^P, λ^P) is usually quite close to the target point (x^+, λ^+)), the steps will have the form

$$\begin{bmatrix} \dfrac{\partial H}{\partial x} & \dfrac{\partial H}{\partial \lambda} \\ \multicolumn{2}{c}{e_i} \end{bmatrix} \begin{bmatrix} \delta x \\ \delta \lambda \end{bmatrix} = \begin{bmatrix} -H \\ 0 \end{bmatrix},$$

where the quantities $\partial H/\partial x$, $\partial H/\partial \lambda$, and H are evaluated at the latest point of the corrector process. The last row of this system serves to fix the ith component of $(\delta x, \delta \lambda)$ at zero; the vector $e_i \in \mathbf{R}^{n+1}$ is a vector of length $n+1$ containing all zeros, except for a 1 in the location i that corresponds to the fixed component. Note that in Figure 11.5 the λ component is chosen to be fixed on the current iteration. On the following iteration, it may be more appropriate to choose x as the fixed component, as we reach the turning point in λ.

The two variants on path-following described above are able to follow curves like those depicted in Figure 11.4 to a solution of the nonlinear system. They rely, however, on the $n \times (n+1)$ matrix in (11.61) having full rank for all (x, λ) along the path, so that the

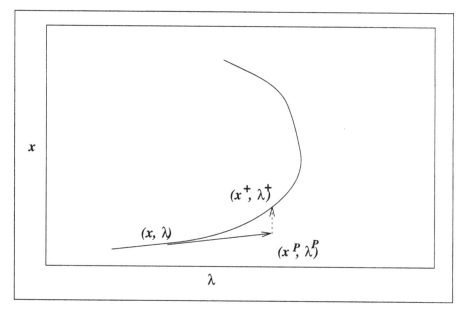

Figure 11.5 The algebraic predictor–corrector procedure, using λ as the fixed variable in the correction process.

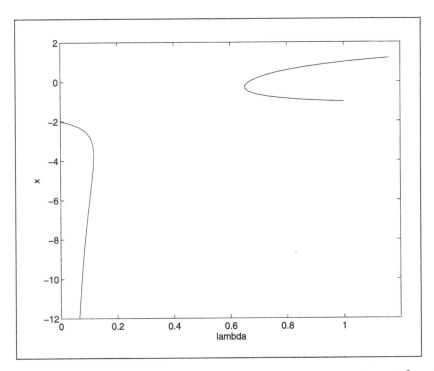

Figure 11.6 Continuation curves for the example in which $H(x, \lambda) = \lambda(x^2 - 1) + (1 - \lambda)(x + 2)$. There is no path from $\lambda = 0$ to $\lambda = 1$.

tangent vector is well-defined. Fortunately, it can be shown that full rank is guaranteed under certain assumptions, as in the following theorem.

Theorem 11.11 (Watson [244]).
Suppose that r is twice continuously differentiable. Then for almost all vectors $a \in \mathbf{R}^n$, there is a zero path emanating from $(0, a)$ along which the $n \times (n + 1)$ matrix (11.61) has full rank. If this path is bounded for $\lambda \in [0, 1)$, then it has an accumulation point $(\bar{x}, 1)$ such that $r(\bar{x}) = 0$. Furthermore, if the Jacobian $J(\bar{x})$ is nonsingular, the zero path between $(a, 0)$ and $(\bar{x}, 1)$ has finite arc length.

The theorem assures us that unless we are unfortunate in the choice of a, the algorithms described above can be applied to obtain a path that either diverges or else leads to a point $\bar{x}$ that is a solution of the original nonlinear system if $J(\bar{x})$ is nonsingular. More detailed convergence results can be found in Watson [244] and the references therein.

We conclude with an example to show that divergence of the zero path (the less desirable outcome of Theorem 11.11) can happen even for innocent-looking problems.

❏ EXAMPLE 11.2

Consider the system $r(x) = x^2 - 1$, for which there are two nondegenerate solutions $+1$ and -1. Suppose we choose $a = -2$ and attempt to apply a continuation method to the function

$$H(x, \lambda) = \lambda(x^2 - 1) + (1 - \lambda)(x + 2) = \lambda x^2 + (1 - \lambda)x + (2 - 3\lambda), \quad (11.63)$$

obtained by substituting into (11.59). The zero paths for this function are plotted in Figure 11.6. As can be seen from that diagram, there is no zero path that joins $(-2, 0)$ to either $(1, 1)$ or $(-1, 1)$, so the continuation methods fail on this example. We can find the values of λ for which no solution exists by fixing λ in (11.63) and using the formula for a quadratic root to obtain

$$x = \frac{-(1-\lambda) \pm \sqrt{(1-\lambda)^2 - 4\lambda(2-3\lambda)}}{2\lambda}.$$

Now, when the term in the square root is negative, the corresponding values of x are complex, that is, there are no real roots x. It is easy to verify that this is the case when

$$\lambda \in \left(\frac{5 - 2\sqrt{3}}{13}, \frac{5 + 2\sqrt{3}}{13} \right) \approx (0.118, 0.651).$$

Note that the zero path starting from $(-2, 0)$ becomes unbounded, which is one of the possible outcomes of Theorem 11.11.

❏

This example indicates that continuation methods may fail to produce a solution even to a fairly simple system of nonlinear equations. However, it is generally true that they are more reliable than the merit-function methods described earlier in the chapter. The extra robustness comes at a price, since continuation methods typically require significantly more function and derivative evaluations and linear algebra operations than the merit-function methods.

NOTES AND REFERENCES

Nonlinear differential equations and integral equations are a rich source of nonlinear equations. When these problems are discretized, we obtain a variable vector $x \in \mathbf{R}^n$ whose components can be used to construct an approximation to the function that solves the problem. In other applications, the vector x is intrinsically finite-dimensional—it may represent the quantities of materials to be transported between pairs of cities in a distribution network, for instance. In all cases, the equations r_i enforce consistency, conservation, and optimality principles in the model. Moré [168] and Averick et al. [5] discuss a number of interesting practical applications.

For analysis of the convergence of Broyden's method, including proofs of Theorem 11.5, see Dennis and Schnabel [69, Chapter 8] and Kelley [141, Chapter 6]. Details on a limited-memory implementation of Broyden's method are given by Kelley [141, Section 7.3].

The trust-region approach (11.51) was proposed by Duff, Nocedal, and Reid [73].

✎ EXERCISES

✎ 11.1 Show that for any vector $s \in \mathbf{R}^n$, we have

$$\left\| \frac{ss^T}{s^Ts} \right\| = 1,$$

where $\|\cdot\|$ denotes the Euclidean matrix norm. (Hint: Use (A.37) and (A.38).)

✎ 11.2 Consider the function $r : \mathbf{R} \to \mathbf{R}$ defined by $r(x) = x^q$, where q is an integer greater than 2. Note that $x^* = 0$ is the sole root of this function and that it is degenerate. Show that Newton's method converges Q-linearly, and find the value of the convergence ratio r in (2.21).

✎ 11.3 Show that Newton's method applied to the function $r(x) = -x^5 + x^3 + 4x$ starting from $x_0 = 1$ produces the cyclic behavior described in the text. Find the roots of this function, and check that they are nondegenerate.

✎ 11.4 For the scalar function $r(x) = \sin(5x) - x$, show that the sum-of-squares merit function has infinitely many local minima, and find a general formula for such points.

✎ 11.5 When $r : \mathbf{R}^n \to \mathbf{R}^n$, show that the function

$$\phi(\lambda) = \left\| (J^T J + \lambda I)^{-1} J^T r \right\|$$

is monotonically decreasing in λ unless $J^T r = 0$. (Hint: Use the singular-value decomposition of J.)

✎ 11.6 Show that $\nabla^2 f(x)$ is Lipschitz continuous under the assumptions of Theorem 11.7.

✎ 11.7 Prove part (iii) of Theorem 11.3.

✎ 11.8 Suppose that Theorem 11.7 is modified to allow search directions that approach the Newton direction only in the limit, that is, $\|p_k - (-J_k^{-1} r_k)\| = o(\|J_k^{-1} r_k\|)$. By modifying the proof, show that x_k converges superlinearly to x^*.

◇ **11.9** Consider a line search Newton method in which the step length α_k is chosen to be the exact minimizer of the merit function $f(\cdot)$; that is,

$$\alpha_k = \arg\min_\alpha f(x_k - \alpha J_k^{-1} r_k).$$

Show that $\alpha_k \to 1$ as $x_k \to x^*$.

◇ **11.10** Let $J \in \mathbf{R}^{n \times m}$ and $r \in \mathbf{R}^n$ and suppose that $JJ^T r = 0$. Show that $J^T r = 0$. (Hint: This doesn't even take one line!)

◇ * **11.11** Suppose we replace the assumption of $x_k \to x^*$ in Theorem 11.10 by an assumption that the nondegenerate solution x^* is a limit point of x^*. By adding some logic to the proof of this result, show that in fact x^* is the only possible limit point of the sequence. (Hint: Show that $\|J_{k+1}^{-1} r_{k+1}\| \leq \frac{1}{2} \|J_k^{-1} r_k\|$ for all k sufficiently large, and hence that for any constant $\epsilon > 0$, the sequence $\{x_k\}$ satisfies $\|x_k - x^*\| \leq \epsilon$ for all k sufficiently large. Since ϵ can be made arbitrarily small, we can conclude that x^* is the only possible limit point of $\{x_k\}$.)

◇ **11.12** Consider the following modification of our example of failure of continuation methods:

$$r(x) = x^2 - 1, \quad a = \tfrac{1}{2}.$$

Show that for this example there is a zero path for $H(x, \lambda) = \lambda(x^2 - 1 + (1 - \lambda)(x - a)$ that connects $(\tfrac{1}{2}, 0)$ to $(1, 0)$, so that continuation methods should work for this choice of starting point.

Chapter 12

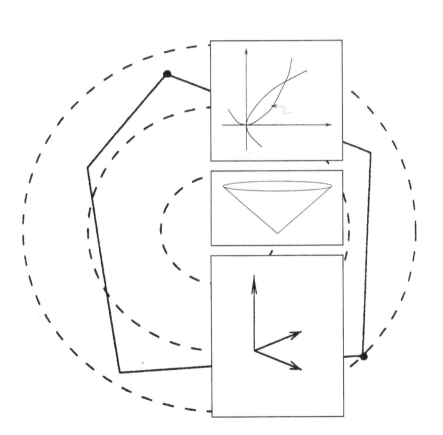

Theory of Constrained Optimization

The second part of this book is about minimizing functions subject to constraints on the variables. A general formulation for these problems is

$$\min_{x \in R^n} f(x) \quad \text{subject to} \quad \begin{cases} c_i(x) = 0, & i \in \mathcal{E}, \\ c_i(x) \geq 0, & i \in \mathcal{I}, \end{cases} \quad (12.1)$$

where f and the functions c_i are all smooth, real-valued functions on a subset of $\mathbf{R}^n$, and $\mathcal{I}$ and $\mathcal{E}$ are two finite sets of indices. As before, we call f the *objective* function, while c_i, $i \in \mathcal{E}$ are the *equality constraints* and c_i, $i \in \mathcal{I}$ are the *inequality constraints*. We define the *feasible set* Ω to be the set of points x that satisfy the constraints; that is,

$$\Omega = \{x \mid c_i(x) = 0, \ i \in \mathcal{E}; \ c_i(x) \geq 0, \ i \in \mathcal{I}\}, \quad (12.2)$$

so that we can rewrite (12.1) more compactly as

$$\min_{x \in \Omega} f(x). \quad (12.3)$$

In this chapter we derive mathematical characterizations of the solutions of (12.3). Recall that for the unconstrained optimization problem of Chapter 2, we characterized solution points x^* in the following way:

Necessary conditions: Local minima of unconstrained problems have $\nabla f(x^*) = 0$ and $\nabla^2 f(x^*)$ positive semidefinite.

Sufficient conditions: Any point x^* at which $\nabla f(x^*) = 0$ and $\nabla^2 f(x^*)$ is positive definite is a strong local minimizer of f.

Our aim in this chapter is to derive similar conditions to characterize the solutions of constrained optimization problems.

LOCAL AND GLOBAL SOLUTIONS

We have seen already that global solutions are difficult to find even when there are no constraints. The situation may be improved when we add constraints, since the feasible set might exclude many of the local minima and it may be comparatively easy to pick the global minimum from those that remain. However, constraints can also make things much more difficult. As an example, consider the problem

$$\min_{x \in R^n} \|x\|_2^2, \quad \text{subject to } \|x\|_2^2 \geq 1.$$

Without the constraint, this is a convex quadratic problem with unique minimizer $x = 0$. When the constraint is added, any vector x with $\|x\|_2 = 1$ solves the problem. There are infinitely many such vectors (hence, infinitely many local minima) whenever $n \geq 2$.

A second example shows how addition of a constraint produces a large number of local solutions that do not form a connected set. Consider

$$\min (x_2 + 100)^2 + 0.01 x_1^2, \quad \text{subject to } x_2 - \cos x_1 \geq 0, \qquad (12.4)$$

illustrated in Figure 12.1. Without the constraint, the problem has the unique solution $(-100, 0)$. With the constraint there are local solutions near the points

$$(x_1, x_2) = (k\pi, -1), \quad \text{for } k = \pm 1, \pm 3, \pm 5, \ldots.$$

Definitions of the different types of local solutions are simple extensions of the corresponding definitions for the unconstrained case, except that now we restrict consideration to the *feasible* points in the neighborhood of x^*. We have the following definition.

A vector x^* is a *local solution* of the problem (12.3) if $x^* \in \Omega$ and there is a neighborhood $\mathcal{N}$ of x^* such that $f(x) \geq f(x^*)$ for $x \in \mathcal{N} \cap \Omega$.

Similarly, we can make the following definitions:

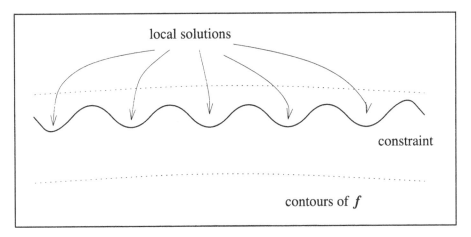

Figure 12.1 Constrained problem with many isolated local solutions.

A vector x^* is a *strict local solution* (also called a *strong local solution*) if $x^* \in \Omega$ and there is a neighborhood $\mathcal{N}$ of x^* such that $f(x) > f(x^*)$ for all $x \in \mathcal{N} \cap \Omega$ with $x \neq x^*$.

A point x^* is an *isolated local solution* if $x^* \in \Omega$ and there is a neighborhood $\mathcal{N}$ of x^* such that x^* is the only local minimizer in $\mathcal{N} \cap \Omega$.

At times, we replace the word "solution" by "minimizer" in our discussion. This alternative is frequently used in the literature, but it is slightly less satisfying because it does not account for the role of the constraints in defining the point in question.

SMOOTHNESS

Smoothness of objective functions and constraints is an important issue in characterizing solutions, just as in the unconstrained case. It ensures that the objective function and the constraints all behave in a reasonably predictable way and therefore allows algorithms to make good choices for search directions.

We saw in Chapter 2 that graphs of nonsmooth functions contain "kinks" or "jumps" where the smoothness breaks down. If we plot the feasible region for any given constrained optimization problem, we usually observe many kinks and sharp edges. Does this mean that the constraint functions that describe these regions are nonsmooth? The answer is often no, because the nonsmooth boundaries can often be described by a collection of smooth constraint functions. Figure 12.2 shows a diamond-shaped feasible region in $\mathbf{R}^2$ that could be described by the single nonsmooth constraint

$$\|x\|_1 = |x_1| + |x_2| \leq 1. \tag{12.5}$$

CHAPTER 12. THEORY OF CONSTRAINED OPTIMIZATION

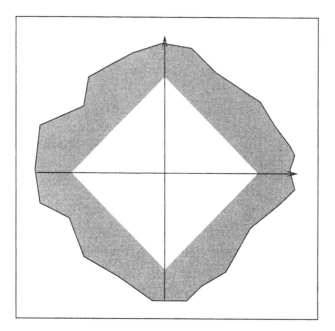

Figure 12.2 A feasible region with a nonsmooth boundary can be described by smooth constraints.

It can also be described by the following set of smooth (in fact, linear) constraints:

$$x_1 + x_2 \leq 1, \quad x_1 - x_2 \leq 1, \quad -x_1 + x_2 \leq 1, \quad -x_1 - x_2 \leq 1. \qquad (12.6)$$

Each of the four constraints represents one edge of the feasible polytope. In general, the constraint functions are chosen so that each one represents a smooth piece of the boundary of Ω.

Nonsmooth, unconstrained optimization problems can sometimes be reformulated as smooth constrained problems. An example is given by the unconstrained scalar problem of minimizing a nonsmooth function $f(x)$ defined by

$$f(x) = \max(x^2, x), \qquad (12.7)$$

which has kinks at $x = 0$ and $x = 1$, and the solution at $x^* = 0$. We obtain a smooth, constrained formulation of this problem by adding an artificial variable t and writing

$$\min t \quad \text{s.t.} \quad t \geq x, \quad t \geq x^2. \qquad (12.8)$$

Reformulation techniques such as (12.6) and (12.8) are used often in cases where f is a maximum of a collection of functions or when f is a 1-norm or ∞-norm of a vector function.

In the examples above we expressed inequality constraints in a slightly different way from the form $c_i(x) \geq 0$ that appears in the definition (12.1). However, any collection of inequality constraints with $\geq$ and $\leq$ and nonzero right-hand-sides can be expressed in the form $c_i(x) \geq 0$ by simple rearrangement of the inequality. In general, it is good practice to state the constraint in a way that is intuitive and easy to understand.

12.1 EXAMPLES

To introduce the basic principles behind the characterization of solutions of constrained optimization problems, we work through three simple examples. The ideas discussed here will be made rigorous in the sections that follow.

We start by noting one item of terminology that recurs throughout the rest of the book: At a feasible point x, the inequality constraint $i \in \mathcal{I}$ is said to be *active* if $c_i(x) = 0$ and *inactive* if the strict inequality $c_i(x) > 0$ is satisfied.

A SINGLE EQUALITY CONSTRAINT

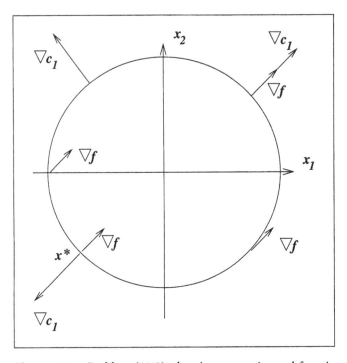

Figure 12.3 Problem (12.9), showing constraint and function gradients at various feasible points.

❑ EXAMPLE 12.1

Our first example is a two-variable problem with a single equality constraint:

$$\min x_1 + x_2 \quad \text{s.t.} \quad x_1^2 + x_2^2 - 2 = 0 \tag{12.9}$$

(see Figure 12.3). In the language of (12.1), we have $f(x) = x_1 + x_2$, $\mathcal{I} = \emptyset$, $\mathcal{E} = \{1\}$, and $c_1(x) = x_1^2 + x_2^2 - 2$. We can see by inspection that the feasible set for this problem is the circle of radius $\sqrt{2}$ centered at the origin—just the boundary of this circle, not its interior. The solution x^* is obviously $(-1, -1)^T$. From any other point on the circle, it is easy to find a way to move that *stays feasible* (that is, remains on the circle) while *decreasing f*. For instance, from the point $x = (\sqrt{2}, 0)^T$ any move in the clockwise direction around the circle has the desired effect.

We also see from Figure 12.3 that at the solution x^*, the *constraint normal* $\nabla c_1(x^*)$ is parallel to $\nabla f(x^*)$. That is, there is a scalar λ_1^* such that

$$\nabla f(x^*) = \lambda_1^* \nabla c_1(x^*). \tag{12.10}$$

(In this particular case, we have $\lambda_1^* = -\frac{1}{2}$.)

❑

We can derive (12.10) by examining first-order Taylor series approximations to the objective and constraint functions. To retain feasibility with respect to the function $c_1(x) = 0$, we require that $c_1(x + d) = 0$; that is,

$$0 = c_1(x + d) \approx c_1(x) + \nabla c_1(x)^T d = \nabla c_1(x)^T d. \tag{12.11}$$

Hence, the direction d retains feasibility with respect to c_1, to first order, when it satisfies

$$\nabla c_1(x)^T d = 0. \tag{12.12}$$

Similarly, a direction of improvement must produce a decrease in f, so that

$$0 > f(x + d) - f(x) \approx \nabla f(x)^T d,$$

or, to first order,

$$\nabla f(x)^T d < 0. \tag{12.13}$$

If there exists a direction d that satisfies both (12.12) and (12.13), we conclude that improvement on our current point x is possible. It follows that a *necessary condition* for optimality for the problem (12.9) is that there exist *no direction d satisfying both (12.12) and (12.13).*

By drawing a picture, the reader can check that the only way that such a direction cannot exist is if $\nabla f(x)$ and $\nabla c_1(x)$ are parallel, that is, if the condition $\nabla f(x) = \lambda_1 \nabla c_1(x)$ holds at x, for some scalar λ_1. If this condition is *not* satisfied, the direction defined by

$$d = -\left(I - \frac{\nabla c_1(x) \nabla c_1(x)^T}{\|\nabla c_1(x)\|^2}\right) \nabla f(x) \tag{12.14}$$

satisfies both conditions (12.12) and (12.13) (see the exercises).

By introducing the *Lagrangian function*

$$\mathcal{L}(x, \lambda_1) = f(x) - \lambda_1 c_1(x), \tag{12.15}$$

and noting that $\nabla_x \mathcal{L}(x, \lambda_1) = \nabla f(x) - \lambda_1 \nabla c_1(x)$, we can state the condition (12.10) equivalently as follows: At the solution x^*, there is a scalar λ_1^* such that

$$\nabla_x \mathcal{L}(x^*, \lambda_1^*) = 0. \tag{12.16}$$

This observation suggests that we can search for solutions of the equality-constrained problem (12.9) by searching for stationary points of the Lagrangian function. The scalar quantity λ_1 in (12.15) is called a *Lagrange multiplier* for the constraint $c_1(x) = 0$.

Though the condition (12.10) (equivalently, (12.16)) appears to be *necessary* for an optimal solution of the problem (12.9), it is clearly not *sufficient*. For instance, in Example 12.1, (12.10) is satisfied at the point $x = (1, 1)$ (with $\lambda_1 = \frac{1}{2}$), but this point is obviously not a solution—in fact, it *maximizes* the function f on the circle. Moreover, in the case of equality-constrained problems, we cannot turn the condition (12.10) into a sufficient condition simply by placing some restriction on the sign of λ_1. To see this, consider replacing the constraint $x_1^2 + x_2^2 - 2 = 0$ by its negative $2 - x_1^2 - x_2^2 = 0$ in Example 12.1. The solution of the problem is not affected, but the value of λ_1^* that satisfies the condition (12.10) changes from $\lambda_1^* = -\frac{1}{2}$ to $\lambda_1^* = \frac{1}{2}$.

A SINGLE INEQUALITY CONSTRAINT

❑ EXAMPLE 12.2

This is a slight modification of Example 12.1, in which the equality constraint is replaced by an inequality. Consider

$$\min x_1 + x_2 \quad \text{s.t.} \quad 2 - x_1^2 - x_2^2 \geq 0, \tag{12.17}$$

for which the feasible region consists of the circle of problem (12.9) and its interior (see Figure 12.4). Note that the constraint normal ∇c_1 points toward the interior of the feasible region at each point on the boundary of the circle. By inspection, we see that the solution is still $(-1, -1)$ and that the condition (12.10) holds for the value $\lambda_1^* = \frac{1}{2}$. However, this inequality-constrained problem differs from the equality-constrained problem (12.9) of Example 12.1 in that the sign of the Lagrange multiplier plays a significant role, as we now argue.

□

As before, we conjecture that a given feasible point x is *not* optimal if we can find a step d that both retains feasibility and decreases the objective function f to first order. The main difference between problems (12.9) and (12.17) comes in the handling of the feasibility condition. As in (12.13), the direction d improves the objective function, to first order, if $\nabla f(x)^T d < 0$. Meanwhile, the direction d retains feasibility if

$$0 \leq c_1(x+d) \approx c_1(x) + \nabla c_1(x)^T d,$$

so, to first order, feasibility is retained if

$$c_1(x) + \nabla c_1(x)^T d \geq 0. \tag{12.18}$$

In determining whether a direction d exists that satisfies both (12.13) and (12.18), we consider the following two cases, which are illustrated in Figure 12.4.

Case I: Consider first the case in which x lies *strictly inside* the circle, so that the strict inequality $c_1(x) > 0$ holds. In this case, *any* vector d satisfies the condition (12.18), provided only that its length is sufficiently small. In particular, whenever $\nabla f(x^*) \neq 0$, we can obtain a direction d that satisfies both (12.13) and (12.18) by setting

$$d = -c_1(x) \frac{\nabla f(x)}{\|\nabla f(x)\|}.$$

The only situation in which such a direction fails to exist is when

$$\nabla f(x) = 0. \tag{12.19}$$

Case II: Consider now the case in which x lies on the boundary of the circle, so that $c_1(x) = 0$. The conditions (12.13) and (12.18) therefore become

$$\nabla f(x)^T d < 0, \quad \nabla c_1(x)^T d \geq 0.$$

The first of these conditions defines an open half-space, while the second defines a closed half-space, as illustrated in Figure 12.5. It is clear from this figure that the two regions fail to intersect only when $\nabla f(x)$ and $\nabla c_1(x)$ point in the same direction, that is, when

$$\nabla f(x) = \lambda_1 \nabla c_1(x), \quad \text{for some } \lambda_1 \geq 0. \tag{12.20}$$

12.1. EXAMPLES   323

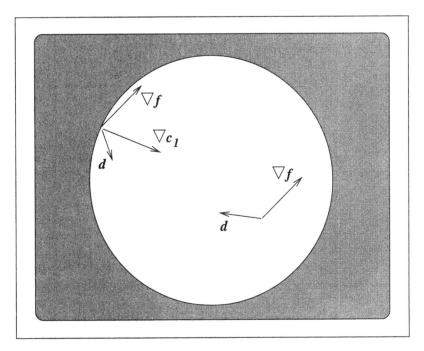

Figure 12.4 Improvement directions from two feasible points for the problem (12.17) at which the constraint is active and inactive, respectively.

Note that the sign of the multiplier is significant here. If (12.10) were satisfied with a *negative* value of λ_1, then $\nabla f(x)$ and $\nabla c_1(x)$ would point in opposite directions, and we see from Figure 12.5 that the set of directions that satisfy both (12.13) and (12.18) would make up an entire open half-plane.

The optimality conditions for both cases I and II can again be summarized neatly with reference to the Lagrangian function. When no first-order feasible descent direction exists at some point x^*, we have that

$$\nabla_x \mathcal{L}(x^*, \lambda_1^*) = 0, \quad \text{for some } \lambda_1^* \geq 0, \tag{12.21}$$

where we also require that

$$\lambda_1^* c_1(x^*) = 0. \tag{12.22}$$

This condition is known as a *complementarity condition*; it implies that the Lagrange multiplier λ_1 can be strictly positive *only when the corresponding constraint c_1 is active*. Conditions of this type play a central role in constrained optimization, as we see in the sections that follow. In case I, we have that $c_1(x^*) > 0$, so (12.22) requires that $\lambda_1^* = 0$. Hence, (12.21)

324 CHAPTER 12. THEORY OF CONSTRAINED OPTIMIZATION

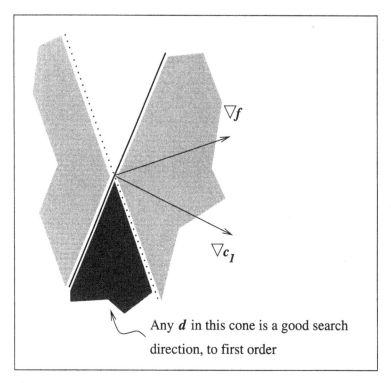

Any d in this cone is a good search direction, to first order

Figure 12.5 A direction d that satisfies both (12.13) and (12.18) lies in the intersection of a closed half-plane and an open half-plane.

reduces to $\nabla f(x^*) = 0$, as required by (12.19). In case II, (12.22) allows λ_1^* to take on a nonnegative value, so (12.21) becomes equivalent to (12.20).

TWO INEQUALITY CONSTRAINTS

❑ EXAMPLE 12.3

Suppose we add an extra constraint to the problem (12.17) to obtain

$$\min x_1 + x_2 \quad \text{s.t.} \quad 2 - x_1^2 - x_2^2 \geq 0, \quad x_2 \geq 0, \tag{12.23}$$

for which the feasible region is the half-disk illustrated in Figure 12.6. It is easy to see that the solution lies at $(-\sqrt{2}, 0)^T$, a point at which both constraints are active. By repeating the arguments for the previous examples, we conclude that a direction d is a feasible descent direction, to first order, if it satisfies the following conditions:

$$\nabla c_i(x)^T d \geq 0, \quad i \in \mathcal{I} = \{1, 2\}, \quad \nabla f(x)^T d < 0. \tag{12.24}$$

However, it is clear from Figure 12.6 that no such direction can exist when $x = (-\sqrt{2}, 0)^T$. The conditions $\nabla c_i(x)^T d \geq 0$, $i = 1, 2$, are both satisfied only if d lies in the quadrant defined by $\nabla c_1(x)$ and $\nabla c_2(x)$, but it is clear by inspection that all vectors d in this quadrant satisfy $\nabla f(x)^T d \geq 0$.

Let us see how the Lagrangian and its derivatives behave for the problem (12.23) and the solution point $(-\sqrt{2}, 0)^T$. First, we include an additional term $\lambda_i c_i(x)$ in the Lagrangian for each additional constraint, so we have

$$\mathcal{L}(x, \lambda) = f(x) - \lambda_1 c_1(x) - \lambda_2 c_2(x),$$

where $\lambda = (\lambda_1, \lambda_2)^T$ is the vector of Lagrange multipliers. The extension of condition (12.21) to this case is

$$\nabla_x \mathcal{L}(x^*, \lambda^*) = 0, \quad \text{for some } \lambda^* \geq 0, \tag{12.25}$$

where the inequality $\lambda^* \geq 0$ means that all components of λ^* are required to be nonnegative. By applying the complementarity condition (12.22) to both inequality constraints, we obtain

$$\lambda_1^* c_1(x^*) = 0, \quad \lambda_2^* c_2(x^*) = 0. \tag{12.26}$$

When $x^* = (-\sqrt{2}, 0)^T$, we have

$$\nabla f(x^*) = \begin{bmatrix} 1 \\ 1 \end{bmatrix}, \quad \nabla c_1(x^*) = \begin{bmatrix} 2\sqrt{2} \\ 0 \end{bmatrix}, \quad \nabla c_2(x^*) = \begin{bmatrix} 0 \\ 1 \end{bmatrix},$$

so that it is easy to verify that $\nabla_x \mathcal{L}(x^*, \lambda^*) = 0$ when we select λ^* as follows:

$$\lambda^* = \begin{bmatrix} 1/(2\sqrt{2}) \\ 1 \end{bmatrix}.$$

Note that both components of λ^* are positive.

We consider now some other feasible points that are *not* solutions of (12.23), and examine the properties of the Lagrangian and its gradient at these points.

For the point $x = (\sqrt{2}, 0)^T$, we again have that both constraints are active. However, the objective gradient $\nabla f(x)$ no longer lies in the quadrant defined by the conditions $\nabla c_i(x)^T d \geq 0$, $i = 1, 2$ (see Figure 12.7). One first-order feasible descent direction from this point—a vector d that satisfies (12.24)—is simply $d = (-1, 0)^T$; there are many others (see the exercises). For this value of x it is easy to verify that the condition $\nabla_x \mathcal{L}(x, \lambda) = 0$

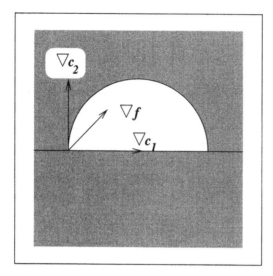

Figure 12.6
Problem (12.23), illustrating the gradients of the active constraints and objective at the solution.

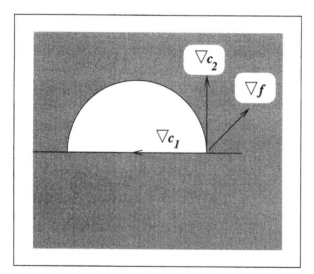

Figure 12.7
Problem (12.23), illustrating the gradients of the active constraints and objective at a nonoptimal point.

is satisfied only when $\lambda = (-1/(2\sqrt{2}), 1)$. Note that the first component λ_1 is negative, so that the conditions (12.25) are not satisfied at this point.

Finally, let us consider the point $x = (1, 0)^T$, at which only the second constraint c_2 is active. At this point, linearization of f and c as in Example 12.2 gives the following conditions, which must be satisfied for d to be a feasible descent direction, to first order:

$$1 + \nabla c_1(x)^T d \geq 0, \quad \nabla c_2(x)^T d \geq 0, \quad \nabla f(x)^T d < 0. \qquad (12.27)$$

In fact, we need worry only about satisfying the second and third conditions, since we can always satisfy the first condition by multiplying d by a sufficiently small positive quantity. By noting that

$$\nabla f(x) = \begin{bmatrix} 1 \\ 1 \end{bmatrix}, \quad \nabla c_2(x) = \begin{bmatrix} 0 \\ 1 \end{bmatrix},$$

it is easy to verify that the vector $d = \left(-\frac{1}{2}, \frac{1}{4}\right)$ satisfies (12.27) and is therefore a descent direction.

To show that optimality conditions (12.25) and (12.26) fail, we note first from (12.26) that since $c_1(x) > 0$, we must have $\lambda_1 = 0$. Therefore, in trying to satisfy $\nabla_x \mathcal{L}(x, \lambda) = 0$, we are left to search for a value λ_2 such that $\nabla f(x) - \lambda_2 \nabla c_2(x) = 0$. No such λ_2 exists, and thus this point fails to satisfy the optimality conditions. □

12.2 FIRST-ORDER OPTIMALITY CONDITIONS

STATEMENT OF FIRST-ORDER NECESSARY CONDITIONS

The three examples above suggest that a number of conditions are important in the characterization of solutions for (12.1). These include the relation $\nabla_x \mathcal{L}(x, \lambda) = 0$, the nonnegativity of λ_i for all inequality constraints $c_i(x)$, and the complementarity condition $\lambda_i c_i(x) = 0$ that is required for all the inequality constraints. We now generalize the observations made in these examples and state the first-order optimality conditions in a rigorous fashion.

In general, the Lagrangian for the constrained optimization problem (12.1) is defined as

$$\mathcal{L}(x, \lambda) = f(x) - \sum_{i \in \mathcal{E} \cup \mathcal{I}} \lambda_i c_i(x). \qquad (12.28)$$

The *active set* $\mathcal{A}(x)$ at any feasible x is the union of the set $\mathcal{E}$ with the indices of the active inequality constraints; that is,

$$\mathcal{A}(x) = \mathcal{E} \cup \{i \in \mathcal{I} \mid c_i(x) = 0\}. \qquad (12.29)$$

Next, we need to give more attention to the properties of the constraint gradients. The vector $\nabla c_i(x)$ is often called the *normal* to the constraint c_i at the point x, because it is usually a vector that is perpendicular to the contours of the constraint c_i at x, and in the case of an inequality constraint, it points toward the feasible side of this constraint. It is possible, however, that $\nabla c_i(x)$ vanishes due to the algebraic representation of c_i, so that the term

$\lambda_i \nabla c_i(x)$ vanishes for all values of λ_i and does not play a role in the Lagrangian gradient $\nabla_x \mathcal{L}$. For instance, if we replaced the constraint in (12.9) by the equivalent condition

$$c_1(x) = (x_1^2 + x_2^2 - 2)^2 = 0,$$

we would have that $\nabla c_1(x) = 0$ for all feasible points x, and in particular that the condition $\nabla f(x) = \lambda_1 \nabla c_1(x)$ no longer holds at the optimal point $(-1, -1)^T$. We usually make an assumption called a *constraint qualification* to ensure that such degenerate behavior does not occur at the value of x in question. One such constraint qualification—probably the one most often used in the design of algorithms—is the one defined as follows:

Definition 12.1 (LICQ).

Given the point x^* and the active set $\mathcal{A}(x^*)$ defined by (12.29), we say that the *linear independence constraint qualification* (LICQ) holds if the set of active constraint gradients $\{\nabla c_i(x^*), i \in \mathcal{A}(x^*)\}$ is linearly independent.

Note that if this condition holds, none of the active constraint gradients can be zero.

This condition allows us to state the following optimality conditions for a general nonlinear programming problem (12.1). These conditions provide the foundation for many of the algorithms described in the remaining chapters of the book. They are called *first-order conditions* because they concern themselves with properties of the gradients (first-derivative vectors) of the objective and constraint functions.

Theorem 12.1 (First-Order Necessary Conditions).

Suppose that x^* is a local solution of (12.1) and that the LICQ holds at x^*. Then there is a Lagrange multiplier vector λ^*, with components λ_i^*, $i \in \mathcal{E} \cup \mathcal{I}$, such that the following conditions are satisfied at (x^*, λ^*)

$$\nabla_x \mathcal{L}(x^*, \lambda^*) = 0, \tag{12.30a}$$
$$c_i(x^*) = 0, \quad \text{for all } i \in \mathcal{E}, \tag{12.30b}$$
$$c_i(x^*) \geq 0, \quad \text{for all } i \in \mathcal{I}, \tag{12.30c}$$
$$\lambda_i^* \geq 0, \quad \text{for all } i \in \mathcal{I}, \tag{12.30d}$$
$$\lambda_i^* c_i(x^*) = 0, \quad \text{for all } i \in \mathcal{E} \cup \mathcal{I}. \tag{12.30e}$$

The conditions (12.30) are often known as the *Karush–Kuhn–Tucker conditions*, or *KKT conditions* for short. Because the complementarity condition implies that the Lagrange multipliers corresponding to inactive inequality constraints are zero, we can omit the terms for indices $i \notin \mathcal{A}(x^*)$ from (12.30a) and rewrite this condition as

$$0 = \nabla_x \mathcal{L}(x^*, \lambda^*) = \nabla f(x^*) - \sum_{i \in \mathcal{A}(x^*)} \lambda_i^* \nabla c_i(x^*). \tag{12.31}$$

A special case of complementarity is important and deserves its own definition:

Definition 12.2 (Strict Complementarity).

Given a local solution x^* of (12.1) and a vector λ^* satisfying (12.30), we say that the *strict complementarity condition holds if* exactly one of λ_i^* and $c_i(x^*)$ is zero for each index $i \in \mathcal{I}$. In other words, we have that $\lambda_i^* > 0$ for each $i \in \mathcal{I} \cap \mathcal{A}(x^*)$.

For a given problem (12.1) and solution point x^*, there may be many vectors λ^* for which the conditions (12.30) are satisfied. When the LICQ holds, however, the optimal λ^* is unique (see the exercises).

The proof of Theorem 12.1 is quite complex, but it is important to our understanding of constrained optimization, so we present it in the next section. First, we illustrate the KKT conditions with another example.

❑ EXAMPLE 12.4

Consider the feasible region illustrated in Figure 12.2 and described by the four constraints (12.6). By restating the constraints in the standard form of (12.1) and including an objective function, the problem becomes

$$\min_x \left(x_1 - \frac{3}{2}\right)^2 + \left(x_2 - \frac{1}{8}\right)^4 \quad \text{s.t.} \quad \begin{bmatrix} 1 - x_1 - x_2 \\ 1 - x_1 + x_2 \\ 1 + x_1 - x_2 \\ 1 + x_1 + x_2 \end{bmatrix} \geq 0. \quad (12.32)$$

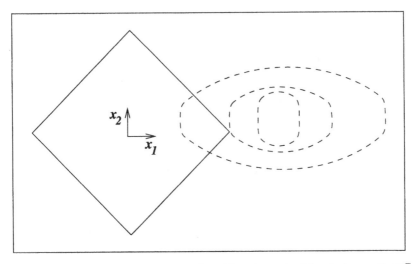

Figure 12.8 Inequality-constrained problem (12.32) with solution at $(1, 0)^T$.

It is fairly clear from Figure 12.8 that the solution is $x^* = (1, 0)$. The first and second constraints in (12.32) are active at this point. Denoting them by c_1 and c_2 (and the inactive constraints by c_3 and c_4), we have

$$\nabla f(x^*) = \begin{bmatrix} -1 \\ 1 \\ -\frac{1}{2} \end{bmatrix}, \quad \nabla c_1(x^*) = \begin{bmatrix} -1 \\ -1 \end{bmatrix}, \quad \nabla c_2(x^*) = \begin{bmatrix} -1 \\ 1 \end{bmatrix}.$$

Therefore, the KKT conditions (12.30a)-(12.30e) are satisfied when we set

$$\lambda^* = \left(\tfrac{3}{4}, \tfrac{1}{4}, 0, 0\right)^T.$$

□

SENSITIVITY

The convenience of using Lagrange multipliers should now be clear, but what of their intuitive significance? The value of each Lagrange multiplier λ_i^* tells us something about the *sensitivity* of the optimal objective value $f(x^*)$ to the presence of constraint c_i. To put it another way, λ_i^* indicates how hard f is "pushing" or "pulling" against the particular constraint c_i. We illustrate this point with a little analysis. When we choose an inactive constraint $i \notin \mathcal{A}(x^*)$ such that $c_i(x^*) > 0$, the solution x^* and function value $f(x^*)$ are quite indifferent to whether this constraint is present or not. If we perturb c_i by a tiny amount, it will still be inactive and x^* will still be a local solution of the optimization problem. Since $\lambda_i^* = 0$ from (12.30e), the Lagrange multiplier indicates accurately that constraint i is not significant.

Suppose instead that constraint i is active, and let us perturb the right-hand-side of this constraint a little, requiring, say, that $c_i(x) \geq -\epsilon \|\nabla c_i(x^*)\|$ instead of $c_i(x) \geq 0$. Suppose that ϵ is sufficiently small that the perturbed solution $x^*(\epsilon)$ still has the same set of active constraints, and that the Lagrange multipliers are not much affected by the perturbation. (These conditions can be made more rigorous with the help of strict complementarity and second-order conditions, as discussed later in the chapter.) We then find that

$$-\epsilon \|\nabla c_i(x^*)\| = c_i(x^*(\epsilon)) - c_i(x^*) \approx (x^*(\epsilon) - x^*)^T \nabla c_i(x^*),$$
$$0 = c_j(x^*(\epsilon)) - c_j(x^*) \approx (x^*(\epsilon) - x^*)^T \nabla c_j(x^*),$$

for all $j \in \mathcal{A}(x^*)$ with $j \neq i$.

The value of $f(x^*(\epsilon))$, meanwhile, can be estimated with the help of (12.30a). We have

$$f(x^*(\epsilon)) - f(x^*) \approx (x^*(\epsilon) - x^*)^T \nabla f(x^*)$$

$$= \sum_{j \in \mathcal{A}(x^*)} \lambda_j^*(x^*(\epsilon) - x^*)^T \nabla c_j(x^*)$$

$$\approx -\epsilon \|\nabla c_i(x^*)\| \lambda_i^*.$$

By taking limits, we see that the family of solutions $x^*(\epsilon)$ satisfies

$$\frac{df(x^*(\epsilon))}{d\epsilon} = -\lambda_i^* \|\nabla c_i(x^*)\|. \qquad (12.33)$$

A sensitivity analysis of this problem would conclude that if $\lambda_i^* \|\nabla c_i(x^*)\|$ is large, then the optimal value is sensitive to the placement of the ith constraint, while if this quantity is small, the dependence is not too strong. If λ_i^* is exactly zero for some active constraint, small perturbations to c_i in some directions will hardly affect the optimal objective value at all; the change is zero, to first order.

This discussion motivates the definition below, which classifies constraints according to whether or not their corresponding Lagrange multiplier is zero.

Definition 12.3.
 Let x^* be a solution of the problem (12.1), and suppose that the KKT conditions (12.30) are satisfied. We say that an inequality constraint c_i is strongly active or binding if $i \in \mathcal{A}(x^*)$ and $\lambda_i^* > 0$ for some Lagrange multiplier λ^* satisfying (12.30). We say that c_i is weakly active if $i \in \mathcal{A}(x^*)$ and $\lambda_i^* = 0$ for all λ^* satisfying (12.30).

Note that the analysis above is independent of scaling of the individual constraints. For instance, we might change the formulation of the problem by replacing some active constraint c_i by $10 c_i$. The new problem will actually be equivalent (that is, it has the same feasible set and same solution), but the optimal multiplier λ_i^* corresponding to c_i will be replaced by $\lambda_i^*/10$. However, since $\|\nabla c_i(x^*)\|$ is replaced by $10\|\nabla c_i(x^*)\|$, the product $\lambda_i^* \|\nabla c_i(x^*)\|$ does not change. If, on the other hand, we replace the objective function f by $10f$, the multipliers λ_i^* in (12.30) all will need to be replaced by $10\lambda_i^*$. Hence in (12.33) we see that the sensitivity of f to perturbations has increased by a factor of 10, which is exactly what we would expect.

12.3 DERIVATION OF THE FIRST-ORDER CONDITIONS

Having studied some motivating examples, observed the characteristics of optimal and nonoptimal points, and stated the KKT conditions, we now describe a complete proof of Theorem 12.1. This analysis is not just of esoteric interest, but is rather the key to understanding all constrained optimization algorithms.

FEASIBLE SEQUENCES

The first concept we introduce is that of a *feasible sequence*. Given a feasible point x^*, a sequence $\{z_k\}_{k=0}^{\infty}$ with $z_k \in \mathbf{R}^n$ is a feasible sequence if the following properties hold:

(i) $z_k \neq x^*$ for all k;

(ii) $\lim_{k \to \infty} z_k = x^*$;

(iii) z_k is feasible for all sufficiently large values of k.

For later reference, we denote the set of all possible feasible sequences approaching x by $\mathcal{T}(x)$.

We characterize a local solution of (12.1) as *a point x at which all feasible sequences have the property that $f(z_k) \geq f(x)$ for all k sufficiently large*. We derive practical, verifiable conditions under which this property holds. To do so we will make use of the concept of a *limiting direction* of a feasible sequence.

Limiting directions of a feasible sequence are vectors d such that we have

$$\lim_{z_k \in \mathcal{S}_d} \frac{z_k - x}{\|z_k - x\|} \to d, \tag{12.34}$$

where $\mathcal{S}_d$ is some subsequence of $\{z_k\}_{k=0}^{\infty}$. In general, a feasible sequence has at least one limiting direction and may have more than one. To see this, note that the sequence of vectors defined by

$$d_k = \frac{z_k - x}{\|z_k - x\|}$$

lies on the surface of the unit sphere, which is a compact set, and thus there is at least one limit point d. Moreover, all such points are limiting directions by the definition (12.34). If we have some sequence $\{z_k\}$ with limiting direction d and corresponding subsequence $\mathcal{S}_d$, we can construct another feasible sequence $\{\bar{z}_k\}$ such that

$$\lim_{k \to \infty} \frac{\bar{z}_k - x}{\|\bar{z}_k - x\|} = d$$

(that is, with a unique limit point) by simply defining each $\bar{z}_k$ to be an element from the subsequence $\mathcal{S}_d$.

We illustrate these concepts by revisiting Example 12.1

❑ **EXAMPLE 12.5** (EXAMPLE 12.1, REVISITED)

Figure 12.9 shows a closeup of the problem (12.9), the equality-constrained problem in which the feasible set is a circle of radius $\sqrt{2}$, near the nonoptimal point $x = (-\sqrt{2}, 0)^T$.

12.3. DERIVATION OF THE FIRST-ORDER CONDITIONS

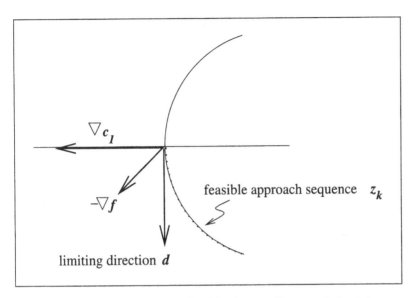

Figure 12.9 Constraint normal, objective gradient, and feasible sequence for problem (12.9).

The figure also shows a feasible sequence approaching x. This sequence could be defined analytically by the formula

$$z_k = \begin{bmatrix} -\sqrt{2 - 1/k^2} \\ -1/k \end{bmatrix}. \tag{12.35}$$

The vector $d = (0, -1)$ plotted in Figure 12.9 is a limiting direction of this feasible sequence. Note that d is tangent to the feasible sequence at x but points in the opposite direction. The objective function $f(x) = x_1 + x_2$ increases as we move along the sequence (12.35); in fact, we have $f(z_{k+1}) > f(z_k)$ for all $k = 2, 3, \ldots$ (see the exercises). It follows that $f(z_k) < f(x)$ for $k = 2, 3, \ldots$. Hence, x cannot be a solution of (12.9).

Another feasible sequence is one that approaches $x^* = (-\sqrt{2}, 0)$ from the opposite direction. Its elements are defined by

$$z_k = \begin{bmatrix} -\sqrt{2 - 1/k^2} \\ 1/k \end{bmatrix}.$$

It is easy to show that f *decreases* along this sequence and that its limiting direction is $d = (0, 1)^T$. Other feasible sequences are obtained by combining elements from the two sequences already discussed, for instance

$$z_k = \begin{cases} (-\sqrt{2-1/k^2}, 1/k), & \text{when } k \text{ is a multiple of 3,} \\ (-\sqrt{2-1/k^2}, -1/k), & \text{otherwise.} \end{cases}$$

In general, feasible sequences of points approaching $(-\sqrt{2}, 0)^T$ will have two limiting directions, $(0, 1)^T$ and $(0, -1)^T$.

□

We now consider feasible sequences and limiting directions for an example that involves inequality constraints.

❑ EXAMPLE 12.6   (EXAMPLE 12.2, REVISITED)

We now reconsider problem (12.17) in Example 12.2. The solution $x^* = (-1, -1)^T$ is the same as in the equality-constrained case, but there is a much more extensive collection of feasible sequences that converge to any given feasible point (see Figure 12.10). From the point $x = (-\sqrt{2}, 0)^T$, the various feasible sequences defined above for the equality-constrained problem are still feasible for (12.17). There are also infinitely many feasible sequences that converge to $x = (-\sqrt{2}, 0)^T$ along a straight line from the interior of the circle. These are defined by

$$z_k = (-1, 0)^T + (1/k)w, \qquad (12.36)$$

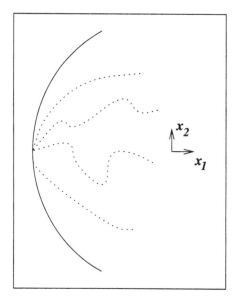

Figure 12.10
Feasible sequences converging to a particular feasible point for the region defined by $x_1^2 + x_2^2 \leq 2$.

where w is any vector whose first component is positive ($w_1 > 0$). Now, z_k is feasible in (12.36), provided that $\|z_k\| \leq 1$, that is,

$$(-1 + w_1/k)^2 + (w_2/k)^2 \leq 1,$$

a condition that is satisfied, provided that $k > (2w_1)/(w_1^2 + w_2^2)$. In addition to these straight-line feasible sequences, we can also define an infinite variety of sequences that approach $(-\sqrt{2}, 0)$ along a curve from the interior of the circle or that make the approach in a seemingly random fashion. □

Given a point x, if it is possible to choose a feasible sequence from $T(x)$ such that the first-order approximation to the objective function actually *increases* monotonically along the sequence, then x must not be optimal. This condition is the fundamental first-order necessary condition, and we state it formally in the following theorem.

Theorem 12.2.
If x^* is a local solution of (12.1), then all feasible sequences $\{z_k\}$ in $T(x^*)$ must satisfy

$$\nabla f(x^*)^T d \geq 0, \qquad (12.37)$$

where d is any limiting direction of the feasible sequence.

PROOF. Suppose that there is a feasible sequence $\{z_k\}$ with the property $\nabla f(x^*)^T d < 0$, for some limiting direction d, and let S_d be the subsequence of $\{z_k\}$ that approaches x^*. By Taylor's theorem (Theorem 2.1), we have for any $z_k \in S_d$ that

$$f(z_k) = f(x^*) + (z_k - x^*)^T \nabla f(x^*) + o(\|z_k - x^*\|)$$
$$= f(x^*) + \|z_k - x^*\| d^T \nabla f(x^*) + o(\|z_k - x^*\|).$$

Since $d^T \nabla f(x^*) < 0$, we have that the remainder term is eventually dominated by the first-order term, that is,

$$f(z_k) < f(x^*) + \tfrac{1}{2}\|z_k - x^*\| d^T \nabla f(x^*), \quad \text{for all } k \text{ sufficiently large.}$$

Hence, given any open neighborhood of x^*, we can choose k sufficiently large that z_k lies within this neighborhood and has a lower value of the objective f. Therefore, x^* is not a local solution. □

This theorem tells us why we can ignore constraints that are *strictly inactive* (that is, constraints for which $c_i(x) > 0$) in formulating optimality conditions. The theorem does not use the whole range of properties of the feasible sequence, but rather one specific property: the limiting directions of $\{z_k\}$. Because of the way in which the limiting directions are defined, it is clear that only the asymptotic behavior of the sequence is relevant, that is, its behavior for large values of the index k. If some constraint $i \in \mathcal{I}$ is inactive at x, then we have $c_i(z_k) > 0$ for all k sufficiently large, so that a constraint that is inactive at x is also inactive at all sufficiently advanced elements of the feasible sequence $\{z_k\}$.

CHARACTERIZING LIMITING DIRECTIONS: CONSTRAINT QUALIFICATIONS

Theorem 12.2 is quite general, but it is not very useful as stated, because it seems to require knowledge of all possible limiting directions for all feasible sequences $T(x^*)$. In this section we show that *constraint qualifications* allow us to characterize the salient properties of $T(x^*)$, and therefore make the condition (12.37) easier to verify.

One frequently used constraint qualification is the *linear independence constrained qualification* (LICQ) given in Definition 12.1. The following lemma shows that when LICQ holds, there is a neat way to characterize the set of all possible limiting directions d in terms of the gradients $\nabla c_i(x^*)$ of the active constraints at x^*.

In subsequent results we introduce the notation A to represent the matrix whose rows are the active constraint gradients at the optimal point, that is,

$$\nabla c_i^* = \nabla c_i(x^*), \quad A^T = [\nabla c_i^*]_{i \in \mathcal{A}(x^*)}, \quad \nabla f^* = \nabla f(x^*), \tag{12.38}$$

where the active set $\mathcal{A}(x^*)$ is defined as in (12.29).

Lemma 12.3.
The following two statements are true.

(i) *If $d \in \mathbf{R}^n$ is a limiting direction of a feasible sequence, then*

$$d^T \nabla c_i^* = 0, \quad \text{for all } i \in \mathcal{E}, \quad d^T \nabla c_i^* \geq 0, \quad \text{for all } i \in \mathcal{A}(x^*) \cap \mathcal{I}. \tag{12.39}$$

(ii) *If (12.39) holds with $\|d\| = 1$ and the LICQ condition is satisfied, then $d \in \mathbf{R}^n$ is a limiting direction of some feasible sequence.*

PROOF. Without loss of generality, let us assume that all the constraints $c_i(\cdot)$, $i = 1, 2, \ldots, m$, are active. (We can arrive at this convenient ordering by simply dropping all inactive constraints—which are irrelevant in some neighborhood of x^*—and renumbering the active constraints that remain.)

To prove (i), let $\{z_k\} \in T(x^*)$ be some feasible sequence for which d is a limiting direction, and assume (by taking a subsequence if necessary) that

$$\lim_{k \to \infty} \frac{z_k - x^*}{\|z_k - x^*\|} = d.$$

12.3. DERIVATION OF THE FIRST-ORDER CONDITIONS

From this definition, we have that

$$z_k = x^* + \|z_k - x^*\| d + o(\|z_k - x^*\|).$$

By taking $i \in \mathcal{E}$ and using Taylor's theorem, we have that

$$\begin{aligned} 0 &= \frac{1}{\|z_k - x^*\|} c_i(z_k) \\ &= \frac{1}{\|z_k - x^*\|} \left[c_i(x^*) + \|z_k - x^*\| \nabla c_i^T d + o(\|z_k - x^*\|) \right] \\ &= \nabla c_i^T d + \frac{o(\|z_k - x^*\|)}{\|z_k - x^*\|}. \end{aligned}$$

By taking the limit as $k \to \infty$, the last term in this expression vanishes, and we have $\nabla c_i^T d = 0$, as required. For the active inequality constraints $i \in \mathcal{A}(x^*) \cap \mathcal{I}$, we have similarly that

$$\begin{aligned} 0 &\leq \frac{1}{\|z_k - x^*\|} c_i(z_k) \\ &= \frac{1}{\|z_k - x^*\|} \left[c_i(x^*) + \|z_k - x^*\| \nabla c_i^T d + o(\|z_k - x^*\|) \right] \\ &= \nabla c_i^T d + \frac{o(\|z_k - x^*\|)}{\|z_k - x^*\|}. \end{aligned}$$

Hence, by a similar limiting argument, we have that $\nabla c_i^T d \geq 0$, as required.

For (ii), we use the implicit function theorem (see the Appendix or Lang [147, p. 131] for a statement of this result). First, since the LICQ holds, we have from Definition 12.1 that the $m \times n$ matrix A of active constraint gradients has full row rank m. Let Z be a matrix whose columns are a basis for the null space of A; that is,

$$Z \in \mathbf{R}^{n \times (n-m)}, \quad Z \text{ has full column rank}, \quad AZ = 0. \quad (12.40)$$

Let d have the properties (12.39), and suppose that $\{t_k\}_{k=0}^{\infty}$ is any sequence of positive scalars such $\lim_{k \to \infty} t_k = 0$. Define the parametrized system of equations $R : \mathbf{R}^n \times \mathbf{R} \to \mathbf{R}^n$ by

$$R(z, t) = \begin{bmatrix} c(z) - t A d \\ Z^T (z - x^* - t d) \end{bmatrix} = \begin{bmatrix} 0 \\ 0 \end{bmatrix}. \quad (12.41)$$

We claim that for each $t = t_k$, the solutions $z = z_k$ of this system for small $t > 0$ give a feasible sequence that approaches x^*.

Clearly, for $t = 0$, the solution of (12.41) is $z = x^*$, and the Jacobian of R at this point is

$$\nabla_z R(x^*, 0) = \begin{bmatrix} A \\ Z^T \end{bmatrix}, \tag{12.42}$$

which is nonsingular by construction of Z. Hence, according to the implicit function theorem, the system (12.41) has a unique solution z_k for all values of t_k sufficiently small. Moreover, we have from (12.41) and (12.39) that

$$i \in \mathcal{E} \Rightarrow c_i(z_k) = t_k \nabla c_i^T d = 0, \tag{12.43a}$$
$$i \in \mathcal{A}(x^*) \cap \mathcal{I} \Rightarrow c_i(z_k) = t_k \nabla c_i^T d \geq 0, \tag{12.43b}$$

so that z_k is indeed feasible. Also, for any positive value $t = \bar{t} > 0$, we cannot have $z(t) = x^*$, since otherwise by substituting $(z, t) = (x^*, \bar{t})$ into (12.41), we obtain

$$\begin{bmatrix} c(x^*) - \bar{t} A d \\ -Z^T(\bar{t} d) \end{bmatrix} = \begin{bmatrix} 0 \\ 0 \end{bmatrix}.$$

Since $c(x^*) = 0$ (we have assumed that all constraints are active) and $\bar{t} > 0$, we have from the full rank of the matrix in (12.42) that $d = 0$, which contradicts $\|d\| = 1$. It follows that $z_k = z(t_k) \neq x^*$ for all k.

It remains to show that d is a limiting direction of $\{z_k\}$. Using the fact that $R(z_k, t_k) = 0$ for all k together with Taylor's theorem, we find that

$$0 = R(z_k, t_k) = \begin{bmatrix} c(z_k) - t_k A d \\ Z^T(z_k - x^* - t_k d) \end{bmatrix}$$
$$= \begin{bmatrix} A(z_k - x^*) + o(\|z_k - x^*\|) - t_k A d \\ Z^T(z_k - x^* - t_k d) \end{bmatrix}$$
$$= \begin{bmatrix} A \\ Z^T \end{bmatrix} (z_k - x^* - t_k d) + o(\|z_k - x^*\|).$$

By dividing this expression by $\|z_k - x^*\|$ and using nonsingularity of the coefficient matrix in the first term, we obtain

$$\lim_{k \to \infty} d_k - \frac{t_k}{\|z_k - x^*\|} d = 0, \quad \text{where } d_k = \frac{z_k - x^*}{\|z_k - x^*\|}. \tag{12.44}$$

Since $\|d_k\| = 1$ for all k and since $\|d\| = 1$, we must have

$$\lim_{k \to \infty} \frac{t_k}{\|z_k - x^*\|} = 1. \tag{12.45}$$

12.3. DERIVATION OF THE FIRST-ORDER CONDITIONS

(We leave the simple proof by contradiction of this statement as an exercise.) Hence, from (12.44), we have $\lim_{k\to\infty} d_k = d$, as required. □

The set of directions defined by (12.39) plays a central role in the optimality conditions, so for future reference we give this set a name and define it formally.

Definition 12.4.
Given a point x^* and the active constraint set $\mathcal{A}(x^*)$ defined by (12.29), the set F_1 is defined by

$$F_1 = \left\{ \alpha d \,\Big|\, \alpha > 0, \begin{array}{ll} d^T \nabla c_i^* = 0, & \text{for all } i \in \mathcal{E}, \\ d^T \nabla c_i^* \geq 0, & \text{for all } i \in \mathcal{A}(x^*) \cap \mathcal{I} \end{array} \right\}.$$

Note that F_1 is a cone. In fact, when a constraint qualification is satisfied, F_1 is the tangent cone to the feasible set at x^*. (See Section 12.6 and the definition (A.1) in the Appendix.)

INTRODUCING LAGRANGE MULTIPLIERS

Lemma 12.3 tells us that when the LICQ holds, the cone F_1 is simply the set of all positive multiples of all limiting directions of all possible feasible sequences. Therefore, the condition (12.37) of Theorem 12.2 holds if $\nabla f(x^*)^T d < 0$ for all $d \in F_1$. This condition, too, would appear to be impossible to check, since the set F_1 contains infinitely many vectors in general. The next lemma gives an alternative, practical way to check this condition that makes use of the Lagrange multipliers, the variables λ_i that were introduced in the definition (12.28) of the Lagrangian $\mathcal{L}$.

Lemma 12.4.
There is no direction $d \in F_1$ for which $d^T \nabla f^* < 0$ if and only if there exists a vector $\lambda \in \mathbb{R}^m$ with

$$\nabla f^* = \sum_{i \in \mathcal{A}(x^*)} \lambda_i \nabla c_i^* = A(x^*)^T \lambda, \quad \lambda_i \geq 0 \text{ for } i \in \mathcal{A}(x^*) \cap \mathcal{I}. \tag{12.46}$$

PROOF. If we define the cone N by

$$N = \left\{ s \,\Big|\, s = \sum_{i \in \mathcal{A}(x^*)} \lambda_i \nabla c_i^*, \quad \lambda_i \geq 0 \text{ for } i \in \mathcal{A}(x^*) \cap \mathcal{I} \right\}, \tag{12.47}$$

then the condition (12.46) is equivalent to $\nabla f^* \in N$. We note first that the set N is closed—a fact that is intuitively clear but nontrivial to prove rigorously. (An intriguing proof of this claim appears in the Notes and References at the end of this chapter.)

We prove the forward implication by supposing that (12.46) holds and choosing d to be any vector satisfying (12.39). We then have that

$$d^T \nabla f^* = \sum_{i \in \mathcal{E}} \lambda_i (d^T \nabla c_i^*) + \sum_{i \in \mathcal{A}(x^*) \cap \mathcal{I}} \lambda_i (d^T \nabla c_i^*).$$

The first summation is zero because $d^T \nabla c_i^* = 0$ for $i \in \mathcal{E}$, while the second term is nonnegative because $\lambda_i \geq 0$ and $d^T \nabla c_i^* \geq 0$ for $i \in \mathcal{A}(x^*) \cap \mathcal{I}$. Hence $d^T \nabla f^* \geq 0$.

For the reverse implication, we show that if ∇f^* does *not* satisfy (12.46) (that is, $\nabla f^* \notin N$), then we can find a vector d for which $d^T \nabla f^* < 0$ and (12.39) holds.

Let $\hat{s}$ be the vector in N that is closest to ∇f^*. Because N is closed, $\hat{s}$ is well-defined. In fact, $\hat{s}$ solves the constrained optimization problem

$$\min \; \|s - \nabla f^*\|_2^2 \quad \text{subject to } s \in N. \tag{12.48}$$

Since $\hat{s} \in N$, we also have $t\hat{s} \in N$ for all scalars $t \geq 0$. Since $\|t\hat{s} - \nabla f^*\|_2^2$ is minimized at $t = 1$, we have

$$\left. \frac{d}{dt} \|t\hat{s} - \nabla f^*\|_2^2 \right|_{t=1} = 0 \Rightarrow \left. (-2\hat{s}^T \nabla f^* + 2t\hat{s}^T \hat{s}) \right|_{t=1} = 0$$

$$\Rightarrow \hat{s}^T (\hat{s} - \nabla f^*) = 0. \tag{12.49}$$

Now, let s be any other vector in N. Since N is convex (check!), we have by the minimizing property of $\hat{s}$ that

$$\|\hat{s} + \theta(s - \hat{s}) - \nabla f^*\|_2^2 \geq \|\hat{s} - \nabla f^*\|_2^2 \quad \text{for all } \theta \in [0, 1],$$

and hence

$$2\theta (s - \hat{s})^T (\hat{s} - \nabla f^*) + \theta^2 \|s - \hat{s}\|_2^2 \geq 0.$$

By dividing this expression by θ and taking the limit as $\theta \downarrow 0$, we have $(s - \hat{s})^T (\hat{s} - \nabla f^*) \geq 0$. Therefore, because of (12.49),

$$s^T (\hat{s} - \nabla f^*) \geq 0, \quad \text{for all } s \in N. \tag{12.50}$$

We claim now that the vector

$$d = \hat{s} - \nabla f^*$$

satisfies both (12.39) and $d^T \nabla f^* < 0$. Note that $d \neq 0$ because ∇f^* does not belong to the cone N. We have from (12.49) that

$$d^T \nabla f^* = d^T(\hat{s} - d) = (\hat{s} - \nabla f^*)^T \hat{s} - d^T d = -\|d\|_2^2 < 0,$$

so that d satisfies the descent property.

By making appropriate choices of coefficients λ_i, $i = 1, 2, \ldots, m$, it is easy to see that

$$i \in \mathcal{E} \Rightarrow \nabla c_i^* \in N \quad \text{and} \quad -\nabla c_i^* \in N;$$
$$i \in \mathcal{A}(x^*) \cap \mathcal{I} \Rightarrow \nabla c_i^* \in N.$$

Hence, from (12.50), we have by substituting $d = \hat{s} - \nabla f^*$ and the particular choices $s = \nabla c_i^*$ and $s = -\nabla c_i^*$ that

$$i \in \mathcal{E} \Rightarrow d^T \nabla c_i^* \geq 0 \quad \text{and} \quad -d^T \nabla c_i^* \geq 0 \Rightarrow d^T \nabla c_i^* = 0;$$
$$i \in \mathcal{A}(x^*) \cap \mathcal{I} \Rightarrow d^T \nabla c_i^* \geq 0.$$

Therefore, d also satisfies (12.39), so the reverse implication is proved. □

PROOF OF THEOREM 12.1

Lemmas 12.3 and 12.4 can be combined to give the KKT conditions described in Theorem 12.1. Suppose that $x^* \in \mathbf{R}^n$ is a feasible point at which the LICQ holds. The theorem claims that if x^* is a local solution for (12.1), then there is a vector $\lambda^* \in \mathbf{R}^m$ that satisfies the conditions (12.30).

We show first that there are multipliers λ_i, $i \in \mathcal{A}(x^*)$, such that (12.46) is satisfied. Theorem 12.2 tells us that $d^T \nabla f^* \geq 0$ for all vectors d that are limiting directions of feasible sequences. From Lemma 12.3, we know that when LICQ holds, the set of all possible limiting directions is exactly the set of vectors that satisfy the conditions (12.39). By putting these two statements together, we find that all directions d that satisfy (12.39) must also have $d^T \nabla f^* \geq 0$. Hence, from Lemma 12.4, we have that there is a vector λ for which (12.46) holds, as claimed.

We now define the vector λ^* by

$$\lambda_i^* = \begin{cases} \lambda_i, & i \in \mathcal{A}(x^*), \\ 0, & \text{otherwise}, \end{cases} \quad (12.51)$$

and show that this choice of λ^*, together with our local solution x^*, satisfies the conditions (12.30). We check these conditions in turn.

- The condition (12.30a) follows immediately from (12.46) and the definitions (12.28) of the Lagrangian function and (12.51) of λ^*.

- Since x^* is feasible, the conditions (12.30b) and (12.30c) are satisfied.

- We have from (12.46) that $\lambda_i^* \geq 0$ for $i \in \mathcal{A}(x^*) \cap \mathcal{I}$, while from (12.51), $\lambda_i^* = 0$ for $i \in \mathcal{I} \setminus \mathcal{A}(x^*)$. Hence, $\lambda_i^* \geq 0$ for $i \in \mathcal{I}$, so that (12.30d) holds.

- We have for $i \in \mathcal{A}(x^*) \cap \mathcal{I}$ that $c_i(x^*) = 0$, while for $i \in \mathcal{I} \setminus \mathcal{A}(x^*)$, we have $\lambda_i^* = 0$. Hence $\lambda_i^* c_i(x^*) = 0$ for $i \in \mathcal{I}$, so that (12.30e) is satisfied as well.

This completes the proof.

12.4 SECOND-ORDER CONDITIONS

So far, we have described the first-order conditions—the KKT conditions—which tell us how the first derivatives of f and the active constraints c_i are related at x^*. When these conditions are satisfied, a move along any vector w from $\mathcal{F}_1$ either increases the first-order approximation to the objective function (that is, $w^T \nabla f(x^*) > 0$), or else keeps this value the same (that is, $w^T \nabla f(x^*) = 0$).

What implications does optimality have for the *second* derivatives of f and the constraints c_i? We see in this section that these derivatives play a "tiebreaking" role. For the directions $w \in \mathcal{F}_1$ for which $w^T \nabla f(x^*) = 0$, we cannot determine from first derivative information alone whether a move along this direction will increase or decrease the objective function f. Second-order conditions examine the second derivative terms in the Taylor series expansions of f and c_i, to see whether this extra information resolves the issue of increase or decrease in f. Essentially, the second-order conditions concern the curvature of the Lagrangian function in the "undecided" directions—the directions $w \in \mathcal{F}_1$ for which $w^T \nabla f(x^*) = 0$.

Since we are discussing second derivatives, stronger smoothness assumptions are needed here than in the previous sections. For the purpose of this section, f and $c_i, i \in \mathcal{E} \cup \mathcal{I}$, are all assumed to be twice continuously differentiable.

Given $\mathcal{F}_1$ from Definition 12.4 and some Lagrange multiplier vector λ^* satisfying the KKT conditions (12.30), we define a subset $\mathcal{F}_2(\lambda^*)$ of $\mathcal{F}_1$ by

$$\mathcal{F}_2(\lambda^*) = \{w \in \mathcal{F}_1 \mid \nabla c_i(x^*)^T w = 0, \text{ all } i \in \mathcal{A}(x^*) \cap \mathcal{I} \text{ with } \lambda_i^* > 0\}.$$

Equivalently,

$$w \in \mathcal{F}_2(\lambda^*) \Leftrightarrow \begin{cases} \nabla c_i(x^*)^T w = 0, & \text{for all } i \in \mathcal{E}, \\ \nabla c_i(x^*)^T w = 0, & \text{for all } i \in \mathcal{A}(x^*) \cap \mathcal{I} \text{ with } \lambda_i^* > 0, \\ \nabla c_i(x^*)^T w \geq 0, & \text{for all } i \in \mathcal{A}(x^*) \cap \mathcal{I} \text{ with } \lambda_i^* = 0. \end{cases} \quad (12.52)$$

The subset $F_2(\lambda^*)$ contains the directions w that tend to "adhere" to the active inequality constraints for which the Lagrange multiplier component λ_i^* is positive, as well as to the equality constraints. From the definition (12.52) and the fact that $\lambda_i^* = 0$ for all inactive components $i \in \mathcal{I} \setminus \mathcal{A}(x^*)$, it follows immediately that

$$w \in F_2(\lambda^*) \quad \Rightarrow \quad \lambda_i^* \nabla c_i(x^*)^T w = 0 \text{ for all } i \in \mathcal{E} \cup \mathcal{I}. \tag{12.53}$$

Hence, from the first KKT condition (12.30a) and the definition (12.28) of the Lagrangian function, we have that

$$w \in F_2(\lambda^*) \quad \Rightarrow \quad w^T \nabla f(x^*) = \sum_{i \in \mathcal{E} \cup \mathcal{I}} \lambda_i^* w^T \nabla c_i(x^*) = 0. \tag{12.54}$$

Hence the set $F_2(\lambda^*)$ contains directions from F_1 for which it is not clear from first derivative information alone whether f will increase or decrease.

The first theorem defines a *necessary* condition involving the second derivatives: If x^* is a local solution, then the curvature of the Lagrangian along directions in $F_2(\lambda^*)$ must be nonnegative.

Theorem 12.5 (Second-Order Necessary Conditions).
Suppose that x^* is a local solution of (12.1) and that the LICQ condition is satisfied. Let λ^* be a Lagrange multiplier vector such that the KKT conditions (12.30) are satisfied, and let $F_2(\lambda^*)$ be defined as above. Then

$$w^T \nabla_{xx} \mathcal{L}(x^*, \lambda^*) w \geq 0, \quad \text{for all } w \in F_2(\lambda^*). \tag{12.55}$$

PROOF. Since x^* is a local solution, all feasible sequences $\{z_k\}$ approaching x^* must have $f(z_k) \geq f(x^*)$ for all k sufficiently large. Our approach in this proof is to construct a feasible sequence whose limiting direction is $w/\|w\|$ and show that the property $f(z_k) \geq f(x^*)$ implies that (12.55) holds.

Since $w \in F_2(\lambda^*) \subset F_1$, we can use the technique in the proof of Lemma 12.3 to construct a feasible sequence $\{z_k\}$ such that

$$\lim_{k \to \infty} \frac{z_k - x^*}{\|z_k - x^*\|} = \frac{w}{\|w\|}.$$

In particular, we have from formula (12.43) that

$$c_i(z_k) = \frac{t_k}{\|w\|} \nabla c_i(x^*)^T w, \quad \text{for all } i \in \mathcal{A}(x^*), \tag{12.56}$$

where $\{t_k\}$ is some sequence of positive scalars decreasing to zero. Moreover, we have from (12.45) that

$$\|z_k - x^*\| = t_k + o(t_k) \tag{12.57}$$

(this is an alternative way of stating the limit $\lim_{k \to \infty} t_k / \|z_k - x^*\| = 1$), and so by substitution into (12.44), we obtain

$$z_k - x^* = \frac{t_k}{\|w\|} w + o(t_k). \tag{12.58}$$

From (12.28), (12.30), and (12.56), we have as in (12.31) that

$$\mathcal{L}(z_k, \lambda^*) = f(z_k) - \sum_{i \in \mathcal{E} \cup \mathcal{I}} \lambda_i^* c_i(z_k)$$

$$= f(z_k) - \frac{t_k}{\|w\|} \sum_{i \in \mathcal{A}(x^*)} \lambda_i^* \nabla c_i(x^*)^T w$$

$$= f(z_k), \tag{12.59}$$

where the last equality follows from the critical property (12.53). On the other hand, we can perform a Taylor series expansion to obtain an estimate of $\mathcal{L}(z_k, \lambda^*)$ near x^*. By using the Taylor's theorem expression (2.6) and continuity of the Hessians $\nabla^2 f$ and $\nabla^2 c_i, i \in \mathcal{E} \cup \mathcal{I}$, we obtain

$$\mathcal{L}(z_k, \lambda^*) = \mathcal{L}(x^*, \lambda^*) + (z_k - x^*)^T \nabla_x \mathcal{L}(x^*, \lambda^*) \tag{12.60}$$
$$+ \tfrac{1}{2}(z_k - x^*)^T \nabla_{xx} \mathcal{L}(x^*, \lambda^*)(z_k - x^*) + o(\|z_k - x^*\|^2).$$

By the complementarity conditions (12.30e) the first term on the right-hand-side of this expression is equal to $f(x^*)$. From (12.30a), the second term is zero. Hence we can rewrite (12.60) as

$$\mathcal{L}(z_k, \lambda^*) = f(x^*) + \tfrac{1}{2}(z_k - x^*)^T \nabla_{xx} \mathcal{L}(x^*, \lambda^*)(z_k - x^*) + o(\|z_k - x^*\|^2). \tag{12.61}$$

By using (12.57) and (12.58), we have for the second-order term and the remainder term that

$$\tfrac{1}{2}(z_k - x^*)^T \nabla_{xx} \mathcal{L}(x^*, \lambda^*)(z_k - x^*) + o(\|z_k - x^*\|^2)$$
$$= \tfrac{1}{2}(t_k/\|w\|)^2 w^T \nabla_{xx} \mathcal{L}(x^*, \lambda^*) w + o(t_k^2).$$

Hence, by substituting this expression together with (12.59) into (12.61), we obtain

$$f(z_k) = f(x^*) + \tfrac{1}{2}(t_k/\|w\|)^2 w^T \nabla_{xx} \mathcal{L}(x^*, \lambda^*) w + o(t_k^2). \tag{12.62}$$

If $w^T \nabla_{xx}\mathcal{L}(x^*, \lambda^*)w < 0$, then (12.62) would imply that $f(z_k) < f(x^*)$ for all k sufficiently large, contradicting the fact that x^* is a local solution. Hence, the condition (12.55) must hold, as claimed. □

Sufficient conditions are conditions on f and c_i, $i \in \mathcal{E} \cup \mathcal{I}$, that ensure that x^* is a local solution of the problem (12.1). (They take the opposite tack to necessary conditions, which assume that x^* is a local solution and deduce properties of f and c_i.) The second-order sufficient condition stated in the next theorem looks very much like the necessary condition just discussed, but it differs in that the constraint qualification is not required, and the inequality in (12.55) is replaced by a strict inequality.

Theorem 12.6 (Second-Order Sufficient Conditions).
Suppose that for some feasible point $x^* \in \mathbf{R}^n$ there is a Lagrange multiplier vector λ^* such that the KKT conditions (12.30) are satisfied. Suppose also that

$$w^T \nabla_{xx}\mathcal{L}(x^*, \lambda^*)w > 0, \quad \text{for all } w \in F_2(\lambda^*), w \neq 0. \tag{12.63}$$

Then x^* is a strict local solution for (12.1).

PROOF. The result is proved if we can show that for any feasible sequence $\{z_k\}$ approaching x^*, we have that $f(z_k) > f(x^*)$ for all k sufficiently large.

Given any feasible sequence, we have from Lemma 12.3(i) and Definition 12.4 that all its limiting directions d satisfy $d \in F_1$. Choose a particular limiting direction d whose associated subsequence $\mathcal{S}_d$ satisfies (12.34). In other words, we have for all $k \in \mathcal{S}_d$ that

$$z_k - x^* = \|z_k - x^*\|d + o(\|z_k - x^*\|). \tag{12.64}$$

From (12.28), we have that

$$\mathcal{L}(z_k, \lambda^*) = f(z_k) - \sum_{i \in \mathcal{A}(x^*)} \lambda_i^* c_i(z_k) \leq f(z_k), \tag{12.65}$$

while the Taylor series approximation (12.61) from the proof of Theorem 12.5 continues to hold.

We know that $d \in F_1$, but suppose first that it is *not* in $F_2(\lambda^*)$. We can then identify some index $j \in \mathcal{A}(x^*) \cap \mathcal{I}$ such that the strict positivity condition

$$\lambda_j^* \nabla c_j(x^*)^T d > 0 \tag{12.66}$$

is satisfied, while for the remaining indices $i \in \mathcal{A}(x^*)$, we have

$$\lambda_i^* \nabla c_i(x^*)^T d \geq 0.$$

From Taylor's theorem and (12.64), we have for all $k \in \mathcal{S}_d$ and for this particular value of j that

$$\lambda_j^* c_j(z_k) = \lambda_j^* c_j(x^*) + \lambda_j^* \nabla c_j(x^*)^T (z_k - x^*) + o(\|z_k - x^*\|)$$
$$= \|z_k - x^*\| \lambda_j^* \nabla c_j(x^*)^T d + o(\|z_k - x^*\|).$$

Hence, from (12.65), we have for $k \in \mathcal{S}_d$ that

$$\mathcal{L}(z_k, \lambda^*) = f(z_k) - \sum_{i \in \mathcal{A}(x^*)} \lambda_i^* c_i(z_k)$$
$$\leq f(z_k) - \|z_k - x^*\| \lambda_j^* \nabla c_j(x^*)^T d + o(\|z_k - x^*\|). \qquad (12.67)$$

From the Taylor series estimate (12.61), we have meanwhile that

$$\mathcal{L}(z_k, \lambda^*) = f(x^*) + O(\|z_k - x^*\|^2),$$

and by combining with (12.67), we obtain

$$f(z_k) \geq f(x^*) + \|z_k - x^*\| \lambda_j^* \nabla c_j(x^*)^T d + o(\|z_k - x^*\|).$$

Therefore, because of (12.66), we have $f(z_k) > f(x^*)$ for all $k \in \mathcal{S}_d$ sufficiently large.
For the other case of $d \in F_2(\lambda^*)$, we use (12.64), (12.65), and (12.61) to write

$$f(z_k) \geq f(x^*) + \frac{1}{2}(z_k - x^*)^T \nabla_{xx} \mathcal{L}(x^*, \lambda^*)(z_k - x^*) + o(\|z_k - x^*\|^2)$$
$$= \frac{1}{2} \|z_k - x^*\|^2 d^T \nabla_{xx} \mathcal{L}(x^*, \lambda^*) d + o(\|z_k - x^*\|^2).$$

Because of (12.63), we again have $f(z_k) > f(x^*)$ for all $k \in \mathcal{S}_d$ sufficiently large.

Since this reasoning applies to *all* limiting directions of $\{z_k\}$, and since each element z_k of the sequence can be assigned to one of the subsequences $\mathcal{S}_d$ that converge to one of these limiting directions, we conclude that $f(z_k) > f(x^*)$ for all k sufficiently large. □

❑ **EXAMPLE 12.7** (EXAMPLE 12.2, ONE MORE TIME)

We now return to Example 12.2 to check the second-order conditions for problem (12.17). In this problem we have $f(x) = x_1 + x_2$, $c_1(x) = 2 - x_1^2 - x_2^2$, $\mathcal{E} = \emptyset$, and $\mathcal{I} = \{1\}$. The Lagrangian is

$$\mathcal{L}(x, \lambda) = (x_1 + x_2) - \lambda_1(2 - x_1^2 - x_2^2),$$

and it is easy to show that the KKT conditions (12.30) are satisfied by $x^* = (-1, -1)^T$, with $\lambda_1^* = \frac{1}{2}$. The Lagrangian Hessian at this point is

$$\nabla_{xx}\mathcal{L}(x^*, \lambda^*) = \begin{bmatrix} 2\lambda_1^* & 0 \\ 0 & 2\lambda_1^* \end{bmatrix} = \begin{bmatrix} 1 & 0 \\ 0 & 1 \end{bmatrix}.$$

This matrix is positive definite, that is, it satisfies $w^T \nabla_{xx}\mathcal{L}(x^*, \lambda^*)w > 0$ for all $w \neq 0$, so it certainly satisfies the conditions of Theorem 12.6. We conclude that $x^* = (-1, -1)^T$ is a strict local solution for (12.17). (In fact, it is the global solution of this problem, since, as we note in a later section, this problem is a convex programming problem.) ☐

❏ EXAMPLE 12.8

For an example in which the issues are more complex, consider the problem

$$\min\ -0.1(x_1 - 4)^2 + x_2^2 \quad \text{s.t.} \quad x_1^2 + x_2^2 - 1 \geq 0, \tag{12.68}$$

in which we seek to minimize a nonconvex function over the *exterior* of the unit circle. Obviously, the objective function is not bounded below on the feasible region, since we can take the feasible sequence

$$\begin{bmatrix} 10 \\ 0 \end{bmatrix}, \begin{bmatrix} 20 \\ 0 \end{bmatrix}, \begin{bmatrix} 30 \\ 0 \end{bmatrix}, \begin{bmatrix} 40 \\ 0 \end{bmatrix},$$

and note that $f(x)$ approaches $-\infty$ along this sequence. Therefore, no global solution exists, but it may still be possible to identify a strict local solution on the boundary of the constraint. We search for such a solution by using the KKT conditions (12.30) and the second-order conditions of Theorem 12.6.

By defining the Lagrangian for (12.68) in the usual way, it is easy to verify that

$$\nabla_x \mathcal{L}(x, \lambda) = \begin{bmatrix} -0.2(x_1 - 4) - 2\lambda x_1 \\ 2x_2 - 2\lambda x_2 \end{bmatrix}, \tag{12.69a}$$

$$\nabla_{xx}\mathcal{L}(x, \lambda) = \begin{bmatrix} -0.2 - 2\lambda & 0 \\ 0 & 2 - 2\lambda \end{bmatrix}. \tag{12.69b}$$

The point $x^* = (1, 0)^T$ satisfies the KKT conditions with $\lambda_1^* = 0.3$ and the active set $\mathcal{A}(x^*) = \{1\}$. To check that the second-order sufficient conditions are satisfied at this point,

we note that
$$\nabla c_1(x^*) = \begin{bmatrix} 2 \\ 0 \end{bmatrix},$$

so that the space F_2 defined in (12.52) is simply
$$F_2(\lambda^*) = \{w \mid w_1 = 0\} = \{(0, w_2)^T \mid w_2 \in \mathbf{R}\}.$$

Now, by substituting x^* and λ^* into (12.69b), we have for any $w \in F_2$ with $w \neq 0$ that
$$w^T \nabla_{xx} \mathcal{L}(x^*, \lambda^*) w = \begin{bmatrix} 0 \\ w_2 \end{bmatrix}^T \begin{bmatrix} -0.4 & 0 \\ 0 & 1.4 \end{bmatrix} \begin{bmatrix} 0 \\ w_2 \end{bmatrix} = 1.4 w_2^2 > 0.$$

Hence, the second-order sufficient conditions are satisfied, and we conclude from Theorem 12.6 that $(1, 0)$ is a strict local solution for (12.68). □

SECOND-ORDER CONDITIONS AND PROJECTED HESSIANS

The second-order conditions are sometimes stated in a form that is weaker but easier to verify than (12.55) and (12.63). This form uses a two-sided projection of the Lagrangian Hessian $\nabla_{xx}\mathcal{L}(x^*, \lambda^*)$ onto subspaces that are related to $F_2(\lambda^*)$.

The simplest case is obtained when the multiplier λ^* that satisfies the KKT conditions (12.30) is unique (as happens, for example, when the LICQ condition holds) and strict complementarity holds. In this case, the definition (12.52) of $F_2(\lambda^*)$ reduces to
$$F_2(\lambda^*) = \mathrm{Null}\, \left[\nabla c_i(x^*)^T\right]_{i \in \mathcal{A}(x^*)} = \mathrm{Null}\, A,$$

where A is defined as in (12.38). In other words, $F_2(\lambda^*)$ is the null space of the matrix whose rows are the active constraint gradients at x^*. As in (12.40), we can define the matrix Z with full column rank whose columns span the space $F_2(\lambda^*)$. Any vector $w \in F_2(\lambda^*)$ can be written as $w = Zu$ for some vector u, and conversely, we have that $Zu \in F_2(\lambda^*)$ for all u. Hence, the condition (12.55) in Theorem 12.5 can be restated as
$$u^T Z^T \nabla_{xx} \mathcal{L}(x^*, \lambda^*) Z u \geq 0 \quad \text{for all } u,$$

or, more succinctly,
$$Z^T \nabla_{xx} \mathcal{L}(x^*, \lambda^*) Z \quad \text{is positive semidefinite.}$$

Similarly, the condition (12.63) in Theorem 12.6 can be restated as

$$Z^T \nabla_{xx} \mathcal{L}(x^*, \lambda^*) Z \text{ is positive definite.}$$

We see at the end of this section that Z can be computed numerically, so that the positive (semi)definiteness conditions can actually be checked by forming these matrices and finding their eigenvalues.

When the optimal multiplier λ^* is unique but the strict complementarity condition is not satisfied, $F_2(\lambda^*)$ is no longer a subspace. Instead, it is an intersection of planes (defined by the first two conditions in (12.52)) and half-spaces (defined by the third condition in (12.52)). We can still, however, define two subspaces $\overline{F}_2$ and $\underline{F}_2$ that "bound" F_2 above and below, in the sense that $\overline{F}_2$ is the smallest-dimensional subspace that *contains* $F_2(\lambda^*)$, while $\underline{F}_2$ is the largest-dimensional subspace *contained in* $F_2(\lambda^*)$. To be precise, we have

$$\underline{F}_2 = \{d \in F_1 \mid \nabla c_i(x^*)^T d = 0, \text{ all } i \in \mathcal{A}(x^*)\},$$
$$\overline{F}_2 = \{d \in F_1 \mid \nabla c_i(x^*)^T d = 0, \text{ all } i \in \mathcal{A}(x^*) \text{ with } i \in \mathcal{E} \text{ or } \lambda_i^* > 0\},$$

so that

$$\underline{F}_2 \subset F_2(\lambda^*) \subset \overline{F}_2. \tag{12.70}$$

As in the previous case, we can construct matrices $\underline{Z}$ and $\overline{Z}$ whose columns span the subspaces $\underline{F}_2$ and $\overline{F}_2$, respectively. If the condition (12.55) of Theorem 12.5 holds, we can be sure that

$$w^T \nabla_{xx} \mathcal{L}(x^*, \lambda^*) w \geq 0, \quad \text{for all } w \in \underline{F}_2,$$

because $\underline{F}_2 \subset F_2(\lambda^*)$. Therefore, an immediate consequence of (12.55) is that the matrix $\underline{Z}^T \nabla_{xx} \mathcal{L}(x^*, \lambda^*) \underline{Z}$ is positive semidefinite.

Analogously, we have from $F_2(\lambda^*) \subset \overline{F}_2$ that condition (12.63) is implied by the condition

$$w^T \nabla_{xx} \mathcal{L}(x^*, \lambda^*) w > 0, \quad \text{for all } w \in \overline{F}_2.$$

Hence, given that the λ^* satisfying the KKT conditions is unique, a sufficient condition for (12.63) is that the matrix $\overline{Z}^T \nabla_{xx} \mathcal{L}(x^*, \lambda^*) \overline{Z}$ be positive definite. Again, this condition provides a practical way to check the second-order sufficient condition.

The matrices $\underline{Z}^T \nabla_{xx} \mathcal{L}(x^*, \lambda^*) \underline{Z}$ and $\overline{Z}^T \nabla_{xx} \mathcal{L}(x^*, \lambda^*) \overline{Z}$ are sometimes called *two-sided projected Hessian* matrices, or simply *projected Hessians* for short.

One way to compute the matrix Z (and its counterparts $\underline{Z}$ and $\overline{Z}$) is to apply a QR factorization to the matrix of active constraint gradients whose null space we seek. In the simplest case above (in which the multiplier λ^* is unique and strictly complementary), we define A as in (12.38) and write the QR factorization of A^T as

$$A^T = Q \begin{bmatrix} R \\ 0 \end{bmatrix} = [\, Q_1 \;\; Q_2 \,] \begin{bmatrix} R \\ 0 \end{bmatrix} = Q_1 R, \qquad (12.71)$$

where R is a square upper triangular matrix, and Q is $n \times n$ orthogonal. If R is nonsingular, we can set $Z = Q_2$. If R is singular (indicating that the active constraint gradients are linearly dependent), a slight enhancement of this procedure that makes use of column pivoting during the QR procedure can be used to identify Z. For more details, see Section 15.2.

CONVEX PROGRAMS

Convex programming problems are constrained optimization problems in which the objective function f is a convex function and the feasible set Ω defined by (12.2) is a convex set. As for unconstrained optimization (see Chapter 2), the convex case is special—all local solutions are also global solutions, and the set of global minima is itself convex. (The proof of these statements is similar to the proof of Theorem 2.5 and is left as an exercise.) Convex programming problems often can be recognized by the algebraic properties of the constraint functions c_i, as we see in the following theorem.

Theorem 12.7.

A sufficient condition for the feasible region Ω defined by the constraints in (12.1) to be convex is that the functions c_i be linear for $i \in \mathcal{E}$ and that $-c_i$ be convex functions for $i \in \mathcal{I}$.

PROOF. Let x_0 and x_1 be two feasible points, that is,

$$c_i(x_0) = c_i(x_1) = 0, \quad i \in \mathcal{E},$$
$$c_i(x_0) \geq 0, \quad c_i(x_1) \geq 0, \quad i \in \mathcal{I},$$

and define $x_\tau = (1 - \tau)x_0 + \tau x_1$ for any $\tau \in [0, 1]$. To prove the theorem, it is sufficient to show that x_τ is feasible. By linearity of the equality constraints, we have

$$c_i(x_\tau) = (1 - \tau)c_i(x_0) + \tau c_i(x_1) = 0, \quad i \in \mathcal{E},$$

so x_τ satisfies the equality constraints. By convexity of $-c_i$ for $i \in \mathcal{I}$, we have

$$-c_i(x_\tau) \leq (1 - \tau)(-c_i(x_0)) + \tau(-c_i(x_1)) \leq 0, \quad i \in \mathcal{I},$$

so x_τ also satisfies the inequality constraints. Hence $x_\tau \in \Omega$. $\square$

It follows from Theorem 12.7 that linear programming is a special case of convex programming.

Of our examples in this chapter, Examples 12.2, 12.3, and 12.4 were all convex programs, while Example 12.1 was not, because its feasible region was not convex.

12.5 OTHER CONSTRAINT QUALIFICATIONS

We now reconsider the constraint qualification, the "extra" assumption that is needed to make necessary conditions of the KKT conditions of Theorem 12.1. In general, constraint qualifications allow us to show that the set F_1 of Definition 12.4 is identical to the set of multiples of limiting feasible directions. (The LICQ of Definition 12.1 was used in this way in the proof of Lemma 12.3, for instance.) To put it another way, a constraint qualification is a condition that ensures that the linear approximation F_1 to the feasible region at x^* captures the essential geometric features of the true feasible set in some neighborhood of x^*.

This point is perhaps best illustrated by an example in which F_1 contains directions that are obviously infeasible with respect to the actual constraint set. Consider the feasible region defined by

$$x_2 \leq x_1^3, \quad x_2 \geq 0, \tag{12.72}$$

as illustrated in Figure 12.11. Supposing that $x^* = (0, 0)$, so that both constraints are active, we have from Definition 12.4 that

$$F_1 = \{d \mid -d_2 \geq 0, \quad d_2 \geq 0\} = \{d \mid d_2 = 0\}.$$

This set includes the direction $(-1, 0)$, which points along the negative direction of the horizontal axis and is clearly *not* a limiting direction for any feasible sequence. In fact, the only possible limiting direction for this constraint set at $x^* = (0, 0)$ is the vector $(1, 0)$. We conclude that the linear approximation to the feasible set at $(0, 0)$ fails to capture the geometry of the true feasible set, so constraint qualifications do not hold in this case.

One situation in which the linear approximation F_1 is *obviously* an adequate representation of the actual feasible set occurs when all the active constraints are already linear! It is not difficult to prove a version of Lemma 12.3 for this situation.

Lemma 12.8.
 Suppose that all active constraints $c_i(\cdot)$, $i \in \mathcal{A}(x^*)$, are linear functions. Then the vector $w \in \mathbf{R}^n$ belongs to F_1 if and only if $w/\|w\|$ is a limiting direction of some feasible sequence.

PROOF.   The reverse implication holds, because Lemma 12.3 shows that all multiples of limiting directions of feasible sequences belong to F_1.

352 CHAPTER 12. THEORY OF CONSTRAINED OPTIMIZATION

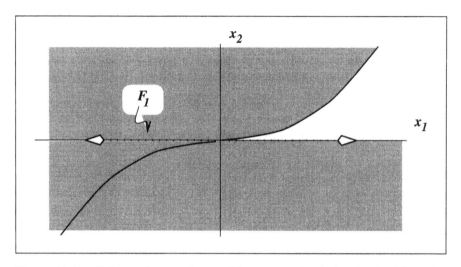

Figure 12.11 Failure of constraint qualifications at $(0, 0)$ for the feasible region defined by (12.72).

To prove the forward implication, we choose $w \in F_1$ arbitrarily and construct a feasible sequence for which $w/\|w\|$ is a limiting direction. First note that we can choose a positive scalar T such that

$$c_i(x^* + tw) > 0, \quad \text{for all } i \in \mathcal{I} \setminus \mathcal{A}(x^*) \text{ and all } t \in [0, T],$$

that is, the inactive constraints stay inactive at least for short steps along w. Now define the sequence z_k by

$$z_k = x^* + (T/k)w, \quad k = 1, 2, \ldots.$$

Since $a_i^T w \geq 0$ for all $i \in \mathcal{I} \cup \mathcal{A}(x^*)$, we have

$$c_i(z_k) = c_i(z_k) - c_i(x^*) = a_i^T(z_k - x^*) = \frac{T}{k} a_i^T w \geq 0, \quad \text{for all } i \in \mathcal{I} \cap \mathcal{A}(x^*),$$

so that z_k is feasible with respect to the active inequality constraints c_i, $i \in \mathcal{I} \cap \mathcal{A}(x^*)$. By the choice of T, we find that z_k is also feasible with respect to the inactive inequality constraints $i \in \mathcal{I} \setminus \mathcal{A}(x^*)$, and it is easy to show that $c_i(z_k) = 0$ for the equality constraints $i \in \mathcal{E}$. Hence, z_k is feasible for each $k = 1, 2, \ldots$. In addition, we have that

$$\frac{z_k - x^*}{\|z_k - x^*\|} = \frac{(T/k)w}{(T/k)\|w\|} = \frac{w}{\|w\|},$$

so that indeed the direction $w/\|w\|$ is the limiting feasible direction of the sequence $\{z_k\}$. □

We conclude from this result that the condition that all active constraints be linear is another possible constraint qualification. It is neither weaker nor stronger than the LICQ condition, that is, there are situations in which one condition is satisfied but not the other (see the exercises).

Another useful generalization of the LICQ is the Mangasarian–Fromovitz constraint qualification (MFCQ), which we define formally as follows.

Definition 12.5 (MFCQ).

Given the point x^* and the active set $\mathcal{A}(x^*)$ defined by (12.29), we say that the Mangasarian–Fromovitz constraint qualification (MFCQ) holds if there exists a vector $w \in \mathbb{R}^n$ such that

$$\nabla c_i(x^*)^T w > 0, \quad \text{for all } i \in \mathcal{A}(x^*) \cap \mathcal{I},$$
$$\nabla c_i(x^*)^T w = 0, \quad \text{for all } i \in \mathcal{E},$$

and the set of equality constraint gradients $\{\nabla c_i(x^*), i \in \mathcal{E}\}$ is linearly independent.

Note the *strict* inequality involving the active inequality constraints.

The MFCQ is a weaker condition than LICQ. If LICQ is satisfied, then the system of equalities defined by

$$\nabla c_i(x^*)^T w = 1, \quad \text{for all } i \in \mathcal{A}(x^*) \cap \mathcal{I},$$
$$\nabla c_i(x^*)^T w = 0, \quad \text{for all } i \in \mathcal{E},$$

has a solution w, by full rank of the active constraint gradients. Hence, we can choose the w of Definition 12.5 to be precisely this vector. On the other hand, it is easy to construct examples in which the MFCQ is satisfied but the LICQ is not; see the exercises.

It is possible to prove a version of the first-order necessary condition result, Theorem 12.1, in which MFCQ replaces LICQ in the assumptions. MFCQ has the particularly nice property that it is equivalent to boundedness of the set of Lagrange multiplier vectors λ^* for which the KKT conditions (12.30) are satisfied. (In the case of LICQ, this set consists of a unique vector λ^*, and so is trivially bounded.)

Note that constraint qualifications are *sufficient conditions* for the linear approximation to be adequate, not necessary conditions. For instance, consider the set defined by $x_2 \geq -x_1^2$ and $x_2 \leq x_1^2$ and the feasible point $x^* = (0, 0)$. None of the constraint qualifications we have discussed are satisfied, but the linear approximation $F_1 = \{w \mid w_2 = 0\}$ accurately reflects the geometry of the feasible set near x^*.

12.6 A GEOMETRIC VIEWPOINT

Finally, we mention an alternative first-order optimality condition that depends only on the geometry of the feasible set Ω and not on its particular algebraic description in terms of the functions $c_i, i \in \mathcal{E} \cup \mathcal{I}$. In geometric terms, our problem (12.1) can be stated as

$$\min f(x) \quad \text{subject to } x \in \Omega, \tag{12.73}$$

where Ω is the feasible set defined by the conditions (12.2).

The tools we need to state this condition are discussed in the Appendix; see in particular the Definition A.1 for the *tangent cone* $T_\Omega(x^*)$ and formula (A.22), which defines the *normal cone* $N_\Omega(x^*)$.

The first-order necessary condition for (12.73) is delightfully simple.

Theorem 12.9.
Suppose that x^* is a local minimizer of f in Ω. Then

$$-\nabla f(x^*) \in N_\Omega(x^*). \tag{12.74}$$

PROOF. Given any $d \in T_\Omega(x^*)$, we have from Definition A.1 that for any decreasing positive sequence $t_j \downarrow 0$, there is a sequence of vectors d_j such that

$$x^* + t_j d_j \in \Omega, \quad d_j \to d. \tag{12.75}$$

(We simply take $x_j \equiv x^*$ in Definition A.1.) Since x^* is a local solution, we must have

$$f(x^* + t_j d_j) \geq f(x^*)$$

for all j sufficiently large. Hence, since f is continuously differentiable, we have from Taylor's theorem (2.4) that

$$t_j \nabla f(x^*)^T d_j + o(t_j) \geq 0.$$

By taking the limits of this inequality as $j \to \infty$, we have

$$\nabla f(x^*)^T d \geq 0.$$

Recall that d was an arbitrary member of $T_\Omega(x^*)$, so we have $-\nabla f(x^*)^T d \leq 0$ for all $d \in T_\Omega(x^*)$. We conclude from (A.22) that $-\nabla f(x^*) \in N_\Omega(x^*)$. □

This result, together with some observations we made earlier in this chapter, suggests that there is a close relationship between $N_\Omega(x^*)$ and the convex combination of active constraint gradients given by (12.47). When the linear independence constraint qualification holds, we can prove this result formally by using an argument based on the implicit function theorem.

Lemma 12.10.
Suppose that the LICQ assumption (Definition 12.1) holds at x^*. Then the tangent cone $T_\Omega(x^*)$ is identical to F_1 from Definition 12.4, while the normal cone $N_\Omega(x^*)$ is simply $-N$, where N is the set defined in (12.47).

PROOF. Suppose that $d \in T_\Omega(x^*)$. Then by Definition A.1, we have for any positive decreasing sequence $t_j \downarrow 0$ that there is a sequence $\{d_j\}$ with the property (12.75). In particular,

$$c_i(x^* + t_j d_j) = 0, \ i \in \mathcal{E}, \quad c_i(x^* + t_j d_j) \geq 0, \ i \in \mathcal{A}(x^*) \cap \mathcal{I}.$$

For $i \in \mathcal{E}$, we have by Taylor's theorem that

$$0 = c_i(x^*) + t_j a_i^T d_j + o(t_j) = t_j a_i^T d_j + o(t_j).$$

By taking the limit as $j \to \infty$, we obtain $a_i^T d = 0$. Similarly, we obtain $a_i^T d \geq 0$ for $i \in \mathcal{A}(x^*) \cap \mathcal{I}$. Hence $d \in F_1$.

For the converse, choose $d \in F_1$ and let x_j and t_j be given sequences as in Definition A.1. For simplicity, we assume as in Lemma 12.3 that all constraints $c_i, i = 1, 2, \ldots, m$, are active at the solution. Similarly to (12.41), and using the notation

$$A(x)^T = [\nabla c_i(x)]_{i=1,2,\ldots,m}, \quad A^* = A(x^*), \quad A_j = A(x_j),$$

we construct the following parametrized system of equations:

$$R(z, t; x_j) = \begin{bmatrix} c(z) - c(x_j) - t A_j d \\ Z^T(z - x_j - td) \end{bmatrix} = \begin{bmatrix} 0 \\ 0 \end{bmatrix}. \quad (12.76)$$

Clearly, the solution for $t = 0$ is $z(0) = x_j$. Further, we have

$$\nabla_z R(z, t; x_j)\big|_{t=0} = \begin{bmatrix} A_j \\ Z \end{bmatrix},$$

which is nonsingular for j sufficiently large, since $A(x_j) \to A(x^*)$, so the solution $z(t)$ of (12.76) is well-defined for all values of t sufficiently small. Moreover, from Definition 12.4, we have for such solutions z that

$$c_i(z(t)) = c(x_j) + t A_j d = 0 \ \text{for} \ i \in \mathcal{E};$$

$$c_i(z(t)) = c(x_j) + tA_j d \geq 0 \text{ for } i \in \mathcal{A} \cap \mathcal{I},$$

so that $z(t)$ is feasible with respect to the constraints in (12.2).

We now define

$$d_j = \frac{z(t_j) - x_j}{t_j}.$$

Since $x_j + t_j d_j = z(t_j) \in \Omega$, the sequence $\{d_j\}$ satisfies one of the properties of Definition A.1. To verify the other property, we rearrange (12.76) to obtain

$$\begin{bmatrix} A_j \\ Z^T \end{bmatrix} d = \frac{1}{t_j} \begin{bmatrix} c(z(t_j)) - c(x_j) \\ z(t_j) - x_j \end{bmatrix} \qquad (12.77)$$

$$= \begin{bmatrix} A_j \\ Z^T \end{bmatrix} d_j + \frac{1}{t_j} o(\|c(z(t_j)) - c(x_j) - t_j A_j d_j\|).$$

By using Taylor's theorem, we have that

$$\begin{aligned} c(z(t_j)) - c(x_j) &= A_j(z(t_j) - x_j) + o(\|z(t_j) - x_j\|) \\ &= A^*(z(t_j) - x_j) + (A_j - A^*)(z(t_j) - x_j) + o(\|z(t_j) - x_j\|) \\ &= A^*(z(t_j) - x_j) + o(\|z(t_j) - x_j\|), \end{aligned}$$

and therefore from the definition of d_j, we have

$$\frac{1}{t_j}[c(z(t_j)) - c(x_j) - t_j A^* d_j] = o(\|z(t_j) - x_j\|/t_j) = o(\|d_j\|).$$

By substituting into (12.77), we obtain

$$\begin{bmatrix} A^* \\ Z^T \end{bmatrix} d = \begin{bmatrix} A^* \\ Z^T \end{bmatrix} d_j + o(\|d_j\|),$$

which implies, by nonsingularity of the matrix $[(A^*)^T \mid Z]$, that

$$d = d_j + o(\|d_j\|).$$

Hence, we have $d_j \to d$, and so $d \in T_\Omega(x^*)$, completing the proof of equivalence between F_1 and $T_\Omega(x^*)$.

The proof of equivalence between $-N$ and $N_\Omega(x^*)$ follows from Lemma 12.4 and the definition (A.22) of $N_\Omega(x^*)$. From Lemma 12.4, we have that

$$g \in N \implies g^T d \geq 0 \text{ for all } d \in F_1.$$

Since we have $F_1 = T_\Omega(x^*)$, it follows by switching the sign of this expression that

$$g \in -N \;\Rightarrow\; g^T d \leq 0 \text{ for all } d \in T_\Omega(x^*).$$

By comparing with (A.22), we see that this is precisely the definition of $N_\Omega(x^*)$, so we conclude that $N_\Omega(x^*) = -N$, as claimed. $\qquad\square$

NOTES AND REFERENCES

We return to our claim that the set N defined by (12.47) is a closed set. This fact is needed in the proof of Lemma 12.4 to ensure that the solution of the projection subproblem (12.48) is well-defined. Note that N has the following form:

$$N = \{s \mid s = A\lambda,\; C\lambda \geq 0\}, \tag{12.78}$$

for appropriately defined matrices A and C.

We outline a proof of closedness of N for the case in which LICQ holds (Definition 12.1), that is, the matrix A has full column rank. It suffices to show that whenever $\{s_k\}$ is a sequence of vectors satisfying $s_k \to s^*$, then $s^* \in N$. Because A has full column rank, for each s_k there is a unique vector λ_k such that $s_k = A\lambda_k$; in fact, we have $\lambda_k = (A^T A)^{-1} A^T s_k$. Because $s_k \in N$, we must have $C\lambda_k \geq 0$, and since $s_k \to s^*$, we also have that

$$\lim_k \lambda_k = \lim_k (A^T A)^{-1} A^T s_k = (A^T A)^{-1} A^T s^* \stackrel{\text{def}}{=} \lambda^*.$$

The inequality $C\lambda^* \geq 0$ is a consequence of $C\lambda_k \geq 0$ for all k, whereas $s^* = A\lambda^*$ follows from the facts that $s_k - A\lambda_k = 0$ for all k, and $s_k - A\lambda_k \to s^* - A\lambda^*$. Therefore we have identified a vector λ^* such that $s^* = A\lambda^*$ and $C\lambda^* \geq 0$, so $s^* \in N$, as claimed.

There is a more general proof, using an argument due to Mangasarian and Schumaker [155] and appealing to Theorem 13.2 (iii), that does not require LICQ. We omit the details here.

The theory of constrained optimization is discussed in many books on numerical optimization. The discussion in Fletcher [83, Chapter 9] is similar to ours, though a little terser, and includes additional material on duality. Bertsekas [9, Chapter 3] emphasizes the role of duality and discusses sensitivity of the solution with respect to the active constraints in some detail. The classic treatment of Mangasarian [154] is particularly notable for its thorough description of constraint qualifications.

The KKT conditions were described in a 1951 paper of Kuhn and Tucker [145], though they were derived earlier (and independently) in an unpublished 1939 master's thesis of W. Karush. Lagrange multipliers and optimality conditions for general problems (including nonsmooth problems) are described in the article of Rockafellar [217], which is both wide-ranging and deep.

Exercises

12.1 Does problem (12.4) have a finite or infinite number of local solutions? Use the first-order optimality conditions (12.30) to justify your answer.

12.2 Show by drawing a picture that the sufficient conditions for Ω to be convex in Theorem 12.7 are not necessary. That is, it is possible to have a convex feasible region Ω even if the constraint functions do not satisfy the conditions of the theorem.

12.3 If f is convex and the feasible region Ω is convex, show that local solutions of the problem (12.3) are also global solutions. Show that the set of global solutions is convex.

12.4 Let $v : \mathbf{R}^n \to \mathbf{R}^m$ be a smooth vector function and consider the unconstrained optimization problems of minimizing $f(x)$ where

$$f(x) = \|v(x)\|_\infty, \qquad f(x) = \max_{i=1,2,\ldots,m} v_i(x).$$

Reformulate these (generally nonsmooth) problems as smooth constrained optimization problems.

* 12.5 Can you perform a smooth reformulation as in the previous question when f is defined by

$$f(x) = \min_{i=1,2,\ldots,m} f_i(x)?$$

(N.B. "min" not "max.") Why or why not?

12.6 We demonstrated in this chapter how a feasible region with nonsmooth boundary could sometimes be described by a set of smooth constraints. Conversely, it is possible that a single smooth constraint describes a feasible set with nonsmooth boundary. As an example, consider the cone defined by

$$\{(x_1, x_2, x_3) \mid x_3^2 \geq x_1^2 + x_2^2\}$$

(draw this set). Can you think of other examples?

12.7 Show that the vector defined by (12.14) satisfies both (12.12) and (12.13) when the first-order optimality condition (12.10) is not satisfied. (Hint: Use the Hölder inequality (A.37).)

12.8 Verify that for the sequence $\{z_k\}$ defined by (12.35), the function $f(x) = x_1 + x_2$ satisfies $f(z_{k+1}) > f(z_k)$ for $k = 2, 3, \ldots$. (Hint: Consider the trajectory $z(s) = (-\sqrt{2 - 1/s^2}, -1/s)^T$ and show that the function $h(s) \stackrel{\text{def}}{=} f(z(s))$ has $h'(s) > 0$ for all $s \geq 2$.)

12.9 Consider the problem (12.9). Specify *two* feasible sequences that approach the maximizing point (1, 1), and show that neither sequence is a decreasing sequence for f.

12.10 Verify that neither the LICQ nor the MFCQ holds for the constraint set defined by (12.72) at $x^* = (0, 0)$.

12.11 Consider the feasible set in $\mathbf{R}^2$ defined by $x_2 \geq 0$, $x_2 \leq x_1^2$. Does the linear approximation F_1 of Definition 12.4 adequately capture the geometry of the feasible set near $x^* = (0, 0)$? Is the LICQ or MFCQ satisfied?

12.12 It is trivial to construct an example of a feasible set and a feasible point x^* at which the LICQ is satisfied but the constraints are nonlinear. Give an example of the reverse situation, that is, where the active constraints are linear but the LICQ is not satisfied.

12.13 Show that for the feasible region defined by

$$(x_1 - 1)^2 + (x_2 - 1)^2 \leq 2,$$
$$(x_1 - 1)^2 + (x_2 + 1)^2 \leq 2,$$
$$x_1 \geq 0,$$

the MFCQ is satisfied at $x^* = (0, 0)$ but the LICQ is not satisfied.

12.14 Verify (12.70).

12.15 Consider the half space defined by $H = \{x \in \mathbf{R}^n \mid a^T x + \alpha \geq 0\}$ where $a \in \mathbf{R}^n$ and $\alpha \in \mathbf{R}$ are given. Formulate and solve the optimization problem for finding the point x in H that has the smallest Euclidean norm.

12.16 (Fletcher [83]) Solve the problem

$$\min_x x_1 + x_2 \text{ subject to } x_1^2 + x_2^2 = 1$$

by eliminating the variable x_2. Show that the choice of sign for a square root operation during the elimination process is critical; the "wrong" choice leads to an incorrect answer.

12.17 Prove that when the KKT conditions (12.30) and the LICQ are satisfied at a point x^*, the Lagrange multiplier λ^* in (12.30) is unique.

12.18 Consider the problem of finding the point on the parabola $y = \frac{1}{5}(x - 1)^2$ that is closest to $(x, y) = (1, 2)$, in the Euclidean norm sense. We can formulate this problem as

$$\min \; f(x, y) = (x - 1)^2 + (y - 2)^2 \quad \text{subject to } (x - 1)^2 = 5y.$$

(a) Find all the KKT points for this problem. Is the LICQ satisfied?

(b) Which of these points are solutions?

(c) By directly substituting the constraint into the objective function and eliminating the variable x, we obtain an unconstrained optimization problem. Show that the solutions of this problem cannot be solutions of the original problem.

12.19 Consider the problem

$$\min_{x \in \mathbb{R}^2} f(x) = -2x_1 + x_2 \quad \text{subject to} \quad \begin{cases} (1 - x_1)^3 - x_2 & \geq \quad 0 \\ x_2 + 0.25x_1^2 - 1 & \geq \quad 0. \end{cases}$$

The optimal solution is $x^* = (0, 1)^T$, where both constraints are active. Does the LICQ hold at this point? Are the KKT conditions satisfied?

12.20 Find the minima of the function $f(x) = x_1 x_2$ on the unit circle $x_1^2 + x_2^2 = 1$. Illustrate this problem geometrically.

12.21 Find the *maxima* of $f(x) = x_1 x_2$ over the unit disk defined by the inequality constraint $1 - x_1^2 - x_2^2 \geq 0$.

Chapter 13

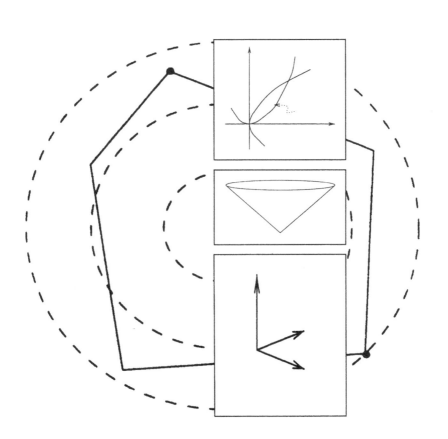

Linear Programming: The Simplex Method

Dantzig's development of the simplex method in the late 1940s marks the start of the modern era in optimization. This method made it possible for economists to formulate large models and analyze them in a systematic and efficient way. Dantzig's discovery coincided with the development of the first digital computers, and the simplex method became one of the earliest important applications of this new and revolutionary technology. From those days to the present, computer implementations of the simplex method have been continually improved and refined. They have benefited particularly from interactions with numerical analysis, a branch of mathematics that also came into its own with the appearance of digital computers, and have now reached a high level of sophistication.

Today, linear programming and the simplex method continue to hold sway as the most widely used of all optimization tools. Since 1950, generations of workers in management, economics, finance, and engineering have been trained in the business of formulating linear models and solving them with simplex-based software. Often, the situations they model are actually nonlinear, but linear programming is appealing because of the advanced state of the software, guaranteed convergence to a global minimum, and the fact that uncertainty in the data can make complicated nonlinear models look like overkill. Nonlinear programming may make inroads as software develops, and a new class of methods known as interior-point

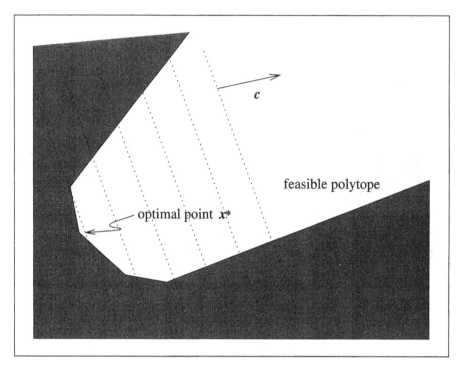

Figure 13.1 A linear program in two dimensions with solution at x^*.

methods (see Chapter 14) has proved to be faster for some linear programming problems, but the continued importance of the simplex method is assured for the foreseeable future.

LINEAR PROGRAMMING

Linear programs have a linear objective function and linear constraints, which may include both equalities and inequalities. The feasible set is a polytope, that is, a convex, connected set with flat, polygonal faces. The contours of the objective function are planar. Figure 13.1 depicts a linear program in two-dimensional space. Contours of the objective function are indicated by the dotted lines. The solution in this case is unique—a single vertex. A simple reorientation of the polytope or the objective gradient c could, however, make the solution nonunique; the optimal value $c^T x$ could be the same on an entire edge. In higher dimensions, the set of optimal points can be a single vertex, an edge or face, or even the entire feasible set!

Linear programs are usually stated and analyzed in the following *standard form*:

$$\min c^T x, \quad \text{subject to } Ax = b, x \geq 0, \tag{13.1}$$

where c and x are vectors in $\mathbf{R}^n$, b is a vector in $\mathbf{R}^m$, and A is an $m \times n$ matrix. Simple devices can be used to transform any linear program to this form. For instance, given the problem

$$\min c^T x, \quad \text{subject to } Ax \geq b$$

(without any bounds on x), we can convert the inequality constraints to equalities by introducing a vector of *surplus variables* z and writing

$$\min c^T x, \quad \text{subject to } Ax - z = b, z \geq 0. \tag{13.2}$$

This form is still not quite standard, since not all the variables are constrained to be nonnegative. We deal with this by *splitting* x into its nonnegative and nonpositive parts, $x = x^+ - x^-$, where $x^+ = \max(x, 0) \geq 0$ and $x^- = \max(-x, 0) \geq 0$. The problem (13.2) can now be written as

$$\min \begin{bmatrix} c \\ -c \\ 0 \end{bmatrix}^T \begin{bmatrix} x^+ \\ x^- \\ z \end{bmatrix}, \quad \text{s.t.} \quad \begin{bmatrix} A & -A & -I \end{bmatrix} \begin{bmatrix} x^+ \\ x^- \\ z \end{bmatrix} = b, \quad \begin{bmatrix} x^+ \\ x^- \\ z \end{bmatrix} \geq 0,$$

which clearly has the same form as (13.1). Inequality constraints of the form $x \leq u$ or $Ax \leq b$ can be dealt with by adding *slack variables* to make up the difference between the left- and right-hand-sides. Hence

$$x \leq u \Leftrightarrow x + w = u, \ w \geq 0,$$
$$Ax \leq b \Leftrightarrow Ax + y = b, \ y \geq 0.$$

We can also convert a "maximize" objective $\max c^T x$ into the "minimize" form of (13.1) by simply negating c to obtain $\min(-c)^T x$.

Many linear programs arise from models of transshipment and distribution networks. These problems have much additional structure in their constraints; special-purpose simplex algorithms that exploit this structure are highly efficient. We do not discuss these *network-flow problems* further in this book, except to note that the subject is important and complex, and that a number of excellent texts are available (see, for example, Ahuja, Magnanti, and Orlin [1]).

For the standard formulation (13.1), we will assume throughout that $m < n$. Otherwise, the system $Ax = b$ contains redundant rows, is infeasible, or defines a unique point. When $m \geq n$, factorizations such as the QR or LU factorization (see the Appendix) can be used to transform the system $Ax = b$ to one with a coefficient matrix of full row rank and, in some cases, decide that the feasible region is empty or consists of a single point.

13.1 OPTIMALITY AND DUALITY

OPTIMALITY CONDITIONS

Optimality conditions for the problem (13.1) can be derived from the theory of Chapter 12. Only the first-order conditions—the Karush–Kuhn–Tucker (KKT) conditions—are needed. Convexity of the problem ensures that these conditions are sufficient for a global minimum, as we show below by a simple argument. (We do not need to refer to the second-order conditions from Chapter 12, which are not informative in any case because the Hessian of the Lagrangian for (13.1) is zero.)

The tools we developed in Chapter 12 make derivation of optimality and duality theory for linear programming much easier than in other treatments of the subject, where this theory has to be developed more or less from scratch.

The KKT conditions follow from Theorem 12.1. As stated in Chapter 12, this theorem requires linear independence of the active constraint gradients (LICQ). However, as we showed in the section on constraint qualifications, the result continues to hold for *dependent* constraints, provided that they are linear, as is the case here.

We partition the Lagrange multipliers for the problem (13.1) into two vectors π and s, where $\pi \in \mathbf{R}^m$ is the multiplier vector for the equality constraints $Ax = b$, while $s \in \mathbf{R}^n$ is the multiplier vector for the bound constraints $x \geq 0$. Using the definition (12.28), we can write the Lagrangian function for (13.1) as

$$\mathcal{L}(x, \pi, s) = c^T x - \pi^T (Ax - b) - s^T x. \tag{13.3}$$

Applying Theorem 12.1, we find that the first-order necessary conditions for x^* to be a solution of (13.1) are that there exist vectors π and s such that

$$A^T \pi + s = c, \tag{13.4a}$$
$$Ax = b, \tag{13.4b}$$
$$x \geq 0, \tag{13.4c}$$
$$s \geq 0, \tag{13.4d}$$
$$x_i s_i = 0, \quad i = 1, 2, \ldots, n. \tag{13.4e}$$

The complementarity condition (13.4e), which essentially says that at least one of the components x_i and s_i must be zero for each $i = 1, 2, \ldots, n$, is often written in the alternative form $x^T s = 0$. Because of the nonnegativity conditions (13.4c), (13.4d), the two forms are identical.

Let (x^*, π^*, s^*) denote a vector triple that satisfies (13.4). By combining the three equalities (13.4a), (13.4d), and (13.4e), we find that

$$c^T x^* = (A^T \pi^* + s^*)^T x^* = (Ax^*)^T \pi^* = b^T \pi^*. \tag{13.5}$$

As we shall see in a moment, $b^T \pi$ is the objective function for the dual problem to (13.1), so (13.5) indicates that the primal and dual objectives are equal for vector triples (x, π, s) that satisfy (13.4).

It is easy to show directly that the conditions (13.4) are sufficient for x^* to be a global solution of (13.1). Let $\bar{x}$ be any other feasible point, so that $A\bar{x} = b$ and $\bar{x} \geq 0$. Then

$$c^T \bar{x} = (A\pi^* + s^*)^T \bar{x} = b^T \pi^* + \bar{x}^T s^* \geq b^T \pi^* = c^T x^*. \tag{13.6}$$

We have used (13.4) and (13.5) here; the inequality relation follows trivially from $\bar{x} \geq 0$ and $s^* \geq 0$. The inequality (13.6) tells us that no other feasible point can have a lower objective value than $c^T x^*$. We can say more: The feasible point $\bar{x}$ is optimal if and only if

$$\bar{x}^T s^* = 0,$$

since otherwise the inequality in (13.6) is strict. In other words, when $s_i^* > 0$, then we must have $\bar{x}_i = 0$ for all solutions $\bar{x}$ of (13.1).

THE DUAL PROBLEM

Given the data c, b, and A, which define the problem (13.1), we can define another, closely related, problem as follows:

$$\max b^T \pi, \quad \text{subject to } A^T \pi \leq c. \tag{13.7}$$

This problem is called the *dual problem* for (13.1). In contrast, (13.1) is often referred to as the *primal*.

The primal and dual problems are two sides of the same coin, as we see when we write down the KKT conditions for (13.7). Let us first rewrite (13.7) in the form

$$\min -b^T \pi \quad \text{subject to } c - A^T \pi \geq 0,$$

to fit the formulation (12.1) from Chapter 12. By using $x \in \mathbf{R}^n$ to denote the Lagrange multipliers for the constraints $A^T \pi \leq c$, we write the Lagrangian function as

$$\bar{\mathcal{L}}(\pi, x) = -b^T \pi - x^T (c - A^T \pi).$$

Noting again that the conclusions of Theorem 12.1 continue to hold if the linear independence assumption is replaced by linearity of all constraints, we find the first-order necessary condition for π to be optimal for (13.7) to be that there exist a vector x such that

$$Ax = b, \tag{13.8a}$$
$$A^T \pi \leq c, \tag{13.8b}$$

$$x \geq 0, \tag{13.8c}$$
$$x_i(c - A^T\pi)_i = 0, \quad i = 1, 2, \ldots, n. \tag{13.8d}$$

If we define $s = c - A^T\pi$ and substitute in (13.8), we find that the conditions (13.8) and (13.4) are identical! The optimal Lagrange multipliers π in the primal problem are the optimal variables in the dual problem, while the optimal Lagrange multipliers x in the dual problem are the optimal variables in the primal problem.

The primal–dual relationship is symmetric; by taking the dual of the dual, we recover the primal. To see this, we restate (13.7) in standard form by introducing the slack vector s (so that $A^T\pi + s = c$) and splitting the unbounded variables π as $\pi = \pi^+ - \pi^-$, where $\pi^+ \geq 0$, and $\pi^- \geq 0$. We can now write the dual as

$$\min \begin{bmatrix} -b \\ b \\ 0 \end{bmatrix}^T \begin{bmatrix} \pi^+ \\ \pi^- \\ s \end{bmatrix} \text{ s.t. } \begin{bmatrix} A^T & -A^T & I \end{bmatrix} \begin{bmatrix} \pi^+ \\ \pi^- \\ s \end{bmatrix} = c, \quad \begin{bmatrix} \pi^+ \\ \pi^- \\ s \end{bmatrix} \geq 0, \tag{13.9}$$

which clearly has the standard form (13.1). The dual of (13.9) is now

$$\max c^T z \text{ subject to } \begin{bmatrix} A \\ -A \\ I \end{bmatrix} z \leq \begin{bmatrix} -b \\ b \\ 0 \end{bmatrix}.$$

Now $Az \leq -b$ and $-Az \leq b$ together imply that $Az = -b$, so we obtain the equivalent problem

$$\min -c^T z \text{ subject to } Az = -b, \ z \leq 0.$$

By making the identification $z = -x$, we recover (13.1), as claimed.

Given a feasible vector x for the primal—that is, $Ax = b$ and $x \geq 0$—and a feasible point (π, s) for the dual—that is, $A^T\pi + s = c, s \geq 0$—we have as in (13.6) that

$$0 \leq x^T s = x^T (c - A^T\pi) = c^T x - b^T \pi. \tag{13.10}$$

Therefore, we have $c^T x \geq b^T \pi$ when both the primal and dual variables are feasible—the dual objective function is a lower bound on the primal objective function. At a solution, the gap between primal and dual shrinks to zero, as we show in the following theorem.

Theorem 13.1 (Duality Theorem of Linear Programming).
  (i) *If either problem (13.1) or (13.7) has a solution with finite optimal objective value, then so does the other, and the objective values are equal.*

(ii) *If either problem (13.1) or (13.7) has an unbounded objective, then the other problem has no feasible points.*

PROOF. For (i), suppose that (13.1) has a finite optimal solution. Then because of Theorem 12.1, there are vectors π and s such that (x, π, s) satisfies (13.4). Since (13.4) and (13.8) are equivalent, it follows that π is a solution of the dual problem (13.7), since there exists a vector x that satisfies (13.8). Because $x^T s = 0$, it follows from (13.10) that $c^T x = b^T \pi$, so the optimal objective values are equal.

We can make a symmetric argument if we start by assuming that the dual problem (13.7) has a solution.

For (ii), suppose that the primal objective value is unbounded below. Then there must exist a direction $d \in \mathbb{R}^n$ along which $c^T x$ decreases without violating feasibility. That is,

$$c^T d < 0, \quad Ad = 0, \quad d \geq 0.$$

Suppose now that there *does* exist a feasible point π for the dual problem (13.7), that is $A^T \pi \leq c$. Multiplying from the left by d^T, using the nonnegativity of d, we obtain

$$0 = d^T A^T \pi \leq d^T c < 0,$$

giving a contradiction.

Again, we can make a symmetric argument to prove (ii) if we start by assuming that the dual objective is unbounded below. □

As we showed in the discussion following Theorem 12.1, the multiplier values π and s for (13.1) tell us how sensitive the optimal objective value is to perturbations in the constraints. In fact, the process of finding (π, s) for a given optimal x is often called *sensitivity analysis*. We can make a simple direct argument to illustrate this dependence. If a small change Δb is made to the vector b (the right-hand-side in (13.1) and objective gradient in (13.7)), then we would usually expect small perturbations in the primal and dual solutions. If these perturbations $(\Delta x, \Delta \pi, \Delta s)$ are small enough, we know that provided the problem is not *degenerate* (defined below), the vectors Δs and Δx have zeros in the same locations as s and x, respectively. Since x and s are complementary (see (13.4e)), it follows that

$$x^T s = x^T \Delta s = (\Delta x)^T s = (\Delta x)^T \Delta s = 0.$$

Now we have from the duality theorem that

$$c^T (x + \Delta x) = (b + \Delta b)^T (\pi + \Delta \pi).$$

Since

$$c^T x = b^T \pi, \quad A(x + \Delta x) = b + \Delta b, \quad A^T \Delta \pi = -\Delta s,$$

we have

$$c^T \Delta x = (b + \Delta b)^T \Delta \pi + \Delta b^T \pi$$
$$= (x + \Delta x)^T A^T \Delta \pi + \Delta b^T \pi$$
$$= -(x + \Delta x)^T \Delta s + \Delta b^T \pi = \Delta b^T \pi.$$

In particular, if $\Delta b = \epsilon e_j$, where e_j is the jth unit vector in $\mathbf{R}^m$, we have for all ϵ sufficiently small that

$$c^T \Delta x = \epsilon \pi_j. \tag{13.11}$$

That is, the change in primal objective is linear in the value of π_j (cf. (12.33)) for small perturbations in the components of the right-hand-side b_j.

13.2 GEOMETRY OF THE FEASIBLE SET

BASIC FEASIBLE POINTS

We assume for the remainder of the chapter that

$$\text{The matrix } A \text{ in (13.1) has full row rank.} \tag{13.12}$$

In practice, a preprocessing phase is applied to the user-supplied data to remove some redundancies from the given constraints and eliminate some of the variables. Reformulation by adding slack, surplus, and artificial variables is also used to force A to satisfy the property (13.12).

Suppose that x is a feasible point with at most m nonzero components. Suppose, too, that we can identify a subset $\mathcal{B}(x)$ of the index set $\{1, 2, \ldots, n\}$ such that

- $\mathcal{B}(x)$ contains exactly m indices;
- $i \notin \mathcal{B}(x) \Rightarrow x_i = 0$;
- the $m \times m$ matrix B defined by

$$B = [A_i]_{i \in \mathcal{B}(x)} \tag{13.13}$$

is nonsingular, where A_i is the ith column of A.

If all these conditions are true, we call x a *basic feasible point* for (13.1).

The simplex method generates a sequence of iterates x^k all of which are basic feasible points. Since we want the iterates to converge to a solution of (13.1), the simplex strategy will make sense only if

(a) the problem *has* basic feasible points; and

(b) at least one such point is a *basic optimal point*, that is, a solution of (13.1) that is also a basic feasible point.

Happily, both (a) and (b) are true under minimal assumptions.

Theorem 13.2 (Fundamental Theorem of Linear Programming).

(i) *If there is a feasible point for (13.1), then there is a basic feasible point.*

(ii) *If (13.1) has solutions, then at least one such solution is a basic optimal point.*

(iii) *If (13.1) is feasible and bounded, then it has an optimal solution.*

PROOF. Among all feasible vectors x, choose one with the minimal number of nonzero components, and denote this number by p. Without loss of generality, assume that the nonzeros are $x_1, x_2, \ldots, x_p$, so we have

$$\sum_{i=1}^{p} A_i x_i = b.$$

Suppose first that the columns $A_1, A_2, \ldots, A_p$ are linearly dependent. Then we can express one of them (A_p, say) in terms of the others, and write

$$A_i = \sum_{i=1}^{p-1} A_i z_i, \quad (13.14)$$

for some scalars $z_1, z_2, \ldots, z_{p-1}$. It is easy to check that the vector

$$x(\epsilon) = x + \epsilon(z_1, z_2, \ldots, z_{p-1}, -1, 0, 0, \ldots, 0)^T = x + \epsilon z \quad (13.15)$$

satisfies $Ax(\epsilon) = b$ for any scalar ϵ. In addition, since $x_i > 0$ for $i = 1, 2, \ldots, p$, we also have $x_i(\epsilon) > 0$ for the same indices $i = 1, 2, \ldots, p$ and all ϵ sufficiently small in magnitude. However, there is a value $\bar{\epsilon} \in (0, x_p]$ such that $x_i(\bar{\epsilon}) = 0$ for some $i = 1, 2, \ldots, p$. Hence, $x(\bar{\epsilon})$ is feasible and has at most $p - 1$ nonzero components, contradicting our choice of p as the minimal number of nonzeros.

Therefore, columns $A_1, A_2, \ldots, A_p$ must be linearly independent, and so $p \leq m$. If $p = m$, we are done, since then x is a basic feasible point and $\mathcal{B}(x)$ is simply $\{1, 2, \ldots, m\}$. Otherwise, $p < m$, and because A has full row rank, we can choose $m - p$ columns from among $A_{p+1}, A_{p+2}, \ldots, A_n$ to build up a set of m linearly independent vectors. We construct $\mathcal{B}(x)$ by adding the corresponding indices to $\{1, 2, \ldots, p\}$. The proof of (i) is complete.

The proof of (ii) is quite similar. Let x^* be a solution with a minimal number of nonzero components p, and assume again that $x_1^*, x_2^*, \ldots, x_p^*$ are the nonzeros. If the columns $A_1, A_2, \ldots, A_p$ are linearly dependent, we define

$$x^*(\epsilon) = x^* + \epsilon z,$$

where z is chosen exactly as in (13.14), (13.15). It is easy to check that $x^*(\epsilon)$ will be feasible for all ϵ sufficiently small, both positive and negative. Hence, since x^* is optimal, we must have

$$c^T(x^* + \epsilon z) \geq c^T x^* \quad \Rightarrow \quad \epsilon c^T z \geq 0$$

for all $|\epsilon|$ sufficiently small. Therefore, $c^T z = 0$, and so $c^T x^*(\epsilon) = c^T x^*$ for all ϵ. The same logic as in the proof of (i) can be applied to find $\bar{\epsilon} > 0$ such that $x^*(\bar{\epsilon})$ is feasible and optimal, with at most $p - 1$ nonzero components. This contradicts our choice of p as the minimal number of nonzeros, so the columns $A_1, A_2, \ldots, A_p$ must be linearly independent. We can now apply the same logic as above to conclude that x^* is already a basic feasible point and therefore a basic optimal point.

The final statement (iii) is a consequence of finite termination of the simplex method. We comment on the latter property in the next section. □

The terminology we use here is not standard, as the following table shows:

our terminology	standard terminology
basic feasible point	basic feasible solution
basic optimal point	optimal basic feasible solution

The standard terms arose because "solution" and "feasible solution" were originally used as synonyms for "feasible point." However, as the discipline of optimization developed, the word "solution" took on a more specific and intuitive meaning (as in "solution to the problem ..."). We keep the terminology of this chapter consistent with the rest of the book by sticking to this more modern usage.

VERTICES OF THE FEASIBLE POLYTOPE

The feasible set defined by the linear constraints is a polytope, and the *vertices* of this polytope are the points that do not lie on a straight line between two other points in the set. Geometrically, they are easily recognizable; see Figure 13.2.

Algebraically, the vertices are exactly the basic feasible points that we described above. We therefore have an important relationship between the algebraic and geometric viewpoints and a useful aid to understanding how the simplex method works.

Theorem 13.3.

All basic feasible points for (13.1) are vertices of the feasible polytope $\{x \mid Ax = b, x \geq 0\}$, and vice versa.

PROOF. Let x be a basic feasible point and assume without loss of generality that $\mathcal{B}(x) = \{1, 2, \ldots, m\}$. The matrix $B = [A_i]_{i=1,2,\ldots,m}$ is therefore nonsingular, and

$$x_{m+1} = x_{m+2} = \cdots = x_n = 0. \tag{13.16}$$

13.2. GEOMETRY OF THE FEASIBLE SET

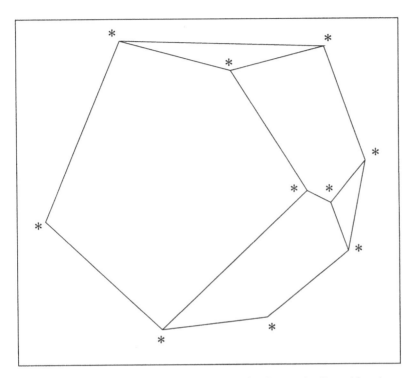

Figure 13.2 Vertices of a three-dimensional polytope (indicated by ∗).

Suppose that x lies on a straight line between two other feasible points y and z. Then we can find $\alpha \in (0, 1)$ such that $x = \alpha y + (1 - \alpha)z$. Because of (13.16) and the fact that α and $1 - \alpha$ are both positive, we must have $y_i = z_i = 0$ for $i = m + 1, m + 2, \ldots, n$. Writing $x_B = (x_1, x_2, \ldots, x_m)^T$ and defining y_B and z_B likewise, we have from $Ax = Ay = Az = b$ that

$$Bx_B = By_B = Bz_B = b,$$

and so, by nonsingularity of B, we have $x_B = y_B = z_B$. Therefore, $x = y = z$, contradicting our assertion that y and z are two feasible points *other* than x. Therefore, x is a vertex.

Conversely, let x be a vertex of the feasible polytope, and suppose that the nonzero components of x are $x_1, x_2, \ldots, x_p$. If the corresponding columns $A_1, A_2, \ldots, A_p$ are linearly dependent, then we can construct the vector $x(\epsilon) = x + \epsilon z$ as in (13.15). Since $x(\epsilon)$ is feasible for all ϵ with sufficiently small magnitude, we can define $\hat{\epsilon} > 0$ such that $x(\hat{\epsilon})$ and $x(-\hat{\epsilon})$ are both feasible. Since $x = x(0)$ obviously lies on a straight line between these two points, it cannot be a vertex. Hence our assertion that $A_1, A_2, \ldots, A_p$ are linearly dependent must be incorrect, so these columns must be linearly independent and $p \leq m$. The same arguments as in the proof of Theorem 13.2 can now be used to show that x is a basic feasible point, completing our proof. □

We conclude this discussion of the geometry of the feasible set with a definition of *degeneracy*. This term has a variety of meanings in optimization, as we discuss in Chapter 16. For the purposes of this chapter, we use the following definition.

Definition 13.1 (Degenerate Linear Program).
A linear program (13.1) is said to be degenerate *if there exists at least one basic feasible point that has fewer than m nonzero components.*

Naturally, *nondegenerate* linear programs are those for which this definition is not satisfied.

13.3 THE SIMPLEX METHOD

OUTLINE OF THE METHOD

As we just described, all iterates of the simplex method are basic feasible points for (13.1) and therefore vertices of the feasible polytope. Most steps consist of a move from one vertex to an adjacent one for which the set of basic indices $\mathcal{B}(x)$ differs in exactly one component. On most steps (but not all), the value of the primal objective function $c^T x$ is decreased. Another type of step occurs when the problem is unbounded: The step is an edge along which the objective function is reduced, and along which we can move infinitely far without ever reaching a vertex.

The major issue at each simplex iteration is to decide which index to change in the basis set $\mathcal{B}$. Unless the step is a direction of unboundedness, one index must be removed from $\mathcal{B}$ and replaced by another from outside $\mathcal{B}$. We can get some insight into how this decision is made by looking again at the KKT conditions (13.4) to see how they relate to the algorithm.

From $\mathcal{B}$ and (13.4), we can derive values for not just the primal variable x but also the dual variables π and s, as we now show. We define the index set $\mathcal{N}$ as the complement of $\mathcal{B}$, that is,

$$\mathcal{N} = \{1, 2, \ldots, n\} \backslash \mathcal{B}. \tag{13.17}$$

Just as B is the column submatrix of A that corresponds to the indices $i \in \mathcal{B}$, we use N to denote the submatrix $N = [A_i]_{i \in \mathcal{N}}$. We also partition the n-element vectors x, s, and c according to the index sets $\mathcal{B}$ and $\mathcal{N}$, using the notation

$$x_{\text{B}} = [x_i]_{i \in \mathcal{B}}, \quad x_{\text{N}} = [x_i]_{i \in \mathcal{N}}, \quad s_{\text{B}} = [s_i]_{i \in \mathcal{B}}, \quad s_{\text{N}} = [s_i]_{i \in \mathcal{N}}.$$

From the KKT condition (13.4b), we have that

$$Ax = Bx_{\text{B}} + Nx_{\text{N}} = b.$$

The primal variable x for this simplex iterate is defined as

$$x_B = B^{-1}b, \quad x_N = 0. \tag{13.18}$$

Since we are dealing only with basic feasible points, we know that B is nonsingular and that $x_B \geq 0$, so this choice of x satisfies two of the KKT conditions: the equality constraints (13.4b) and the nonnegativity condition (13.4c).

We choose s to satisfy the complementarity condition (13.4e) by setting $s_B = 0$. The remaining components π and s_N can be found by partitioning this condition into $\mathcal{B}$ and $\mathcal{N}$ components and using $s_B = 0$ to obtain

$$B^T \pi = c_B, \quad N^T \pi + s_N = c_N. \tag{13.19}$$

Since B is square and nonsingular, the first equation uniquely defines π as

$$\pi = B^{-T} c_B. \tag{13.20}$$

The second equation in (13.19) implies a value for s_N:

$$s_N = c_N - N^T \pi = c_N - (B^{-1}N)^T c_B. \tag{13.21}$$

Computation of the vector s_N is often referred to as *pricing*. The components of s_N are often called the *reduced costs* of the nonbasic variables x_N.

The only KKT condition that we have not enforced explicitly is the nonnegativity condition $s \geq 0$. The basic components s_B certainly satisfy this condition, by our choice $s_B = 0$. If the vector s_N defined by (13.21) also satisfies $s_N \geq 0$, we have found an optimal vector triple (x, π, s), so the algorithm can terminate and declare success. The usual case, however, is that one or more of the components of s_N are negative, so the condition $s \geq 0$ is violated. The new index to enter the basic index set $\mathcal{B}$—the *entering index*—is now chosen to be *one of the indices* $q \in \mathcal{N}$ *for which* $s_q < 0$. As we show below, the objective $c^T x$ will decrease when we allow x_q to become positive if and only if q has the property that $s_q < 0$. Our procedure for altering $\mathcal{B}$ and changing x and s accordingly is as follows:

- allow x_q to increase from zero during the next step;
- fix all other components of x_N at zero;
- figure out the effect of increasing x_q on the current basic vector x_B, given that we want to stay feasible with respect to the equality constraints $Ax = b$;
- keep increasing x_q until one of the components of x_B (corresponding to x_p, say) is driven to zero, or determining that no such component exists (the unbounded case);
- remove index p (known as the *leaving index*) from $\mathcal{B}$ and replace it with the entering index q.

It is easy to formalize this procedure in algebraic terms. Since we want both the new iterate x^+ and the current iterate x to satisfy $Ax = b$, and since $x_N = 0$ and $x_i^+ = 0$ for $i \in \mathcal{N}\setminus\{q\}$, we have

$$Ax^+ = Bx_B^+ + A_q x_q^+ = Bx_B = Ax.$$

By multiplying this expression by B^{-1} and rearranging, we obtain

$$x_B^+ = x_B - B^{-1} A_q x_q^+. \tag{13.22}$$

We show in a moment that the direction $-B^{-1}A_q$ is a descent direction for $c^T x$. Geometrically speaking, (13.22) is a move along an edge of the feasible polytope that decreases $c^T x$. We continue to move along this edge until a new vertex is encountered. We have to stop at this vertex, since by definition we cannot move any further without leaving the feasible region. At the new vertex, a new constraint $x_i \geq 0$ must have become active, that is, one of the components x_i, $i \in \mathcal{B}$, has decreased to zero. This index i is the one that is removed from the basis.

Figure 13.3 shows a path traversed by the simplex method for a problem in $\mathbf{R}^2$. In this example the function $c^T x$ is decreased at each iteration, and the optimal vertex x^* is found in three steps.

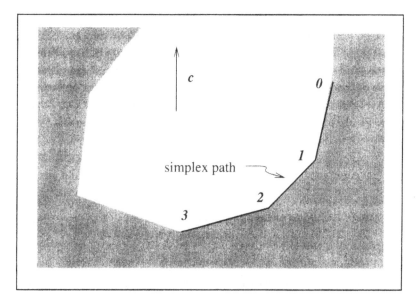

Figure 13.3 Simplex iterates for a two-dimensional problem.

FINITE TERMINATION OF THE SIMPLEX METHOD

Let us now verify that the step defined by (13.22) leads to a decrease in $c^T x$. By using the definition (13.22) of x_B^+ together with

$$x_N^+ = (0, \ldots, 0, x_q^+, 0, \ldots, 0)^T,$$

we have

$$\begin{aligned} c^T x^+ &= c_B^T x_B^+ + c_N^T x_N^+ \\ &= c_B^T x_B^+ + c_q x_q^+ \\ &= c_B^T x_B - c_B^T B^{-1} A_q x_q^+ + c_q x_q^+. \end{aligned} \quad (13.23)$$

Now, from (13.20) we have $c_B^T B^{-1} = \pi^T$, while from (13.19) we have $A_q^T \pi = c_q - s_q$, since A_q is a column of N. Therefore,

$$c_B^T B^{-1} A_q x_q^+ = \pi^T A_q x_q^+ = (c_q - s_q) x_q^+,$$

so by substituting in (13.23) we obtain

$$c^T x^+ = c_B^T x_B - (c_q - s_q) x_q^+ + c_q x_q^+ = c_B^T x_B - s_q x_q^+.$$

Since $x_N = 0$, we have $c^T x = c_B^T x_B$ and therefore

$$c^T x^+ = c^T x - s_q x_q^+. \quad (13.24)$$

Since we chose q such that $s_q < 0$, and since $x_q^+ > 0$ if we are able to move at all along the edge, it follows from (13.24) that the step (13.22) produces a decrease in the primal objective function $c^T x$.

If the problem is *nondegenerate* (see Definition 13.1), then we are guaranteed that $x_q^+ > 0$, so we can be assured of a strict decrease in the objective function $c^T x$ at every simplex step. We can therefore prove the following result concerning termination of the simplex method.

Theorem 13.4.

Provided that the linear program (13.1) is nondegenerate and bounded, the simplex method terminates at a basic optimal point.

PROOF. The simplex method cannot visit the same basic feasible point x at two different iterations, because it attains a strict decrease at each iteration. Since each subset of m indices drawn from the set $\{1, 2, \ldots, n\}$ is associated with at most one basic feasible point, it follows that no basis B can be visited at two different simplex iterations. The number of possible bases is at most $\binom{n}{m}$ (which is the number of ways to choose the m elements of a basis $\mathcal{B}$ from among the n possible indices), so there can be only a finite number of iterations. Since the method is always able to take a step away from a nonoptimal basic feasible point, the point of termination must be a basic optimal point. $\square$

Note that this result gives us a proof of Theorem 13.2 (iii) for the nondegenerate case.

A SINGLE STEP OF THE METHOD

We have covered most of the mechanics of taking a single step of the simplex method. To make subsequent discussions easier to follow, we summarize our description in a semiformal way.

Procedure 13.1 (One Step of Simplex).
Given $\mathcal{B}, \mathcal{N}, x_{\mathcal{B}} = B^{-1}b \geq 0, x_{\mathcal{N}} = 0$;
    Solve $B^T \pi = c_{\mathcal{B}}$ for π,
    Compute $s_{\mathcal{N}} = c_{\mathcal{N}} - N^T \pi$;
    if $s_{\mathcal{N}} \geq 0$
        STOP; (* optimal point found *)
    Select $q \in \mathcal{N}$ with $s_q < 0$ as the entering index;
    Solve $Bt = A_q$ for t;
    if $t \leq 0$
        STOP; (* problem is unbounded *)
    Calculate $x_q^+ = \min_{i \mid t_i > 0} (x_{\mathcal{B}})_i / t_i$, and use p to denote the index of the basic
        variable for which this minimum is achieved;
    Update $x_{\mathcal{B}}^+ = x_{\mathcal{B}} - t x_q^+, x_{\mathcal{N}}^+ = (0, \ldots, 0, x_q^+, 0, \ldots, 0)^T$;
    Change $\mathcal{B}$ by adding q and removing p.

We need to flesh out this description with specifics of three important points. These are as follows.

- Linear algebra issues—maintaining an LU factorization of B that can be used to solve for π and t.

- Selection of the entering index q from among the negative components of $s_{\mathcal{N}}$. (In general, there are many such components.)

- Handling of *degenerate steps*, in which $x_q^+ = 0$, so that x is not changed.

Proper handling of these issues is crucial to the efficiency of a simplex implementation. We give some details in the next three sections.

13.4 LINEAR ALGEBRA IN THE SIMPLEX METHOD

We have to solve two linear systems involving the matrix B at each step; namely,

$$B^T \pi = c_B, \quad Bt = A_q. \qquad (13.25)$$

We never calculate the inverse matrix B^{-1} explicitly just to solve these systems, since this calculation is unnecessarily expensive. Instead, we calculate or maintain some factorization of B—usually an LU factorization—and use triangular substitutions with the factors to recover π and t. It is less expensive to update the factorization than to calculate it afresh at each iteration, because the matrix B changes by just a single column between iterations.

The standard factorization/updating procedures start with an LU factorization of B at the first iteration of the simplex algorithm. Since in practical applications B is large and sparse, its rows and columns are rearranged during the factorization to maintain both numerical stability and sparsity of the L and U factors. One successful pivot strategy that trades off between these two aims was proposed by Markowitz in 1957 [158]; it is still used as the basis of many practical sparse LU algorithms. Other considerations may also enter into our choice of pivot. As we discuss below, it may help to improve the efficiency of the updating procedures if as many as possible of the leading columns of U contain just a single nonzero: the diagonal element. Many heuristics have been devised for choosing row and column permutations that produce this and other desirable structural features.

Let us assume for simplicity that row and column permutations are already incorporated in B, so that we write the initial LU factorization as

$$LU = B, \qquad (13.26)$$

(L is unit lower triangular, U is upper triangular). The system $Bt = A_q$ can then be solved by the following two-step procedure:

$$L\bar{t} = A_q, \quad Ut = \bar{t}. \qquad (13.27)$$

Similarly, the system $B^T \pi = c_B$ is solved by performing the following two triangular substitutions:

$$U^T \bar{\pi} = c_B, \quad L^T \pi = \bar{\pi}.$$

We now discuss a procedure for updating the factors L and U after one step of the simplex method, when the index q is removed from $\mathcal{B}$ and replaced by the index p. The corresponding change to the basis matrix B is that the column A_q is removed from B and replaced by A_p. We call the resulting matrix B^+ and note that if we rewrite (13.26) as $U = L^{-1}B$, it is easy to see that the modified matrix $L^{-1}B^+$ will be upper triangular except in the column occupied by A_p. That is, $L^{-1}B^+$ has the form shown in Figure 13.4.

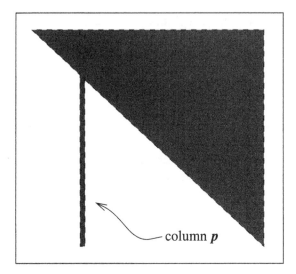

Figure 13.4
$L^{-1}B^+$, which is upper triangular except for the column occupied by A_p.

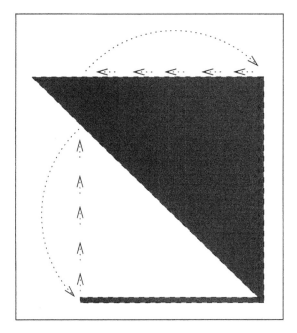

Figure 13.5
After cyclic row and column permutation P_1, the non–upper triangular part of $P_1 L^{-1} B^+ P_1^T$ appears in the last row.

We now perform a cyclic permutation that moves column p to the last column position m and moves columns $p+1, p+2, \ldots, m$ one position to the left to make room for it. If we apply the same permutation to rows p through m, the net effect is to move the non–upper triangular part to the last row of the matrix, as shown in Figure 13.5. If we denote the permutation matrix by P_1, the matrix of Figure 13.5 is $P_1 L^{-1} B^+ P_1^T$.

13.4. LINEAR ALGEBRA IN THE SIMPLEX METHOD

Finally, we perform sparse Gaussian elimination on the matrix $P_1 L^{-1} B^+ P_1^T$ to restore upper triangular form. That is, we find L_1 and U_1 (lower and upper triangular, respectively) such that

$$P_1 L^{-1} B^+ P_1^T = L_1 U_1. \tag{13.28}$$

It is easy to show that L_1 and U_1 have a simple form. The lower triangular matrix L_1 differs from the identity only in the last row, while U_1 is identical to $P_1 L^{-1} B^+ P_1^T$ except that the (m, m) element is changed and the off-diagonal elements in the last row have been eliminated.

We give details of this process for the case of $m = 5$. Using the notation

$$L^{-1} B = U = \begin{bmatrix} u_{11} & u_{12} & u_{13} & u_{14} & u_{15} \\ & u_{22} & u_{23} & u_{24} & u_{25} \\ & & u_{33} & u_{34} & u_{35} \\ & & & u_{44} & u_{45} \\ & & & & u_{55} \end{bmatrix}, \quad L^{-1} A_q = \begin{bmatrix} w_1 \\ w_2 \\ w_3 \\ w_4 \\ w_5 \end{bmatrix},$$

and supposing that A_p occurs in the second column of B (so that the second column is replaced by $L^{-1} A_q$), we have

$$L^{-1} B^+ = \begin{bmatrix} u_{11} & w_1 & u_{13} & u_{14} & u_{15} \\ & w_2 & u_{23} & u_{24} & u_{25} \\ & w_3 & u_{33} & u_{34} & u_{35} \\ & w_4 & & u_{44} & u_{45} \\ & w_5 & & & u_{55} \end{bmatrix}.$$

After the cyclic permutation P_1, we have

$$P_1 L^{-1} B^+ P_1^T = \begin{bmatrix} u_{11} & u_{13} & u_{14} & u_{15} & w_1 \\ & u_{33} & u_{34} & u_{35} & w_3 \\ & & u_{44} & u_{45} & w_4 \\ & & & u_{55} & w_5 \\ & u_{23} & u_{24} & u_{25} & w_2 \end{bmatrix}. \tag{13.29}$$

The factors L_1 and U_1 are now as follows:

$$L_1 = \begin{bmatrix} 1 & & & & \\ & 1 & & & \\ & & 1 & & \\ & & & 1 & \\ 0 & l_{52} & l_{53} & l_{54} & 1 \end{bmatrix}, \quad U_1 = \begin{bmatrix} u_{11} & u_{13} & u_{14} & u_{15} & w_1 \\ & u_{33} & u_{34} & u_{35} & w_3 \\ & & u_{44} & u_{45} & w_4 \\ & & & u_{55} & w_5 \\ & & & & \hat{w}_2 \end{bmatrix}, \quad (13.30)$$

for some values of l_{52}, l_{53}, l_{54}, and $\hat{w}_2$ (see the exercises).

The result of this updating process is the factorization (13.28), which we can rewrite as follows:

$$B^+ = L^+ U^+, \quad \text{where} \quad L^+ = L P_1^T L_1, \quad U^+ = U_1 P_1. \quad (13.31)$$

There is no need to calculate L^+ and U^+ explicitly. Rather, the nonzero elements in L_1 and the last column of U_1, and the permutation information in P_1, can be stored in compact form, so that triangular substitutions involving L^+ and U^+ can be performed by applying a number of permutations and sparse triangular substitutions involving these various factors.

Subsequent simplex iterates give rise to similar updating procedures. At each step, we need to store the same type of information: the permutation matrix and the nonzero elements in the last row of L_1 and the last column of U_1.

The procedure we have just outlined is due to Forrest and Tomlin. It is quite efficient, because it requires the storage of little data at each update and does not require much movement of data in memory. Its major disadvantage is possible numerical instability. Large elements in the factors of a matrix are a sure indicator of instability, and the multipliers in the L_1 factor (l_{52} in (13.30), for example) may be very large. An earlier scheme of Bartels and Golub allowed swapping of rows to avoid these problems. For instance, if $|u_{33}| < |u_{23}|$ in (13.29), we could swap rows 2 and 5 to ensure that the subsequent multiplier l_{52} in the L_1 factor does not exceed 1 in magnitude.

The Bartels–Golub procedure pays a price for its superior stability properties. The lower right corner of the upper triangular factor is generally altered and possibly filled in during each updating step. The fill-in can be reduced by choosing a permutation of B during the initial factorization to achieve a certain structure in the initial upper triangular factor U. Specifically, we choose the permutation such that U has the structure

$$U = \begin{bmatrix} D & U_{12} \\ 0 & U_{22} \end{bmatrix}, \quad (13.32)$$

where D is square (say, $\hat{m} \times \hat{m}$) and diagonal. Then, at the first update, fill-in can only appear in the last $m - \hat{m}$ columns of U. Since these columns are moved one place to the left during the update, the updated factor U_1 also has the form of (13.32), but with D one dimension smaller (that is, $\hat{m} - 1 \times \hat{m} - 1$ and diagonal). Consequently, during the second update, we may get fill-in in the last $m - \hat{m} + 1$ columns of the U factor, and so on. This procedure (due to Saunders) can sharply restrict the amount of fill-in and produce an algorithm that is both efficient and stable.

All these updating schemes require us to keep a record of permutations and of certain nonzero elements of the sparse triangular factors. Although this information can often be stored in a highly compact form—just a few bytes of storage for each update—the total amount of space may build up to unreasonable levels after many such updates have been performed. As the number of updates builds up, so does the time needed to solve for the vectors t and π in Procedure 13.1. If an unstable updating procedure is used, numerical errors may also come into play, blocking further progress by the simplex algorithm. For all these reasons, most simplex implementations periodically calculate a fresh LU factorization of the current basis matrix B and discard the updates. The new factorization uses the same permutation strategies that we apply to the very first factorization, which balance the requirements of stability, sparsity, and structure.

13.5 OTHER (IMPORTANT) DETAILS

PRICING AND SELECTION OF THE ENTERING INDEX

There are usually many negative components of s_N at each step. We need to select one of these, s_q, as the entering variable. How do we make this selection? Ideally, we would like to choose the sequence of entering variables that gets us to the solution x^* in the fewest possible steps, but we rarely have the global perspective needed to implement this strategy. Instead, we use more shortsighted but practical strategies that obtain a significant decrease in $c^T x$ on just the present iteration. There is usually a tradeoff between the effort spent searching for a good entering variable q that achieves this aim and the decrease in $c^T x$ to be obtained from pivoting on q. Different pivot strategies resolve the tradeoff in different ways.

Dantzig's original selection rule is one of the simplest. It calculates $s_N = N^T \pi$ and selects the most negative s_q as the entering component. This rule is motivated by (13.24). It gives the maximum improvement in $c^T x$ per unit step in the entering variable x_q. A large reduction in $c^T x$ is not guaranteed, however. It could be that we can increase x_q^+ only a tiny amount from zero before reaching the next vertex. As we see later, it is even possible that we cannot move at all!

To avoid calculation of the entire vector s_N, *partial pricing* strategies are a popular option. At each iteration, these methods evaluate just a subvector of s_N by multiplying π by a column submatrix of N. The most negative s_q from this subvector is chosen as entering pivot. To give all the components of N a chance at entering the basis, partial pricing strategies cycle through the nonbasic elements, changing the subvector of s_N they evaluate at each iteration so that no component is ignored for too long.

Neither of these strategies guarantees that we can make a substantial move along the chosen edge before finding a new vertex. *Multiple pricing* strategies are more thorough: For a small subset of indices from $\mathcal{N}$, they evaluate not just the prices s_q but the distance that can be moved along the resulting edge and the consequent reduction in objective function, namely $s_q x_q^+$. The index subset is initially chosen from the most negative components of s_N; ten such components is a typical number. For each component, we compute $t = B^{-1}A_q$ and x_q^+ as in the algorithm outline above, and choose the entering variable as the index that minimizes $s_q x_q^+$. For subsequent iterations we deal just with this same index set, essentially fixing all the components of x_N outside this set at zero and solving a reduced linear program. Eventually, we find that all the prices s_q for indices in the subset are nonnegative, so we choose a new index subset by computing the whole vector s_N and repeat the process. This approach has the advantage that we operate only on a few columns A_q of the nonbasic matrix N at a time; the remainder of N does not even need to be in main memory at all. After the great improvements in price and speed of computer memory in recent years, however, savings such as this are less important than they once were.

There is plenty of scope to define heuristics that combine ideas from multiple and partial pricing. At every iteration we can update the subset for which we compute the entering variable value x_q^+ by

- retaining the most promising indices from the previous iteration (i.e., the most negative s_q's); and

- pricing a new part of the vector s_N and choosing the most negative components s_q from this part.

Strategies like this can combine the thoroughness of multiple pricing with the advantages of partial pricing, which brings in "new blood" in the form of new parts of $\mathcal{N}$ at each iteration, at relatively low cost.

A sophisticated rule known as *steepest edge* chooses the most downhill direction from among all the candidates—the one that produces the largest decrease in $c^T x$ per unit distance moved along the edge. By contrast, Dantzig's rule maximizes the decrease in $c^T x$ per unit change in x_q^+, which is *not* the same thing. (A small change in x_q^+ can correspond to a large distance moved along the edge.) During the pivoting step, the overall change in x is

$$x^+ = \begin{bmatrix} x_B^+ \\ x_N^+ \end{bmatrix} = \begin{bmatrix} x_B \\ x_N \end{bmatrix} + \begin{bmatrix} -B^{-1}A_q \\ e_q \end{bmatrix} x_q^+ = x + \eta_q x_q^+,$$

where e_q is the unit vector with a 1 in the position corresponding to the index $q \in \mathcal{N}$ and zeros elsewhere, and

$$\eta_q = \begin{bmatrix} -B^{-1}A_q \\ e_q \end{bmatrix} = \begin{bmatrix} -t \\ e_q \end{bmatrix}, \tag{13.33}$$

where we have used the fact that $t = B^{-1}A_q$; see Procedure 13.1. The change in $c^T x$ per unit step along η_q is given by

$$\frac{c^T \eta_q}{\|\eta_q\|}. \tag{13.34}$$

The steepest-edge rule chooses q to minimize this quantity.

If we had to compute each η_q explicitly to perform the minimization in (13.34), the steepest-edge strategy would be prohibitively expensive—as expensive as the strategy of seeking the best possible reduction in $c^T x$. Goldfarb and Reid showed that edge steepness for all indices $i \in \mathcal{N}$ can, in fact, be maintained quite economically. Instead of performing one linear system solution procedure with B^T (to find π from $B^T \pi = c_B$ in Procedure 13.1), their method requires two such procedures with this same coefficient matrix. In both cases, a number of inner products involving the columns A_i for $i \in \mathcal{N}$ are required. We now outline their steepest-edge procedure.

First, note that we know already the numerator $c^T \eta_q$ in (13.34) without calculating η_q, because we proved in the discussion leading up to (13.24) that $c^T \eta_q = s_q$. For the denominator, there is a simple recurrence for keeping track of each norm $\|\eta_q\|$, which is updated from step to step. To derive this recurrence, suppose that A_p is the column to be removed from the basis matrix B to make way for A_q. Assuming without loss of generality that A_p occupies the first column of B, we can express the pivot algebraically as follows:

$$B^+ = B + (A_q - A_p)e_1^T = B + (A_q - Be_1)e_1^T, \tag{13.35}$$

where $e_1 = (1, 0, 0, \ldots, 0)^T$. Choosing an index $i \in \mathcal{N}$ with $i \neq q$, we want to see how $\|\eta_i\|$ is affected by the update (13.35).

For convenience, we use γ_i to denote $\|\eta_i\|^2$, and note from (13.33) that

$$\gamma_i = \|\eta_i\|^2 = \|B^{-1}A_i\|^2 + 1,$$
$$\gamma_i^+ = \|\eta_i^+\|^2 = \|(B^+)^{-1}A_i\|^2 + 1.$$

By applying the Sherman–Morrison formula (A.55) to the rank-one update formula in (13.35), we obtain

$$(B^+)^{-1} = B^{-1} - \frac{(B^{-1}A_q - e_1)e_1^T B^{-1}}{1 + e_1^T (B^{-1}A_q - e_1)} = B^{-1} - \frac{(t - e_1)e_1^T B^{-1}}{e_1^T t},$$

where again we have used the fact that $t = B^{-1}A_q$. Therefore, we have that

$$(B^+)^{-1}A_i = B^{-1}A_i - \frac{e_1^T B^{-1}A_i}{e_1^T t}(t - e_1).$$

By substituting for $(B^+)^{-1}A_i$ in (13.35) and performing some simple algebra (see the exercises), we obtain

$$\gamma_i^+ = \gamma_i - 2\left(\frac{e_1^T B^{-1} A_i}{e_1^T t}\right) A_i^T B^{-T} t + \left(\frac{e_1^T B^{-1} A_i}{e_1^T t}\right)^2 \gamma_q. \tag{13.36}$$

We now solve the following two linear systems to obtain $\hat{t}$ and r:

$$B^T \hat{t} = t, \qquad B^T r = e_1. \tag{13.37}$$

The formula (13.36) then becomes

$$\gamma_i^+ = \gamma_j - 2\left(\frac{r^T A_i}{r^T A_q}\right) \hat{t}^T A_i + \left(\frac{r^T A_i}{r^T A_q}\right)^2 \gamma_q. \tag{13.38}$$

Hence, once the systems (13.37) have been solved, each γ_i^+ can be updated at the cost of evaluating the two inner products $r^T A_i$ and $\hat{t}^T A_i$.

The steepest-edge strategy does not guarantee that we can take a long step before reaching another vertex, but it has proved to be highly effective in practice. Recent testing (Goldfarb and Forrest [114]) has shown it to be superior even to interior-point methods on some very large problems.

STARTING THE SIMPLEX METHOD

The simplex method requires a basic feasible starting point $\bar{x}$ and a corresponding initial basic index set $\mathcal{B} \subset \{1, 2, \ldots, n\}$ with $|\mathcal{B}| = m$ for which

- the $m \times m$ matrix B defined by (13.13) is nonsingular; and
- $\bar{x}_B \geq 0$ and $\bar{x}_N = 0$, where $\mathcal{N}$ is the complement of $\mathcal{B}$ (13.17).

The problem of finding this initial point and basis may itself be nontrivial—in fact, its difficulty is equivalent to that of actually finding the solution of a linear program. One approach for dealing with this difficulty is the *two-phase* or *Phase-I/Phase-II* approach, which proceeds as follows. In Phase I, we set up an auxiliary problem, a linear program different from (13.1), and solve it with the simplex method. The Phase-I problem is constructed so that an initial basic feasible point is trivial to determine, and so that its solution gives a basic feasible initial point for Phase II. In Phase II, a second linear program very similar to the original problem (13.1) is solved, starting from the Phase-I solution. The solution of the original problem (13.1) can be extracted easily from the solution of the Phase-II problem.

In Phase I we introduce artificial variables z into (13.1) and redefine the objective function to be the sum of these artificial variables. To be specific, the *Phase-I problem* is

$$\min e^T z, \quad \text{subject to } Ax + Ez = b, (x, z) \geq 0, \tag{13.39}$$

where $z \in \mathbf{R}^m$, $e = (1, 1, \ldots, 1)^T$, and E is a diagonal matrix whose diagonal elements are

$$E_{jj} = +1 \text{ if } b_j \geq 0, \quad E_{jj} = -1 \text{ if } b_j < 0.$$

It is easy to see that the point (x, z) defined by

$$x = 0, \quad z_j = |b_j|, \quad j = 1, 2, \ldots, m, \quad (13.40)$$

is a basic feasible point for (13.39). Obviously, this point satisfies the constraints in (13.39), while the initial basis matrix B is simply the matrix E, which is clearly nonsingular.

At any feasible point for (13.39), the artificial variables z represent the amounts by which the constraints $Ax = b$ are violated by the x component. The objective function is simply the sum of these violations, so by minimizing this sum we are forcing x to become feasible for the original problem (13.1). In fact, it is not difficult to see that the Phase-I problem (13.39) has an optimal objective value of zero if and only if the original problem (13.1) has feasible points by using the following argument: If there exists a vector $(\tilde{x}, \tilde{z})$ that is feasible for (13.39) such that $e^T \tilde{z} = 0$, we must have $\tilde{z} = 0$, and therefore $A\tilde{x} = b$ and $\tilde{x} \geq 0$, so $\tilde{x}$ is feasible for the original problem (13.1). Conversely, if $\tilde{x}$ is feasible for (13.1), then the point $(\tilde{x}, 0)$ is feasible for (13.39) with an objective value of 0. Since the objective in (13.39) is obviously nonnegative at all feasible points, then $(\tilde{x}, 0)$ must be optimal for (13.39), verifying our claim.

We can now apply the simplex method to (13.39) from the initial point (13.40). If it terminates at an optimal solution for which the objective $e^T z$ is positive, we conclude by the argument above that the original problem (13.1) is infeasible. Otherwise, the simplex method identifies a point $(\tilde{x}, \tilde{z})$ with $e^T \tilde{z} = 0$, which is a basic feasible point for the following linear program, which is the *Phase-II problem*:

$$\min c^T x \text{ subject to } Ax + z = b, \quad x \geq 0, \quad 0 \geq z \geq 0. \quad (13.41)$$

We can note immediately that this problem is equivalent to (13.1), because any solution (and indeed any feasible point) must have $z = 0$. We need to retain the artificial variables z in Phase II, however, since some components of z may be present in the optimal basis from Phase I that we are using as the initial basis for (13.41), though of course the values $\tilde{z}_j$ of these components must be zero. In fact, we can modify (13.41) to include *only* those components of z that are present in the optimal basis for (13.39), omitting the others.

The problem (13.41) is not quite in standard form because of the two-sided bounds on z. However, it is easy to modify the simplex method described above to handle these additional bounds (we omit the details). We can customize the simplex algorithm slightly by deleting each component of z from the problem (13.41) as soon as it is swapped out of the basis. This strategy ensures that components of z do not repeatedly enter and leave the basis, and therefore avoids unnecessary simplex iterations with little added complication to the algorithm.

If (x^*, z^*) is a basic solution of (13.41), it must have $z^* = 0$, and so x^* is a solution of (13.1). In fact, x^* is a *basic* solution of (13.1), though this claim is not completely obvious because the final basis $\mathcal{B}$ for the Phase-II problem may still contain components of z^*, making it unsuitable as an optimal basis for (13.1). Since we know that A has full row rank, however, we can construct an optimal basis for (13.1) in a postprocessing phase: We can extract the indices that correspond to x components from $\mathcal{B}$ and add extra indices to $\mathcal{B}$ in a way that maintains nonsingularity of the submatrix B defined by (13.13).

A final point to note is that in many problems we do not need to add a complete set of m artificial variables to form the Phase-I problem. This observation is particularly relevant when slack and surplus variables have already been added to the problem formulation, as in (13.2), to obtain a linear program with inequality constraints in standard form (13.1). Some of these slack/surplus variables can play the roles of artificial variables, making it unnecessary to include such variables explicitly.

We illustrate this point with the following example.

❑ Example 13.1

Consider the inequality-constrained linear program defined by

$$\min 3x_1 + x_2 + x_3 \quad \text{subject to}$$
$$2x_1 + x_2 + x_3 \leq 2,$$
$$x_1 - x_2 - x_3 \leq -1, \quad x \geq 0.$$

By adding slack variables to both inequality constraints, we obtain the following equivalent problem in standard form:

$$\min 3x_1 + x_2 + x_3 \quad \text{subject to}$$
$$2x_1 + x_2 + x_3 + x_4 = 2,$$
$$x_1 - x_2 - x_3 + x_5 = -1, \quad x \geq 0.$$

By inspection, it is easy to see that the vector $x = (0, 0, 0, 2, 0)$ is feasible with respect to the first linear constraint and the lower bound $x \geq 0$, though it does not satisfy the second constraint. Hence, in forming the Phase-I problem, we add just a single artificial variable z_2 to take care of the infeasibility in this second constraint, and obtain

$$\min z_2 \quad \text{subject to} \tag{13.42}$$
$$2x_1 + x_2 + x_3 + x_4 = 2,$$
$$x_1 - x_2 - x_3 + x_5 - z_2 = -1, \quad (x, z_2) \geq 0.$$

It is easy to see that the vector

$$(x, z_2) = ((0, 0, 0, 2, 0), 1)$$

is feasible with respect to (13.42). In fact, it is a basic feasible point, since the matrix B corresponding to the initial basis is

$$B = \begin{bmatrix} 1 & 0 \\ 0 & -1 \end{bmatrix}.$$

In this example, the variable x_4 plays the role of artificial variable for the first constraint. There was no need to add an explicit artificial variable z_1.

DEGENERATE STEPS AND CYCLING

The third issue arises from the fact that the step x_q^+ computed by the procedure above may be zero. There may well be an index i such that $(x_B)_i = 0$ while $t_i > 0$, so the algorithm cannot move any distance at all along the feasible direction without leaving the feasible set. A step of this kind is known as a *degenerate step*. Although we have not moved anywhere, the step may still be useful because we have changed one component of the basis $\mathcal{B}$. The new basis may be closer to the optimal basis, so in that sense the algorithm may have made progress towards the solution. Although no reduction in $c^T x$ is made at a degenerate step, it may be laying the groundwork for reductions in $c^T x$ at subsequent steps.

However, a problem known as *cycling* can occur when we take a number of consecutive degenerate steps. After making a number of alterations to the basis set without changing the objective value, it is possible that we return to an earlier basis set $\mathcal{B}$! If we continue to apply our algorithm from this point, we will repeat the same cycle, returning to $\mathcal{B}$ ad infinitum and never terminating.

Cycling was thought originally to be a rare phenomenon, but it has been observed with increasing frequency in the large linear programs that arise as relaxations of integer programming problems. Since integer programs are an important source of linear programs, it is essential for practical simplex codes to incorporate a cycling avoidance strategy.

We describe the *perturbation strategy* by first showing how a perturbation to the right-hand-side b of the constraints in (13.1) affects the basic solutions and the pivot strategy. Then, we outline how these perturbations can be used to avoid cycling.

Suppose we perturb the right-hand-side vector b to $b(\epsilon)$ defined by

$$b(\epsilon) = b + E \begin{bmatrix} \epsilon \\ \epsilon^2 \\ \vdots \\ \epsilon^m \end{bmatrix},$$

where E is a nonsingular matrix and $\epsilon > 0$ is a small constant. The perturbation in b induces a perturbation in each basic solution x_B, so that by defining $x_B(\epsilon)$ as the solution of

$Bx_B(\epsilon) = b(\epsilon)$, we obtain

$$x_B(\epsilon) = x_B + B^{-1}E \begin{bmatrix} \epsilon \\ \epsilon^2 \\ \vdots \\ \epsilon^m \end{bmatrix} = x_B + \sum_{k=1}^{m} (B^{-1}E)_{\cdot k} \epsilon^k, \quad (13.43)$$

where $(B^{-1}E)_{\cdot k}$ denotes the kth column of $B^{-1}E$. We choose E so that the perturbation vector $\sum_{k=1}^{m} (B^{-1}E)_{\cdot k} \epsilon^k$ is strictly positive for all sufficiently small positive values of ϵ.

With this perturbation, we claim that ties cannot arise in the choice of variable to leave the basis. From Procedure 13.1 above, we have a tie if there are two indices $i, l \in \mathcal{B}$ such that

$$\frac{(x_B(\epsilon))_i}{t_i} = \frac{(x_B(\epsilon))_l}{t_l}, \quad \text{for all } \epsilon > 0. \quad (13.44)$$

This equality can occur only if all the coefficients in the expansion (13.43) are identical for the two rows i and l. In particular, we must have

$$\frac{(B^{-1}E)_{ik}}{t_i} = \frac{(B^{-1}E)_{lk}}{t_l}, \quad k = 1, 2, \ldots, m.$$

Hence, rows i and l of the matrix $B^{-1}E$ are multiples of each other, so $B^{-1}E$ is singular. This contradicts nonsingularity of B and E, so ties cannot occur, and there is a unique minimizing index p. Moreover, if a step of simplex is taken with this leaving index p, there is a range of $\epsilon > 0$ for which all variables remaining in the basis $(x_B(\epsilon))_i$, $i \neq p$, remain strictly positive.

How can this device be used to avoid cycling? Suppose we are at a basis $\mathcal{B}$ for which there are zero components of x_B. At this point, we choose the perturbation matrix E so that the perturbation (13.43) is positive for some range of $\epsilon > 0$. For instance, if we choose E to be the current basis B, we have from (13.43) that

$$x_B(\epsilon) = x_B + B^{-1}E \begin{bmatrix} \epsilon \\ \epsilon^2 \\ \vdots \\ \epsilon^m \end{bmatrix} = x_B + \begin{bmatrix} \epsilon \\ \epsilon^2 \\ \vdots \\ \epsilon^m \end{bmatrix} > x_B,$$

for *all* $\epsilon > 0$. We now perform a step of simplex for the perturbed problem, choosing the unique leaving index p as above.

After the simplex step, the basis components $(x_B(\epsilon))_i$, $i \neq p$, carried over from the previous iteration are strictly positive for sufficiently small ϵ, as is the entering component

x_q. Hence, it will be possible to take a nonzero step again at the next simplex iteration, and because ties (13.44) are not possible at this iteration either for sufficiently small ϵ, the leaving variable will be uniquely determined as well.

The reasoning of the previous paragraph holds only if the value of ϵ is sufficiently small. There are robust strategies, such as the *lexicographic* strategy, that avoid an explicit choice of ϵ but rather ask questions of the "what if?" variety to break ties. Given a set of tied indices $\bar{B} \subset B$ with

$$\frac{(x_B)_i}{t_i} = \frac{(x_B)_l}{t_l}, \quad \text{for all } i, l \in \bar{B},$$

such a strategy would ask, "What if we made a perturbation of the type (13.43)? Which index from $\bar{B}$ would be preferred then?" Implementation of this idea is not difficult: We evaluate successive coefficients of ϵ^k, $k = 1, 2, \ldots$, in the expansion (13.43) for the tying elements. If there is an element $i \in \bar{B}$ with

$$(B^{-1}E)_{i1} > (B^{-1}E)_{l1}, \quad \text{for all } l \in \bar{B}, \quad l \neq i,$$

Then clearly $(x_B(\epsilon))_i > (x_B(\epsilon))_l$ for all ϵ sufficiently small, so we choose i as the leaving index. Otherwise, we retain in $\bar{B}$ the indices i that tie for the largest value of $(B^{-1}E)_{i1}$ and evaluate the second column of $B^{-1}E$. If one value of $(B^{-1}E)_{i2}$ dominates, we choose it as the leaving index. Otherwise, we consider coefficients $(B^{-1}E)_{ik}$ for successively higher-order k until just one candidate index remains.

13.6 WHERE DOES THE SIMPLEX METHOD FIT?

In linear programming, as in all optimization problems in which inequality constraints (including bounds) are present, the fundamental issue is to partition the inequality constraints into those that are *active* at the solution and those that are *inactive*. (The inactive constraints are, of course, those for which a strict inequality holds at the solution.) The simplex method belongs to a general class of algorithms for constrained optimization known as *active set methods*, which explicitly maintain estimates of the active and inactive index sets that are updated at each step of the algorithm. (In the case of simplex and linear programming, B is the set of "probably inactive" indices—those for which the bound $x_i \geq 0$ is inactive—while $\mathcal{N}$ is the set of "probably active" indices.) Like most active set methods, the simplex method makes only modest changes to these index sets at each step: A single index is transferred from B into $\mathcal{N}$, and vice versa.

Active set algorithms for quadratic programming, bound-constrained optimization, and nonlinear programming use the same basic strategy as simplex of defining a set of "probably active" constraint indices and taking a step toward the solution of a reduced problem in which these constraints are satisfied as equalities. When nonlinearity enters the problem, many of the features that make the simplex method so effective no longer apply. For instance, it is no longer true in general that at least $n - m$ of the bounds $x \geq 0$ are active at the solution. The specialized linear algebra techniques no longer apply, and it may no longer be possible to obtain an exact solution to the reduced problem at each iteration. Nevertheless, the simplex method is rightly viewed as the antecedent of the active set class.

One undesirable feature of the simplex method attracted attention from its earliest days. Though highly efficient on almost all practical problems (the method generally requires at most $2m$ to $3m$ iterations, where m is the row dimension of the constraint matrix in (13.1)), there are pathological problems on which the algorithm performs very poorly. Klee and Minty [144] presented an n-dimensional problem whose feasible polytope has 2^n vertices, for which the simplex method visits every single vertex before reaching the optimal point! This example verified that the complexity of the simplex method is *exponential*; roughly speaking, its running time can be an exponential function of the dimension of the problem. For many years, theoreticians searched for a linear programming algorithm that has *polynomial complexity*, that is, an algorithm in which the running time is bounded by a polynomial function of the size of the input. In the late 1970s, Khachiyan [142] described an *ellipsoid method* that indeed has polynomial complexity but turned out to be much too slow in practice. In the mid-1980s, Karmarkar [140] described a polynomial algorithm that approaches the solution through the interior of the feasible polytope rather than working its way around the boundary as the simplex method does. Karmarkar's announcement marked the start of intense research in the field of *interior-point methods*, which are the subject of the next chapter.

NOTES AND REFERENCES

An alternative procedure for performing the Phase-I calculation of an initial basis was described by Wolfe [247]. This technique does not require artificial variables to be introduced in the problem formulation, but rather starts at any point x that satisfies $Ax = b$ with at most m nonzero components in x. (Note that we do not require the basic part x_B to consist of all positive components.) Phase I then consists in solving the problem

$$\min_{x} \sum_{x_i < 0} -x_i \quad \text{subject to } Ax = b,$$

and terminating when an objective value of 0 is attained. This problem is *not* a linear program—its objective is only piecewise linear—but it can be solved by the simplex method nonetheless. The key is to *redefine* the cost vector f at each iteration x such that $f_i = -1$ for $x_i < 0$ and $f_i = 0$ otherwise.

EXERCISES

13.1 Consider the overdetermined linear system $Ax = b$ with m rows and n columns ($m > n$). When we apply Gaussian elimination with complete pivoting to A, we obtain

$$PAQ = L \begin{bmatrix} U_{11} & U_{12} \\ 0 & 0 \end{bmatrix},$$

where P and Q are permutation matrices, L is $m \times m$ lower triangular, U_{11} is $\bar{m} \times \bar{m}$ upper triangular and nonsingular, U_{12} is $\bar{m} \times (n - \bar{m})$, and $\bar{m} \leq n$ is the rank of A.

(a) Show that the system $Ax = b$ is feasible if the last $m - \bar{m}$ components of $L^{-1}Pb$ are zero, and infeasible otherwise.

(b) When $\bar{m} = n$, find the unique solution of $Ax = b$.

(c) Show that the reduced system formed from the first $\bar{m}$ rows of PA and the first $\bar{m}$ components of Pb is equivalent to $Ax = b$ (i.e., a solution of one system also solves the other).

13.2 Convert the following linear program to standard form:

$$\max_{x,y} c^T x + d^T y \text{ subject to } A_1 x = b_1, \; A_2 x + B_2 y \leq b_2, \; l \leq y \leq u,$$

where there are no explicit bounds on x.

13.3 Show that the dual of the linear program

$$\min c^T x \text{ subject to } Ax \geq b, \; x \geq 0,$$

is

$$\max b^T \pi \text{ subject to } A^T \pi \leq c, \; \pi \geq 0.$$

13.4 Show that when $m \leq n$ and the rows of A are linearly dependent in (13.1), then the matrix B in (13.13) is singular, and therefore there are no basic feasible points.

13.5 Verify formula (13.36).

13.6 Calculate the values of l_{52}, l_{53}, l_{54}, and $\hat{w}_2$ in (13.30), by equating the last row of $L_1 U_1$ to the last row of the matrix in (13.29).

13.7 By extending the procedure (13.27) appropriately, show how the factorization (13.31) can be used to solve linear systems with coefficient matrix B^+ efficiently.

CHAPTER 14

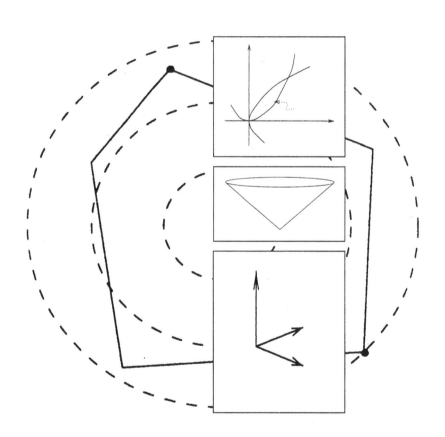

Linear Programming: Interior-Point Methods

In the 1980s it was discovered that many large linear programs could be solved efficiently by formulating them as nonlinear problems and solving them with various modifications of nonlinear algorithms such as Newton's method. One characteristic of these methods was that they required all iterates to satisfy the inequality constraints in the problem *strictly*, so they soon became known as interior-point methods. By the early 1990s, one class—primal–dual methods—had distinguished itself as the most efficient practical approach and proved to be a strong competitor to the simplex method on large problems. These methods are the focus of this chapter.

The motivation for interior-point methods arose from the desire to find algorithms with better theoretical properties than the simplex method. As we mentioned in Chapter 13, the simplex method can be quite inefficient on certain problems. Roughly speaking, the time required to solve a linear program may be exponential in the size of the problem, as measured by the number of unknowns and the amount of storage needed for the problem data. In practice, the simplex method is much more efficient than this bound would suggest, but its poor worst-case complexity motivated the development of new algorithms with better guaranteed performance. Among them is the *ellipsoid method* proposed by Khachiyan [142], which finds a solution in time that is at worst polynomial in the problem size. Unfortunately, this method approaches its worst-case bound on *all* problems and is not competitive with the simplex method.

Karmarkar's *projective algorithm* [140], announced in 1984, also has the polynomial complexity property, but it came with the added inducement of good practical behavior. The initial claims of excellent performance on large linear programs were never fully borne out, but the announcement prompted a great deal of research activity and a wide array of methods described by such labels as "affine-scaling," "logarithmic barrier," "potential-reduction," "path-following," "primal–dual," and "infeasible interior-point." All are related to Karmarkar's original algorithm, and to the log-barrier approach of Section 17.2, but many of the approaches can be motivated and analyzed independently of the earlier methods.

Interior-point methods share common features that distinguish them from the simplex method. Each interior-point iteration is expensive to compute and can make significant progress towards the solution, while the simplex method usually requires a larger number of inexpensive iterations. The simplex method works its way around the boundary of the feasible polytope, testing a sequence of vertices in turn until it finds the optimal one. Interior-point methods approach the boundary of the feasible set only in the limit. They may approach the solution either from the interior or the exterior of the feasible region, but they never actually lie on the boundary of this region.

In this chapter we outline some of the basic ideas behind primal–dual interior-point methods, including the relationship to Newton's method and homotopy methods and the ideas of the central path and central neighborhoods. We sketch the important methods in this class, with an emphasis on path-following methods. We describe in detail a practical predictor–corrector algorithm proposed by Mehrotra, which is the basis of much of the current generation of software.

14.1 PRIMAL–DUAL METHODS

OUTLINE

We consider the linear programming problem in standard form; that is,

$$\min c^T x, \quad \text{subject to } Ax = b, x \geq 0, \tag{14.1}$$

where c and x are vectors in $\mathbf{R}^n$, b is a vector in $\mathbf{R}^m$, and A is an $m \times n$ matrix. The dual problem for (14.1) is

$$\max b^T \lambda, \quad \text{subject to } A^T \lambda + s = c, s \geq 0, \tag{14.2}$$

where λ is a vector in $\mathbf{R}^m$ and s is a vector in $\mathbf{R}^n$. As shown in Chapter 13, primal–dual solutions of (14.1), (14.2) are characterized by the Karush–Kuhn–Tucker conditions (13.4), which we restate here as follows:

$$A^T \lambda + s = c, \tag{14.3a}$$
$$Ax = b, \tag{14.3b}$$
$$x_i s_i = 0, \quad i = 1, 2, \ldots, n, \tag{14.3c}$$
$$(x, s) \geq 0. \tag{14.3d}$$

Primal–dual methods find solutions (x^*, λ^*, s^*) of this system by applying variants of Newton's method to the three equalities in (14.3) and modifying the search directions and step lengths so that the inequalities $(x, s) \geq 0$ are satisfied *strictly* at every iteration. The equations (14.3a), (14.3b), (14.3c) are only mildly nonlinear and so are not difficult to solve by themselves. However, the problem becomes much more difficult when we add the nonnegativity requirement (14.3d). The nonnegativity condition is the source of all the complications in the design and analysis of interior-point methods.

To derive primal–dual interior-point methods, we restate the optimality conditions (14.3) in a slightly different form by means of a mapping F from $\mathbf{R}^{2n+m}$ to $\mathbf{R}^{2n+m}$:

$$F(x, \lambda, s) = \begin{bmatrix} A^T \lambda + s - c \\ Ax - b \\ XSe \end{bmatrix} = 0, \tag{14.4a}$$
$$(x, s) \geq 0, \tag{14.4b}$$

where

$$X = \mathrm{diag}(x_1, x_2, \ldots, x_n), \qquad S = \mathrm{diag}(s_1, s_2, \ldots, s_n), \tag{14.5}$$

and $e = (1, 1, \ldots, 1)^T$. Primal–dual methods generate iterates (x^k, λ^k, s^k) that satisfy the bounds (14.4b) strictly, that is, $x^k > 0$ and $s^k > 0$. This property is the origin of the term *interior-point*. By respecting these bounds, the methods avoid spurious solutions, that is, points that satisfy $F(x, \lambda, s) = 0$ but not $(x, s) \geq 0$. Spurious solutions abound, and do not provide useful information about solutions of (14.1) or (14.2), so it makes sense to exclude them altogether from the region of search.

Many interior-point methods actually require the iterates to be *strictly feasible*; that is, each (x^k, λ^k, s^k) must satisfy the linear equality constraints for the primal and dual problems. If we define the primal–dual *feasible set* $\mathcal{F}$ and *strictly feasible set* $\mathcal{F}^o$ by

$$\mathcal{F} = \{(x, \lambda, s) \mid Ax = b, A^T \lambda + s = c, (x, s) \geq 0\}, \tag{14.6a}$$
$$\mathcal{F}^o = \{(x, \lambda, s) \mid Ax = b, A^T \lambda + s = c, (x, s) > 0\}, \tag{14.6b}$$

the strict feasibility condition can be written concisely as

$$(x^k, \lambda^k, s^k) \in \mathcal{F}^o.$$

Like most iterative algorithms in optimization, primal–dual interior-point methods have two basic ingredients: a procedure for determining the step and a measure of the desirability of each point in the search space. As mentioned above, the search direction procedure has its origins in Newton's method for the nonlinear equations (14.4a). Newton's method forms a linear model for F around the current point and obtains the search direction $(\Delta x, \Delta \lambda, \Delta s)$ by solving the following system of linear equations:

$$J(x, \lambda, s) \begin{bmatrix} \Delta x \\ \Delta s \\ \Delta \lambda \end{bmatrix} = -F(x, \lambda, s),$$

where J is the Jacobian of F. (See Chapter 11 for a detailed discussion of Newton's method for nonlinear systems.) If the current point is strictly feasible (that is, $(x, \lambda, s) \in \mathcal{F}^o$), the Newton step equations become

$$\begin{bmatrix} 0 & A^T & I \\ A & 0 & 0 \\ S & 0 & X \end{bmatrix} \begin{bmatrix} \Delta x \\ \Delta \lambda \\ \Delta s \end{bmatrix} = \begin{bmatrix} 0 \\ 0 \\ -XSe \end{bmatrix}. \tag{14.7}$$

A full step along this direction usually is not permissible, since it would violate the bound $(x, s) \geq 0$. To avoid this difficulty, we perform a line search along the Newton direction so that the new iterate is

$$(x, \lambda, s) + \alpha(\Delta x, \Delta \lambda, \Delta s),$$

for some line search parameter $\alpha \in (0, 1]$. Unfortunately, we often can take only a small step along the direction ($\alpha \ll 1$) before violating the condition $(x, s) > 0$; hence, the pure Newton direction (14.7), which is known as the *affine scaling direction*, often does not allow us to make much progress toward a solution.

Primal–dual methods modify the basic Newton procedure in two important ways:

1. They bias the search direction toward the interior of the nonnegative orthant $(x, s) \geq 0$, so that we can move further along the direction before one of the components of (x, s) becomes negative.

2. They keep the components of (x, s) from moving "too close" to the boundary of the nonnegative orthant.

We consider these modifications in turn.

THE CENTRAL PATH

The central path $\mathcal{C}$ is an arc of strictly feasible points that plays a vital role in primal–dual algorithms. It is parametrized by a scalar $\tau > 0$, and each point $(x_\tau, \lambda_\tau, s_\tau) \in \mathcal{C}$ solves the following system:

$$A^T \lambda + s = c, \qquad (14.8\text{a})$$
$$Ax = b, \qquad (14.8\text{b})$$
$$x_i s_i = \tau, \quad i = 1, 2, \ldots, n, \qquad (14.8\text{c})$$
$$(x, s) > 0. \qquad (14.8\text{d})$$

These conditions differ from the KKT conditions only in the term τ on the right-hand-side of (14.8c). Instead of the complementarity condition (14.3c), we require that the pairwise products $x_i s_i$ have the same value for all indices i. From (14.8), we can define the central path as

$$\mathcal{C} = \{(x_\tau, \lambda_\tau, s_\tau) \mid \tau > 0\}.$$

It can be shown that $(x_\tau, \lambda_\tau, s_\tau)$ is defined uniquely for each $\tau > 0$ if and only if $\mathcal{F}^o$ is nonempty. A plot of $\mathcal{C}$ for a typical problem, projected into the space of primal variables x, is shown in Figure 14.1.

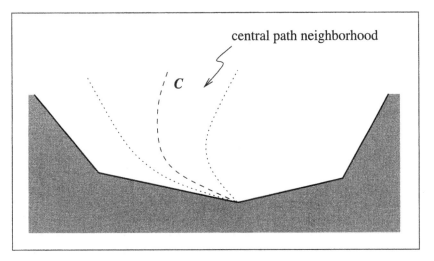

Figure 14.1 Central path, projected into space of primal variables x, showing a typical neighborhood $\mathcal{N}$.

Another way of defining $\mathcal{C}$ is to use the mapping F defined in (14.4) and write

$$F(x_\tau, \lambda_\tau, s_\tau) = \begin{bmatrix} 0 \\ 0 \\ \tau e \end{bmatrix}, \quad (x_\tau, s_\tau) > 0. \tag{14.9}$$

The equations (14.8) approximate (14.3) more and more closely as τ goes to zero. If $\mathcal{C}$ converges to anything as $\tau \downarrow 0$, it must converge to a primal–dual solution of the linear program. The central path thus guides us to a solution along a route that steers clear of spurious solutions by keeping all x and s components strictly positive and decreasing the pairwise products $x_i s_i$, $i = 1, 2, \ldots, n$, to zero at roughly the same rate.

Primal–dual algorithms take Newton steps toward points on $\mathcal{C}$ for which $\tau > 0$, rather than pure Newton steps for F. Since these steps are biased toward the interior of the nonnegative orthant defined by $(x, s) \geq 0$, it usually is possible to take longer steps along them than along the pure Newton steps for F, before violating the positivity condition.

To describe the biased search direction, we introduce a *centering parameter* $\sigma \in [0, 1]$ and a *duality measure* μ defined by

$$\mu = \frac{1}{n} \sum_{i=1}^{n} x_i s_i = \frac{x^T s}{n}, \tag{14.10}$$

which measures the average value of the pairwise products $x_i s_i$. By writing $\tau = \sigma \mu$ and applying Newton's method to the system (14.9), we obtain

$$\begin{bmatrix} 0 & A^T & I \\ A & 0 & 0 \\ S & 0 & X \end{bmatrix} \begin{bmatrix} \Delta x \\ \Delta \lambda \\ \Delta s \end{bmatrix} = \begin{bmatrix} 0 \\ 0 \\ -XSe + \sigma\mu e \end{bmatrix}. \tag{14.11}$$

The step $(\Delta x, \Delta \lambda, \Delta s)$ is a Newton step toward the point $(x_{\sigma\mu}, \lambda_{\sigma\mu}, s_{\sigma\mu}) \in \mathcal{C}$, at which the pairwise products $x_i s_i$ are all equal to $\sigma\mu$. In contrast, the step (14.7) aims directly for the point at which the KKT conditions (14.3) are satisfied.

If $\sigma = 1$, the equations (14.11) define a *centering direction*, a Newton step toward the point $(x_\mu, \lambda_\mu, s_\mu) \in \mathcal{C}$, at which all the pairwise products $x_i s_i$ are identical to μ. Centering directions are usually biased strongly toward the interior of the nonnegative orthant and make little, if any, progress in reducing the duality measure μ. However, by moving closer to $\mathcal{C}$, they set the scene for substantial progress on the next iteration. (Since the next iteration starts near $\mathcal{C}$, it will be able to take a relatively long step without leaving the nonnegative orthant.) At the other extreme, the value $\sigma = 0$ gives the standard Newton step (14.7), sometimes known as the *affine-scaling direction*. Many algorithms use intermediate values of σ from the open interval $(0, 1)$ to trade off between the twin goals of reducing μ and improving centrality.

A PRIMAL–DUAL FRAMEWORK

With these basic concepts in hand, we can define a general framework for primal–dual algorithms.

Framework 14.1 (Primal–Dual).
 Given $(x^0, \lambda^0, s^0) \in \mathcal{F}^o$
 for $k = 0, 1, 2, \ldots$
  Solve

$$\begin{bmatrix} 0 & A^T & I \\ A & 0 & 0 \\ S^k & 0 & X^k \end{bmatrix} \begin{bmatrix} \Delta x^k \\ \Delta \lambda^k \\ \Delta s^k \end{bmatrix} = \begin{bmatrix} 0 \\ 0 \\ -X^k S^k e + \sigma_k \mu_k e \end{bmatrix}, \quad (14.12)$$

  where $\sigma_k \in [0, 1]$ and $\mu_k = (x^k)^T s^k / n$;
  Set

$$(x^{k+1}, \lambda^{k+1}, s^{k+1}) = (x^k, \lambda^k, s^k) + \alpha_k(\Delta x^k, \Delta \lambda^k, \Delta s^k), \quad (14.13)$$

  choosing α_k such that $(x^{k+1}, s^{k+1}) > 0$.
 end (for).

The choices of centering parameter σ_k and step length α_k are crucial to the performance of the method. Techniques for controlling these parameters, directly and indirectly, give rise to a wide variety of methods with varying theoretical properties.

So far, we have assumed that the starting point (x^0, λ^0, s^0) is strictly feasible and, in particular, that it satisfies the linear equations $Ax^0 = b$, $A^T \lambda^0 + s^0 = c$. All subsequent iterates also respect these constraints, because of the zero right-hand-side terms in (14.12).

For most problems, however, a strictly feasible starting point is difficult to find. *Infeasible-interior-point methods* require only that the components of x^0 and s^0 be strictly positive. The search direction needs to be modified so that it improves feasibility as well as centrality at each iteration, but this requirement entails only a slight change to the step equation (14.11). If we define the residuals for the two linear equations as

$$r_b = Ax - b, \qquad r_c = A^T \lambda + s - c, \quad (14.14)$$

the modified step equation is

$$\begin{bmatrix} 0 & A^T & I \\ A & 0 & 0 \\ S & 0 & X \end{bmatrix} \begin{bmatrix} \Delta x \\ \Delta \lambda \\ \Delta s \end{bmatrix} = \begin{bmatrix} -r_c \\ -r_b \\ -XSe + \sigma \mu e \end{bmatrix}. \quad (14.15)$$

The search direction is still a Newton step toward the point $(x_{\sigma\mu}, \lambda_{\sigma\mu}, s_{\sigma\mu}) \in \mathcal{C}$. It tries to correct all the infeasibility in the equality constraints in a single step. If a full step is taken at any iteration (that is, $\alpha_k = 1$ for some k), the residuals r_b and r_c become zero, and all subsequent iterates remain strictly feasible.

PATH-FOLLOWING METHODS

Path-following algorithms explicitly restrict the iterates to a neighborhood of the central path $\mathcal{C}$ and follow $\mathcal{C}$ to a solution of the linear program. By preventing the iterates from coming too close to the boundary of the nonnegative orthant, they ensure that search directions calculated from each iterate make at least some minimal amount of progress toward the solution.

A key ingredient of any optimization algorithm is a measure of the desirability of each point in the search space. In path-following algorithms, the duality measure μ defined by (14.10) fills this role. The duality measure μ_k is forced to zero as $k \to \infty$, so the iterates (x^k, λ^k, s^k) come closer and closer to satisfying the KKT conditions (14.3).

The two most interesting neighborhoods of $\mathcal{C}$ are the so-called 2-norm neighborhood $\mathcal{N}_2(\theta)$ defined by

$$\mathcal{N}_2(\theta) = \{(x, \lambda, s) \in \mathcal{F}^o \mid \|XSe - \mu e\|_2 \leq \theta\mu\}, \tag{14.16}$$

for some $\theta \in [0, 1)$, and the one-sided ∞-norm neighborhood $\mathcal{N}_{-\infty}(\gamma)$ defined by

$$\mathcal{N}_{-\infty}(\gamma) = \{(x, \lambda, s) \in \mathcal{F}^o \mid x_i s_i \geq \gamma\mu \text{ all } i = 1, 2, \ldots, n\}, \tag{14.17}$$

for some $\gamma \in (0, 1]$. (Typical values of the parameters are $\theta = 0.5$ and $\gamma = 10^{-3}$.) If a point lies in $\mathcal{N}_{-\infty}(\gamma)$, each pairwise product $x_i s_i$ must be at least some small multiple γ of their average value μ. This requirement is actually quite modest, and we can make $\mathcal{N}_{-\infty}(\gamma)$ encompass most of the feasible region $\mathcal{F}$ by choosing γ close to zero. The $\mathcal{N}_2(\theta)$ neighborhood is more restrictive, since certain points in $\mathcal{F}^o$ do not belong to $\mathcal{N}_2(\theta)$ no matter how close θ is chosen to its upper bound of 1.

The projection of the neighborhood $\mathcal{N}$ onto the space of x variables for a typical problem is shown as the region between the dotted lines in Figure 14.1.

By keeping all iterates inside one or another of these neighborhoods, path-following methods reduce all the pairwise products $x_i s_i$ to zero at more or less the same rate.

Path-following methods are akin to homotopy methods for general nonlinear equations, which also define a path to be followed to the solution. Traditional homotopy methods stay in a tight tubular neighborhood of their path, making incremental changes to the parameter and chasing the homotopy path all the way to a solution. For primal–dual methods, this neighborhood is conical rather than tubular, and it tends to be broad and loose for larger values of the duality measure μ. It narrows as $\mu \to 0$, however, because of the positivity requirement $(x, s) > 0$.

The algorithm we specify below, known as a *long-step path-following* algorithm, can make rapid progress because of its use of the wide neighborhood $\mathcal{N}_{-\infty}(\gamma)$, for γ close to zero. It depends on two parameters $\sigma_{\min}$ and $\sigma_{\max}$, which are upper and lower bounds on the centering parameter σ_k. The search direction is, as usual, obtained by solving (14.12), and we choose the step length α_k to be as large as possible, subject to staying inside $\mathcal{N}_{-\infty}(\gamma)$.

Here and in later analysis, we use the notation

$$(x^k(\alpha), \lambda^k(\alpha), s^k(\alpha)) \stackrel{\text{def}}{=} (x^k, \lambda^k, s^k) + \alpha(\Delta x^k, \Delta \lambda^k, \Delta s^k), \qquad (14.18a)$$

$$\mu_k(\alpha) \stackrel{\text{def}}{=} x^k(\alpha)^T s^k(\alpha)/n. \qquad (14.18b)$$

Algorithm 14.2 (Long-Step Path-Following).
Given $\gamma, \sigma_{\min}, \sigma_{\max}$ with $\gamma \in (0, 1), 0 < \sigma_{\min} < \sigma_{\max} < 1$,
and $(x^0, \lambda^0, s^0) \in \mathcal{N}_{-\infty}(\gamma)$;
for $k = 0, 1, 2, \ldots$
    Choose $\sigma_k \in [\sigma_{\min}, \sigma_{\max}]$;
    Solve (14.12) to obtain $(\Delta x^k, \Delta \lambda^k, \Delta s^k)$;
    Choose α_k as the largest value of α in $[0, 1]$ such that

$$(x^k(\alpha), \lambda^k(\alpha), s^k(\alpha)) \in \mathcal{N}_{-\infty}(\gamma); \qquad (14.19)$$

    Set $(x^{k+1}, \lambda^{k+1}, s^{k+1}) = (x^k(\alpha_k), \lambda^k(\alpha_k), s^k(\alpha_k))$;
end (for).

Typical behavior of the algorithm is illustrated in Figure 14.2 for the case of $n = 2$. The horizontal and vertical axes in this figure represent the pairwise products $x_1 s_1$ and $x_2 s_2$, so the central path $\mathcal{C}$ is the line emanating from the origin at an angle of $45°$. (A point at the origin of this illustration is a primal–dual solution if it also satisfies the feasibility conditions (14.3a), (14.3b), and (14.3d).) In the unusual geometry of Figure 14.2, the search directions $(\Delta x^k, \Delta \lambda^k, \Delta s^k)$ transform to curves rather than straight lines.

As Figure 14.2 shows (and the analysis confirms), the lower bound $\sigma_{\min}$ on the centering parameter ensures that each search direction starts out by moving away from the boundary of $\mathcal{N}_{-\infty}(\gamma)$ and into the relative interior of this neighborhood. That is, small steps along the search direction improve the centrality. Larger values of α take us outside the neighborhood again, since the error in approximating the nonlinear system (14.9) by the linear step equations (14.11) becomes more pronounced as α increases. Still, we are guaranteed that a certain minimum step can be taken before we reach the boundary of $\mathcal{N}_{-\infty}(\gamma)$, as we show in the analysis below.

We provide a complete analysis of Algorithm 14.2 in Section 14.4. It shows that the theory of primal–dual methods can be understood without recourse to profound mathematics. This algorithm is fairly efficient in practice, but with a few more changes it becomes the basis of a truly competitive method, which we discuss in the next section.

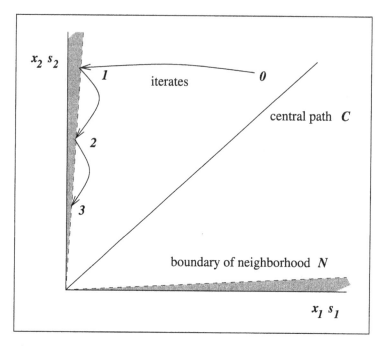

Figure 14.2 Iterates of Algorithm 14.2, plotted in (xs) space.

An infeasible-interior-point variant of Algorithm 14.2 can be constructed by generalizing the definition of $\mathcal{N}_{-\infty}(\gamma)$ to allow violation of the feasibility conditions. In this extended neighborhood, the residual norms $\|r_b\|$ and $\|r_c\|$ are bounded by a constant multiple of the duality measure μ. By squeezing μ to zero, we also force r_b and r_c to zero, so that the iterates approach complementarity and feasibility simultaneously.

14.2 A PRACTICAL PRIMAL–DUAL ALGORITHM

Most existing interior-point codes for general-purpose linear programming problems are based on a predictor–corrector algorithm proposed by Mehrotra [164]. The two key features of this algorithm are

  a. addition of a *corrector step* to the search direction of Framework 14.1, so that the algorithm more closely follows a trajectory to the primal–dual solution set; and

  b. adaptive choice of the centering parameter σ.

The method can be motivated by "shifting" the central path C so that it starts at our current iterate (x, λ, s) and ends, as before, at the set of solution points Ω. We denote this

14.2. A Practical Primal–Dual Algorithm

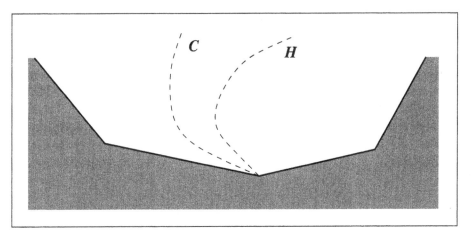

Figure 14.3 Central path C, and a trajectory $\mathcal{H}$ from the current (noncentral) point (x, λ, s) to the solution set Ω.

modified trajectory by $\mathcal{H}$ and parametrize it by the parameter $\tau \in [0, 1)$, so that

$$\mathcal{H} = \{(\hat{x}(\tau), \hat{\lambda}(\tau), \hat{s}(\tau)) \mid \tau \in [0, 1)\},$$

with $(\hat{x}(0), \hat{\lambda}(0), \hat{s}(0)) = (x, \lambda, s)$ and, if the limit exists as $\tau \uparrow 1$, we have

$$\lim_{\tau \uparrow 1} (\hat{x}(\tau), \hat{\lambda}(\tau), \hat{s}(\tau)) \in \Omega.$$

The relationship of $\mathcal{H}$ to the central path C is depicted in Figure 14.3.

Algorithms from Framework 14.1 can be thought of as first-order methods, in that they find the tangent to a trajectory like $\mathcal{H}$ and perform a line search along it. This tangent is known as the *predictor step*. Mehrotra's algorithm takes the next logical step of calculating the *curvature* of $\mathcal{H}$ at the current point, thereby obtaining a second-order approximation to this trajectory. The curvature is used to define the *corrector step*, and it can be obtained at a low marginal cost, since it reuses the matrix factors from the calculation of the predictor step.

The second important feature of Mehrotra's algorithm is that it chooses the centering parameter σ *adaptively*, in contrast to algorithms from Framework 14.1, which assign a value to σ_k prior to calculating the search direction. At each iteration, Mehrotra's algorithm first calculates the affine-scaling direction (the predictor step) and assesses its usefulness as a search direction. If this direction yields a large reduction in μ without violating the positivity condition $(x, s) > 0$, the algorithm concludes that little centering is needed, so it chooses σ close to zero and calculates a centered search direction with this small value. If the affine-scaling direction is not so productive, the algorithm enforces a larger amount of centering by choosing a value of σ closer to 1.

The algorithm thus combines three steps to form the search direction: a *predictor step*, which allows us to determine the centering parameter σ_k, a corrector step using second-order information of the path $\mathcal{H}$ leading toward the solution, and a centering step in which the chosen value of σ_k is substituted in (14.15). We will see that computation of the centered direction and the corrector step can be combined, so adaptive centering does not add further to the cost of each iteration.

In concrete terms, the computation of the search direction $(\Delta x, \Delta \lambda, \Delta s)$ proceeds as follows. First, we calculate the predictor step $(\Delta x^{\text{aff}}, \Delta \lambda^{\text{aff}}, \Delta s^{\text{aff}})$ by setting $\sigma = 0$ in (14.15), that is,

$$\begin{bmatrix} 0 & A^T & I \\ A & 0 & 0 \\ S & 0 & X \end{bmatrix} \begin{bmatrix} \Delta x^{\text{aff}} \\ \Delta \lambda^{\text{aff}} \\ \Delta s^{\text{aff}} \end{bmatrix} = \begin{bmatrix} -r_c \\ -r_b \\ -XSe \end{bmatrix}. \quad (14.20)$$

To measure the effectiveness of this direction, we find $\alpha_{\text{aff}}^{\text{pri}}$ and $\alpha_{\text{aff}}^{\text{dual}}$ to be the longest step lengths that can be taken along this direction before violating the nonnegativity conditions $(x, s) \geq 0$, with an upper bound of 1. Explicit formulae for these values are as follows:

$$\alpha_{\text{aff}}^{\text{pri}} \stackrel{\text{def}}{=} \min\left(1, \min_{i:\Delta x_i^{\text{aff}} < 0} -\frac{x_i}{\Delta x_i^{\text{aff}}}\right), \quad (14.21a)$$

$$\alpha_{\text{aff}}^{\text{dual}} \stackrel{\text{def}}{=} \min\left(1, \min_{i:\Delta s_i^{\text{aff}} < 0} -\frac{s_i}{\Delta s_i^{\text{aff}}}\right). \quad (14.21b)$$

We define μ_{aff} to be the value of μ that would be obtained by a full step to the boundary, that is,

$$\mu_{\text{aff}} = (x + \alpha_{\text{aff}}^{\text{pri}} \Delta x^{\text{aff}})^T (s + \alpha_{\text{aff}}^{\text{pri}} \Delta s^{\text{aff}})/n, \quad (14.22)$$

and set the damping parameter σ to be

$$\sigma = \left(\frac{\mu_{\text{aff}}}{\mu}\right)^3.$$

It is easy to see that this choice has the effect mentioned above: When good progress is made along the predictor direction, we have $\mu_{\text{aff}} \ll \mu$, so the σ obtained from this formula is small; and conversely.

The corrector step is obtained by replacing the right-hand-side in (14.20) by $(0, 0, -\Delta X^{\text{aff}} \Delta S^{\text{aff}} e)$, while the centering step requires a right-hand-side of $(0, 0, \sigma \mu e)$. We can obtain the complete Mehrotra step, which includes the predictor, corrector, and centering step components, by adding the right-hand-sides for these three components and solving the following system:

$$\begin{bmatrix} 0 & A^T & I \\ A & 0 & 0 \\ S & 0 & X \end{bmatrix} \begin{bmatrix} \Delta x \\ \Delta \lambda \\ \Delta s \end{bmatrix} = \begin{bmatrix} -r_c \\ -r_b \\ -XSe - \Delta X^{\text{aff}} \Delta S^{\text{aff}} e + \sigma \mu e \end{bmatrix}. \quad (14.23)$$

We calculate the maximum steps that can be taken along these directions before violating the nonnegativity condition $(x, s) > 0$ by formulae similar to (14.21); namely,

$$\alpha_{\max}^{\text{pri}} \stackrel{\text{def}}{=} \min\left(1, \min_{i:\Delta x_i < 0} -\frac{x_i^k}{\Delta x_i}\right), \quad (14.24a)$$

$$\alpha_{\max}^{\text{dual}} \stackrel{\text{def}}{=} \min\left(1, \min_{i:\Delta s_i < 0} -\frac{s_i^k}{\Delta s_i}\right), \quad (14.24b)$$

and then choose the primal and dual step lengths as follows:

$$\alpha_k^{\text{pri}} = \min(1, \eta \alpha_{\max}^{\text{pri}}), \quad \alpha_k^{\text{dual}} = \min(1, \eta \alpha_{\max}^{\text{dual}}), \quad (14.25)$$

where $\eta \in [0.9, 1.0)$ is chosen so that $\eta \to 1$ near the solution, to accelerate the asymptotic convergence. (For details of the choice of η and other elements of the algorithm such as the choice of starting point (x^0, λ^0, s^0) see Mehrotra [164].)

We summarize this discussion by specifying Mehrotra's algorithm in the usual format.

Algorithm 14.3 (Mehrotra Predictor–Corrector Algorithm).
Given (x^0, λ^0, s^0) with $(x^0, s^0) > 0$;
for $k = 0, 1, 2, \ldots$
    Set $(x, \lambda, s) = (x^k, \lambda^k, s^k)$ and solve (14.20) for $(\Delta x^{\text{aff}}, \Delta \lambda^{\text{aff}}, \Delta s^{\text{aff}})$;
    Calculate $\alpha_{\text{aff}}^{\text{pri}}, \alpha_{\text{aff}}^{\text{dual}}$, and μ_{aff} as in (14.21) and (14.22);
    Set centering parameter to $\sigma = (\mu_{\text{aff}}/\mu)^3$;
    Solve (14.23) for $(\Delta x, \Delta \lambda, \Delta s)$;
    Calculate α_k^{pri} and α_k^{dual} from (14.25);
    Set

$$x^{k+1} = x^k + \alpha_k^{\text{pri}} \Delta x,$$
$$(\lambda^{k+1}, s^{k+1}) = (\lambda^k, s^k) + \alpha_k^{\text{dual}}(\Delta \lambda, \Delta s);$$

end (for).

It is important to note that no convergence theory is available for Mehrotra's algorithm, at least in the form in which it is described above. In fact, there are examples for which the algorithm diverges. Simple safeguards could be incorporated into the method to force it into the convergence framework of existing methods. However, most programs do not implement these safeguards, because the good practical performance of Mehrotra's algorithm makes them unnecessary.

SOLVING THE LINEAR SYSTEMS

Most of the computational effort in primal–dual methods is taken up in solving linear systems such as (14.15), (14.20), and (14.23). The coefficient matrix in these systems is usually large and sparse, since the constraint matrix A is itself large and sparse in most applications. The special structure in the step equations allows us to reformulate them as systems with more compact symmetric coefficient matrices, which are easier and cheaper to factor than the original form.

The reformulation procedures are simple, as we show by applying them to the system (14.15). Since the current point (x, λ, s) has x and s strictly positive, the diagonal matrices X and S are nonsingular. Hence, by eliminating Δs from (14.15), we obtain the following equivalent system:

$$\begin{bmatrix} 0 & A \\ A^T & -D^{-2} \end{bmatrix} \begin{bmatrix} \Delta \lambda \\ \Delta x \end{bmatrix} = \begin{bmatrix} -r_b \\ -r_c + s - \sigma\mu X^{-1}e \end{bmatrix}, \quad (14.26a)$$

$$\Delta s = -s + \sigma\mu X^{-1}e - X^{-1}S\Delta x, \quad (14.26b)$$

where we have introduced the notation

$$D = S^{-1/2}X^{1/2}. \quad (14.27)$$

This form of the step equations usually is known as the *augmented system*. Since the matrix $X^{-1}S$ is also diagonal and nonsingular, we can go a step further, eliminating Δx from (14.26a) to obtain another equivalent form:

$$AD^2A^T \Delta\lambda = -r_b + A(-S^{-1}Xr_c + x - \sigma\mu S^{-1}e), \quad (14.28a)$$

$$\Delta s = -r_c - A^T \Delta\lambda, \quad (14.28b)$$

$$\Delta x = -x + \sigma\mu S^{-1}e - S^{-1}X\Delta s. \quad (14.28c)$$

This form often is called the *normal-equations* form because the system (14.28a) can be viewed as the normal equations (10.11) for a linear least-squares problem with coefficient matrix DA^T.

Most implementations of primal–dual methods are based on the system (14.28). They use direct sparse Cholesky algorithms to factor the matrix AD^2A^T, and then perform triangular solves with the resulting sparse factors to obtain the step $\Delta\lambda$ from (14.28a). The steps Δs and Δx are recovered from (14.28b) and (14.28c). General-purpose sparse Cholesky software can be applied to AD^2A^T, but a few modifications are needed because AD^2A^T

may be ill-conditioned or singular. Ill conditioning is often observed during the final stages of a primal–dual algorithm, when the elements of the diagonal weighting matrix D^2 take on both huge and tiny values.

The formulation (14.26) has received less attention than (14.28), mainly because algorithms and software for factoring sparse symmetric indefinite matrices are more complicated, slower, and less prevalent than sparse Cholesky algorithms. Nevertheless, the formulation (14.26) is cleaner and more flexible than (14.28) in a number of respects. Fill-in of the coefficient matrix in (14.28) caused by dense columns in A does not occur, and free variables (that is, components of x with no explicit lower or upper bounds) can be handled without resorting to the various artificial devices needed in the normal-equations form.

14.3 OTHER PRIMAL–DUAL ALGORITHMS AND EXTENSIONS

OTHER PATH-FOLLOWING METHODS

Framework 14.1 is the basis of a number of other algorithms of the path-following variety. They are less important from a practical viewpoint, but we mention them here because of their elegance and their strong theoretical properties.

Some path-following methods choose conservative values for the centering parameter σ (that is, σ only slightly less than 1) so that unit steps $\alpha = 1$ can be taken along the resulting direction from (14.11) without leaving the chosen neighborhood. These methods, which are known as *short-step* path-following methods, make only slow progress toward the solution because they require the iterates to stay inside a restrictive $\mathcal{N}_2$ neighborhood (14.16).

Better results are obtained with the *predictor–corrector* method, due to Mizuno, Todd, and Ye [165], which uses two $\mathcal{N}_2$ neighborhoods, nested one inside the other. (Despite the similar terminology, this algorithm is quite distinct from Algorithm 14.3 of Section 14.2.) Every second step of this method is a predictor step, which starts in the inner neighborhood and moves along the affine-scaling direction (computed by setting $\sigma = 0$ in (14.11)) to the boundary of the outer neighborhood. The gap between neighborhood boundaries is wide enough to allow this step to make significant progress in reducing μ. Alternating with the predictor steps are corrector steps (computed with $\sigma = 1$ and $\alpha = 1$), which take the next iterate back inside the inner neighborhood in preparation for the subsequent predictor step. The predictor–corrector algorithm produces a sequence of duality measures μ_k that converge superlinearly to zero, in contrast to the linear convergence that characterizes most methods.

POTENTIAL-REDUCTION METHODS

Potential-reduction methods take steps of the same form as path-following methods, but they do not explicitly follow the central path $\mathcal{C}$ and can be motivated independently of it. They use a logarithmic potential function to measure the worth of each point in $\mathcal{F}^o$ and aim to achieve a certain fixed reduction in this function at each iteration. The primal–dual potential function, which we denote generically by Φ, usually has two important properties:

$$\Phi \to \infty \quad \text{if } x_i s_i \to 0 \text{ for some } i, \text{ while } \mu = x^T s / n \not\to 0, \tag{14.29a}$$

$$\Phi \to -\infty \quad \text{if and only if } (x, \lambda, s) \to \Omega. \tag{14.29b}$$

The first property (14.29a) stops any one of the pairwise products $x_i s_i$ from approaching zero independently of the others, and therefore keeps the iterates away from the boundary of the nonnegative orthant. The second property (14.29b) relates Φ to the solution set Ω. If our algorithm forces Φ to $-\infty$, then (14.29b) ensures that the sequence approaches the solution set.

An interesting primal–dual potential function is the Tanabe–Todd–Ye function Φ_ρ defined by

$$\Phi_\rho(x, s) = \rho \log x^T s - \sum_{i=1}^{n} \log x_i s_i, \tag{14.30}$$

for some parameter $\rho > n$ (see [233], [236]). Like all algorithms based on Framework 14.1, potential-reduction algorithms obtain their search directions by solving (14.12), for some $\sigma_k \in (0, 1)$, and they take steps of length α_k along these directions. For instance, the step length α_k may be chosen to approximately minimize Φ_ρ along the computed direction. The "cookbook" value $\sigma_k \equiv n/(n + \sqrt{n})$ is sufficient to guarantee constant reduction in Φ_ρ at every iteration. Hence, Φ_ρ will approach $-\infty$, forcing convergence. Adaptive and heuristic choices of σ_k and α_k are also covered by the theory, provided that they at least match the reduction in Φ_ρ obtained from the conservative theoretical values of these parameters.

EXTENSIONS

Primal–dual methods for linear programming can be extended to wider classes of problems. There are simple extensions of the algorithm to the monotone linear complementarity problem (LCP) and convex quadratic programming problems for which the convergence and polynomial complexity properties of the linear programming algorithms are retained. The LCP is the problem of finding vectors x and s in $\mathbf{R}^n$ that satisfy the following conditions:

$$s = Mx + q, \quad (x, s) \geq 0, \quad x^T s = 0, \tag{14.31}$$

where M is a positive semidefinite $n \times n$ matrix and $q \in \mathbf{R}^n$. The similarity between (14.31) and the KKT conditions (14.3) is obvious: The last two conditions in (14.31) correspond to (14.3d) and (14.3c), respectively, while the condition $s = Mx + q$ is similar to the equations (14.3a) and (14.3b). For practical instances of the problem (14.31), see Cottle, Pang, and Stone [59].

In convex quadratic programming, we minimize a convex quadratic objective subject to linear constraints. A convex quadratic generalization of the standard form linear program (14.1) is

$$\min c^T x + \tfrac{1}{2} x^T G x, \quad \text{subject to } Ax = b, \, x \geq 0, \tag{14.32}$$

where G is a symmetric $n \times n$ positive semidefinite matrix. The KKT conditions for this problem are similar to (14.3) and the linear complementarity problem (14.31). In fact, we can show that any LCP can be formulated as a convex quadratic program, and vice versa. See Section 16.7 for further discussion of interior-point methods for (14.32).

Extensions of primal–dual methods to nonlinear programming problems can be devised by writing down the KKT conditions in a form similar to (14.3), adding slack variables where necessary to convert all the inequality conditions to simple bounds of the type (14.3d). As in Framework 14.1, the basic primal–dual step is found by applying Newton's method to the equality KKT conditions, curtailing the step length to ensure that the bounds are satisfied strictly by all iterates. When the nonlinear programming problem is convex (that is, its objective and constraint functions are all convex functions), global convergence of primal–dual methods is not too difficult to prove. Extensions to general nonlinear programming problems are not so straightforward; we discuss some approaches under investigation in Section 17.2.

Interior-point methods are proving to be effective in *semidefinite programming*, a class of problems in which some of the variables are symmetric matrices that are constrained to be positive semidefinite. This class, which has been the topic of concentrated research since the early 1990s, has applications in many areas, including control theory and combinatorial optimization. Further information on this increasingly important topic can be found in Nesterov and Nemirovskii [181], Boyd et al. [26], and Vandenberghe and Boyd [240].

14.4 ANALYSIS OF ALGORITHM 14.2

We now present a detailed analysis of Algorithm 14.2. As is typical in interior-point methods, the analysis builds from a purely technical lemma to a powerful theorem in just a few pages. We start with the technical lemma—Lemma 14.1—and use it to prove Lemma 14.2, a bound on the vector of pairwise products $\Delta x_i \Delta s_i$, $i = 1, 2, \ldots, n$. Theorem 14.3 finds a lower bound on α_k and a corresponding estimate of the reduction in μ at each iteration. Global convergence is an immediate consequence of this result. Theorem 14.4 goes a step further, proving that $O(n)$ iterations are required to identify a point for which $\mu_k < \epsilon$, for a given tolerance $\epsilon \in (0, 1)$.

Lemma 14.1.
Let u and v be any two vectors in $\mathbf{R}^n$ with $u^T v \geq 0$. Then

$$\|UVe\| \leq 2^{-3/2} \|u + v\|^2,$$

where

$$U = \text{diag}(u_1, u_2, \ldots, u_n), \qquad V = \text{diag}(v_1, v_2, \ldots, v_n).$$

PROOF. First, note that for any two scalars α and β with $\alpha\beta \geq 0$, we have from the algebraic-geometric mean inequality that

$$\sqrt{|\alpha\beta|} \leq \frac{1}{2}|\alpha + \beta|. \tag{14.33}$$

Since $u^T v \geq 0$, we have

$$0 \leq u^T v = \sum_{u_i v_i \geq 0} u_i v_i + \sum_{u_i v_i < 0} u_i v_i = \sum_{i \in \mathcal{P}} |u_i v_i| - \sum_{i \in \mathcal{M}} |u_i v_i|, \tag{14.34}$$

where we partitioned the index set $\{1, 2, \ldots, n\}$ as

$$\mathcal{P} = \{i \mid u_i v_i \geq 0\}, \qquad \mathcal{M} = \{i \mid u_i v_i < 0\}.$$

Now,

$$\begin{aligned}
\|UVe\| &= \left(\|[u_i v_i]_{i\in\mathcal{P}}\|^2 + \|[u_i v_i]_{i\in\mathcal{M}}\|^2\right)^{1/2} \\
&\leq \left(\|[u_i v_i]_{i\in\mathcal{P}}\|_1^2 + \|[u_i v_i]_{i\in\mathcal{M}}\|_1^2\right)^{1/2} \quad \text{since } \|\cdot\|_2 \leq \|\cdot\|_1 \\
&\leq \left(2\|[u_i v_i]_{i\in\mathcal{P}}\|_1^2\right)^{1/2} \quad \text{from (14.34)} \\
&\leq \sqrt{2}\left\|\left[\frac{1}{4}(u_i + v_i)^2\right]_{i\in\mathcal{P}}\right\|_1 \quad \text{from (14.33)} \\
&= 2^{-3/2} \sum_{i\in\mathcal{P}} (u_i + v_i)^2 \\
&\leq 2^{-3/2} \sum_{i=1}^n (u_i + v_i)^2 \\
&\leq 2^{-3/2} \|u + v\|^2,
\end{aligned}$$

completing the proof. $\square$

Lemma 14.2.

If $(x, \lambda, s) \in \mathcal{N}_{-\infty}(\gamma)$, then

$$\|\Delta X \Delta S e\| \leq 2^{-3/2}(1 + 1/\gamma)n\mu.$$

PROOF. It is easy to show using (14.12) that

$$\Delta x^T \Delta s = 0 \tag{14.35}$$

(see the exercises). By multiplying the last block row in (14.12) by $(XS)^{-1/2}$ and using the definition $D = X^{1/2}S^{-1/2}$, we obtain

$$D^{-1}\Delta x + D\Delta s = (XS)^{-1/2}(-XSe + \sigma\mu e). \tag{14.36}$$

Because $(D^{-1}\Delta x)^T(D\Delta s) = \Delta x^T \Delta s = 0$, we can apply Lemma 14.1 with $u = D^{-1}\Delta x$ and $v = D\Delta s$ to obtain

$$\|\Delta X \Delta Se\| = \|(D^{-1}\Delta X)(D\Delta S)e\|$$
$$\leq 2^{-3/2}\|D^{-1}\Delta x + D\Delta s\|^2 \quad \text{from Lemma 14.1}$$
$$= 2^{-3/2}\|(XS)^{-1/2}(-XSe + \sigma\mu e)\|^2 \quad \text{from (14.36)}.$$

Expanding the squared Euclidean norm and using such relationships as $x^T s = n\mu$ and $e^T e = n$, we obtain

$$\|\Delta X \Delta Se\| \leq 2^{-3/2}\left[x^T s - 2\sigma\mu e^T e + \sigma^2\mu^2 \sum_{i=1}^n \frac{1}{x_i s_i}\right]$$
$$\leq 2^{-3/2}\left[x^T s - 2\sigma\mu e^T e + \sigma^2\mu^2 \frac{n}{\gamma\mu}\right] \quad \text{since } x_i s_i \geq \gamma\mu$$
$$\leq 2^{-3/2}\left[1 - 2\sigma + \frac{\sigma^2}{\gamma}\right]n\mu$$
$$\leq 2^{-3/2}(1 + 1/\gamma)n\mu,$$

as claimed. $\square$

Theorem 14.3.
Given the parameters γ, $\sigma_{\min}$, and $\sigma_{\max}$ in Algorithm 14.2, there is a constant δ independent of n such that

$$\mu_{k+1} \leq \left(1 - \frac{\delta}{n}\right)\mu_k, \tag{14.37}$$

for all $k \geq 0$.

PROOF. We start by proving that

$$\left(x^k(\alpha), \lambda^k(\alpha), s^k(\alpha)\right) \in \mathcal{N}_{-\infty}(\gamma) \text{ for all } \alpha \in \left[0, 2^{3/2}\gamma\frac{1-\gamma}{1+\gamma}\frac{\sigma_k}{n}\right], \tag{14.38}$$

where $(x^k(\alpha), \lambda^k(\alpha), s^k(\alpha))$ is defined as in (14.18). It follows that α_k is bounded below as follows:

$$\alpha_k \geq 2^{3/2} \frac{\sigma_k}{n} \gamma \frac{1-\gamma}{1+\gamma}. \tag{14.39}$$

For any $i = 1, 2, \ldots, n$, we have from Lemma 14.2 that

$$|\Delta x_i^k \Delta s_i^k| \leq \|\Delta X^k \Delta S^k e\|_2 \leq 2^{-3/2}(1+1/\gamma)n\mu_k. \tag{14.40}$$

Using (14.12), we have from $x_i^k s_i^k \geq \gamma \mu_k$ and (14.40) that

$$\begin{aligned} x_i^k(\alpha)s_i^k(\alpha) &= \left(x_i^k + \alpha \Delta x_i^k\right)\left(s_i^k + \alpha \Delta s_i^k\right) \\ &= x_i^k s_i^k + \alpha\left(x_i^k \Delta s_i^k + s_i^k \Delta x_i^k\right) + \alpha^2 \Delta x_i^k \Delta s_i^k \\ &\geq x_i^k s_i^k(1-\alpha) + \alpha \sigma_k \mu_k - \alpha^2|\Delta x_i^k \Delta s_i^k| \\ &\geq \gamma(1-\alpha)\mu_k + \alpha \sigma_k \mu_k - \alpha^2 2^{-3/2}(1+1/\gamma)n\mu_k. \end{aligned}$$

By summing the n components of the equation $S^k \Delta x^k + X^k \Delta s^k = -X^k S^k e + \sigma_k \mu_k e$ (the third block row from (14.12)), and using (14.35) and the definition of μ_k and $\mu_k(\alpha)$ (see (14.18)), we obtain

$$\mu_k(\alpha) = (1 - \alpha(1 - \sigma_k))\mu_k.$$

From these last two formulas, we can see that the proximity condition

$$x_i^k(\alpha)s_i^k(\alpha) \geq \gamma \mu_k(\alpha)$$

is satisfied, provided that

$$\gamma(1-\alpha)\mu_k + \alpha\sigma_k\mu_k - \alpha^2 2^{-3/2}(1+1/\gamma)n\mu_k \geq \gamma(1-\alpha+\alpha\sigma_k)\mu_k.$$

Rearranging this expression, we obtain

$$\alpha\sigma_k\mu_k(1-\gamma) \geq \alpha^2 2^{-3/2} n \mu_k(1+1/\gamma),$$

which is true if

$$\alpha \leq \frac{2^{3/2}}{n}\sigma_k\gamma\frac{1-\gamma}{1+\gamma}.$$

We have proved that $(x^k(\alpha), \lambda^k(\alpha), s^k(\alpha))$ satisfies the proximity condition for $\mathcal{N}_{-\infty}(\gamma)$ when α lies in the range stated in (14.38). It is not difficult to show that $(x^k(\alpha), \lambda^k(\alpha), s^k(\alpha)) \in$

$\mathcal{F}^o$ for all α in the given range (see the exercises). Hence, we have proved (14.38) and therefore (14.39).

We complete the proof of the theorem by estimating the reduction in μ on the kth step. Because of (14.35), (14.39), and the last block row of (14.11), we have

$$\mu_{k+1} = x^k(\alpha_k)^T s^k(\alpha_k)/n$$
$$= \left[(x^k)^T s^k + \alpha_k \left((x^k)^T \Delta s^k + (s^k)^T \Delta x^k\right) + \alpha_k^2 (\Delta x^k)^T \Delta s^k\right]/n \quad (14.41)$$
$$= \mu_k + \alpha_k \left(-(x^k)^T s^k/n + \sigma_k \mu_k\right) \quad (14.42)$$
$$= (1 - \alpha_k(1 - \sigma_k))\mu_k \quad (14.43)$$
$$\leq \left(1 - \frac{2^{3/2}}{n}\gamma\frac{1-\gamma}{1+\gamma}\sigma_k(1-\sigma_k)\right)\mu_k. \quad (14.44)$$

Now, the function $\sigma(1 - \sigma)$ is a concave quadratic function of σ, so on any given interval it attains its minimum value at one of the endpoints. Hence, we have

$$\sigma_k(1 - \sigma_k) \geq \min\{\sigma_{\min}(1 - \sigma_{\min}), \sigma_{\max}(1 - \sigma_{\max})\}, \quad \text{for all } \sigma_k \in [\sigma_{\min}, \sigma_{\max}].$$

The proof is completed by substituting this estimate into (14.44) and setting

$$\delta = 2^{3/2}\gamma\frac{1-\gamma}{1+\gamma}\min\{\sigma_{\min}(1-\sigma_{\min}), \sigma_{\max}(1-\sigma_{\max})\}. \qquad \square$$

We conclude with the complexity result.

Theorem 14.4.
Given $\epsilon > 0$ and $\gamma \in (0, 1)$, suppose the starting point $(x^0, \lambda^0, s^0) \in \mathcal{N}_{-\infty}(\gamma)$ in Algorithm 14.2 has

$$\mu_0 \leq 1/\epsilon^\kappa \quad (14.45)$$

for some positive constant κ. Then there is an index K with $K = O(n \log 1/\epsilon)$ such that

$$\mu_k \leq \epsilon, \quad \text{for all } k \geq K.$$

PROOF. By taking logarithms of both sides in (14.37), we obtain

$$\log \mu_{k+1} \leq \log\left(1 - \frac{\delta}{n}\right) + \log \mu_k.$$

By repeatedly applying this formula and using (14.45), we have

$$\log \mu_k \leq k \log\left(1 - \frac{\delta}{n}\right) + \log \mu_0 \leq k \log\left(1 - \frac{\delta}{n}\right) + \kappa \log\frac{1}{\epsilon}.$$

The following well-known estimate for the log function,

$$\log(1+\beta) \leq \beta, \quad \text{for all } \beta > -1,$$

implies that

$$\log \mu_k \leq k\left(-\frac{\delta}{n}\right) + \kappa \log \frac{1}{\epsilon}.$$

Therefore, the convergence criterion $\mu_k \leq \epsilon$ is satisfied if we have

$$k\left(-\frac{\delta}{n^\omega}\right) + \kappa \log \frac{1}{\epsilon} \leq \log \epsilon.$$

This inequality holds for all k that satisfy

$$k \geq K = (1+\kappa)\frac{n^\omega}{\delta}\log \frac{1}{\epsilon},$$

so the proof is complete. □

NOTES AND REFERENCES

For more details on the material of this chapter, see the book by Wright [255].

As noted earlier, Karmarkar's method arose from a search for linear programming algorithms with better worst-case behavior than the simplex method. The first algorithm with polynomial complexity, Khachiyan's ellipsoid algorithm [142], was a computational disappointment, but the execution times required by Karmarkar's method were not too much greater than simplex codes at the time of its introduction, particularly for large linear programs. Karmarkar's is a *primal* algorithm; that is, it is described, motivated, and implemented purely in terms of the primal problem (14.1) without reference to the dual. At each iteration, Karmarkar's algorithm performs a projective transformation on the primal feasible set that maps the current iterate x^k to the center of the set and takes a step in the feasible steepest descent direction for the transformed space. Progress toward optimality is measured by a logarithmic potential function. Nice descriptions of the algorithm can be found in Karmarkar's original paper [140] and in Fletcher [83].

Karmarkar's method falls outside the scope of this chapter, and in any case, its practical performance does not appear to match the most efficient primal–dual methods. The algorithms we discussed in this chapter—path-following, potential-reduction—have polynomial complexity, like Karmarkar's method.

Historians of interior-point methods point out that many of the ideas that have been examined since 1984 actually had their genesis in three works that preceded Karmarkar's paper. The first of these is the book of Fiacco and McCormick [79] on logarithmic barrier functions, which proves existence of the central path, among many other results. Further analysis of the central path was carried out by McLinden [162], in the context of nonlinear complementarity problems. Finally, there is Dikin's paper [72], in which an interior-point method known as primal affine-scaling was originally proposed. The outburst of research on primal–dual methods, which culminated in the efficient software packages available today, dates to the seminal paper of Megiddo [163].

Todd gives an excellent survey of potential reduction methods in [235]. He relates the primal–dual potential reduction method mentioned above to pure primal potential reduction methods, including Karmarkar's original algorithm, and discusses extensions to special classes of nonlinear problems.

For an introduction to complexity theory and its relationship to optimization, see the book by Vavasis [241].

Interior-point software for linear programming is now widely available. Most of the implementations are based on Algorithm 14.3, with additional "higher-order correction" steps proposed by Gonzdio [116]. Because this software is typically less complex than simplex-based codes, some of it is freely available for research and even for commercial purposes. See the web page for this book for further information.

✐ EXERCISES

✐ 14.1 This exercise illustrates the fact that the bounds $(x, s) \geq 0$ are essential in relating solutions of the system (14.4a) to solutions of the linear program (14.1) and its dual. Consider the following linear program in $\mathbf{R}^2$:

$$\min x_1, \quad \text{subject to } x_1 + x_2 = 1, \quad (x_1, x_2) \geq 0.$$

Show that the primal–dual solution is

$$x^* = \begin{pmatrix} 0 \\ 1 \end{pmatrix}, \quad \lambda^* = 0, \quad s^* = \begin{pmatrix} 1 \\ 0 \end{pmatrix}.$$

Also verify that the system $F(x, \lambda, s) = 0$ has the spurious solution

$$x = \begin{pmatrix} 1 \\ 0 \end{pmatrix}, \quad \lambda = 1, \quad s = \begin{pmatrix} 0 \\ -1 \end{pmatrix},$$

which has no relation to the solution of the linear program.

14.2

(i) Show that $\mathcal{N}_2(\theta_1) \subset \mathcal{N}_2(\theta_2)$ when $0 \leq \theta_1 < \theta_2 < 1$ and that $\mathcal{N}_{-\infty}(\gamma_1) \subset \mathcal{N}_{-\infty}(\gamma_2)$ for $0 < \gamma_2 \leq \gamma_1 \leq 1$.

(ii) Show that $\mathcal{N}_2(\theta) \subset \mathcal{N}_{-\infty}(\gamma)$ if $\gamma \leq 1 - \theta$.

14.3
Given an arbitrary point $(x, \lambda, s) \in \mathcal{F}^o$, find the range of γ values for which $(x, \lambda, s) \in \mathcal{N}_{-\infty}(\gamma)$. (The range depends on x and s.)

14.4
For $n = 2$, find a point $(x, s) > 0$ for which the condition

$$\|XSe - \mu e\|_2 \leq \theta \mu$$

is *not* satisfied for any $\theta \in [0, 1)$.

14.5
Prove that the neighborhoods $\mathcal{N}_{-\infty}(1)$ (see (14.17)) and $\mathcal{N}_2(0)$ (see (14.16)) coincide with the central path $\mathcal{C}$.

14.6
Show that Φ_ρ defined by (14.30) has the property (14.29a).

14.7
Prove that the coefficient matrix in (14.11) is nonsingular if and only if A has full row rank.

14.8
Given $(\Delta x, \Delta \lambda, \Delta s)$ satisfying (14.12), prove (14.35).

14.9
Given that X and S are diagonal with positive diagonal elements, show that the coefficient matrix in (14.28a) is symmetric and positive definite if and only if A has full row rank. Does this result continue to hold if we replace D by a diagonal matrix in which exactly m of the diagonal elements are positive and the remainder are zero? (Here m is the number of rows of A.)

14.10
Given a point (x, λ, s) with $(x, s) > 0$, consider the trajectory $\mathcal{H}$ defined by

$$F\left(\hat{x}(\tau), \hat{\lambda}(\tau), \hat{s}(\tau)\right) = \begin{bmatrix} (1-\tau)(A^T\lambda + s - c) \\ (1-\tau)(Ax - b) \\ (1-\tau)XSe \end{bmatrix}, \quad (\hat{x}(\tau), \hat{s}(\tau)) \geq 0,$$

for $\tau \in [0, 1]$, and note that $\left(\hat{x}(0), \hat{\lambda}(0), \hat{s}(0)\right) = (x, \lambda, s)$, while the limit of $\left(\hat{x}(\tau), \hat{\lambda}(\tau), \hat{s}(\tau)\right)$ as $\tau \uparrow 1$ will lie in the primal–dual solution set of the linear program. Find equations for the first, second, and third derivatives of $\mathcal{H}$ with respect to τ at $\tau = 0$. Hence, write down a Taylor series approximation to $\mathcal{H}$ near the point (x, λ, s).

14.11
Consider the following linear program, which contains "free variables" denoted by y:

$$\min c^T x + d^T y, \quad \text{subject to } A_1 x + A_2 y = b, x \geq 0.$$

By introducing Lagrange multipliers λ for the equality constraints and s for the bounds $x \geq 0$, write down optimality conditions for this problem in an analogous fashion to (14.3). Following (14.4) and (14.11), use these conditions to derive the general step equations for a primal–dual interior-point method. Express these equations in augmented system form analogously to (14.26) and explain why it is not possible to reduce further to a formulation like (14.28) in which the coefficient matrix is symmetric positive definite.

✏ **14.12** Program Algorithm 14.3 in Matlab. Choose $\eta = 0.99$ uniformly in (14.25). Test your code on a linear programming problem (14.1) generated by choosing A randomly, and then setting $x, s, b,$ and c as follows:

$$x_i = \begin{cases} \text{random positive number} & i = 1, 2, \ldots, m, \\ 0 & i = m+1, m+2, \ldots, n, \end{cases}$$

$$s_i = \begin{cases} \text{random positive number} & i = m+1, m+2, \ldots, n \\ 0 & i = 1, 2, \ldots, m, \end{cases}$$

$\lambda = $ random vector,

$c = A^T \lambda + s,$

$b = Ax.$

Choose the starting point (x^0, λ^0, s^0) with the components of x^0 and s^0 set to large positive values.

CHAPTER 15

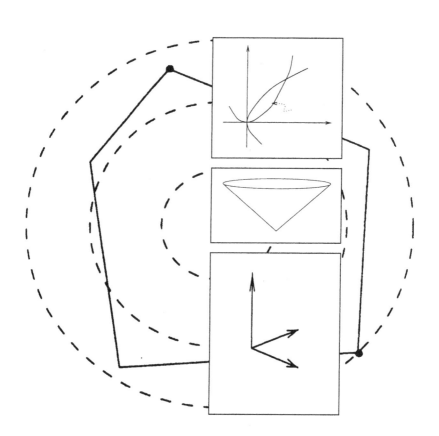

Fundamentals of Algorithms for Nonlinear Constrained Optimization

We now begin our discussion of algorithms for solving the general constrained optimization problem

$$\min_{x \in \mathbf{R}^n} f(x) \quad \text{subject to} \quad \begin{aligned} c_i(x) &= 0, \quad i \in \mathcal{E}, \\ c_i(x) &\geq 0, \quad i \in \mathcal{I}, \end{aligned} \tag{15.1}$$

where the objective function f and the constraint functions c_i are all smooth, real-valued functions on a subset of $\mathbf{R}^n$, and $\mathcal{I}$ and $\mathcal{E}$ are finite index sets of inequality and equality constraints, respectively. In Chapter 12 we used this general statement of the problem to derive optimality conditions that characterize its solutions. This theory is useful for motivating algorithms, but to be truly efficient, the algorithm must take account of the particular properties and structure of the objective and constraint functions. There are many important special cases of (15.1) for which specialized algorithms are available. They include the following:

Linear programming, where the objective function f and all the constraints c_i are linear functions. This problem can be solved by the techniques of Chapters 13 and 14.

Quadratic programming, where the constraints c_i are linear and the objective function f is quadratic. Algorithms for solving this problem are discussed in Chapter 16.

Nonlinear programming, where at least some of the constraints c_i are general nonlinear functions.

Linearly constrained optimization where, as the name suggests, all the constraints c_i are linear.

Bound-constrained optimization, where the only constraints in the problem have the form $x_i \geq l_i$ or $x_i \leq u_i$, where l_i and u_i are lower and upper bounds on the ith component of x.

Convex programming, where the objective function f is convex, the equality constraints $c_i(x) = 0$, $i \in \mathcal{E}$, are linear, and the inequality constraint functions $c_i(x)$, $i \in \mathcal{I}$, are concave.

These categories are neither mutually exclusive nor exhaustive, and some of the classes can be divided into important subclasses. For instance, convex quadratic programming is a subclass of quadratic programming in which the objective function is convex. This finer characterization is relevant to the discussion of algorithms; for example, it is easier to choose a merit function if the problem is convex.

The constrained optimization algorithms described in the following chapters are iterative in nature. They generate a sequence of guesses for the solution x^* that, we hope, tend towards a solution. They may also generate a sequence of guesses for the Lagrange multipliers associated with the constraints. In deciding how to move from one iterate to the next, the methods use information about the objective and constraint functions and their derivatives, possibly combined with information gathered at earlier iterations and earlier stages of the algorithm. They terminate when they have either identified an approximate solution or when further progress seems to be impossible.

In this chapter we discuss some of the fundamental issues in the design of algorithms for nonlinear constrained optimization problems. As in the chapters on unconstrained optimization, we will only study algorithms for finding local minimizers for (15.1); the problem of finding a global minimizer is outside the scope of this book.

INITIAL STUDY OF A PROBLEM

Before solving a constrained optimization problem by means of one of the algorithms described in this book, it is useful first to study the problem to see whether a simplification is possible. In some cases, it is possible to find a solution without use of a computer. For example, an examination of the constraints may show that the feasible region is empty or that the objective function is not bounded in the feasible region (see Exercise 1).

It may also be possible to solve the KKT conditions (12.30) directly, as was done in some examples in Chapter 12, by guessing which of the inequality constraints are active at the solution. Knowledge of the active constraints reduces the KKT conditions to a system of equations that can be solved directly. This approach, however, is rarely practical. Even if the active constraints can be identified—normally the most difficult issue facing a constrained optimization algorithm—we would still need to solve the resulting system of equations numerically. As we see in Chapter 11, algorithms for nonlinear equations are not guaranteed to find a solution from arbitrary starting points. Therefore, we need nonlinear optimization algorithms that directly address general problems of the form (15.1).

Once we are ready to solve the problem using an optimization algorithm, we must decide under which of the categories listed above it falls. If the problem contains discrete variables (for instance, binary variables for which the only permissible values are 0 and 1), we cannot use the techniques of this book but must resort to discrete optimization algorithms. See the algorithms described in Wolsey [249] and Nemhauser and Wolsey [179].

Attention should also be given to the nature of the constraints. Some may be considered "hard" and others "soft" constraints. From the algorithmic point of view, hard constraints are those that must be satisfied in order that the functions and constraints in (15.1) be meaningful. Some of these functions may not even be defined at infeasible points. For instance, a variable is constrained to be positive because its square root is required in the calculation of the objective function. Another example is a problem in which all the variables must sum to zero to satisfy some conservation law.

Constrained optimization problems with soft constraints are sometimes recast by the modeler as an unconstrained problem in which a penalty term including the constraints is added to the objective function. As we will see in the next chapter, this penalty approach usually introduces ill-conditioning, which may or may not be harmful depending on the algorithm used for the unconstrained optimization. The user of optimization algorithms must decide whether it is preferable to use one of the approaches in which the constraints are treated explicitly or whether a penalty approach is adequate.

For problems with hard constraints that must be satisfied at all iterates, we must use *feasible algorithms*. Usually not all the constraints are hard, and therefore these algorithms choose an initial point that satisfies the hard constraints and produce a new iterate that is also feasible for these constraints. Feasible algorithms are usually slower and more expensive than algorithms that allow the iterates to be infeasible, since they cannot follow shortcuts to the solution that cross infeasible territory. However, they have the advantage that the objective function f can be used to judge the merit of each point. Since the constraints are always satisfied, there is no need to introduce a more complex merit function that takes account of the constraint violations.

15.1 CATEGORIZING OPTIMIZATION ALGORITHMS

We now outline the ideas presented in the rest of the book. There is no "standard taxonomy" for nonlinear optimization algorithms; in the remaining chapters we have grouped the various approaches as follows.

❚ In Chapter 17 we discuss *penalty, barrier,* and *augmented Lagrangian* methods, and also provide a brief description of *sequential linearly constrained* methods. We now give a brief overview of each of these approaches.

By combining the objective function and constraints into a *penalty function* we can attack problem (15.1) by solving a sequence of unconstrained problems. For example, if only equality constraints are present in (15.1), we can define the penalty function as

$$f(x) + \frac{1}{2\mu} \sum_{i \in \mathcal{E}} c_i^2(x), \qquad (15.2)$$

where $\mu > 0$ is referred to as a *penalty parameter*. We then minimize this unconstrained function, for a series of increasing values of μ, until the solution of the constrained optimization problem is identified to sufficient accuracy.

If we use an *exact* penalty function, it may be possible to solve (15.1) with one single unconstrained minimization. For the equality constrained problem, an exact penalty function is obtained by setting

$$f(x) + \frac{1}{\mu} \sum_{i \in \mathcal{E}} |c_i(x)|,$$

for some sufficiently small (but positive) choice of μ. Often, however, exact penalty functions are nondifferentiable, and to minimize them requires the solution of a sequence of subproblems.

In *barrier methods*, we add terms to the objective that are insignificant when x is safely in the interior of the feasible set but approach zero as x approaches the boundary. For instance, if only inequality constraints are present in (15.1), the logarithmic barrier function has the form

$$f(x) - \mu \sum_{i \in \mathcal{I}} \log c_i(x),$$

where $\mu > 0$ is now referred to as a *barrier parameter*. The minimizers of this function can be shown to approach solutions of the original constrained problem as $\mu \downarrow 0$, under certain conditions. Again, the usual strategy is to find approximate minimizers for a decreasing sequence of values of μ.

In *augmented Lagrangian methods*, we define a function that combines the properties of the Lagrangian function (12.28) and the quadratic penalty function (15.2). This so-called augmented Lagrangian function has the following form, when only equality constraints are present in the problem (15.1):

$$\mathcal{L}_A(x, \lambda; \mu) = f(x) - \sum_{i \in \mathcal{E}} \lambda_i c_i(x) + \frac{1}{2\mu} \sum_{i \in \mathcal{E}} c_i^2(x).$$

Methods based on this function proceed by fixing λ to some estimate of the optimal Lagrange multipliers and $\mu > 0$ to some positive value, and then finding a value of x that approximately minimizes $\mathcal{L}_A$. This new x iterate is then used to update λ, μ may be decreased, and the process is repeated. We show later that this approach avoids certain numerical difficulties associated with the minimization of the penalty and barrier functions discussed above.

In *sequential linearly constrained* methods we minimize, at every iteration, a certain Lagrangian function subject to a linearization of the constraints. These methods have been used mainly to solve large problems.

II In Chapter 18 we describe methods based on *sequential quadratic programming*. Here the idea is to model (15.1) by a quadratic subproblem at each iterate and to define the search direction as the solution of this subproblem. More specifically, in the case of equality constraints in (15.1), we define the search direction p_k at the iterate (x_k, λ_k) to be the solution of

$$\min_p \tfrac{1}{2} p^T W_k p + \nabla f_k^T p \qquad (15.3\text{a})$$

$$\text{subject to} \quad A_k p + c_k = 0. \qquad (15.3\text{b})$$

The objective in this subproblem is an approximation of the Lagrangian function and the constraints are linearizations of the constraints in (15.1). The new iterate is obtained by searching along this direction until a certain *merit function* is decreased.

Sequential quadratic programming methods have proved to be effective in practice. They are the basis of some of the best software for solving both small and large constrained optimization problems. They typically require fewer function evaluations than some of the other methods, at the expense of solving a relatively complicated quadratic subproblem at each iteration.

III In Chapter 16 we consider algorithms for solving *quadratic programming problems*. We consider this category separately because of its importance and because algorithms can be tailored to its particular characteristics. We discuss active set and interior-point methods. Active set quadratic programming methods are the basis for the sequential quadratic programming methods mentioned above.

The algorithms in categories II and III make use of elimination techniques for the constraints. As a background to those algorithms we now discuss this topic. Later in this chapter we discuss merit functions, which are important components of sequential quadratic programming methods and some other algorithms. These discussions will set the stage for our study of practical algorithms in the remaining chapters of the book.

A Study Note: The concepts that follow constitute background material. The reader may wish to peruse only the next two sections and return to them as needed during study of Chapters 17 and 18.

15.2 ELIMINATION OF VARIABLES

A natural approach for dealing with constrained optimization problems is to attempt to eliminate the constraints so as to obtain an unconstrained problem, or at least to eliminate some constraints and obtain a simpler problem. Elimination techniques must be used with care, however, because they may alter the problem or introduce ill-conditioning.

We begin with an example in which it is safe and convenient to apply elimination of variables. In the problem

$$\min f(x) = f(x_1, x_2, x_3, x_4) \quad \text{subject to} \quad \begin{aligned} x_1 + x_3^2 - x_4 x_5 &= 0, \\ -x_2 + x_4 + x_3^2 &= 0, \end{aligned}$$

there is no risk in setting

$$x_1 = x_4 x_5 - x_3^2, \qquad x_2 = x_4 + x_3^2,$$

and minimizing the unconstrained function of two variables

$$h(x_3, x_4) = f(x_4 x_3 - x_3^2, x_4 + x_3^2, x_3, x_4).$$

This new problem can be solved by any means of one of the unconstrained algorithms described in earlier chapters.

The dangers of nonlinear elimination are illustrated in the following example.

❑ **EXAMPLE 15.1** (FLETCHER [83])

Consider the problem

$$\min x^2 + y^2 \quad \text{subject to } (x-1)^3 = y^2.$$

The contours of the objective function and the constraints are illustrated in Figure 15.1, which shows that the solution is $(x, y) = (1, 0)$.

We attempt to solve this problem by eliminating y. By doing so, we obtain

$$h(x) = x^2 + (x-1)^3.$$

Clearly, $h(x) \to -\infty$ as $x \to -\infty$. By blindly applying this transformation we may conclude that the problem is in fact unbounded, but this view ignores the fact that the constraint $(x-1)^3 = y^2$ implicitly imposes the bound $x \geq 1$ that is active at the solution. This

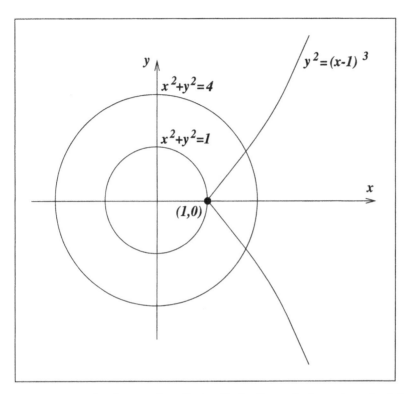

Figure 15.1 The danger of nonlinear elimination—the lower branch of the curve should be excluded from the feasible set

bound should therefore be explicitly introduced into the problem if we are to perform the elimination of y. ☐

This example shows that elimination of nonlinear equations may result in errors that can be difficult to trace. For this reason, nonlinear elimination is not used by most optimization algorithms. Instead, many algorithms linearize the constraints and apply elimination techniques to the simplified problem. We will now describe systematic procedures for performing this elimination of linear constraints.

SIMPLE ELIMINATION FOR LINEAR CONSTRAINTS

Let us consider the minimization of a nonlinear function subject to a set of linear equality constraints,

$$\min\ f(x) \quad \text{subject to } Ax = b, \tag{15.4}$$

where A is an $m \times n$ matrix with $m \leq n$. Suppose for simplicity that A has full row rank. (If such is not the case, we find either that the problem is inconsistent, or that some of the constraints are redundant and can be deleted without affecting the solution of the problem.) Under this assumption, we can find a subset of m columns of A that is linearly independent. If we gather these columns into an $m \times m$ matrix B, and define an $n \times n$ permutation matrix P that swaps these columns to the first m column positions in A, we can write

$$AP = [B \mid N], \tag{15.5}$$

where N denotes the $n-m$ remaining columns of A. (The notation here is consistent with that of Chapter 13, where we discussed similar concepts in the context of linear programming.) We define the subvectors $x_B \in \mathbf{R}^m$ and $x_N \in \mathbf{R}^{n-m}$ in such a way that

$$P^T x = \begin{bmatrix} x_B \\ x_N \end{bmatrix}, \tag{15.6}$$

and call x_B the *basic variables* and B the *basis matrix*. Noting that $PP^T = I$, we can rewrite the constraint $Ax = b$ as

$$b = Ax = AP(P^T x) = B x_B + N x_N.$$

From this formula we deduce that the basic variables are given by

$$x_B = B^{-1} b - B^{-1} N x_N. \tag{15.7}$$

We can therefore compute a feasible point for the constraints $Ax = b$ by choosing *any* value of x_N, and then setting x_B according to the formula (15.7). The problem (15.4) is therefore equivalent to the unconstrained problem

$$\min_{x_N} h(x_N) \stackrel{\text{def}}{=} f\left(P \begin{bmatrix} B^{-1} b - B^{-1} N x_N \\ x_N \end{bmatrix} \right). \tag{15.8}$$

We refer to (15.7) as *simple elimination of variables*.

This discussion shows that a nonlinear optimization problem with linear equality constraints is, from a mathematical point of view, the same as an unconstrained problem.

❑ **EXAMPLE 15.2**

Consider the problem

$$\min \sin(x_1 + x_2) + x_3^2 + \frac{1}{3}(x_4 + x_5^4 + x_6/2) \tag{15.9}$$

15.2. ELIMINATION OF VARIABLES

subject to

$$8x_1 - 6x_2 + x_3 + 9x_4 + 4x_5 = 6$$
$$3x_1 + 2x_2 - x_4 + 6x_5 + 4x_6 = -4. \tag{15.10}$$

By defining the permutation matrix P so as to reorder the components of x as $x^T = (x_3, x_6, x_1, x_2, x_4, x_5)^T$, we find that the coefficient matrix AP is

$$AP = \begin{bmatrix} 1 & 0 & | & 8 & -6 & 9 & 4 \\ 0 & 4 & | & 3 & 2 & -1 & 6 \end{bmatrix}.$$

The basis matrix B is diagonal, and therefore easy to invert. We obtain from (15.7) that

$$\begin{bmatrix} x_3 \\ x_6 \end{bmatrix} = -\begin{bmatrix} 8 & -6 & 9 & 4 \\ \frac{3}{4} & \frac{1}{2} & \frac{-1}{4} & \frac{3}{2} \end{bmatrix} \begin{bmatrix} x_1 \\ x_2 \\ x_4 \\ x_5 \end{bmatrix} + \begin{bmatrix} 6 \\ -1 \end{bmatrix}. \tag{15.11}$$

By substituting for x_3 and x_6 in (15.9), the problem becomes

$$\min_{x_1, x_2, x_4, x_5} \sin(x_1 + x_2) + (8x_1 - 6x_2 + 9x_4 + 4x_5 - 6)^2 \tag{15.12}$$

$$+ \frac{1}{3}(x_4 + x_5^4 - [(1/2) + (3/8)x_1 + (1/4)x_2 - (1/8)x_4 + (3/4)x_5]).$$

We could have chosen two other columns of the coefficient matrix A (that is, two variables other than x_3 and x_6) as the basis for elimination in the system (15.10). For those choices, however, the matrix $B^{-1}N$ would have been more complicated. □

Selecting a set of m independent columns can be done, in general, by means of Gaussian elimination. In the parlance of linear algebra, we can compute the row echelon form of the matrix and choose the pivot columns as the columns of the basis B. Ideally, we would like B to be easy to factorize and well conditioned. For this purpose, we can use a sparse Gaussian elimination algorithm that attempts to preserve sparsity while keeping rounding errors under control. A well-known implementation of this algorithm is MA48 from the Harwell library [133]. As we discuss below, however, there is no guarantee that the Gaussian elimination process will identify the best choice of basis matrix.

There is an interesting interpretation of the elimination-of-variables approach that we have just described. To simplify the notation, we will assume from now on that the coefficient matrix is already given to us so that the basic columns appear in the first m positions, that is, $P = I$. (It is straightforward, but cumbersome, to adapt the arguments that follow to the case of $P \neq I$.)

From (15.6) and (15.7) we see that any feasible point x for the linear constraints in (15.4) can be written as

$$x = Yb + Zx_N, \qquad (15.13)$$

where

$$Y = \begin{bmatrix} B^{-1} \\ 0 \end{bmatrix}, \quad Z = \begin{bmatrix} -B^{-1}N \\ I \end{bmatrix}. \qquad (15.14)$$

Note that Z has $n - m$ linearly independent columns (due to the presence of the identity matrix in the lower block) and that it satisfies $AZ = 0$. Therefore, Z is a *basis for the null space* of A. In addition, the columns of Y and the columns of Z form a linearly independent set, which implies that Y is a *basis for the range space* of A^T. We note also from (15.14),(15.5) that Yb is a particular solution of the linear constraints $Ax = b$.

In other words, the simple elimination technique expresses feasible points as the sum of a particular solution of $Ax = b$, the first term in (15.13), plus a displacement along the null space (or tangent) space of the constraints—the second term in (15.13). The relations (15.13), (15.14) indicate that the particular Yb solution is obtained by holding $n - m$ components of x at zero while relaxing the other m components until they reach the constraints; see Figure 15.2. The displacement Yb is sometimes known as the coordinate relaxation step. A different choice of basis would correspond in this figure to a particular solution along the x_2 axis.

Simple elimination is inexpensive but can give rise to numerical instabilities. If the feasible set in Figure 15.2 consisted of a line that was almost parallel to the x_1 axis, Then a particular solution along this axis would be very large in magnitude. Since the total displacement (15.13) is not large in general, we would compute x as the difference of very large vectors, giving rise to numerical cancellation. In that situation it would be preferable to choose a particular solution along the x_2 axis, that is, to select a different basis. Selection of the best basis is, however, not a straightforward task in general. Numerical errors can also occur in the definition of Z when the basis matrix B is poorly conditioned.

To overcome this danger we could define the particular solution Yb as the minimum-norm step to the constraints. This approach is a special case of more general elimination strategies, which we now describe.

GENERAL REDUCTION STRATEGIES FOR LINEAR CONSTRAINTS

In analogy with (15.13), we choose matrices $Y \in \mathbf{R}^{n \times m}$ and $Z \in \mathbf{R}^{n \times (n-m)}$ whose columns form a linearly independent set, and express any solution of the linear constraints

$Ax = b$ as

$$x = Yx_Y + Zx_Z, \qquad (15.15)$$

for some vectors x_Y and x_Z of dimensions m and $n - m$, respectively. We also require that Y and Z be chosen so that Z has linearly independent columns and so that

$$AY \quad \text{is nonsingular, and} \quad AZ = 0. \qquad (15.16)$$

(Note that AY is an $m \times m$ matrix.) As in simple elimination, we define Z to be a basis for the null space of the constraints, but we now leave the choice of Y unspecified; see Figure 15.3.

By substituting (15.15) into the constraints $Ax = b$, we obtain

$$Ax = (AY)x_Y = b,$$

so by nonsingularity of AY, x_Y can be written explicitly as follows:

$$x_Y = (AY)^{-1}b. \qquad (15.17)$$

By substituting this expression into (15.15), we conclude that any vector x of the form

$$x = Y(AY)^{-1}b + Zx_Z \qquad (15.18)$$

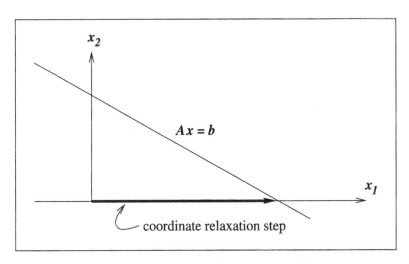

Figure 15.2 Simple elimination—we fix x_2 to zero and choose x_1 to ensure that the constraint is satisfied.

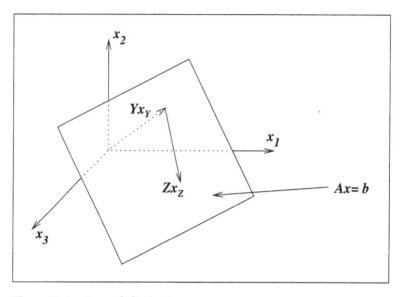

Figure 15.3 General elimination.

will satisfy the constraints $Ax = b$ for any choice of $x_z \in \mathbf{R}^{n-m}$. Therefore, the problem (15.4) can be restated equivalently as the unconstrained problem

$$\min_{x_z} \; f(Y(AY)^{-1}b + Zx_z). \tag{15.19}$$

Ideally, we would like to choose Y in such a way that the matrix AY is as well conditioned as possible, since it needs to be factorized to give the particular solution $Y(AY)^{-1}b$. We can do this by computing Y and Z by means of a QR factorization of A^T, which has the form

$$A^T \Pi = [\; Q_1 \;\; Q_2 \;] \begin{bmatrix} R \\ 0 \end{bmatrix}, \tag{15.20}$$

where $[\; Q_1 \;\; Q_2 \;]$ is orthogonal. The submatrices Q_1 and Q_2 have orthonormal columns and are of dimension $n \times m$ and $n \times (n-m)$, while R is $m \times m$ upper triangular and nonsingular and Π is an $m \times m$ permutation matrix. (See the discussion following (A.54) in the Appendix for more details.)

We now define

$$[\; Y \;\; Z \;] = [\; Q_1 \;\; Q_2 \;], \tag{15.21}$$

so that Y and Z form an *orthonormal basis* of $\mathbf{R}^n$. If we expand (15.20) and do a little rearrangement, we obtain

$$AY = \Pi R^T, \quad AZ = 0.$$

Therefore, Y and Z have the desired properties, and the condition number of AY is the same as that of R, which in turn is the same as that of A itself. From (15.18) we see that any solution of $Ax = b$ can be expressed as

$$x = Q_1 R^{-T} \Pi^T b + Q_2 x_z,$$

for some vector x_z. The computation $R^{-T} \Pi^T b$ can be carried out inexpensively, at the cost of a single triangular substitution.

A simple computation shows that the particular solution $Q_1 R^{-T} \Pi^T b$ can also be written as

$$x_p = A^T (AA^T)^{-1} b, \tag{15.22}$$

and is therefore the minimum-norm solution of the constraints $Ax = b$, that is, the solution of

$$\min \|Ax - b\|_2.$$

See Figure 15.4 for an illustration of this step.

Elimination via the orthogonal basis (15.21) is ideal from the point of view of numerical stability. The main cost associated with this reduction strategy is in the computation of the

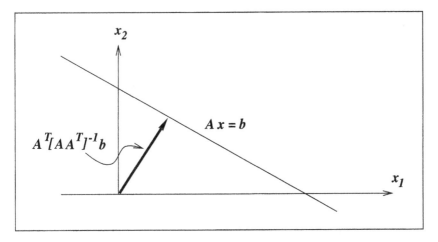

Figure 15.4 The minimum-norm step

QR factorization (15.20). Unfortunately, for problems in which A is large and sparse, a sparse QR factorization can be much more costly to compute than the application of sparse Gaussian elimination used in the simple elimination approach. Therefore, other elimination strategies have been developed that seek a compromise between these two techniques; see Exercise 6.

THE EFFECT OF INEQUALITY CONSTRAINTS

Elimination of variables is not always beneficial if inequality constraints are present alongside the equalities. For instance, if problem (15.9), (15.10) had the additional constraint $x \geq 0$, then after eliminating the variables x_3 and x_6, we would be left with the problem of minimizing the function in (15.12) subject to the constraints

$$(x_1, x_2, x_4, x_5) \geq 0,$$
$$8x_1 - 6x_2 + 9x_4 + 4x_5 \leq 6,$$
$$(3/4)x_1 + (1/2)x_2 - (1/4)x_4 + (3/2)x_5 \leq -1.$$

Hence, the cost of eliminating the equality constraints (15.10) is to make the inequalities more complicated than the simple bounds $x \geq 0$. For many algorithms, this transformation will not yield any benefit.

If, however, we added the general inequality constraint $3x_1 + 2x_3 \geq 1$ to the problem (15.9), (15.10), the elimination (15.11) transforms the problem into one of minimizing the function in (15.12) subject to the inequality constraint

$$-13x_1 + 12x_2 - 18x_4 - 8x_5 \geq -11. \tag{15.23}$$

In this case, the inequality constraint does not become much more complicated after elimination of the equality constraints, so it is probably worthwhile to perform the elimination.

15.3 MEASURING PROGRESS: MERIT FUNCTIONS

Suppose that an algorithm for solving the nonlinear programming problem (15.1) generates a step that gives a substantial reduction in the objective function but leads us farther away from the feasible region than the current iterate. Should we accept this step?

This question is not easy to answer. In constrained optimization we are confronted with the often conflicting goals of reducing the objective function and satisfying the constraints, and we must look for a measure that strikes the right balance between these two goals. Merit functions are designed to quantify this balance and control the algorithm: A step p will be accepted only if it leads to a sufficient reduction in the merit function ϕ.

15.3. MEASURING PROGRESS: MERIT FUNCTIONS

In unconstrained optimization, the objective function f is the natural choice for the merit function. All the unconstrained optimization methods described in this book require that f be decreased at each step (or at least after a certain number of iterations). In feasible methods, in which the starting point and all subsequent iterates satisfy all the constraints in the problem, the objective function is still an appropriate merit function. For example, when all the constraints are linear, some active set methods (Chapters 16 and 18) use a phase-1 iteration to compute a feasible starting point and define each step of the algorithm so that feasibility is retained at all subsequent iterates.

On the other hand, algorithms that allow iterates to violate at least some constraints require some way to assess the quality of the steps and iterates. The most common way to make this assessment is with a merit function.

A widely used merit function for the general nonlinear programming problem (15.1) is the ℓ_1 *exact function* defined by

$$\phi_1(x; \mu) = f(x) + \frac{1}{\mu} \sum_{i \in \mathcal{E}} |c_i(x)| + \frac{1}{\mu} \sum_{i \in \mathcal{I}} [c_i(x)]^-, \tag{15.24}$$

where we use the notation $[x]^- = \max\{0, -x\}$. The positive scalar μ is the *penalty parameter*, which determines the weight that we assign to constraint satisfaction relative to minimization of the objective. This merit function is called *exact* because for a range of values of the penalty parameter μ, the solution to the nonlinear programming problem (15.1) is a local minimizer of $\phi(x; \mu)$. Note that $\phi(x; \mu)$ is not differentiable due to the presence of the absolute value and $[\cdot]^-$ functions.

Another useful merit function is *Fletcher's augmented Lagrangian*. For the case where only equality constraints are present in (15.1), this merit function is defined to be

$$\phi_F(x; \mu) = f(x) - \lambda(x)^T c(x) + \frac{1}{2\mu} \sum_{i \in \mathcal{E}} c_i(x)^2, \tag{15.25}$$

where $\mu > 0$ is the penalty parameter, $\|\cdot\|$ denotes the ℓ_2 norm, and

$$\lambda(x) = [A(x)A(x)^T]^{-1} A(x) \nabla f(x) \tag{15.26}$$

are called the least-squares multiplier estimates. This merit function is differentiable, and is also exact. To define this function for problems including inequality constraints we can make use of slack variables.

We now give a more precise definition of exactness of a merit function. Since the previous discussion suggests that merit functions always make use of a scalar penalty parameter $\mu > 0$, we will write a general merit function as $\phi(x; \mu)$, and assume that constraints are penalized by decreasing μ.

Definition 15.1 (Exact Merit Function).

The merit function $\phi(x; \mu)$ is said to be exact if there is a positive scalar μ^* such that for any $\mu \in (0, \mu^*]$, any local solution of the nonlinear programming problem (15.1) is a local minimizer of $\phi(x; \mu)$.

One can show that the ℓ_1 merit function $\phi(x; \nu)$ is exact for all $\mu < \mu^*$, where

$$\frac{1}{\mu^*} = \max\{|\lambda_i^*|, \ i \in \mathcal{E}; \lambda_i^*, \ i \in \mathcal{I}\},$$

and where the λ_i^* denote the Lagrange multipliers associated with an optimal solution x^*. Many algorithms based on the ℓ_1 merit function contain heuristics for adjusting the penalty parameter whenever the algorithm determines that the current value probably does not satisfy $\mu > \mu^*$. (This decision could be based on an approximation to μ^* obtained from the current Lagrange multiplier estimates.) Precise rules for setting and changing μ are described in Chapter 18.

Fletcher's augmented Lagrangian merit function ϕ_F is also exact. Since the threshold value μ^* is not as simple to write as in the case of the ℓ_1 merit function—because it involves bounds on certain derivatives—we discuss the choice of μ in the context of specific algorithms. See Section 18.5 for further discussion.

We now contrast some of the other properties of these two merit functions. The ℓ_1 merit function is inexpensive to evaluate, since function and constraint values are available at every iteration of constrained optimization algorithms. One of its potential drawbacks is that it may reject steps that make good progress toward the solution—a phenomenon known as the Maratos effect, which is described in Chapter 18. Several strategies have been devised that seem to successfully remove the damaging effects of the Maratos effect, but they introduce a certain degree of complexity into the algorithms.

Fletcher's augmented Lagrangian merit function ϕ_F is differentiable and does not suffer from the Maratos effect. Its main drawback is the expense of evaluating it at trial points—if a line search is needed to ensure that ϕ_F decreases at each step—since (15.26) requires the solution of a linear system. To circumvent this problem we can replace $\lambda(x_k + \alpha p)$, after the first trial value $\lambda(x_k + p)$ has been evaluated, by the interpolant

$$\lambda(x_k) + \alpha(\lambda(x_k + p) - \lambda(x_k)),$$

and use this interpolant in the line search. Other potential drawbacks of Fletcher's function are that λ is not uniquely defined by (15.26) when A loses rank, and that λ can be excessively large when A is nearly rank deficient.

As noted above, ϕ_1 has a simpler form than ϕ_F, but is not differentiable. In fact, as we now show, exact penalty functions that have the form of ϕ_1 *must* be nondifferentiable.

For simplicity, we restrict our attention to the case where only equality constraints are present, and assemble the constraint functions $c_i(x), i \in \mathcal{E}$, into a vector $c(x)$. Consider a merit function of the form

$$\phi(x; \mu) = f(x) + \frac{1}{\mu} h(c(x)), \qquad (15.27)$$

where $h : \mathbf{R}^m \to \mathbf{R}$ is a function satisfying the properties $h(y) \geq 0$ for all $y \in \mathbf{R}^m$ and $h(0) = 0$. Suppose for contradiction that h is differentiable. Since h has a minimizer at zero, we have from Theorem 2.2 that $\nabla h(0) = 0$. Now, if x^* is a local solution of the problem (15.1), we have $c(x^*) = 0$ and therefore $\nabla h(c(x^*)) = 0$. Hence, since x^* is a local minimizer of Ψ, we have that

$$0 = \Psi'(x^*) = \nabla f(x^*) + \frac{1}{\mu} \nabla c(x^*) \nabla h(c(x^*)) = \nabla f(x^*).$$

However, it is not generally true that the gradient of f vanishes at the solution of a constrained optimization problem, so our original assumption that h is differentiable must be incorrect, and our claim is proved.

The ℓ_1 merit function ϕ_1 is a special case of (15.27) in which $h(c) = \|c\|_1$. Other exact merit functions used in practice define $h(x) = \|x\|_2$ (not squared) and $h(x) = \|x\|_\infty$. To obtain exact, differentiable merit functions, we must extend the form (15.27) by including additional terms in the merit function. (The second term in (15.25) serves this purpose.)

We conclude by describing a merit function that is used in several popular programs for nonlinear programming, but that is quite different in nature. Supposing once again, for simplicity, that the problem contains only equality constraints, we define the *augmented Lagrangian in x and* λ as follows:

$$\mathcal{L}_A(x, \lambda; \mu) = f(x) - \lambda^T c(x) + \frac{1}{2\mu} \|c(x)\|_2^2. \qquad (15.28)$$

If at the current point (x_k, λ_k) the algorithm generates a search direction (p_x, p_λ) (that is, a step both in the primal and dual variables), we assess the acceptability of the proposed new iterate by substituting $(x, \lambda) = (x_k + p_x, \lambda_k + p_\lambda)$ into (15.28) and comparing it with $\mathcal{L}_A(x_k, \lambda_k; \mu)$.

Unlike the merit functions ϕ_1 and ϕ_F described above, $\mathcal{L}_A$ depends on the dual variables as well as the primals. A solution (x^*, λ^*) of the nonlinear programming problem is a stationary point for $\mathcal{L}_A(x, \lambda; \mu)$, but in general not a minimizer. Nevertheless, some sequential quadratic programming programs use $\mathcal{L}_A$ successfully as a merit function by adaptively modifying μ and λ.

NOTES AND REFERENCES

General elimination techniques are described in Fletcher [83].

Merit functions have received much attention. Boggs and Tolle [23] survey much of the work and provide numerous references. The ℓ_1 function was first suggested as a merit function for sequential quadratic programming methods by Han [132]. The merit augmented Lagrangian function (15.25) was proposed by Fletcher [81], whereas the primal–dual function (15.28) was proposed by Wright [250] and Schittkowski [222].

EXERCISES

15.1 Do the following problems have solutions? Explain.

$$\min x_1 + x_2 \quad \text{subject to } x_1^2 + x_2^2 = 2, \ 0 \le x_1 \le 1, \ 0 \le x_2 \le 1;$$
$$\min x_1 + x_2 \quad \text{subject to } x_1^2 + x_2^2 \le 1, \ x_1 + x_2 = 3;$$
$$\min x_1 x_2 \quad \text{subject to } x_1 + x_2 = 2.$$

15.2 Show that if in Example 15.1 we eliminate x in terms of y, then the correct solution of the problem is obtained by performing unconstrained minimization.

15.3 Explain why we can be sure that the matrix Z in (15.14) has linearly independent columns, regardless of the choice of B and N.

15.4 Show that the basis matrices (15.14) are linearly independent.

15.5 Show that the particular solution $Q_1 R^{-T} \Pi^T b$ can be written as (15.22).

15.6 In this exercise we compute basis matrices that attempt to be a compromise between the orthonormal basis (15.21) and simple elimination (15.14). Let us assume that the basis matrix is given by the first m columns of A, so that $P = I$ in (15.5), and define

$$Y = \begin{bmatrix} I \\ (B^{-1}N)^T \end{bmatrix}, \quad Z = \begin{bmatrix} -B^{-1}N \\ I \end{bmatrix}.$$

(a) Show that the columns of Y and Z are no longer of norm 1 and that the relations $AZ = 0$ and $Y^T Z = 0$ hold. Therefore, the columns of Y and Z form an independent set, showing that this is a valid choice of the basis matrices. (b) Show that the particular solution $Y(AY)^{-1}b$ defined by this choice of Y is, as in the orthogonal factorization approach, the minimum-norm solution (15.22) of $Ax = b$. More specifically, show that

$$Y(AY)^{-1} = A^T (AA^T)^{-1}.$$

It follows that the matrix $Y(AY)^{-1}$ is independent of the choice of basis matrix B in (15.5), and its conditioning is determined by that of A alone. Note, however, that the matrix Z still depends explicitly on B, so a careful choice of B is needed to ensure well conditioning in this part of the computation.

◈ **15.7** Verify that by adding the inequality constraint $3x_1 + 2x_3 \geq 1$ to the problem (15.9), (15.10), the elimination (15.11) transforms the problem into one of minimizing the function (15.12) subject to the inequality constraint (15.23).

CHAPTER 16

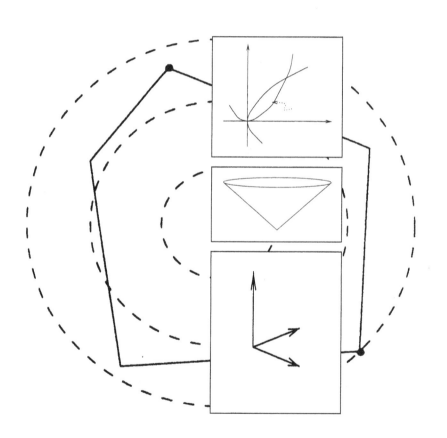

Quadratic Programming

An optimization problem with a quadratic objective function and linear constraints is called a quadratic program. Problems of this type are important in their own right, and they also arise as subproblems in methods for general constrained optimization, such as sequential quadratic programming (Chapter 18) and augmented Lagrangian methods (Chapter 17).

The general quadratic program (QP) can be stated as

$$\min_x \ q(x) = \tfrac{1}{2} x^T G x + x^T d \qquad (16.1a)$$

$$\text{subject to } a_i^T x = b_i, \qquad i \in \mathcal{E}, \qquad (16.1b)$$

$$a_i^T x \geq b_i, \qquad i \in \mathcal{I}, \qquad (16.1c)$$

where G is a symmetric $n \times n$ matrix, $\mathcal{E}$ and $\mathcal{I}$ are finite sets of indices, and d, x, and $\{a_i\}, i \in \mathcal{E} \cup \mathcal{I}$, are vectors with n elements. Quadratic programs can always be solved (or can be shown to be infeasible) in a finite number of iterations, but the effort required to find a solution depends strongly on the characteristics of the objective function and the number of inequality constraints. If the Hessian matrix G is positive semidefinite, we say that (16.1) is a *convex QP*, and in this case the problem is sometimes not much more difficult to solve than a linear program. *Nonconvex* QPs, in which G is an indefinite matrix, can be more challenging, since they can have several stationary points and local minima.

In this chapter we limit ourselves to studying algorithms that find the solution of a convex quadratic program or a stationary point of a general (nonconvex) quadratic program. We start by considering an interesting application of quadratic programming.

AN EXAMPLE: PORTFOLIO OPTIMIZATION

Every investor knows that there is a tradeoff between risk and return: To increase the expected return on investment, an investor must be willing to tolerate greater risks. Portfolio theory studies how to model this tradeoff given a collection of n possible investments with returns r_i, $i = 1, 2, \ldots, n$. The returns r_i are usually not known in advance, and are often assumed to be random variables that follow a normal distribution. We can characterize these variables by their expected value $\mu_i = E[r_i]$ and their variance $\sigma_i^2 = E[(r_i - \mu_i)^2]$. The variance measures the fluctuations of the variable r_i about its mean, so that larger values of σ_i indicate riskier investments.

An investor constructs a portfolio by putting a fraction x_i of the available funds into investment i, for $i = 1, 2, \ldots, n$. Assuming that all available funds are invested and that short-selling is not allowed, the constraints are $\sum_{i=1}^{n} x_i = 1$ and $x \geq 0$. The return on the portfolio is given by

$$R = \sum_{i=1}^{n} x_i r_i. \tag{16.2}$$

To measure the desirability of the portfolio, we need to obtain measures of its expected return variance. The expected return is simply

$$E(R) = E\left[\sum_{i=1}^{n} x_i r_i\right] = \sum_{i=1}^{n} x_i E[r_i] = x^T \mu.$$

The variance, too, can be calculated from elementary laws of statistics. It depends on the *covariances* between each pair of investments, which are defined by

$$\rho_{ij} = \frac{E[(r_i - \mu_i)(r_j - \mu_j)]}{\sigma_i \sigma_j}, \quad \text{for } i, j = 1, 2, \ldots, n.$$

The correlation measures the tendency of the return on investments i and j to move in the same direction. Two investments whose returns tend to rise and fall together have a positive covariance; the nearer ρ_{ij} is to 1, the more closely the two investments track each other. Investments whose returns tend to move in opposite directions have negative covariance.

The variance of the total portfolio R is then given by

$$E[(R - E[R])^2] = \sum_{i=1}^{n}\sum_{j=1}^{n} x_i x_j \sigma_i \sigma_j \rho_{ij} = x^T G x,$$

where we have defined the $n \times n$ symmetric matrix G by

$$G_{ij} = \rho_{ij}\sigma_i\sigma_j.$$

It can be shown that G is positive semidefinite.

We are interested in portfolios for which the expected return $x^T\mu$ is large while the variance $x^T G x$ is small. In the model proposed by Markowitz [157], we combine these two aims into a single objective function with the aid of a "risk tolerance parameter" denoted by κ, and solve the following problem to find the "optimal" portfolio:

$$\max \; x^T\mu - \kappa x^T G x, \quad \text{subject to} \; \sum_{i=1}^{n} x_i = 1, \; x \geq 0.$$

The parameter κ lies in the range $[0, \infty)$, and its chosen value depends on the preferences of the individual investor. Conservative investors would place more emphasis on minimizing risk in their portfolio, so they would choose a large value of κ to increase the weight of the variance measure in the objective function. More daring investors are prepared to take on more risk in the hope of a higher expected return, so their value of κ would be closer to zero.

The difficulty in applying this portfolio optimization technique to real-life investing lies in defining the expected returns, variances, and covariances for the investments in question. One possibility is to use historical data, defining the quantities μ_i, σ_i, and ρ_{ij} to be equal to their historical values between the present day and, say, five years ago. It is not wise to assume that future performance will mirror the past, of course. Moreover, historical data will not be available for many interesting investments (such as start-up companies based on new technology). Financial professionals often combine the historical data with their own insights and expectations to produce values of μ_i, σ_i, and ρ_{ij}.

16.1 EQUALITY--CONSTRAINED QUADRATIC PROGRAMS

We begin our discussion of algorithms for quadratic programming by considering the case where only equality constraints are present. As we will see in this chapter, active set methods for general quadratic programming solve an equality–constrained QP at each iteration.

PROPERTIES OF EQUALITY-CONSTRAINED QPS

Let us denote the number of constraints by m, assume that $m \leq n$, and write the quadratic program as

$$\min_x q(x) \stackrel{\text{def}}{=} \tfrac{1}{2} x^T G x + x^T d \tag{16.3a}$$

$$\text{subject to} \quad Ax = b, \tag{16.3b}$$

where A is the $m \times n$ Jacobian of constraints defined by

$$A = [a_i]_{i \in \mathcal{E}}^T.$$

For the present, we assume that A has full row rank (rank m), and that the constraints (16.3b) are consistent. (Toward the end of the chapter we discuss the case in which A is rank deficient.)

The first-order necessary conditions for x^* to be a solution of (16.3) state that there is a vector λ^* such that the following system of equations is satisfied:

$$\begin{bmatrix} G & -A^T \\ A & 0 \end{bmatrix} \begin{bmatrix} x^* \\ \lambda^* \end{bmatrix} = \begin{bmatrix} -d \\ b \end{bmatrix}. \tag{16.4}$$

It is easy to derive these conditions as a consequence of the general result for first-order optimality conditions, Theorem 12.1. As in Chapter 12, we call λ^* the vector of Lagrange multipliers. The system (16.4) can be rewritten in a form that is useful for computation by expressing x^* as $x^* = x + p$, where x is some estimate of the solution and p is the desired step. By introducing this notation and rearranging the equations, we obtain

$$\begin{bmatrix} G & A^T \\ A & 0 \end{bmatrix} \begin{bmatrix} -p \\ \lambda^* \end{bmatrix} = \begin{bmatrix} g \\ c \end{bmatrix}, \tag{16.5}$$

where

$$c = Ax - b, \qquad g = d + Gx, \qquad p = x^* - x. \tag{16.6}$$

The matrix in (16.5) is called the Karush–Kuhn–Tucker (KKT) matrix, and the following result gives conditions under which it is nonsingular. As in Chapter 15, we use Z to denote the $n \times (n - m)$ matrix whose columns are a basis for the null space of A. That is, Z has full rank and $AZ = 0$.

16.1. EQUALITY--CONSTRAINED QUADRATIC PROGRAMS

Lemma 16.1.
Let A have full row rank, and assume that the reduced-Hessian matrix $Z^T G Z$ is positive definite. Then the KKT matrix

$$K = \begin{bmatrix} G & A^T \\ A & 0 \end{bmatrix} \quad (16.7)$$

is nonsingular, and there is a unique vector pair (x^*, λ^*) satisfying (16.4).

PROOF. Suppose there are vectors p and v such that

$$\begin{bmatrix} G & A^T \\ A & 0 \end{bmatrix} \begin{bmatrix} p \\ v \end{bmatrix} = 0. \quad (16.8)$$

Since $Ap = 0$, we have from (16.8) that

$$0 = \begin{bmatrix} p \\ v \end{bmatrix}^T \begin{bmatrix} G & A^T \\ A & 0 \end{bmatrix} \begin{bmatrix} p \\ v \end{bmatrix} = p^T G p.$$

Since p lies in the null space of A, it can be written as $p = Zu$ for some vector $u \in \mathbf{R}^{n-m}$. Therefore, we have

$$0 = p^T G p = u^T Z^T G Z u,$$

which by positive definiteness of $Z^T G Z$ implies that $u = 0$. Therefore, $p = 0$, and by (16.8), $A^T v = 0$; and the full row rank of A implies that $v = 0$. We conclude that equation (16.8) is satisfied only if $p = 0$ and $v = 0$, so the matrix is nonsingular, as claimed. □

❑ **EXAMPLE 16.1**

Consider the quadratic programming problem

$$\min q(x) = 3x_1^2 + 2x_1x_2 + x_1x_3 + 2.5x_2^2 + 2x_2x_3 + 2x_3^2 - 8x_1 - 3x_2 - 3x_3,$$
$$\text{subject to} \quad x_1 + x_3 = 3, \quad x_2 + x_3 = 0. \quad (16.9)$$

We can write this problem in the form (16.3) by defining

$$G = \begin{bmatrix} 6 & 2 & 1 \\ 2 & 5 & 2 \\ 1 & 2 & 4 \end{bmatrix}, \quad d = \begin{bmatrix} -8 \\ -3 \\ -3 \end{bmatrix}, \quad A = \begin{bmatrix} 1 & 0 & 1 \\ 0 & 1 & 1 \end{bmatrix}, \quad b = \begin{bmatrix} 3 \\ 0 \end{bmatrix}.$$

The solution x^* and optimal Lagrange multiplier vector λ^* are given by

$$x^* = \begin{bmatrix} 2 \\ -1 \\ 1 \end{bmatrix}, \quad \lambda^* = \begin{bmatrix} 3 \\ -2 \end{bmatrix}.$$

In this example, the matrix G is positive definite, and the null-space basis matrix can be defined as

$$Z = (-1, -1, 1)^T. \tag{16.10}$$

□

We have seen that when the conditions of Lemma 16.1 are satisfied, then there is a unique vector pair (x^*, λ^*) that satisfies the first-order necessary conditions for (16.3). In fact, the second-order sufficient conditions (see Theorem 12.6) are also satisfied at (x^*, λ^*), so x^* is a strict local minimizer of (16.3). However, we can use a direct argument to show that x^* is actually a *global* solution of (16.3).

Theorem 16.2.
Suppose that the conditions of Lemma 16.1 are satisfied. Then the vector x^* satisfying (16.4) is the unique global solution of (16.3).

PROOF. Let x be any other feasible point (satisfying $Ax = b$), and as before, we use p to denote the difference $x^* - x$. Since $Ax^* = Ax = b$, we have that $Ap = 0$. By substituting into the objective function (16.3a), we obtain

$$q(x) = \tfrac{1}{2}(x^* - p)^T G(x^* - p) + d^T(x^* - p)$$
$$= \tfrac{1}{2} p^T G p - p^T G x^* - d^T p + q(x^*). \tag{16.11}$$

From (16.4) we have that $Gx^* = -d + A^T\lambda^*$, so from $Ap = 0$ we have that

$$p^T G x^* = p^T(-d + A^T\lambda^*) = -p^T d.$$

By substituting this relation into (16.11), we obtain

$$q(x) = \tfrac{1}{2} p^T G p + q(x^*).$$

Since p lies in the null space of A, we can write $p = Zu$ for some vector $u \in \mathbf{R}^{n-m}$, so that

$$q(x) = \tfrac{1}{2} u^T Z^T G Z u + q(x^*).$$

By positive definiteness of $Z^T G Z$, we conclude that $q(x) > q(x^*)$ except when $u = 0$, that is, when $x = x^*$. Therefore, x^* is the unique global solution of (16.3). $\square$

When the projected Hessian matrix $Z^T G Z$ has zero or negative eigenvalues, the problem (16.3) does not have a bounded solution, except in a special case. To demonstrate this claim, suppose that there is a vector pair (x^*, λ^*) that satisfies the KKT conditions (16.4). Let u be some vector such that $u^T Z^T G Z u \leq 0$, and set $p = Zu$. Then for any $\alpha > 0$, we have that

$$A(x^* + \alpha p) = b,$$

so that $x^* + \alpha p$ is feasible, while

$$q(x^* + \alpha p) = q(x^*) + \alpha p^T (G x^* + d) + \tfrac{1}{2}\alpha^2 p^T G p$$
$$= q(x^*) + \tfrac{1}{2}\alpha^2 p^T G p \leq q(x^*),$$

where we have used the facts that $G x^* + d = A^T \lambda^*$ from (16.4) and $p^T A^T \lambda^* = u^T Z^T A^T \lambda^* = 0$. Therefore, from any x^* satisfying the KKT conditions, we can find a feasible direction p along which q does not increase. In fact, we can always find a direction of *strict decrease* for q unless $Z^T G Z$ has no negative eigenvalues. The only case in which (16.3) has solutions is the one in which there exists some point x^* for which (16.4) is satisfied, while $Z^T G Z$ is positive semidefinite. Even in this case, the solution is not a strict local minimizer.

16.2 SOLVING THE KKT SYSTEM

In this section we discuss efficient methods for solving the KKT system (16.4) (or, alternatively, (16.5)).

The first important observation is that if $m \geq 1$, the KKT matrix is always indefinite. The following result characterizes the *inertia* of (16.7) under the assumptions of Lemma 16.1. The inertia of a matrix is the scalar triple that indicates the number of its positive, negative, and zero eigenvalues.

Lemma 16.3.
Suppose that A has full row rank and that the reduced Hessian $Z^T G Z$ is positive definite. Then the KKT matrix (16.7) has n positive eigenvalues, m negative eigenvalues, and no zero eigenvalues.

This result follows from Theorem 16.6 given later in this chapter. Knowing that the KKT system is indefinite, we now describe the main techniques developed for its solution.

DIRECT SOLUTION OF THE KKT SYSTEM

One option for solving (16.5) is to perform a triangular factorization on the full KKT matrix and then perform backward and forward substitution with the triangular factors. We cannot use the Cholesky factorization algorithm because the KKT matrix is indefinite. Instead, we could use Gaussian elimination with partial pivoting (or a sparse variant of this algorithm) to obtain the L and U factors, but this approach has the disadvantage that it ignores the symmetry of the system.

The most effective strategy in this case is to use a *symmetric indefinite factorization*. We have described these types of factorizations in Chapter 6. For a general symmetric matrix K, these factorizations have the form

$$P^T K P = LBL^T, \qquad (16.12)$$

where P is a permutation matrix, L is unit lower triangular, and B is block-diagonal with either 1×1 or 2×2 blocks. The symmetric permutations defined by the matrix P are introduced for numerical stability of the computation and, in the case of large sparse K, to maintain sparsity. The computational cost of symmetric indefinite factorization (16.12) is typically about half the cost of sparse Gaussian elimination.

To solve (16.5) we first compute the factorization (16.12), substituting the KKT matrix for K. We then perform the following sequence of operations to arrive at the solution:

$$\text{solve } Ly = P^T \begin{bmatrix} g \\ c \end{bmatrix} \text{ to obtain } y;$$

$$\text{solve } B\hat{y} = y \text{ to obtain } \hat{y};$$

$$\text{solve } L^T \bar{y} = \hat{y} \text{ to obtain } \bar{y};$$

$$\text{set } \begin{bmatrix} -p \\ \lambda^* \end{bmatrix} = P\bar{y}.$$

Note that multiplications with the permutation matrices P and P^T can be performed by rearranging vector components, and are therefore inexpensive. Solution of the system $B\hat{y} = y$ entails solving a number of small 1×1 and 2×2 systems, so the number of operations is a small multiple of the vector length $(m+n)$, again quite inexpensive. Triangular substitutions with L and L^T are more significant. Their precise cost depends on the amount of sparsity, but is usually significantly less than the cost of performing the factorization (16.12).

This approach of factoring the full $(n+m) \times (n+m)$ KKT matrix (16.7) is quite effective on some problems. Difficulties may arise when the heuristics for choosing the permutation matrix P are not able to do a very good job of maintaining sparsity in the L factor, so that L becomes much more dense than the original coefficient matrix.

An alternative to the direct factorization approach for the matrix in (16.5) is to apply an iterative method. The conjugate gradient method is not recommended because it can be unstable on systems that are not positive definite. Therefore, we must consider techniques for general linear systems, or for symmetric indefinite systems. Candidates include the QMR and LSQR methods (see the Notes and References at the end of the chapter).

RANGE-SPACE METHOD

In the range-space method, we use the matrix G to perform block elimination on the system (16.5). Assuming that G is positive definite, we multiply the first equation in (16.5) by AG^{-1} and then subtract the second equation to obtain a linear system in the vector λ^* alone:

$$(AG^{-1}A^T)\lambda^* = (AG^{-1}g - c). \tag{16.13}$$

We solve this symmetric positive definite system for λ^*, and then recover p from the first equation in (16.5) by solving

$$Gp = A^T\lambda^* - g. \tag{16.14}$$

This approach requires us to perform operations with G^{-1}, as well as to compute the factorization of the $m \times m$ matrix $AG^{-1}A^T$. It is therefore most useful when:

- G is well conditioned and easy to invert (for instance, when G is diagonal or block-diagonal);
- G^{-1} is known explicitly through a quasi-Newton updating formula; or
- the number of equality constraints m is small, so that the number of backsolves needed to form the matrix $AG^{-1}A^T$ is not too large.

The range-space approach is, in effect, a special case of the symmetric elimination approach of the previous section. It corresponds to the case in which the first n variables in the system (16.5) are eliminated *before* we eliminate any of the last m variables. In other words, it is obtained if we choose the matrix P in (16.12) as

$$P = \begin{bmatrix} P_1 & 0 \\ 0 & P_2 \end{bmatrix},$$

where P_1 and P_2 are permutation matrices of dimension $n \times n$ and $m \times m$, respectively.

We can actually use a similar approach to the range-space method to derive an explicit inverse formula for the KKT matrix in (16.5). This formula is

$$\begin{bmatrix} G & A^T \\ A & 0 \end{bmatrix}^{-1} = \begin{bmatrix} C & E \\ E^T & F \end{bmatrix}, \tag{16.15}$$

with

$$C = G^{-1} - G^{-1}A^T(AG^{-1}A^T)^{-1}AG^{-1},$$
$$E = G^{-1}A^T(AG^{-1}A^T)^{-1},$$
$$F = -(AG^{-1}A^T)^{-1}.$$

The solution of (16.5) can be obtained by multiplying its right-hand-side by this matrix. If we take advantage of common expressions and group terms appropriately, we recover the approach (16.13), (16.14).

NULL-SPACE METHOD

The null-space method, which we now describe, does not require nonsingularity of G and is therefore of wider applicability than the range-space method. It assumes only that the conditions of Lemma 16.1 hold, namely, that A has full row rank and that $Z^T G Z$ is positive definite. It requires, however, knowledge of the null-space basis matrix Z. Like the range-space method, it exploits the block structure in the KKT system to decouple (16.5) into two smaller systems.

Suppose that we partition the vector p in (16.5) into two components, as follows:

$$p = Yp_Y + Zp_Z, \tag{16.16}$$

where Z is the $n \times (n-m)$ null-space matrix, Y is any $n \times m$ matrix such that $[Y \mid Z]$ is nonsingular, p_Y is an m-vector, and p_Z is an $(n-m)$-vector (see Chapter 15). As illustrated in Figure 15.3, Yp_Y is a particular solution of $Ax = b$, and Zp_Z is a displacement along these constraints.

By substituting p into the second equation of (16.5) and recalling that $AZ = 0$, we obtain

$$(AY)p_Y = -c. \tag{16.17}$$

Since A has rank m and $[Y \mid Z]$ is $n \times n$ nonsingular, the product $A[Y \mid Z] = [AY \mid 0]$ has rank m. Therefore, AY is a nonsingular $m \times m$ matrix, and p_Y is well determined by the equations (16.17). Meanwhile, we can substitute (16.16) into the first equation of (16.5) to obtain

$$-GYp_Y - GZp_Z + A^T\lambda^* = g$$

and multiply by Z^T to obtain

$$(Z^T GZ)p_Z = -[Z^T GYp_Y + Z^T g]. \tag{16.18}$$

This system, which can be solved by means of the Cholesky factorization of the $(n-m) \times (n-m)$ reduced-Hessian matrix $Z^T G Z$, determines p_z, and hence the total displacement $p = Y p_Y + Z p_z$. To obtain the Lagrange multiplier, we multiply the first equation of (16.5) by Y^T to obtain the linear system

$$(AY)^T \lambda^* = Y^T(g + Gp), \tag{16.19}$$

which can be solved for λ^*.

❑ EXAMPLE 16.2

Consider the problem (16.9) given in the previous example. We can choose

$$Y = \begin{pmatrix} 2/3 & -1/3 \\ -1/3 & 2/3 \\ 1/3 & 1/3 \end{pmatrix}$$

and Z is as in (16.10). For this particular choice of Y, we have $AY = I$.
Suppose we have $x = (0, 0, 0)^T$ in (16.5). Then

$$c = Ax - b = -b, \qquad g = d + Gx = d = (-8, -3, -3)^T.$$

Simple calculation shows that

$$p_Y = (3, 0)^T, \qquad p_z = 0,$$

so that

$$p = x^* - x = Y p_Y + Z p_z = (2, -1, 1)^T.$$

After recovering λ^* from (16.19), we conclude that

$$x^* = (2, -1, 1)^T, \qquad \lambda^* = (3, -2)^T.$$

❑

The null-space approach can be effective when the number of degrees of freedom $n - m$ is small. Its main drawback is the need for the null-space matrix Z, which as we have seen in Chapter 15, can be expensive to compute in many large problems. The matrix Z is not uniquely defined, and if it is poorly chosen, the reduced system (16.18) may become ill conditioned. If we choose Z to have orthonormal columns, as is normally done in software for small and medium-sized problems, then the conditioning of $Z^T G Z$ is at least as good as that of G itself. When A is large and sparse, however, this choice of Z is relatively expensive to compute, so for practical reasons we are often forced to use one of the less reliable choices of Z described in Chapter 15.

The reduced system (16.18) also can be solved by means of the conjugate gradient (CG) method. If we adopt this approach, it is not necessary to form the reduced Hessian $Z^T G Z$ explicitly, since the CG method requires only that we compute matrix–vector products involving this matrix. In fact, it is not even necessary to form Z explicitly, as long as we are able to compute products of Z and Z^T with arbitrary vectors. For some choices of Z and for large problems, these products are much cheaper to compute than Z itself, as we have seen in Chapter 15. A drawback of the conjugate gradient approach in the absence of an explicit representation of Z is that standard preconditioning techniques, such as modified Cholesky, cannot be used, and the development of effective preconditioners for this case is still the subject of investigation.

It is difficult to give hard and fast rules about the relative effectiveness of null-space and range-space methods, since factors such as fill-in during computation of Z vary significantly even among problems of the same dimension. In general, we can recommend the range-space method when G is positive definite and $AG^{-1}A^T$ can be computed cheaply (because G is easy to invert or because m is small relative to n). Otherwise, the null-space method is often preferable, in particular when it is much more expensive to compute factors of G than to compute the null-space matrix Z and the factors of $Z^T G Z$.

A METHOD BASED ON CONJUGACY

The range-space and null-space methods described above are useful in a wide variety of settings and applications. The following approach is of more limited use, but has proved to be the basis for efficient methods for convex QP. It is applicable in the case where the Hessian G is positive definite, and can be regarded as a null-space method that makes clever use of conjugacy to provide a particularly simple formula for the step computation. (See Chapter 5 for a definition of conjugacy.)

The key idea is to compute an $n \times n$ nonsingular matrix W with the following properties:

$$W^T G W = I, \qquad AW = \begin{bmatrix} 0 & U \end{bmatrix}, \tag{16.20}$$

where U is an $m \times m$ upper triangular matrix. The first equation in (16.20) states that the columns of W are conjugate with respect to the Hessian G, whereas the second condition implies that the first $n - m$ columns of W lie in the null space of A.

The matrix W can be constructed explicitly by using the QR and Cholesky factorizations. (These factorizations are described in Section A.2.) By using a variant of the QR factorization (see Exercise 6), we can find an orthogonal matrix Q and an $m \times m$ upper triangular matrix $\hat{U}$ such that

$$AQ = \begin{bmatrix} 0 & \hat{U} \end{bmatrix}.$$

Since $Q^T G Q$ is symmetric and positive definite, we can compute its Cholesky decomposition $Q^T G Q = L L^T$. By defining $W = Q L^{-T}$, we find that (16.20) is satisfied with $U = \hat{U} L^{-T}$.

We can partition the columns of W to obtain the matrices Z and Y used in (16.16), by writing

$$W = [\; Z \quad Y \;],$$

where Z is the first $n - m$ columns and Y is the last m columns. By using this definition and (16.20), we find that

$$Z^T G Z = I, \quad Z^T G Y = 0, \quad Y^T G Y = I, \qquad (16.21)$$

and

$$AY = U, \quad AZ = 0. \qquad (16.22)$$

These relations allow us to simplify the equations (16.17), (16.18), and (16.19) that are used to solve for p_Y, p_Z, and λ^*, respectively. We have

$$U p_Y = -c, \qquad (16.23a)$$
$$p_Z = -Z^T g, \qquad (16.23b)$$
$$U^T \lambda^* = Y^T g + p_Y, \qquad (16.23c)$$

so these vectors can be recovered at the cost of two triangular substitutions involving U and matrix–vector products involving Y^T and Z^T. The vector p can then be recovered from (16.16).

16.3 INEQUALITY-CONSTRAINED PROBLEMS

In the remainder of the chapter we discuss several classes of algorithms for solving QPs that contain inequality constraints and possibly equality constraints. *Classical active-set methods* can be applied both to convex and nonconvex problems, and they have been the most widely used methods since the 1970s. *Gradient–projection methods* attempt to accelerate the solution process by allowing rapid changes in the active set, and are most efficient when the only constraints in the problem are bounds on the variables. *Interior-point methods* have recently been shown to be effective for solving large convex quadratic programs.

We discuss these methods in the following sections. We mention also that quadratic programs can also be solved by the augmented Lagrangian methods of Chapter 17, or by means of an exact penalization method such as the Sℓ_1QP approach discussed in Chapter 18.

OPTIMALITY CONDITIONS FOR INEQUALITY-CONSTRAINED PROBLEMS

We begin our discussion with a brief review of the optimality conditions for inequality-constrained quadratic programming, and discuss some of the less obvious properties of the solutions.

The conclusions of Theorem 12.1 can be applied to (16.1) by noting that the Lagrangian for this problem is

$$\mathcal{L}(x, \lambda) = \tfrac{1}{2} x^T G x + x^T d - \sum_{i \in \mathcal{I} \cup \mathcal{E}} \lambda_i (a_i^T x - b_i). \tag{16.24}$$

In addition, we define the active set $\mathcal{A}(x^*)$ at an optimal point x^* as in (12.29) as the indices of the constraints at which equality holds, that is,

$$\mathcal{A}(x^*) = \{i \in \mathcal{E} \cup \mathcal{I} : a_i^T x^* = b_i\}. \tag{16.25}$$

By simplifying the general conditions (12.30), we conclude that any solution x^* of (16.1) satisfies the following first-order conditions:

$$Gx^* + d - \sum_{i \in \mathcal{A}(x^*)} \lambda_i^* a_i = 0, \tag{16.26a}$$

$$a_i^T x^* = b_i, \quad \text{for all } i \in \mathcal{A}(x^*), \tag{16.26b}$$

$$a_i^T x^* \geq b_i, \quad \text{for all } i \in \mathcal{I} \backslash \mathcal{A}(x^*), \tag{16.26c}$$

$$\lambda_i^* \geq 0, \quad \text{for all } i \in \mathcal{I} \cap \mathcal{A}(x^*). \tag{16.26d}$$

A technical point: In Theorem 12.1 we assumed that the linear independence constraint qualification (LICQ) was satisfied. As mentioned in Chapter 12, this theorem still holds if we replace LICQ by other constraint qualifications, such as linearity of the constraints, which is certainly satisfied for quadratic programming. Hence, in the optimality conditions for quadratic programming given above we need not assume that the active constraints are linearly dependent at the solution.

We omit a detailed discussion of the second-order conditions, which follow from the theory of Section 12.4. Second-order sufficient conditions for x^* to be a local minimizer are satisfied if $Z^T G Z$ is positive definite, where Z is defined to be a null-space basis matrix for the active constraint Jacobian matrix

$$[a_i]_{i \in \mathcal{A}(x^*)}^T.$$

As we showed in Theorem 16.2, x^* is actually a global solution for the equality-constrained case when this condition holds. When G is not positive definite, the general problem (16.1)

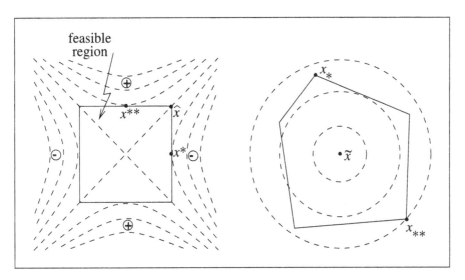

Figure 16.1 Nonconvex quadratic programs

may have more than one strict local minimizer at which the second-order necessary conditions are satisfied. Such problems are referred to as being "nonconvex" or "indefinite," and they cause some complication for algorithms. Examples of indefinite quadratic programs are illustrated in Figure 16.1. On the left we have plotted the contours of a problem in which G has one positive and one negative eigenvalue. We have indicated by $+$ or $-$ that the function tends toward plus or minus infinity in that direction. Note that x^{**} is a local maximizer, x^* a local minimizer, and the center of the box is a stationary point. The picture on the right in Figure 16.1, in which both eigenvalues of G are negative, shows a global maximizer at $\tilde{x}$ and local minimizers at x_* and x_{**}.

DEGENERACY

A second property that causes difficulties for some algorithms is *degeneracy*. Unfortunately, this term has been given a variety of meanings, which can cause confusion. Essentially, it refers to situations in which either

(a) the active constraint gradients a_i, $i \in \mathcal{A}(x^*)$, are linearly dependent at the solution x^*, or

(b) the strict complementarity condition of Definition 12.2 fails to hold, that is, the optimal Lagrange multiplier vector λ^* has $\lambda_i^* = 0$ for some active index $i \in \mathcal{A}(x^*)$. Such constraints are *weakly active* according to Definition 12.3.

Two examples of degeneracy are shown in Figure 16.2. In the left-hand picture, there is a single active constraint at the solution x^*, which is also an unconstrained minimizer of

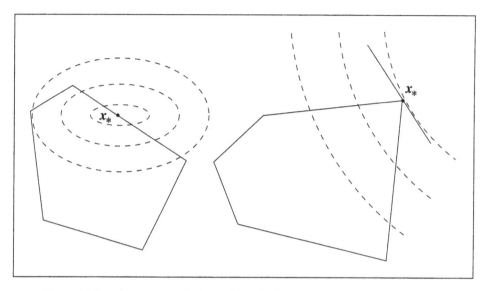

Figure 16.2 Degenerate solutions of quadratic programs.

the objective function. In the notation of (16.26a), we have that $Gx^* + d = 0$, so that the lone Lagrange multiplier must be zero. In the right-hand picture, three constraints are active at the solution. Since each of the three constraint gradients is a vector in $\mathbf{R}^2$, they must be linearly dependent.

A more subtle case of degeneracy is illustrated by the problem

$$\min x_1^2 + (x_2 + 1)^2 \quad \text{s.t.} \quad x \geq 0,$$

which has a solution at $x^* = 0$. The unconstrained minimizer does not lie on one of the constraints, nor are there more than $n = 2$ active constraints at the solution. But this problem is degenerate because the Lagrange multiplier associated with the constraint $x_1 \geq 0$ is zero at the solution.

Degeneracy can cause problems for algorithms for two main reasons. First, linear independence of the active constraint gradients can cause numerical difficulties in computation of the null-space matrix Z, and it causes the matrix $AG^{-1}A^T$ that arises in the range-space method to become singular. (Here, A denotes the matrix whose rows are the active constraints.) Second, when the problem contains weakly active constraints, it is difficult for the algorithm to determine whether or not these constraints are active at the solution or not. In the case of active-set methods and gradient projection methods described below, this "indecisiveness" can cause the algorithm to zigzag, as the iterates move on and off the weakly active constraints on successive iterations. Safeguards must be used in these algorithms to prevent such behavior.

16.4 ACTIVE-SET METHODS FOR CONVEX QP

We now describe active-set methods, which are generally the most effective methods for small- to medium-scale problems. We start by discussing the convex case, in which the matrix G in (16.1a) is positive semidefinite. Since the feasible region defined by (16.1b), (16.1c) is a convex set, any local solution of the QP is a global minimizer. The case in which G is an indefinite matrix raises complications in the algorithms; our discussion of this case appears in the next section.

Recall our definition (16.25) above of the active set $\mathcal{A}(x)$ at the optimal point x^*. We will call it an *optimal* active set.

If $\mathcal{A}(x^*)$ were known in advance, we could find the solution by applying one of the techniques for equality-constrained QP of Section 16.2 to the problem

$$\min_x q(x) = \tfrac{1}{2} x^T G x + x^T d \quad \text{subject to} \quad a_i^T x = b_i, \ i \in \mathcal{A}(x^*).$$

Of course, we usually don't have prior knowledge of $\mathcal{A}(x^*)$, and as we will now see, determination of this set is the main challenge facing algorithms for inequality-constrained QP.

We have already encountered an active-set approach for linear programming in Chapter 13, namely the simplex method. An active-set method starts by making a guess of the optimal active set, and if this guess turns out to be incorrect, it repeatedly uses gradient and Lagrange multiplier information to drop one index from the current estimate of $\mathcal{A}(x^*)$ and add a new index. Active-set methods for QP differ from the simplex method in that the iterates may not move from one vertex of the feasible region to another. Some iterates (and, indeed, the solution of the problem) may lie at other points on the boundary or interior of the feasible region.

Active-set methods for QP come in three varieties, known as *primal, dual,* and *primal–dual*. We restrict our discussion to primal methods, which generate iterates that remain feasible with respect to the primal problem (16.1) while steadily decreasing the primal objective function $q(\cdot)$.

Primal active-set methods usually start by computing a feasible initial iterate x_0, and then ensure that all subsequent iterates remain feasible. They find a step from one iterate to the next by solving a quadratic subproblem in which a subset of the constraints in (16.1b), (16.1c) is imposed as equalities. This subset is referred to as the *working set* and is denoted at the kth iterate x_k by $\mathcal{W}_k$. It consists of all the equality constraints $i \in \mathcal{E}$ (see 16.1b) together with some—but not necessarily all—of the active inequality constraints. An important requirement we impose on $\mathcal{W}_k$ is that the gradients a_i of the constraints in the working set be linearly independent, even when the full set of active constraints at that point has linearly dependent gradients. Later, we discuss how this condition can be enforced without compromising convergence of the algorithm to a solution.

Given an iterate x_k and the working set $\mathcal{W}_k$, we first check whether x_k minimizes the quadratic q in the subspace defined by the working set. If not, we compute a step p

by solving an equality-constrained QP subproblem in which the constraints corresponding to the working set $\mathcal{W}_k$ are regarded as equalities and all other constraints are temporarily disregarded. To express this subproblem in terms of the step p, we define

$$p = x - x_k, \qquad g_k = Gx_k + d,$$

and by substituting for x into the objective function (16.1a), we find that

$$q(x) = q(x_k + p) = \tfrac{1}{2} p^T G p + g_k^T p + c,$$

where $c = \tfrac{1}{2} x_k^T G x_k + d^T x_k$ is a constant term. Since we can drop c from the objective without changing the solution of the problem, we can write the QP subproblem to be solved at the kth iteration as follows:

$$\min_{p} \tfrac{1}{2} p^T G p + g_k^T p \tag{16.27a}$$

$$\text{subject to } a_i^T p = 0 \text{ for all } i \in \mathcal{W}_k. \tag{16.27b}$$

We denote the solution of this subproblem by p_k. Note that for each $i \in \mathcal{W}_k$, the term $a_i^T x$ does not change as we move along p_k, since we have $a_i^T (x_k + p_k) = a_i^T x_k = b_i$. It follows that since the constraints in $\mathcal{W}_k$ were satisfied at x_k, they are also satisfied at $x_k + \alpha p_k$, for any value of α. When G is positive definite, the solution of (16.27b) can be computed by any of the techniques described in Section 16.2.

Suppose for the moment that the optimal p_k from (16.27) is nonzero. We need to decide how far to move along this direction. If $x_k + p_k$ is feasible with respect to all the constraints, we set $x_{k+1} = x_k + p_k$. Otherwise, we set

$$x_{k+1} = x_k + \alpha_k p_k, \tag{16.28}$$

where the step-length parameter α_k is chosen to be the largest value in the range $[0, 1)$ for which all constraints are satisfied. We can derive an explicit definition of α_k by considering what happens to the constraints $i \notin \mathcal{W}_k$, since the constraints $i \in \mathcal{W}_k$ will certainly be satisfied regardless of the choice of α_k. If $a_i^T p_k \geq 0$ for some $i \notin \mathcal{W}_k$, then for all $\alpha_k \geq 0$ we have $a_i^T (x_k + \alpha_k p_k) \geq a_i^T x_k \geq b_i$. Hence, this constraint will be satisfied for all nonnegative choices of the step-length parameter. Whenever $a_i^T p_k < 0$ for some $i \notin \mathcal{W}_k$, however, we have that $a_i^T (x_k + \alpha_k p_k) \geq b_i$ only if

$$\alpha_k \leq \frac{b_i - a_i^T x_k}{a_i^T p_k}.$$

Since we want α_k to be as large as possible in $[0, 1]$ subject to retaining feasibility, we have the following definition:

$$\alpha_k \stackrel{\text{def}}{=} \min\left(1, \min_{i \notin \mathcal{W}_k, a_i^T p_k < 0} \frac{b_i - a_i^T x_k}{a_i^T p_k}\right). \tag{16.29}$$

We call the constraints i for which the minimum in (16.29) is achieved the *blocking constraints*. (If $\alpha_k = 1$ and no new constraints are active at $x_k + \alpha_k p_k$, then there are no blocking constraints on this iteration.) Note that it is quite possible for α_k to be zero, since we could have $a_i^T p_k < 0$ for some constraint i that is active at x_k but not a member of the current working set $\mathcal{W}_k$.

If $\alpha_k < 1$, that is, the step along p_k was blocked by some constraint not in $\mathcal{W}_k$, a new working set $\mathcal{W}_{k+1}$ is constructed by adding one of the blocking constraints to $\mathcal{W}_k$.

We continue to iterate in this manner, adding constraints to the working set until we reach a point $\hat{x}$ that minimizes the quadratic objective function over its current working set $\hat{\mathcal{W}}$. It is easy to recognize such a point because the subproblem (16.27) has solution $p = 0$. Since $p = 0$ satisfies the optimality conditions (16.5) for (16.27), we have that

$$\sum_{i \in \hat{\mathcal{W}}} a_i \hat{\lambda}_i = g = G\hat{x} + d, \tag{16.30}$$

for some Lagrange multipliers $\hat{\lambda}_i$, $i \in \hat{\mathcal{W}}$. It follows that $\hat{x}$ and $\hat{\lambda}$ satisfy the first KKT condition (16.26a), if we define the multipliers corresponding to the inequality constraints that are not in the working set to be zero. Because of the control imposed on the step-length, $\hat{x}$ is also feasible with respect to all the constraints, so the second and third KKT conditions (16.26b) and (16.26c) are satisfied by $\hat{x}$.

We now examine the signs of the multipliers corresponding to the inequality constraints in the working set, that is, the indices $i \in \hat{\mathcal{W}} \cap \mathcal{I}$. If these multipliers are all nonnegative, the fourth KKT condition (16.26d) is also satisfied, so we conclude that $\hat{x}$ is a KKT point for the original problem (16.1). In fact, since G is positive semidefinite, we can show that $\hat{x}$ is a local minimizer. When G is positive definite, $\hat{x}$ is a strict local minimizer.

If, on the other hand, one of the multipliers $\hat{\lambda}_j$, $j \in \hat{\mathcal{W}} \cap \mathcal{I}$, is negative, the condition (16.26d) is not satisfied, and the objective function $q(\cdot)$ may be decreased by dropping this constraint, as shown in Section 12.2. We then remove an index j corresponding to one of the negative multipliers from the working set and solve a new subproblem (16.27) for the new step. We show in the following theorem that this strategy produces a direction p at the next iteration that is feasible with respect to the dropped constraint. We continue to assume that the vectors in the working set are linearly independent, and we defer a discussion of how this can be achieved to the next section when the algorithm has been fully stated.

Theorem 16.4.

Suppose that the point $\hat{x}$ satisfies first-order conditions for the equality-constrained subproblem with working set $\hat{\mathcal{W}}$; that is, equation (16.30) is satisfied along with $a_i^T \hat{x} = b_i$ for all $i \in \hat{\mathcal{W}}$. Suppose, too, that the constraint gradients $a_i, i \in \hat{\mathcal{W}}$, are linearly independent, and that there is an index $j \in \hat{\mathcal{W}}$ such that $\hat{\lambda}_j < 0$. Finally, let p be the solution obtained by dropping the constraint j and solving the following subproblem:

$$\min_p \tfrac{1}{2} p^T G p + (G\hat{x} + d)^T p, \tag{16.31a}$$

$$\text{subject to } a_i^T p = 0, \text{ for all } i \in \hat{\mathcal{W}} \text{ with } i \neq j. \tag{16.31b}$$

(This is the subproblem to be solved at the next iteration of the algorithm.) Then p is a feasible direction for constraint j, that is, $a_j^T p \geq 0$. Moreover, if p satisfies second-order sufficient conditions for (16.31), then we have that $a_j^T p > 0$, and p is a descent direction for $q(\cdot)$.

PROOF. Since p solves (16.31), we have from the results of Section 16.1 that there are multipliers $\tilde{\lambda}_i$, for all $i \in \hat{\mathcal{W}}$ with $i \neq j$, such that

$$\sum_{i \in \hat{\mathcal{W}}, i \neq j} \tilde{\lambda}_i a_i = Gp + (G\hat{x} + d). \tag{16.32}$$

In addition, we have by second-order necessary conditions that if Z is a null-space basis vector for the matrix

$$[a_i]_{i \in \hat{\mathcal{W}}, i \neq j}^T,$$

then $Z^T G Z$ is positive semidefinite. Clearly, p has the form $p = Z p_z$ for some vector p_z, so it follows that $p^T G p \geq 0$.

We have made the assumption that $\hat{x}$ and $\hat{\mathcal{W}}$ satisfy the relation (16.30). By subtracting (16.30) from (16.32), we obtain

$$\sum_{i \in \hat{\mathcal{W}}, i \neq j} (\tilde{\lambda}_i - \hat{\lambda}_i) a_i - \hat{\lambda}_j a_j = Gp. \tag{16.33}$$

By taking inner products of both sides with p and using the fact that $a_i^T p = 0$ for all $i \in \hat{\mathcal{W}}$ with $i \neq j$, we have that

$$-\hat{\lambda}_j a_j^T p = p^T G p. \tag{16.34}$$

Since $p^T G p \geq 0$ and $\hat{\lambda}_j < 0$ by assumption, it follows immediately that $a_j^T p \geq 0$.

If the second-order sufficient conditions are satisfied, we have that $Z^T G Z$ defined above is positive definite. From (16.34) we can have $a_j^T p = 0$ only if $p^T G p = p_z^T Z^T G Z p_z =$

0, which happens only if $p_z = 0$ and $p = 0$. But if $p = 0$, then by substituting into (16.33) and using linear independence of a_i for $i \in \hat{\mathcal{W}}$, we must have that $\hat{\lambda}_j = 0$, which contradicts our choice of j. We conclude that $p^T G p > 0$ in (16.34), and therefore $a_j^T p > 0$ whenever p satisfies the second-order sufficient conditions for (16.31). □

While any index j for which $\hat{\lambda}_j < 0$ usually will give directions along which the algorithm can make progress, the most negative multiplier is often chosen in practice (and in the algorithm specified below). This choice is motivated by the sensitivity analysis given in Chapter 12, which shows that the rate of decrease in the objective function when one constraint is removed is proportional to the magnitude of the Lagrange multiplier for that constraint.

We conclude with a result that shows that whenever p_k obtained from (16.27) is nonzero and satisfies second-order sufficient optimality conditions for the current working set, then it is a direction of strict descent for $q(\cdot)$.

Theorem 16.5.

Suppose that the solution p_k of (16.27) is nonzero and satisfies the second-order sufficient conditions for optimality for that problem. Then the function $q(\cdot)$ is strictly decreasing along the direction p_k.

PROOF. Since p_k satisfies the second-order conditions, that is, $Z^T G Z$ is positive definite for the matrix Z whose columns are a basis of the null space of the constraints (16.27b), we have by applying Theorem 16.2 to (16.27) that p_k is the unique global solution of (16.27). Since $p = 0$ is also a feasible point for (16.27), its objective value in (16.27a) must be larger than that of p_k, so we have

$$\tfrac{1}{2} p_k^T G p_k + g_k^T p_k < 0.$$

Since $p_k^T G p_k \geq 0$ by convexity, this inequality implies that $g_k^T p_k < 0$. Therefore, we have

$$q(x_k + \alpha_k p_k) = q(x_k) + \alpha g_k^T p_k + \tfrac{1}{2} \alpha^2 p_k^T G p_k < q(x_k),$$

for all $\alpha > 0$ sufficiently small. □

A corollary of this result is that when G is positive definite—the *strictly* convex case—the second-order sufficient conditions are satisfied for *all* feasible subproblems of the form (16.27), so that we obtain a strict decrease in $q(\cdot)$ whenever $p_k \neq 0$. This fact is significant in a later section, when we discuss finite termination of the algorithm.

SPECIFICATION OF THE ACTIVE-SET METHOD FOR CONVEX QP

Having given a complete description of the active-set algorithm for convex QP, it is time for the following formal specification:

Algorithm 16.1 (Active-Set Method for Convex QP).
Compute a feasible starting point x_0;
Set $\mathcal{W}_0$ to be a subset of the active constraints at x_0;
   for $k = 0, 1, 2, \ldots$
      Solve (16.27) to find p_k;
      if $\quad p_k = 0$
          Compute Lagrange multipliers $\hat{\lambda}_i$ that satisfy (16.30),
              set $\hat{\mathcal{W}} = \mathcal{W}_k$;
          if $\quad \hat{\lambda}_i \geq 0$ for all $i \in \mathcal{W}_k \cap \mathcal{I}$;
              STOP with solution $x^* = x_k$;
          else
              Set $j = \arg\min_{j \in \mathcal{W}_k \cap \mathcal{I}} \hat{\lambda}_j$;
              $x_{k+1} = x_k$; $\mathcal{W}_{k+1} \leftarrow \mathcal{W}_k \setminus \{j\}$;
      else (* $p_k \neq 0$ *)
          Compute α_k from (16.29);
          $x_{k+1} \leftarrow x_k + \alpha_k p_k$;
          if   there are blocking constraints
              Obtain $\mathcal{W}_{k+1}$ by adding one of the blocking
                    constraints to $\mathcal{W}_{k+1}$;
          else
              $\mathcal{W}_{k+1} \leftarrow \mathcal{W}_k$;
end (for)

Various techniques can be used to determine an initial feasible point. One such is to use the "Phase I" approach described in Chapter 13. Though no significant modifications are needed to generalize this method from linear programming to quadratic programming, we describe a variant here that allows the user to supply an initial estimate $\tilde{x}$ of the vector x. This estimate need not be feasible, but prior knowledge of the QP may be used to select a value of $\tilde{x}$ that is "not too infeasible," which will reduce the work needed to perform the Phase I step.

Given $\tilde{x}$, we define the following feasibility linear program:

$$\min_{(x,z)} \quad e^T z$$
$$\text{subject to } a_i^T x + \gamma_i z_i = b_i, \quad i \in \mathcal{E},$$
$$a_i^T x + \gamma_i z_i \geq b_i, \quad i \in \mathcal{I},$$
$$z \geq 0,$$

where $e = (1, \ldots, 1)^T$, $\gamma_i = -\text{sign}(a_i^T \tilde{x} - b_i)$ for $i \in \mathcal{E}$, while $\gamma_i = 1$ for $i \in \mathcal{I}$. A feasible initial point for this problem is then

$$x = \tilde{x}, \quad z_i = |a_i^T \tilde{x} - b_i| \ (i \in \mathcal{E}), \quad z_i = \max(b_i - a_i^T \tilde{x}, 0) \ (i \in \mathcal{I}).$$

16.4. ACTIVE-SET METHODS FOR CONVEX QP

It is easy to verify that if $\tilde{x}$ is feasible for the original problem (16.1), then $(\tilde{x}, 0)$ is optimal for the feasibility subproblem. In general, if the original problem has feasible points, then the optimal objective value in the subproblem is zero, and any solution of the subproblem yields a feasible point for the original problem. The initial working set $\mathcal{W}_0$ for Algorithm 16.1 can be found by taking a linearly independent subset of the active constraints at the x component of the solution of the feasibility problem.

An alternative approach is the so-called "big M" method, which does away with the "Phase I" and instead includes a measure of infeasibility in the objective that is guaranteed to be zero at the solution. That is, we introduce a scalar artificial variable t into (16.1) to measure the constraint violation, and solve the problem

$$\min_{(x,t)} \tfrac{1}{2} x^T G x + x^T d + M t, \tag{16.35a}$$

$$\text{subject to } t \geq (a_i^T x - b_i), \quad i \in \mathcal{E}, \tag{16.35b}$$

$$t \geq -(a_i^T x - b_i), \quad i \in \mathcal{E}, \tag{16.35c}$$

$$t \geq b_i - a_i^T x, \quad i \in \mathcal{I}, \tag{16.35d}$$

$$t \geq 0, \tag{16.35e}$$

for some large positive value of M. It can be shown by applying the theory of exact penalty functions (see Chapter 15) that whenever there exist feasible points for the original problem (16.1), then for all M sufficiently large, the solution of (16.35) will have $t = 0$, with an x component that is a solution for (16.1).

Our strategy is to use some heuristic to choose a value of M and solve (16.35) by the usual means. If the solution we obtain has a positive value of t, we increase M and try again. Note that a feasible point is easy to obtain for the subproblem (16.35): We set $x = \tilde{x}$ (where, as before, $\tilde{x}$ is the user-supplied initial guess) and choose t large enough that all the constraints in (16.35) are satisfied.

This approach is related to the $S\ell_1$QP method described in Chapter 18; the main difference is that the "big M" method is based on the infinity norm rather than on the ℓ_1 norm. (See (12.7) in Chapter 12.)

AN EXAMPLE

In this section we use subscripts on the vector x to denote its components, while superscripts denote the iteration index. For example, x^4 denotes the fourth iterate of x, while x_1 denotes the first component.

464 CHAPTER 16. QUADRATIC PROGRAMMING

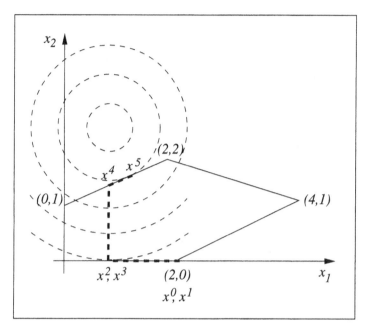

Figure 16.3 Iterates of the active-set method.

❑ EXAMPLE 16.3

We apply Algorithm 16.1 to the following simple 2-dimensional problem, which is illustrated in Figure 16.3.

$$\min_x q(x) = (x_1 - 1)^2 + (x_2 - 2.5)^2 \quad (16.36\text{a})$$
$$\text{subject to} \quad x_1 - 2x_2 + 2 \geq 0, \quad (16.36\text{b})$$
$$-x_1 - 2x_2 + 6 \geq 0, \quad (16.36\text{c})$$
$$-x_1 + 2x_2 + 2 \geq 0, \quad (16.36\text{d})$$
$$x_1 \geq 0, \quad (16.36\text{e})$$
$$x_2 \geq 0. \quad (16.36\text{f})$$

We label the constraints, in order, with the indices 1 through 5. For this problem it is easy to determine a feasible initial point; suppose that we choose $x^0 = (2, 0)$. Constraints 3 and 5 are active at this point, and we set $\mathcal{W}_0 = \{3, 5\}$. (Note that we could just as validly have chosen $\mathcal{W}_0 = \{5\}$ or $\mathcal{W}_0 = \{3\}$ or even $\mathcal{W} = \emptyset$; each would lead the algorithm to perform quite differently.)

Since x^0 lies on a vertex of the feasible region, it is obviously a minimizer of the objective function q with respect to the working set $\mathcal{W}_0$; that is, the solution of (16.27) with $k = 0$ is

16.4. ACTIVE-SET METHODS FOR CONVEX QP

$p = 0$. We can then use (16.30) to find the multipliers $\hat{\lambda}_3$ and $\hat{\lambda}_5$ associated with the active constraints. Substitution of the data from our problem into (16.30) yields

$$\begin{bmatrix} -1 \\ 2 \end{bmatrix} \hat{\lambda}_3 + \begin{bmatrix} 0 \\ 1 \end{bmatrix} \hat{\lambda}_5 = \begin{bmatrix} 2 \\ -5 \end{bmatrix},$$

which has the solution $(\hat{\lambda}_3, \hat{\lambda}_5) = (-2, -1)$.

We now remove constraint 3 from the working set, since it has the most negative multiplier, and set $\mathcal{W}_1 = \{5\}$. We begin iteration 1 by finding the solution of (16.27) for $k = 1$, which is $p^1 = (-1, 0)^T$. The step-length formula (16.29) yields $\alpha_1 = 1$, and the new iterate is $x^2 = (1, 0)$.

There are no blocking constraints, so that $\mathcal{W}_2 = \mathcal{W}_1 = \{5\}$, and we find at the start of iteration 2 that the solution of (16.27) is again $p^2 = 0$. From (16.30) we deduce that the Lagrange multiplier for the lone working constraint is $\hat{\lambda}_5 = -5$, so we drop 5 from the working set to obtain $\mathcal{W}_3 = \emptyset$.

Iteration 3 starts by solving the unconstrained problem, to obtain the solution $p^3 = (0, 2.5)$. The formula (16.29) yields a step length of $\alpha_3 = 0.6$ and a new iterate $x^4 = (1, 1.5)$. There is a single blocking constraint—constraint 1—so we obtain $\mathcal{W}_4 = \{1\}$. The solution of (16.27) for $k = 4$ is then $p^4 = (0.4, 0.2)$, and the new step length is 1. There are no blocking constraints on this step, so the next working set is unchanged: $\mathcal{W}_5 = \{1\}$. The new iterate is $x^5 = (1.4, 1.7)$.

Finally, we solve (16.27) for $k = 5$ to obtain a solution $p^5 = 0$. The formula (16.30) yields a multiplier $\hat{\lambda}_1 = 1.25$, so we have found the solution. We set $x^* = (1.4, 1.7)$ and terminate. □

FURTHER REMARKS ON THE ACTIVE-SET METHOD

We noted above that there is flexibility in the choice of the initial working set, and that each initial choice leads to a different iteration sequence. When the initial active constraints have independent gradients, as above, we can include them all in $\mathcal{W}_0$. Alternatively, we can select a subset. For instance, if in the example above we have chosen $\mathcal{W}_0 = \{3\}$, the first iterate would have yielded $p^0 = (0.2, 0.1)$ and a new iterate of $x^1 = (2.2, 0.1)$. If we had chosen $\mathcal{W}_0 = \{5\}$, we would have moved immediately to the new iterate $x^1 = (1, 0)$, without first performing the operation of dropping the index 3, as is done in the example. Finally, if we had selected $\mathcal{W}_0 = \emptyset$, we would obtain $p^1 = (-1, 2.5)$, $\alpha_1 = \frac{2}{3}$, a new iterate of $x^1 = (\frac{4}{3}, \frac{5}{3})$, and a new working set of $\mathcal{W}_1 = \{1\}$. The solution x^* would have been found on the next iteration.

Even if the initial working set $\mathcal{W}_0$ coincides with the initial active set, the sets $\mathcal{W}_k$ and $\mathcal{A}(x^k)$ may differ at later iterations. For instance, when a particular step encounters more than one blocking constraint, just one of them is added to the working set, so the identification between W_k and $\mathcal{A}(x^k)$ is broken. Moreover, subsequent iterates differ in general according to what choice is made.

We require that the constraint gradients in W_0 be linearly independent, and our strategy for modifying the working set ensures that this same property holds for all subsequent working sets W_k. When we encounter a blocking constraint on a particular step, a constraint that is not in the working set is encountered during the line search, and its constraint normal cannot be a linear combination of the normals a_i in the current working set (see Exercise 15). Hence, linear independence is maintained after the blocking constraint is added to the working set. On the other hand, deletion of an index from the working set certainly does not introduce linear dependence.

The strategy of removing the constraint corresponding to the most negative Lagrange multiplier often works well in practice, but has the disadvantage that it is susceptible to the scaling of the constraints. (By multiplying constraint i by some factor $\beta > 0$ we do not change the geometry of the optimization problem, but we introduce a scaling of $1/\beta$ to the corresponding multiplier λ_i.) Choice of the most negative multiplier is analogous to Dantzig's original pivot rule for the simplex method in linear programming (see Chapter 13), and as we saw there, more sophisticated strategies that were more resistant to scaling often gave better results. We will not discuss these advanced features here.

Finally, we note that the strategy of adding or deleting at most one constraint at each iteration of the Algorithm 16.1 places a natural lower bound on the number of iterations needed to reach optimality. Suppose, for instance, that we have a problem in which m constraints are active at the solution x^*, but that we start from a point x^0 that is strictly feasible with respect to all the inequality constraints. In this case, the algorithm will need at least m iterations to move from x^0 to x^*. Even more iterations would be required if the algorithm adds some constraint j to the working set at some iteration, only to remove it at a later step.

FINITE TERMINATION OF THE CONVEX QP ALGORITHM

It is not difficult to show that Algorithm 16.1 converges for strictly convex QPs under certain assumptions, that is, it identifies the solution x^* in a finite number of iterations. This claim is certainly true if we assume that the method always takes a nonzero step length α_k whenever the direction p_k computed from (16.27) is nonzero. Our argument proceeds as follows:

- If the solution of (16.27) is $p_k = 0$, the current point x_k is the unique global minimizer of $q(\cdot)$ for the working set W_k; see Theorem 16.5. If it is not the solution of the original problem (16.1) (that is, at least one of the Lagrange multipliers is negative), Theorems 16.4 and 16.5 together show that the step p_{k+1} computed after a constraint is dropped will be a strict decrease direction for $q(\cdot)$. Therefore, because of our assumption $\alpha_k > 0$, we have that the value of q is lower than $q(x_k)$ at all subsequent iterations. It follows that the algorithm can never return to the working set W_k, since subsequent iterates have values of q that are lower than the global minimizer for this working set.

- The algorithm encounters an iterate k for which $p_k = 0$ solves (16.27) at least on every nth iteration. To show this claim, note that whenever we have an iteration for which $p_k \neq 0$, either we have $\alpha_k = 1$ (in which case we reach the minimizer of q on the current working set $\mathcal{W}_k$, so that the next iteration will yield $p_{k+1} = 0$), or else a constraint is added to the working set $\mathcal{W}_k$. If the latter situation occurs repeatedly, then after at most n iterations the working set will contain n indices, which correspond to n linearly independent vectors. The solution of (16.27) will then be $p_k = 0$, since only the zero vector will satisfy the constraints (16.27b).

- Taken together, the two statements above indicate that the algorithm finds the global minimum of q on its current working set periodically (at least once every n iterations) and that having done so, it never visits this particular working set again. It follows that since there are only a finite number of possible working sets, the algorithm cannot iterate forever. Eventually, it encounters a minimizer for a current working set that satisfies optimality conditions for (16.1), and it terminates with a solution.

The assumption that we can always take a nonzero step along a nonzero descent direction p_k calculated from (16.27) guarantees that the algorithm does not undergo *cycling*. This term refers to the situation in which a sequence of consecutive iterations results in no movement in iterate x, while the working set $\mathcal{W}_k$ undergoes deletions and additions of indices and eventually repeats itself. That is, for some integers k and $l \geq 1$, we have that $x^k = x^{k+l}$ and $\mathcal{W}_k = \mathcal{W}_{k+l}$. At some point in the cycle, a constraint is dropped (as in Theorem 16.4) but a new constraint $i \notin \mathcal{W}_k$ is encountered immediately without any movement along the computed direction p. Procedures for handling degeneracy and cycling in quadratic programming are similar to those for linear programming discussed in Chapter 13; we will not discuss them here. Most QP implementations simply ignore the possibility of cycling.

UPDATING FACTORIZATIONS

We have seen that the step computation in the active-set method given in Algorithm 16.1 requires the solution of the equality-constrained subproblem (16.27). As mentioned at the beginning of this chapter, this computation amounts to solving the KKT system (16.5). Since the working set can change by just one index at every iteration, the KKT matrix differs in at most one row and one column from the previous iteration's KKT matrix. Indeed, G remains fixed, whereas the matrix A of constraint gradients corresponding to the current working set may change through addition or deletion of a single row.

It follows from this observation that we can compute the matrix factors needed to solve (16.27) at the current iteration by updating the factors computed at the previous iteration, rather than recomputing them from scratch. These updating techniques are crucial to the efficiency of active-set methods.

We will limit our discussion to the case in which the step is computed by using the null-space method (16.16)–(16.19). Suppose that A has m linearly independent rows, and assume that the bases Y and Z are defined by means of a QR factorization of A (see Chapter 15 for details). Thus

$$A^T \Pi = Q \begin{bmatrix} R \\ 0 \end{bmatrix} = \begin{bmatrix} Q_1 & Q_2 \end{bmatrix} \begin{bmatrix} R \\ 0 \end{bmatrix} \qquad (16.37)$$

(see (15.20)), where Π is a permutation matrix; R is square, upper triangular and nonsingular; $Q = \begin{bmatrix} Q_1 & Q_2 \end{bmatrix}$ is $n \times n$ orthogonal; and Q_1 and R both have m columns and Q_2 has $n - m$ columns. As noted in Chapter 15, we can choose Z to be simply the orthonormal matrix Q_2.

Suppose that one constraint is added to the working set at the next iteration, so that the new constraint matrix is $\bar{A}^T = [A^T, a]$, where a is a column vector of length n such that $\bar{A}^T$ retains full column rank. As we now show, there is an economical way to update the Q and R factors in (16.37) to obtain new factors (and hence a new null-space basis matrix $\bar{Z}$, with $n - m - 1$ columns) for the expanded matrix $\bar{A}$. Note first that since $Q_1 Q_1^T + Q_2 Q_2^T = I$,

$$\bar{A}^T \begin{bmatrix} \Pi & 0 \\ 0 & 1 \end{bmatrix} = \begin{bmatrix} A\Pi & | & a \end{bmatrix} = Q \begin{bmatrix} R & | & Q_1^T a \\ 0 & | & Q_2^T a \end{bmatrix}. \qquad (16.38)$$

We can now define an orthogonal matrix $\hat{Q}$ that transforms the vector $Q_2^T a$ to a vector in which all elements except the first are zero. That is, we have

$$\hat{Q}(Q_2^T a) = \begin{bmatrix} \gamma \\ 0 \end{bmatrix},$$

where γ is a scalar. (Since $\hat{Q}$ is orthogonal, we have $\|Q_2^T a\| = |\gamma|$.) From (16.38) we now have

$$\bar{A}^T \begin{bmatrix} \Pi & 0 \\ 0 & 1 \end{bmatrix} = Q \begin{bmatrix} R & | & Q_1^T a \\ 0 & | & \hat{Q}^T \begin{bmatrix} \gamma \\ 0 \end{bmatrix} \end{bmatrix} = Q \begin{bmatrix} I & 0 \\ 0 & \hat{Q}^T \end{bmatrix} \begin{bmatrix} R & | & Q_1^T a \\ 0 & | & \gamma \\ 0 & | & 0 \end{bmatrix}.$$

This factorization has the form

$$\bar{A}^T \bar{\Pi} = \bar{Q} \begin{bmatrix} \bar{R} \\ 0 \end{bmatrix},$$

16.4. ACTIVE-SET METHODS FOR CONVEX QP

where

$$\bar{\Pi} = \begin{bmatrix} \Pi & 0 \\ 0 & 1 \end{bmatrix}, \quad \bar{Q} = Q \begin{bmatrix} I & 0 \\ 0 & \hat{Q}^T \end{bmatrix} = \begin{bmatrix} Q_1 & Q_2 \hat{Q}^T \end{bmatrix}, \quad \bar{R} = \begin{bmatrix} R & Q_1^T a \\ 0 & \gamma \end{bmatrix}.$$

We can therefore choose $\bar{Z}$ to be the last $n - m - 1$ columns of $Q_2 \hat{Q}^T$. If we know Z explicitly and need an explicit representation of $\bar{Z}$, we need to account for the cost of obtaining $\hat{Q}$ and the cost of forming the product $Q_2 \hat{Q}^T = Z \hat{Q}^T$. Because of the special structure of $\hat{Q}$, this cost is of order $n(n - m)$, compared to the cost of computing (16.37) from scratch, which is of order $n^2 m$. The updating strategy is much less expensive, especially when the null space is small (that is, $n - m \ll n$).

An updating technique can also be designed for the case in which a row is removed from A. This operation has the effect of deleting a column from R in (16.37), which disturbs the upper triangular property of this matrix by introducing a number of nonzeros on the diagonal immediately below the main diagonal of the matrix. Upper triangularity can be restored by applying a sequence of plane rotations. These rotations introduce a number of inexpensive transformations into Q, and the updated null-space matrix is obtained by selecting the last $n - m + 1$ columns from this matrix after the transformations are complete. The new null-space basis in this case will have the form

$$\bar{Z} = \begin{bmatrix} Z & \bar{z} \end{bmatrix}, \tag{16.39}$$

that is, the current matrix Z is augmented by a single column. The total cost of this operation varies with the location of the removed column in A, but is in all cases cheaper than recomputing a QR factorization from scratch. For details of these procedures, see Gill et al. [105, Section 5].

Let us now consider the reduced Hessian. Because of the special form of (16.27) we have $c = 0$ in (16.5), and the step p_Y given in (16.17) is zero. Thus from (16.18), the null-space component p_z is the solution of

$$(Z^T G Z) p_z = -Z^T g. \tag{16.40}$$

We can sometimes find ways of updating the factorization of the reduced Hessian $Z^T G Z$ after Z has changed. Suppose that we have the Cholesky factorization of the current reduced Hessian, written as

$$Z^T G Z = L L^T,$$

and that at the next step Z changes as in (16.39), gaining a column after deletion of a constraint. A series of plane rotations can then be used to transform the Cholesky factor L into the new factor $\bar{L}$ for the new reduced Hessian $\bar{Z}^T G \bar{Z}$.

A variety of other simplifications are possible. For example, as discussed in Section 16.6, we can update the reduced gradient $Z^T g$ at the same time as we update Z to $\bar{Z}$.

16.5 ACTIVE-SET METHODS FOR INDEFINITE QP

We now consider the case in which the Hessian matrix G has some negative eigenvalues. Algorithm 16.1, the active-set method for convex QP, can be adapted to this indefinite case by modifying the computation of the search direction and step length in certain situations.

To explain the need for the modification, we consider the computation of a step by a null-space method, that is, $p = Zp_z$, where p_z is given by (16.40). If the reduced Hessian $Z^T G Z$ is positive definite, then this step p points to the minimizer of the subproblem (16.27), and the logic of the iteration need not be changed. If $Z^T G Z$ has negative eigenvalues, however, p points only to a saddle point of (16.27) and is therefore not a suitable step. Instead, we seek an alternative direction s_z that is a direction of *negative curvature* for $Z^T G Z$. We then have that

$$q(x + \alpha Z s_z) \to -\infty \quad \text{as } \alpha \to \infty. \tag{16.41}$$

Additionally, we can choose the sign of s_z so that Zs_z is a non–ascent direction for q at the current point x, that is, $\nabla q(x)^T Z s_z \leq 0$. By moving along the direction Zs_z, we will encounter a constraint that can then be added to the working set for the next iteration. (If we don't find such a constraint, the problem is unbounded.) If the reduced Hessian for the new working set is not positive definite, we can repeat this process until enough constraints have been added to make the reduced Hessian positive definite. A difficulty with this general approach, however, is that if we allow the reduced Hessian to have several negative eigenvalues, we need to compute its spectral factorization or symmetric indefinite factorization in order to obtain appropriate negative curvature directions, but it is difficult to make these methods efficient when the reduced Hessian changes from one working set to the next.

Inertia controlling methods are a practical class of algorithms for indefinite QP that never allow the reduced Hessian to have more than one negative eigenvalue. As in the convex case, there is a preliminary phase in which a feasible starting point x_0 is found. We place the additional demand on x_0 that it be either a vertex (in which case the reduced Hessian is the null matrix) or a constrained stationary point at which the reduced Hessian is positive definite. (We see below how these conditions can be met.) At each iteration, the algorithm will either add or remove a constraint from the working set. If a constraint is added, the reduced Hessian is of smaller dimension and must remain positive definite or be the null matrix (see the exercises). Therefore, an indefinite reduced Hessian can arise only when one of the constraints is removed from the working set, which happens only when the current

16.5. ACTIVE-SET METHODS FOR INDEFINITE QP

point is a minimizer with respect to the current working set. In this case, we will choose the new search direction to be a direction of negative curvature for the reduced Hessian.

There are various algorithms for indefinite QP that differ in the way that indefiniteness is detected, in the computation of the negative curvature direction, and in the handling of the working set. We now discuss an algorithm that makes use of pseudo-constraints (as proposed by Fletcher [80]) and that computes directions of negative curvature by means of the LDL^T factorization (as proposed by Gill and Murray [107]).

Suppose that the current reduced Hessian $Z^T G Z$ is positive definite and that it is factored as $Z^T G Z = LDL^T$, where L is unit lower triangular and D is diagonal with positive diagonal entries. We denote the number of elements in the current working set $\mathcal{W}$ by t. After removing a constraint from the working set, the new null-space basis can be chosen in the form $Z_+ = [Z \mid z]$ (that is, one additional column), so that the new factors have the form

$$L_+ = \begin{bmatrix} L & 0 \\ l^T & 1 \end{bmatrix}, \quad D_+ = \begin{bmatrix} D & 0 \\ 0 & d_{n-t+1} \end{bmatrix},$$

for some vector l and element d_{n-t+1}. If we discover that d_{n-t+1} in D is negative, we know that the reduced Hessian is indefinite on the manifold defined by the new working set. We can then compute a direction s_z of negative curvature for $Z^T G Z$ and a corresponding direction s of negative curvature for G as follows:

$$L_+^T s_z = e_{n-t+1}, \quad s = Z_+ s_z.$$

We can verify that these directions have the desired properties:

$$\begin{aligned} s^T G s &= s_z^T Z_+^T G Z_+ s_z \\ &= s_z^T L_+ D_+ L_+^T s_z \\ &= e_{n-t+1}^T D_+ e_{n-t+1} \\ &= d_{n-t+1} < 0. \end{aligned}$$

Before moving along this direction s, we will define the working set in a special way. Suppose that i is the index of the constraint that is scheduled to be removed from the working set—the index whose removal causes the reduced Hessian to become indefinite and leads to the negative curvature direction s derived above. Rather then removing i explicitly from the working set, as we do in Algorithm 16.1, we leave it in the working set as a *pseudo-constraint*. By doing so, we ensure that the reduced Hessian for this working set remains positive definite. We now move along the negative curvature direction s until we encounter a constraint, and then continue to take steps in the usual manner, adding constraints until the solution of an equality-constrained subproblem is found. If at this point we can safely delete the pseudo-constraint i from the working set while retaining positive definiteness of the reduced Hessian, then we do so. If not, we retain it in the working set until a similar opportunity arises on a later iteration.

We illustrate this strategy with a simple example.

ILLUSTRATION

Consider the following indefinite quadratic program in two variables:

$$\min \tfrac{1}{2} x^T \begin{bmatrix} 1 & 0 \\ 0 & -1 \end{bmatrix} x, \tag{16.42a}$$

$$\text{subject to } x_2 \geq 0, \tag{16.42b}$$

$$x_1 + 2x_2 \geq 2, \tag{16.42c}$$

$$-5x_1 + 4x_2 \leq 10, \tag{16.42d}$$

$$x_1 \leq 3. \tag{16.42e}$$

We use superscripts to number the iterates of the algorithm, and use subscripts to denote components of a vector. We choose the initial point $x^1 = (2, 0)$ and define the working set as $\mathcal{W} = \{1, 2\}$, where we have numbered the constraints in the order they appear in (16.42); see Figure 16.4. Since x^1 is a vertex, it is the solution with respect to the

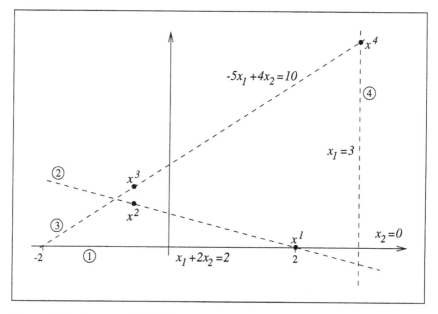

Figure 16.4 Iterates of indefinite QP algorithm.

16.5. ACTIVE-SET METHODS FOR INDEFINITE QP

initial working set $\mathcal{W}$. We compute Lagrange multipliers by solving the system

$$\begin{bmatrix} 0 & 1 \\ 1 & 2 \end{bmatrix} \begin{bmatrix} \lambda_1 \\ \lambda_2 \end{bmatrix} = \begin{bmatrix} 2 \\ 0 \end{bmatrix},$$

giving $(\lambda_1, \lambda_2) = (-4, 2)$. We must therefore remove the first constraint from the working set.

The working set is redefined as $\mathcal{W} = \{2\}$, and we can define the null-space basis as $Z = (2, -1)^T$. Since $Z^T G Z = 3$, we have that $L = 1$ and $D = 3$, which is positive. We solve the equality-constrained subproblem for the current working set, and find that the solution is $x^2 = (-\frac{2}{3}, \frac{4}{3})^T$, and $\lambda^2 = -\frac{2}{3}$.

Since this multiplier is negative, we must remove the second constraint from the working set. We then have $\mathcal{W} = \emptyset$ and could define

$$Z = \begin{bmatrix} 2 & 1 \\ -1 & 2 \end{bmatrix},$$

where the second column has been chosen to be perpendicular to the first. To simplify the computations, however, let us define $Z = I$, so that $L = I$ and $D = G$. This D factor has a negative element in the second diagonal element; see (16.42a). We have now encountered the situation where an indefinite QP algorithm must differ from an algorithm for convex QP.

By solving the system $L^T s_z = e_{n-t} = e_2$ we find that $s_z = (0, 1)^T$ and the direction of negative curvature is given by $s = Z s_z = (0, 1)^T$. The direction s is indeed a descent direction, because we have

$$\nabla q(x^2)^T s = \begin{bmatrix} -2/3 \\ -4/3 \end{bmatrix}^T \begin{bmatrix} 0 \\ 1 \end{bmatrix} < 0.$$

(If this inequality were not true, we could simply change the sign of s.) Before moving along this direction, we include the second constraint as a pseudo-constraint, so that the working set is redefined as $\mathcal{W} = \{2\}$.

We now compute the new iterate $x^3 = x^2 + \alpha s = (-\frac{2}{3}, \frac{5}{3})$, where $\alpha = \frac{1}{3}$ is the step length to the newly encountered third constraint. We add this constraint to the working set, which is now $\mathcal{W} = \{2, 3\}$, and update the LDL^T factorization as we change the working set from $\{2\}$ to $\{2, 3\}$. Since the current iterate is a vertex, it is the solution with respect to the current working set. We also find that both L and D are null matrices.

We will now try to remove the pseudo-constraint. We attempt to solve the equality-constrained subproblem with respect to the working set $\{3\}$, but find that $Z = [4, 5]^T$ and $Z^T G Z = -9$, so that $L = 1$ and $D = -9$. In other words, removal of the pseudo-constraint would yield a reduced Hessian with negative eigenvalues, so we must retain the second constraint as a pseudo-constraint for the time being. However, we can use the hypothetical deletion to calculate a negative-curvature search direction, to move us away from the current iterate. By solving $L^T s_z = e_{2-1} = e_1$ we find that $s_z = 1$, and so the desired direction is $s = Z s_z = (4, 5)^T$. It is easy to verify that s is a descent direction.

We now compute the new iterate $x^4 = x^3 + \alpha s = (3, 25/4)^T$, and find that the step length $\alpha = 11/12$ leads us to the fourth constraint. By adding this constraint to the working set, we obtain $\mathcal{W} = \{2, 3, 4\}$. It is now possible to *drop the pseudo-constraint 2*, since by doing so the working set would become $\{3, 4\}$ and the reduced Hessian is the null matrix, which has no negative eigenvalues. In other words, x^4 solves the subproblem with respect to the working set $\{3, 4\}$.

Continuing with the iteration, we compute Lagrange multipliers at x^4 by solving the following system:

$$\begin{bmatrix} 3 \\ -25/4 \end{bmatrix} = \begin{bmatrix} 5 & -1 \\ -4 & 0 \end{bmatrix} \begin{bmatrix} \lambda_3 \\ \lambda_4 \end{bmatrix}.$$

We obtain $(\lambda_3, \lambda_4) = (25/16, 77/16)$, and we conclude that x^4 is a local solution of the quadratic program (16.42). (In fact, it is also a global solution).

One drawback of this approach is that by including more constraints in the working set, we increase the possibility of degeneracy.

CHOICE OF STARTING POINT

We mentioned above that the initial point must be chosen so as to be a vertex, or at least a point at which the reduced-Hessian matrix is positive definite. To ensure that this property holds, we may need to introduce *artificial constraints* that are active at the initial point and are linearly independent with respect to the other constraints in the working set. For example, consider the quadratic program

$$\min \quad -x_1 x_2$$
$$\text{subject to } -1 \leq x_1 \leq 1,$$
$$-1 \leq x_2 \leq 1.$$

If the starting point is $(0, 0)^T$, no constraints are active, and the reduced Hessian is indefinite for the working set $\mathcal{W}^0 = \emptyset$. By introducing the artificial constraints $x_1 = 0$ and $x_1 - x_2 = 0$, the initial point has the desired properties, so we can start the active-set procedure described above. Once the minimizer for the current working set has been computed, the sign of each artificial constraint normal is chosen so that its multiplier is nonpositive and the constraint may be deleted. Temporary constraints are usually deleted first, if there is a choice.

16.5. ACTIVE-SET METHODS FOR INDEFINITE QP

FAILURE OF THE ACTIVE-SET METHOD

When G is not positive definite, the inertia controlling algorithm just described is not guaranteed to find a local minimizer of the quadratic program. In fact, there may exist nonoptimal points at which *any* active-set method will find it difficult to proceed. These points satisfy the first-order optimality conditions (i.e., they are stationary points), and their reduced Hessian is positive definite, but at least one of the Lagrange multipliers corresponding to the inequality constraints is zero. At such a point, it is not possible to compute a direction that improves the objective function by deleting *only one* constraint at a time. We may have to delete two or more constraints at once in order to make progress.

This phenomenon is illustrated by the problem

$$\min -x_1 x_2 \quad \text{subject to } 0 \leq x_1 \leq 1, \ 0 \leq x_2 \leq 1.$$

Suppose that the starting point is $(0, 0)$ and that both active constraints are included in the initial working set. It is easy to verify that this point is a stationary point, and it is certainly a minimizer on the given working set by virtue of being a vertex. However, if we delete either constraint from the working set, the new reduced Hessian is singular. No feasible direction of negative curvature can therefore be computed by removing only one constraint, so the active-set method will terminate at this point, which is not a local solution of the quadratic program. We can decrease the objective by moving along the direction $(1, 1)$, for instance.

Even though this type of failure is possible, it is not very common, and rounding errors tend to move the algorithm away from such difficult points. Various devices have been proposed to decrease the probability of failure, but none of them can guarantee that a solution will be found in all cases. In the example above, we could move x_1 from zero to 1 while holding x_2 fixed at zero, and this would not affect the value of the objective q. Next we could move x_2 from zero to 1 to obtain the optimal solution.

In general, the cause of this difficulty is that first-order optimality does not lead to second-order optimality for active-set methods.

DETECTING INDEFINITENESS USING THE LBL^T FACTORIZATION

We can also use the symmetric indefinite factorization algorithm (16.12) to determine whether the reduced Hessian $Z^T B Z$ is positive definite. Since this factorization is useful in other optimization methods, such as interior-point methods and sequential quadratic programming methods, we present some of its fundamental properties.

Recall that the inertia of a symmetric matrix K is defined as the triplet formed by the number of positive, negative, and zero eigenvalues of K. We write

$$\text{inertia}(K) = (n_+, n_-, n_0).$$

Sylvester's law of inertia states that if C is a nonsingular matrix, then inertia$(C^T K C) =$ inertia(K); see, for example, [115]. The following result follows from the repeated application of Sylvester's law.

Theorem 16.6.
Let K be defined by (16.7) and suppose that A has rank m. Then

$$\text{inertia}(K) = \text{inertia}(Z^T G Z) + (m, m, 0).$$

This theorem implies that if $Z^T G Z$ is positive definite, then the inertia of the KKT matrix is $(n, m, 0)$, as stated in Lemma 16.3. Note that Theorem 16.6 also shows that K has at least m positive eigenvalues, even if G is negative definite.

Let us suppose that we compute the symmetric indefinite factorization (16.12) of the KKT matrix K. To simplify the discussion, we assume that the permutation matrix P is the identity, and write

$$K = LBL^T.$$

Sylvester's theorem implies that B and K have the same inertia. Moreover, since the 2×2 blocks in B are normally constructed so as to have one positive and one negative eigenvalue, we can compute the inertia of K by examining the blocks of B. In particular, the number of positive eigenvalues of K equals the number of 2×2 blocks plus the number of positive 1×1 blocks.

The symmetric indefinite factorization can therefore be used in inertia controlling methods for indefinite quadratic programming to control the logic of the algorithm.

16.6 THE GRADIENT--PROJECTION METHOD

In the classical active-set method described in the previous two sections, the active set and working set change slowly, usually by a single index at each iteration. As a result, this method may require many iterations to converge on large-scale problems. For instance, if the starting point x^0 has no active constraints, while 200 constraints are active at the solution, then at least 200 iterations of the active-set method will be required to reach the solution.

The gradient projection method is designed to make rapid changes to the active set. It is most efficient when the constraints are simple in form—in particular, when there are only bounds on the variables. Accordingly, we will restrict our attention to the following bound-constrained problem:

$$\min_x q(x) = \tfrac{1}{2} x^T G x + x^T d \qquad (16.43\text{a})$$

$$\text{subject to } l \leq x \leq u, \qquad (16.43\text{b})$$

16.6. THE GRADIENT--PROJECTION METHOD

where G is symmetric and l and u are are vectors of upper and lower bounds on the components of x. (As we see later, the dual of a convex quadratic program is a bound-constrained problem; thus the gradient projection method is of wide applicability.) The feasible region defined by (16.43b) is sometimes called a "box" because of its rectangular shape. Some components of x may lack an upper or a lower bound; we handle these cases by setting the appropriate components of l and u to $-\infty$ and $+\infty$, respectively. We do not make any positive definiteness assumptions on G, since the gradient projection approach can be applied to both convex and nonconvex problems.

Each iteration of the gradient projection algorithm consists of two stages. In the first stage, we search along the steepest descent direction from the current point x, that is, the direction $-g$, where, as in (16.6), $g = Gx + d$. When a bound is encountered, the search direction is "bent" so that it stays feasible. We search along this piecewise path and locate the first local minimizer of q, which we denote by x^c and refer to as the *Cauchy point*, by analogy with our terminology of Chapter 4. The working set $\mathcal{W}$ is now defined to be the set of bound constraints that are active at the Cauchy point, i.e., $\mathcal{W} = \mathcal{A}(x^c)$. In the second stage of each gradient projection iteration, we "explore" the face of the feasible box on which the Cauchy point lies by solving a subproblem in which the active components x_i for $i \in \mathcal{A}(x^c)$ are fixed at the values x_i^c. In this section we denote the iteration number by a superscript (i.e., x^k) and use subscripts to denote the elements of a vector.

CAUCHY POINT COMPUTATION

We now derive an explicit expression for the piecewise linear path obtained by projecting the steepest descent direction onto the feasible box, and outline the search procedure for identifying the first local minimum of q along this path.

We define the projection of an arbitrary point x onto the feasible region (16.43b) as follows: The ith component is given by

$$P(x, l, u)_i = \begin{cases} l_i & \text{if } x_i < l_i, \\ x_i & \text{if } x_i \in [l_i, u_i], \\ u_i & \text{if } x_i > u_i. \end{cases} \quad (16.44)$$

The piecewise linear path $x(t)$ starting at the reference point x^0 and obtained by projecting the steepest descent direction at x^0 onto the feasible region (16.43b) is thus given by

$$x(t) = P(x^0 - tg, l, u), \quad (16.45)$$

where $g = \nabla q(x^0)$; see Figure 16.5. We then compute the Cauchy point x^c, which is defined as the first local minimizer of the univariate, piecewise quadratic $q(x(t))$, for $t \geq 0$. This minimizer is obtained by examining each of the line segments that make up $x(t)$. To perform this search, we need to determine the values of t at which the kinks in $x(t)$, or *breakpoints*, occur. We first identify the values of t for which each component reaches its bound along the chosen direction $-g$. These values $\bar{t}_i$ are given by the following explicit formulae:

$$\bar{t}_i = \begin{cases} (x_i^0 - u_i)/g_i & \text{if } g_i < 0 \text{ and } u_i < +\infty, \\ (x_i^0 - l_i)/g_i & \text{if } g_i > 0 \text{ and } l_i > -\infty, \\ \infty & \text{otherwise.} \end{cases} \quad (16.46)$$

The components of $x(t)$ for any t are therefore

$$x_i(t) = \begin{cases} x_i^0 - tg_i & \text{if } t \leq \bar{t}_i, \\ x_i^0 - \bar{t}_i g_i & \text{otherwise.} \end{cases}$$

This has a simple geometrical interpretation: A component $x_i(t)$ will move at the rate given by the projection of the gradient along this direction, and will remain fixed once this variable reaches one of its bounds.

To search for the first local minimizer along $P(x^0 - tg, l, u)$, we eliminate the duplicate values and zero values of $\bar{t}_i$ from the set $\{\bar{t}_1, \bar{t}_2, \ldots, \bar{t}_n\}$, and sort the remaining numbers into an ordered sequence $t_1, t_2, t_3, \ldots$ such that $0 < t_1 < t_2 < t_3 \leq \cdots$. We now examine the intervals $[0, t_1], [t_1, t_2], [t_2, t_3], \ldots$ in turn. Suppose we have examined the intervals up to

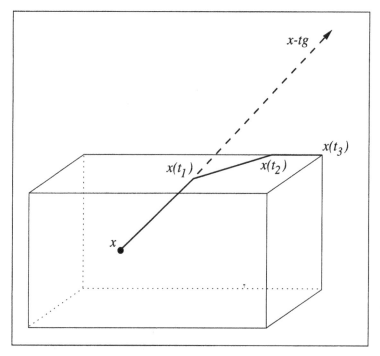

Figure 16.5 The piecewise linear path $x(t)$, for an example in $\mathbf{R}^3$.

16.6. THE GRADIENT--PROJECTION METHOD

$[t_{j-2}, t_{j-1}]$ for some j, and determined that the local minimizer is at some value $t \geq t_{j-1}$. For the interval $[t_{j-1}, t_j]$ between the $(j-1)$st and jth breakpoints, we have that

$$x(t) = x(t_{j-1}) + \Delta t p^{j-1},$$

where

$$\Delta t = t - t_{j-1}, \qquad \Delta t \in [0, t_j - t_{j-1}],$$

and

$$p^{j-1} = \begin{cases} -g_i & \text{if } t_{j-1} < \bar{t}_i, \\ 0 & \text{otherwise.} \end{cases} \tag{16.47}$$

By using this notation, we write the quadratic (16.43a) on the line segment $[x(t_{j-1}), x(t_j)]$ as

$$q(x(t)) = d^T(x(t_{j-1}) + \Delta t p^{j-1}) + \tfrac{1}{2}(x(t_{j-1}) + \Delta t p^{j-1})^T G(x(t_{j-1}) + \Delta t p^{j-1}),$$

where $\Delta t \in [0, t_j - t_{j-1}]$.

By expanding and grouping the coefficients of 1, Δt, and $(\Delta t)^2$, we find that

$$q(x(t)) = f_{j-1} + f'_{j-1} \Delta t + \tfrac{1}{2} f''_{j-1} (\Delta t)^2, \qquad \Delta t \in [0, t_j - t_{j-1}], \tag{16.48}$$

where the coefficients f_{j-1}, f'_{j-1}, and f''_{j-1} are defined by

$$\begin{aligned} f_{j-1} &= d^T x(t_{j-1}) + \tfrac{1}{2} x(t_{j-1})^T G x(t_{j-1}), \\ f'_{j-1} &= d^T p^{j-1} + x(t_{j-1})^T G p^{j-1}, \\ f''_{j-1} &= (p^{j-1})^T G p^{j-1}. \end{aligned}$$

By differentiating (16.48) with respect to Δt and equating to zero, we obtain $\Delta t^* = -f'_{j-1}/f''_{j-1}$. If $\Delta t^* \in [0, t_j - t_{j-1})$ and $f''_{j-1} > 0$, we conclude that there is a local minimizer of $q(x(t))$ at

$$t = t_{j-1} + \Delta t^*.$$

Otherwise, there is a minimizer at $t = t_{j-1}$ when $f'_{j-1} > 0$. In all other cases we move on to the next interval $[t_j, t_{j+1}]$ and continue the search. We need to calculate the new direction p^j from (16.47), and we use this new value to calculate f_j, f'_j, and f''_j. Since p^j differs from p^{j-1} typically in just one component, computational savings can sometimes be made by updating these coefficients rather than computing them from scratch.

SUBSPACE MINIMIZATION

After the Cauchy point x^c has been computed, the components of x^c that are at their lower or upper bounds define the active set $\mathcal{A}(x^c)$. In the second stage of the gradient projection iteration, we approximately solve the QP obtained by fixing these components at the values x_i^c. The subproblem is therefore

$$\min_x q(x) = \tfrac{1}{2} x^T G x + x^T d \tag{16.49a}$$
$$\text{subject to } x_i = x_i^c, \quad i \in \mathcal{A}(x^c), \tag{16.49b}$$
$$l_i \le x_i \le u_i, \quad i \notin \mathcal{A}(x^c). \tag{16.49c}$$

It is not necessary to solve this problem exactly. Nor is it desirable, since the subproblem may be almost as difficult as the original problem (16.43). In fact, to obtain global convergence all we require of the approximate solution x^+ of (16.49) is that $q(x^+) \le q(x^c)$ and that x^+ be feasible with respect to the constraints (16.49b), (16.49c). A strategy that is intermediate between choosing $x^+ = x^c$ as the approximate solution (on the one hand) and solving (16.49) exactly (on the other hand) is to ignore the bound constraints (16.49c) and apply an iterative method to the problem (16.49a), (16.49b). The use of conjugate gradient iterations in combination with the null-space approach for equality-constrained problems, as discussed in Section 16.2, is one possibility. The natural starting point for this process would be x^c. We could terminate the iterations as soon as a bound (16.49c) is encountered by the algorithm or as soon as the conjugate gradient iteration generates a direction of negative curvature; see Algorithm 4.3 in Section 4.1. Note that the null-space basis matrix Z for this problem has a particularly simple form.

We conclude by summarizing the gradient projection algorithm for quadratic programming and stating a convergence theorem.

Algorithm 16.2 (Gradient–Projection Method for QP).
Compute a feasible starting point x^0;
   for $\quad k = 0, 1, 2, \ldots$
      if x^k satisfies the KKT conditions for (16.43)
         STOP with solution $x^* = x^k$;
      Set $x = x^k$ and find the Cauchy point x^c;
      Find an approximate solution x^+ of (16.49) such that $q(x^+) \le q(x^c)$
         and x^+ is feasible;
      $x^{k+1} \leftarrow x^+$;
   end (for)

When applied to problems that satisfy strict complementarity (that is, problems for which all Lagrange multipliers associated with an active bound at the optimizer x^* are nonzero), the active sets $\mathcal{A}(x^c)$ generated by the gradient projection algorithm eventually settle down. That is, constraint indices cannot repeatedly enter and leave the active set on successive iterations. When the problem is degenerate, the active set may not settle down, but various devices have been proposed to prevent this undesirable behavior from taking place.

Finally, we note that while gradient projection methods can be applied in principle to problems with general linear constraints, significant computation is required to apply the projection operator P. If the constraint set is defined as $a_i^T x \geq b_i, i \in \mathcal{I}$, we must solve the following convex quadratic program to compute the projection of a given point $\bar{x}$ onto the feasible set:

$$\min_{x} \|x - \bar{x}\|^2 \quad \text{subject to } a_i^T x \geq b_i \text{ for all } i \in \mathcal{I}.$$

The expense of solving this "projection subproblem" may approach the cost of solving the original quadratic program, so it is usually not economical to apply gradient projection to this case.

16.7 INTERIOR-POINT METHODS

The primal–dual interior-point approach can be applied to convex quadratic programs through a simple extension of the linear-programming algorithms described in Chapter 14. The resulting algorithms are simple to describe, relatively easy to implement, and quite efficient on certain types of problems. Extensions of interior-point methods to nonconvex problems are currently under investigation and will not be discussed here.

For simplicity, we restrict our attention to convex quadratic programs with inequality constraints, which we write as follows:

$$\min_{x} q(x) \stackrel{\text{def}}{=} \tfrac{1}{2} x^T G x + x^T d \quad (16.50\text{a})$$
$$\text{subject to} \quad Ax \geq b, \quad (16.50\text{b})$$

where G is symmetric and positive semidefinite, and where the $m \times n$ matrix A and right-hand-side b are defined by

$$A = [a_i]_{i \in \mathcal{I}}, \quad b = [b_i]_{i \in \mathcal{I}}, \quad \mathcal{I} = \{1, 2, \ldots, m\}.$$

(Equality constraints can be accommodated with simple changes to the approaches described below.) We can specialize the KKT conditions (12.30) to obtain the following set of necessary conditions for (16.50): If x^* is a solution of (16.50), there is a Lagrange multiplier vector λ^* such that the following conditions are satisfied for $(x, \lambda) = (x^*, \lambda^*)$:

$$Gx - A^T \lambda + d = 0,$$
$$Ax - b \geq 0,$$

$$(Ax - b)_i \lambda_i = 0, \quad i = 1, 2, \ldots, m,$$
$$\lambda \geq 0.$$

By introducing the slack vector $y = Ax - b$, we can rewrite these conditions as

$$Gx - A^T\lambda + d = 0, \tag{16.51a}$$
$$Ax - y - b = 0, \tag{16.51b}$$
$$y_i \lambda_i = 0, \quad i = 1, 2, \ldots, m, \tag{16.51c}$$
$$(y, \lambda) \geq 0. \tag{16.51d}$$

It is easy to see the close correspondence between (16.51) and the KKT conditions (14.3) for the linear programming problem (14.1). As in the case of linear programming, the KKT conditions are not only necessary but also sufficient, because the objective function is convex and the feasible region is convex. Hence, we can solve the convex quadratic program (16.50) by finding solutions of the system (16.51).

As in Chapter 14, we can rewrite (16.51) as a constrained system of nonlinear equations and derive primal–dual interior-point algorithms by applying modifications of Newton's method to this system. Analogously to (14.4), we define

$$F(x, y, \lambda) = \begin{bmatrix} Gx - A^T\lambda + d \\ Ax - y - b \\ Y\Lambda e \end{bmatrix}, \quad (y, \lambda) \geq 0,$$

where

$$Y = \operatorname{diag}(y_1, y_2, \ldots, y_m), \quad \Lambda = \operatorname{diag}(\lambda_1, \lambda_2, \ldots, \lambda_m), \quad e = (1, 1, \ldots, 1)^T.$$

Given a current iterate (x, y, λ) that satisfies $(y, \lambda) > 0$, we can define a duality measure μ by

$$\mu = \frac{1}{m} \sum_{i=1}^{m} y_i \lambda_i = \frac{y^T \lambda}{m}, \tag{16.52}$$

similarly to (14.10).

The *central path* $\mathcal{C}$ is the set of points $(x_\tau, y_\tau, \lambda_\tau)$ ($\tau > 0$) such that

$$F(x_\tau, y_\tau, \lambda_\tau) = \begin{bmatrix} 0 \\ 0 \\ \tau e \end{bmatrix}, \quad (y_\tau, \lambda_\tau) > 0$$

(cf. (14.9)). The generic step $(\Delta x, \Delta y, \Delta \lambda)$ is a Newton-like step from the current point (x, y, λ) toward the point $(x_{\sigma\mu}, y_{\sigma\mu}, \lambda_{\sigma\mu})$ on the central path, where $\sigma \in [0, 1]$ is a parameter chosen by the algorithm. As in (14.15), we find that this step satisfies the following linear system:

$$\begin{bmatrix} G & -A^T & 0 \\ A & 0 & -I \\ 0 & Y & \Lambda \end{bmatrix} \begin{bmatrix} \Delta x \\ \Delta s \\ \Delta \lambda \end{bmatrix} = \begin{bmatrix} -r_d \\ -r_b \\ -\Lambda S e + \sigma \mu e \end{bmatrix}, \qquad (16.53)$$

where

$$r_d = Gx - A^T\lambda + d, \qquad r_b = Ax - y - b.$$

As in Framework 14.1 of Chapter 14, we obtain the next iterate by setting

$$(x^+, y^+, \lambda^+) = (x, y, \lambda) + \alpha(\Delta x, \Delta y, \Delta \lambda),$$

where α is chosen to retain the inequality $(y^+, \lambda^+) > 0$ and possibly to satisfy various other conditions.

Mehrotra's predictor–corrector algorithm can also be extended to convex quadratic programming with the exception of one aspect: The step lengths in primal variables (x, y) and dual variables λ cannot be different, as they are in the linear programming case. The reason is that the primal and dual variables are coupled through the matrix G, so different step lengths can disturb feasibility of the equation (16.51a).

The major computational operation is the solution of the system (16.53) at each iteration of the interior-point method. As in Chapter 14, this system may be restated in more compact forms. The augmented system form is

$$\begin{bmatrix} G & -A^T \\ A & \Lambda^{-1}Y \end{bmatrix} \begin{bmatrix} \Delta x \\ \Delta s \end{bmatrix} = \begin{bmatrix} -r_d \\ -r_b + (-y + \sigma\mu\Lambda^{-1}e) \end{bmatrix}, \qquad (16.54)$$

and a symmetric indefinite factorization scheme can be applied to the coefficient matrix. The normal equations form is

$$(G + A^T(Y^{-1}\Lambda)A)\Delta x = -r_d + A^T(Y^{-1}\Lambda)[-r_b - y + \sigma\mu\Lambda^{-1}e],$$

which can be solved by means of a modified Cholesky algorithm. Note that the factorization must be recomputed at each iteration, because the change in y and λ leads to changes in the nonzero components of $A^T(Y^{-1}\Lambda)A$.

EXTENSIONS AND COMPARISON WITH ACTIVE-SET METHODS

It is possible to extend all the algorithms of Chapter 14 to the problem (16.50). To obtain path-following methods, we define the strictly feasible set $\mathcal{F}^o$ by

$$\mathcal{F}^o = \{(x, y, \lambda) \mid Gx - A^T\lambda + d = 0, \ Ax - y - b = 0, \ (y, \lambda) > 0\},$$

and the central path neighborhoods $\mathcal{N}_{-\infty}(\gamma)$ and $\mathcal{N}_2(\theta)$ by

$$\mathcal{N}_2(\theta) = \{(x, y, s) \in \mathcal{F}^o \mid \|YSe - \mu e\| \leq \theta\mu\},$$
$$\mathcal{N}_{-\infty}(\gamma) = \{(x, y, s) \in \mathcal{F}^o \mid y_i s_i \geq \gamma\mu, \ \text{for all } i = 1, 2, \ldots, m\},$$

where $\theta \in [0, 1)$ and $\gamma \in (0, 1]$ (cf. (14.16), (14.17)). As before, all iterates of path-following algorithms are constrained to belong to one or other of these neighborhoods. Potential reduction algorithms are obtained by redefining the Tanabe–Todd–Ye potential function Φ_ρ as

$$\Phi_\rho(y, \lambda) = \rho \log y^T \lambda - \sum_{i=1}^{m} \log y_i \lambda_i,$$

for some $\rho > m$ (cf. (14.30)). As in Chapter 14, iterates of these methods are restricted to $\mathcal{F}^o$, steps are obtained by solving (16.53) with $r_d = 0$ and $r_b = 0$, and the step length α is chosen to force a substantial reduction in Φ_ρ at each iteration.

The basic difference between interior-point and active-set methods for convex QP is that active-set methods generally require a large number of steps in which each search direction is relatively inexpensive to compute, while interior-point methods take a smaller number of more expensive steps. As is the case in linear programming, however, the active-set methods are more complicated to implement, particularly if the program tries to take advantage of sparsity in G and A. In this situation, the factorization updating procedures at each active set are quite difficult to implement efficiently, as the need to maintain sparsity adds complications. By contrast, the nonzero structure of the matrix to be factored at each interior-point iteration does not change (only the numerical values), so standard sparse factorization software can be used to obtain the steps.

16.8 DUALITY

We conclude this chapter with a brief discussion of duality for convex quadratic programming. Duality can be a useful tool in quadratic programming because in some classes of problems we can take advantage of the special structure of the dual to solve the problem more efficiently.

16.8. DUALITY

If G is positive definite, the dual of

$$\min_{x} \tfrac{1}{2} x^T G x + x^T d \tag{16.55a}$$

$$\text{subject to} \quad Ax \geq b \tag{16.55b}$$

is given by

$$\max_{x,\lambda} \tfrac{1}{2} x^T G x + x^T d - \lambda^T (Ax - b)$$

$$\text{subject to} \quad Gx + d - A^T \lambda = 0,$$

$$\lambda \geq 0. \tag{16.56a}$$

By eliminating x from the second equation, we obtain the bound-constrained problem

$$\max_{\lambda} -\tfrac{1}{2} \lambda^T (A G^{-1} A^T) \lambda + \lambda^T (b + A G^{-1} d) - \tfrac{1}{2} d^T G^{-1} d \tag{16.57a}$$

$$\text{subject to} \quad \lambda \geq 0. \tag{16.57b}$$

This problem can be solved by means of the gradient projection method, which normally allows us to identify the active set more rapidly than with a classical active set methods. Moreover, in some areas of application, such as entropy maximization, it is possible to show that the bounds in (16.57b) are not active at the solution. In this case, the problem (16.57) becomes an unconstrained quadratic optimization problem in λ.

For large problems, the complicated form of the objective (16.57a) can be a drawback. If a direct solver is used, then we need to explicitly form $A G^{-1} A^T$. An alternative is to factor G and to apply the conjugate gradient method to $A G^{-1} A^T$.

NOTES AND REFERENCES

If G is not positive definite, the problem of determining whether a feasible point for (16.1) is a global minimizer is an NP-hard problem; see Murty and Kabadi [176].

Markowitz formulated the portfolio optimization problem in 1952 [157]. See his book [159] for further details.

For a discussion on the QMR method see Freund and Nachtigal [94]. The LSQR approach is equivalent to applying the conjugate gradient to the normal equations for the system (16.5); see Paige and Saunders [188].

We have assumed throughout this chapter that all equality-constrained quadratic programs have linearly independent constraints, i.e., that A has rank m. If this is not the case, redundant constraints can be removed from the problem. This can be done, for example, by computing a QR factorization of A^T, which normally (but not always) gives a good indication of the rank of A and of the rows that can be removed. In the large-scale case,

Gaussian elimination techniques are typically used, but this makes it more difficult to decide which constraints should be removed.

The first inertia-controlling method for indefinite quadratic programming was proposed by Fletcher [80]. See also Gill et al. [110] and Gould [120] for a discussion of methods for general quadratic programming.

For further discussion on the gradient projection method see, for example, Conn, Gould, and Toint [51] and Burke and Moré [32].

In some areas of application, the KKT matrix is not just sparse but also contains special structure. For instance, the quadratic programs that arise in optimal control and model predictive control have banded matrices G and A (see Wright [253]). This structure is more easily exploited in the interior-point case, because the matrix in (16.54) can be reordered to have block-banded structure, for which efficient algorithms are available. When active-set methods are applied to this problem, however, the advantages of bandedness and sparsity are lost after just a few updates of the basis.

Further details of interior-point methods for convex quadratic programming can be found in Chapter 8 of Wright [255].

EXERCISES

16.1 Solve the following quadratic program and illustrate it geometrically.

$$\max \ f(x) = 2x_1 + 3x_2 + 4x_1^2 + 2x_1x_2 + x_2^2,$$
$$\text{subject to } x_1 - x_2 \geq 0,$$
$$x_1 + x_2 \leq 4,$$
$$x_1 \leq 3.$$

16.2 The problem of finding the shortest distance from a point x_0 to the hyperplane $\{x \mid Ax = b\}$ where A has full row rank can be formulated as the quadratic program

$$\min \ \tfrac{1}{2}(x - x_0)^T(x - x_0)$$
$$\text{s.t.} \quad Ax = b.$$

Show that the optimal multiplier is

$$\lambda^* = (AA^T)^{-1}(b - Ax_0),$$

and that the solution is

$$x^* = x_0 - A^T(AA^T)^{-1}(b - Ax_0).$$

Show that in the special case where A is a row vector, the shortest distance from x_0 to the solution set of $Ax = b$ is

$$\frac{|b - Ax_0|}{\|A\|}.$$

⬥ **16.3** Use Theorem 12.1 to verify that the first-order necessary conditions for (16.3) are given by (16.4).

⬥ **16.4** Use Theorem 12.6 to show that if the conditions of Lemma 16.1 hold, then the second-order sufficient conditions for (16.3) are satisfied by the vector pair (x^*, λ^*) that satisfies (16.4).

⬥ **16.5** Consider (16.3) and suppose that the projected Hessian matrix $Z^T G Z$ has a negative eigenvalue; that is, $u^T Z^T G Z u < 0$ for some vector u. Show that if there exists any vector pair (x^*, λ^*) that satisfies (16.4), then the point x^* is only a stationary point of (16.3) and not a local minimizer. (Hint: Consider the function $q(x^* + \alpha Z u)$ for $\alpha \neq 0$, and use an expansion like that in the proof of Theorem 16.2.)

⬥ **16.6** By using the QR factorization and a permutation matrix, show that for a full-rank $m \times n$ matrix A (with $m < n$) we can find an orthogonal matrix Q and an $m \times m$ upper triangular matrix $\hat{U}$ such that $AQ = \begin{bmatrix} 0 & \hat{U} \end{bmatrix}$. (Hint: Start by applying the standard QR factorization to A^T.)

⬥ **16.7** Verify that the first-order conditions for optimality of (16.1) are equivalent to (16.26) when we make use of the active set definition (16.25).

⬥ **16.8** For each of the alternative choices of initial working set $\mathcal{W}_0$ in the example (16.36) (that is, $\mathcal{W}_0 = \{3\}$, $\mathcal{W}_0 = \{5\}$, and $\mathcal{W}_0 = \emptyset$) work through the first two iterations of Algorithm 16.1.

⬥ **16.9** Program Algorithm 16.1 and use it to solve

$$\min x_1^2 + 2x_2^2 - 2x_1 - 6x_2 - 2x_1 x_2$$
$$\text{subject to } \tfrac{1}{2} x_1 + \tfrac{1}{2} x_2 \leq 1,$$
$$-x_1 + 2x_2 \leq 2,$$
$$x_1, x_2 \geq 0.$$

Choose three initial starting points: one in the interior of the feasible region, one at a vertex, and one on the boundary of the feasible region (not a vertex).

⬥ **16.10** Consider the problem (16.3a)–(16.3b), and assume that A has full column rank. Prove the following three statements: (a) The quadratic program has a strong local minimizer at a point x^* satisfying the Lagrange equations $Gx^* + d = A\lambda^*$ if and only if

$Z^T G Z$ is positive definite. (b) The quadratic program has an infinite number of solutions if the first equations in (16.4) are compatible and if the reduced Hessian $Z^T G Z$ is positive semidefinite and singular. (c) There are no finite solutions if either $Z^T G Z$ is indefinite or if the first equations in (16.4) are inconsistent.

16.11 Assume that $A \neq 0$. Show that the KKT matrix (16.7) is indefinite.

16.12 Consider the quadratic program

$$\max = 6x_1 + 4x_2 - 13 - x_1^2 - x_2^2,$$
$$\text{subject to } x_1 + x_2 \leq 3,$$
$$x_1 \geq 0, x_2 \geq 0.$$

First solve it graphically and then use your program implementing the active-set method given in Algorithm 16.1.

16.13 Prove that if the KKT matrix (16.7) is nonsingular, then A must have full rank.

16.14 Explain why it is that the size of the Lagrange multipliers is dependent on the scaling of the objective function and constraints.

16.15 Using (16.27) and (16.29), explain briefly why the gradient of each blocking constraint cannot be a linear combination of the constraint gradients in the current working set $\mathcal{W}_k$. More specifically, suppose that the initial working set $\mathcal{W}_0$ in Algorithm 16.1 is chosen so that the gradients of the constraints belonging to this working set are linearly independent. Show that the step selection (16.29) guarantees that constraint gradients of all subsequent working sets remain linearly independent.

16.16 Let W be an $n \times n$ symmetric matrix and suppose that Z is of dimension $n \times t$. Suppose that $Z^T W Z$ is positive definite and that $\bar{Z} = [Z, z]$, i.e., that we have appended a column to Z. Show that $\bar{Z}^T W \bar{Z}$ is positive definite.

16.17 Find a null-space basis matrix Z for the problem (16.49a), (16.49b).

16.18 Write down KKT conditions for the following convex quadratic program with mixed equality and inequality constraints:

$$\min\ q(x) = \tfrac{1}{2} x^T G x + x^T d,$$
$$\text{s.t. } Ax \geq b, \quad \bar{A} x = \bar{b},$$

where G is symmetric and positive semidefinite. Use these conditions to derive an analogue of the generic primal–dual step (16.53) for this problem.

Chapter 17

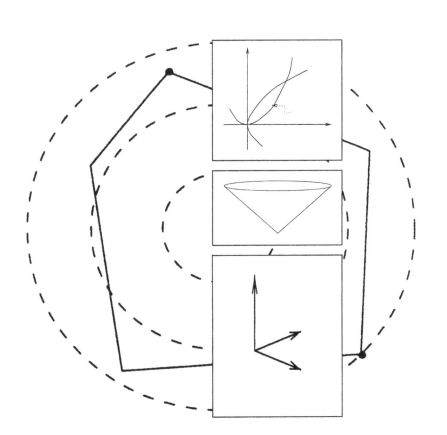

Penalty, Barrier, and Augmented Lagrangian Methods

An important class of methods for constrained optimization seeks the solution by replacing the original constrained problem by a sequence of unconstrained subproblems. In this chapter we describe three important algorithms in this class. The *quadratic penalty* method replaces the constraints by penalty terms in the objective function, where each penalty term is a multiple of the square of the constraint violation. Because of its simplicity and intuitive appeal, this approach is often used in computations, although it has some important disadvantages. This method can be considered a precursor to the *method of multipliers* or *augmented Lagrangian method*, in which explicit Lagrange multiplier estimates are used to avoid the ill-conditioning that is inherent in the quadratic penalty function.

A somewhat different approach is used in the *log-barrier method*, in which logarithmic terms prevent feasible iterates from moving too close to the boundary of the feasible region. This method is interesting in its own right, and it also provides the foundation of primal and primal–dual interior-point methods, which have proved to be important in linear programming (see Chapter 14), convex quadratic programming (see Chapter 16), and semidefinite programming, and may yet make their mark in general constrained optimization as well.

In Section 17.3 we describe exact penalty function approaches, in which a single unconstrained problem (rather than a sequence) takes the place of the original constrained problem. Not surprisingly, the resulting function is often difficult to minimize. Section 17.5 describes *sequential linearly constrained methods*, an important class of practical methods for large constrained optimization problems.

17.1 THE QUADRATIC PENALTY METHOD

MOTIVATION

One fundamental approach to constrained optimization is to replace the original problem by a penalty function that consists of

- the original objective of the constrained optimization problem, *plus*
- one additional term for each constraint, which is positive when the current point x violates that constraint and zero otherwise.

Most approaches define a *sequence* of such penalty functions, in which the penalty terms for the constraint violations are multiplied by some positive coefficient. By making this coefficient larger and larger, we penalize constraint violations more and more severely, thereby forcing the minimizer of the penalty function closer and closer to the feasible region for the constrained problem.

Such approaches are sometimes known as *exterior penalty methods*, because the penalty term for each constraint is nonzero only when x is infeasible with respect to that constraint. Often, the minimizers of the penalty functions are infeasible with respect to the original problem, and approach feasibility only in the limit as the penalty parameter grows increasingly large.

The simplest penalty function of this type is the *quadratic penalty function*, in which the penalty terms are the squares of the constraint violations. We devote most of our discussion in this section to the equality-constrained problem

$$\min_x f(x) \quad \text{subject to } c_i(x) = 0, i \in \mathcal{E}, \tag{17.1}$$

which is a special case of (12.1). The quadratic penalty function $Q(x; \mu)$ for this formulation is

$$Q(x; \mu) \stackrel{\text{def}}{=} f(x) + \frac{1}{2\mu} \sum_{i \in \mathcal{E}} c_i^2(x), \tag{17.2}$$

where $\mu > 0$ is the *penalty parameter*. By driving μ to zero, we penalize the constraint violations with increasing severity. It makes good intuitive sense to consider a sequence of values $\{\mu_k\}$ with $\mu_k \downarrow 0$ as $k \to \infty$, and to seek the approximate minimizer x_k of $Q(x; \mu_k)$ for each k. Because the penalty terms in (17.2) are smooth (each term c_i^2 has at least as many

17.1. THE QUADRATIC PENALTY METHOD 493

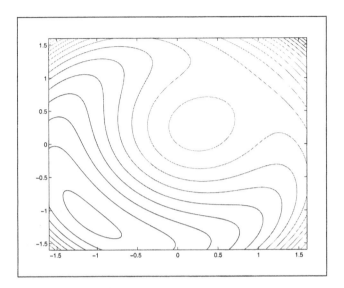

Figure 17.1 Contours of $Q(x;\mu)$ from (17.4) for $\mu = 1$, contour spacing 0.5.

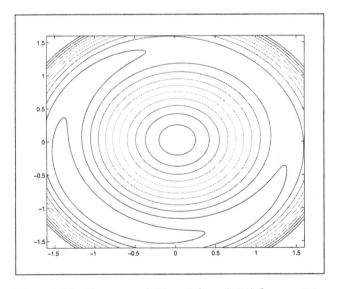

Figure 17.2 Contours of $Q(x;\mu)$ from (17.4) for $\mu = 0.1$, contour spacing 2.

derivatives as c_i itself), we can use techniques from unconstrained optimization to search for x_k. The approximate minimizers x_k, x_{k-1}, etc, can be used to identify a good starting point for the minimization of $Q(\cdot;\mu_{k+1})$ at iteration $k + 1$. By choosing the sequence $\{\mu_k\}$ and the starting points wisely, it may be possible to perform just a few steps of unconstrained minimization for each value μ_k.

❑ EXAMPLE 17.1

Consider the problem (12.9) from Chapter 12, that is,

$$\min x_1 + x_2 \quad \text{s.t.} \quad x_1^2 + x_2^2 - 2 = 0, \tag{17.3}$$

for which the solution is $(-1, -1)^T$ and the quadratic penalty function is

$$Q(x; \mu) = x_1 + x_2 + \frac{1}{2\mu} \left(x_1^2 + x_2^2 - 2 \right)^2. \tag{17.4}$$

We plot the contours of this function in Figures 17.1 and 17.2. In Figure 17.1 we have $\mu = 1$, and we observe a mimizer of Q near the point $(-1.1, -1.1)^T$. (There is also a local maximizer near $x = (0.3, 0.3)^T$.) In Figure 17.2 we have $\mu = 0.1$, so points that do not lie on the feasible circle defined by $x_1^2 + x_2^2 = 2$ suffer a much greater penalty than in the first figure—the "trough" of low values of Q is clearly evident. The minimizer in this figure is much closer to the solution $(-1, -1)^T$ of the problem (17.3). A local maximum lies near $(0, 0)^T$, and Q goes rapidly to ∞ outside the circle $x_1^2 + x_2^2 = 2$.

❏

For the general constrained optimization problem (12.1), which contains inequality constraints as well as equality constraints, we can define Q as

$$Q(x; \mu) \stackrel{\text{def}}{=} f(x) + \frac{1}{2\mu} \sum_{i \in \mathcal{E}} c_i^2(x) + \frac{1}{2\mu} \sum_{i \in \mathcal{I}} \left([c_i(x)]^- \right)^2, \tag{17.5}$$

where $[y]^-$ denotes $\max(-y, 0)$. In this case, Q may be *less* smooth than the objective and constraint functions. For instance, if one of the inequality constraints is $x_1 \geq 0$, then the function $\min(0, x_1)^2$ has a discontinuous second derivative (see the exercises), so that Q is no longer twice continuously differentiable.

ALGORITHMIC FRAMEWORK

A general framework for algorithms based on the penalty function (17.2) can be specified as follows.

Framework 17.1 (Quadratic Penalty).
Given $\mu_0 > 0$, tolerance $\tau_0 > 0$, starting point x_0^s;
for $k = 0, 1, 2, \ldots$
  Find an approximate minimizer x_k of $Q(\cdot; \mu_k)$, starting at x_k^s,
    and terminating when $\|\nabla Q(x; \mu_k)\| \leq \tau_k$;
  if final convergence test satisfied

STOP with approximate solution x_k;
Choose new penalty parameter $\mu_{k+1} \in (0, \mu_k)$;
Choose new starting point x_{k+1}^s;
end (for)

The parameter sequence $\{\mu_k\}$ can be chosen adaptively, based on the difficulty of minimizing the penalty function at each iteration. When minimization of $Q(x; \mu_k)$ proves to be expensive for some k, we choose μ_{k+1} to be only modestly smaller than μ_k; for instance $\mu_{k+1} = 0.7\mu_k$. If we find the approximate minimizer of $Q(x; \mu_k)$ cheaply, we could try a more ambitious reduction, for instance $\mu_{k+1} = 0.1\mu_k$. The convergence theory for Framework 17.1 allows wide latitude in the choice of tolerances τ_k; it requires only that $\lim \tau_k = 0$, to ensure that the minimization is carried out more and more accurately.

When only equality constraints are present, $Q(x; \mu_k)$ is smooth, so the algorithms for unconstrained minimization described in the first chapters of the book can be used to identify the approximate solution x_k. However, the minimization of $Q(x; \mu_k)$ becomes more and more difficult to perform when μ_k becomes small, unless we use special techniques to calculate the search directions. For one thing, the Hessian $\nabla_{xx}^2 Q(x; \mu_k)$ becomes quite ill conditioned near the minimizer. This property alone is enough to make unconstrained minimization algorithms such as quasi-Newton and conjugate gradient perform poorly. Newton's method, on the other hand, is not sensitive to ill conditioning of the Hessian, but it, too, may encounter difficulties for small μ_k for two other reasons. First, ill conditioning of $\nabla_{xx}^2 Q(x; \mu_k)$ might be expected to cause problems when we come to solve the linear equations to calculate the Newton step. We discuss this issue further at the end of the section, where we show that these effects are not so severe and that a reformulation of the Newton equations is possible. Second, even when x is close to the minimizer of $Q(\cdot; \mu_k)$, the quadratic Taylor series approximation to $Q(x; \mu_k)$ about x is a reasonable approximation of the true function only in a small neighborhood of x. This property can be seen in Figure 17.2, where the contours of Q near the minimizer have a banana shape, rather than the elliptical shape that characterizes quadratic functions. Since Newton's method is based on the quadratic model, the steps that it generates may not make rapid progress toward the minimizer of $Q(x; \mu_k)$, unless we are quite close to the minimizer. This difficulty can be overcome partly by judicious choice of the starting point x_{k+1}^s.

Despite the intuitive appeal and simplicity of Framework 17.1, we point out that the method of multipliers of Section 17.4 would generally be more effective, since it tends to avoid the ill conditioning that causes problems in Framework 17.1.

CONVERGENCE OF THE QUADRATIC PENALTY FUNCTION

We describe some convergence properties of this approach in the following two theorems.

Theorem 17.1.

Suppose that each x_k is the exact global minimizer of $Q(x; \mu_k)$ in Framework 17.1 above, and that $\mu_k \downarrow 0$. Then every limit point x^* of the sequence $\{x_k\}$ is a solution of the problem (17.1).

PROOF. Let $\bar{x}$ be a global solution of (17.1), that is,

$$f(\bar{x}) \leq f(x) \quad \text{for all } x \text{ with } c_i(x) = 0, i \in \mathcal{E}.$$

Since x_k minimizes $Q(\cdot; \mu_k)$ for each k, we have that $Q(x_k; \mu_k) \leq Q(\bar{x}; \mu_k)$, which leads to the inequality

$$f(x_k) + \frac{1}{2\mu_k} \sum_{i \in \mathcal{E}} c_i^2(x_k) \leq f(\bar{x}) + \frac{1}{2\mu_k} \sum_{i \in \mathcal{E}} c_i^2(\bar{x}) = f(\bar{x}). \tag{17.6}$$

By rearranging this expression, we obtain

$$\sum_{i \in \mathcal{E}} c_i^2(x_k) \leq 2\mu_k [f(\bar{x}) - f(x_k)]. \tag{17.7}$$

Suppose that x^* is a limit point of $\{x_k\}$, so that there is an infinite subsequence $\mathcal{K}$ such that

$$\lim_{k \in \mathcal{K}} x_k = x^*.$$

By taking the limit as $k \to \infty$, $k \in \mathcal{K}$, on both sides of side of (17.7), we obtain

$$\sum_{i \in \mathcal{E}} c_i^2(x^*) = \lim_{k \in \mathcal{K}} \sum_{i \in \mathcal{E}} c_i^2(x_k) \leq \lim_{k \in \mathcal{K}} 2\mu_k [f(\bar{x}) - f(x_k)] = 0,$$

where the last equality follows from $\mu_k \downarrow 0$. Therefore, we have that $c_i(x^*) = 0$ for all $i \in \mathcal{E}$, so that x^* is feasible. Moreover, by taking the limit as $k \to \infty$ for $k \in \mathcal{K}$ in (17.6), we have by nonnegativity of μ_k and of each $c_i(x_k)^2$ that

$$f(x^*) \leq f(x^*) + \lim_{k \in \mathcal{K}} \frac{1}{2\mu_k} \sum_{i \in \mathcal{E}} c_i^2(x_k) \leq f(\bar{x}).$$

Since x^* is a feasible point whose objective value is no larger than that of the global minimizer $\bar{x}$, we conclude that x^*, too, is a global minimizer, as claimed. □

Since this result requires us to find the *global* minimizer for each subproblem, its very desirable property of convergence to the global solution of (17.1) may be difficult to realize in practice. The next result concerns convergence properties of the sequence $\{x_k\}$ when we allow inexact (but increasingly accurate) minimizations of $Q(\cdot; \mu_k)$. In contrast to Theorem 17.1, it shows that the sequence is attracted to KKT points (that is, points satisfying first-order necessary conditions; see (12.30)), rather than to a global minimizer. It also shows that the quantities $c_i(x_k)/\mu_k$ may be used as estimates of the Lagrange multipliers λ_i^* in certain circumstances. This observation is important for the analysis of Section 17.4.

Theorem 17.2.

If the tolerances τ_k in Framework 17.1 above satisfy

$$\lim_{k \to \infty} \tau_k = 0$$

and the penalty parameters satisfy $\mu_k \downarrow 0$, then for all limit points x^* of the sequence $\{x_k\}$ at which the constraint gradients $\nabla c_i(x^*)$ are linearly independent, we have that x^* is a KKT point for the problem (17.1). For such points, we have for the infinite subsequence $\mathcal{K}$ such that $\lim_{k \in \mathcal{K}} x_k = x^*$ that

$$\lim_{k \in \mathcal{K}} -c_i(x_k)/\mu_k = \lambda_i^*, \quad \text{for all } i \in \mathcal{E}, \tag{17.8}$$

where λ^* is multiplier vector that satisfies the KKT conditions (12.30).

PROOF. By differentiating $Q(x; \mu_k)$ in (17.2), we obtain

$$\nabla_x Q(x_k; \mu_k) = \nabla f(x_k) + \sum_{i \in \mathcal{E}} \frac{c_i(x_k)}{\mu_k} \nabla c_i(x_k). \tag{17.9}$$

By applying the termination criterion for Framework 17.1, we have that

$$\left\| \nabla f(x_k) + \sum_{i \in \mathcal{E}} \frac{c_i(x_k)}{\mu_k} \nabla c_i(x_k) \right\| \leq \tau_k. \tag{17.10}$$

By rearranging this expression (and in particular using the inequality $\|a\| - \|b\| \leq \|a+b\|$), we obtain

$$\left\| \sum_{i \in \mathcal{E}} c_i(x_k) \nabla c_i(x_k) \right\| \leq \mu_k [\tau_k + \|\nabla f(x_k)\|]. \tag{17.11}$$

When we take limits as $k \to \infty$ for $k \in \mathcal{K}$, the bracketed term on the right-hand-side approaches $\|\nabla f(x^*)\|$, so because $\mu_k \downarrow 0$, the right-hand-side approaches zero. From the corresponding limit on the left-hand-side, we obtain

$$\sum_{i \in \mathcal{E}} c_i(x^*) \nabla c_i(x^*) = 0. \tag{17.12}$$

Since by assumption the constraint gradients $\nabla c_i(x^*)$ are linearly independent, we have that $c_i(x^*) = 0$ for all $i \in \mathcal{E}$, so x^* is feasible.

Since x^* is feasible, the second KKT condition (12.30b) is satisfied. We need to check the first KKT condition (12.30a) as well, and to show that the limit (17.8) holds.

By using $A(x)$ to denote the matrix of constraint gradients, that is,

$$A(x)^T = [\nabla c_i(x)]_{i \in \mathcal{E}}, \tag{17.13}$$

and λ_k to denote the vector $-c(x_k)/\mu_k$, we have as in (17.10) that

$$A(x_k)^T \lambda_k = \nabla f(x_k) - \nabla_x Q(x_k; \mu), \quad \|\nabla_x Q(x_k; \mu)\| \leq \tau_k. \tag{17.14}$$

For all $k \in \mathcal{K}$ sufficiently large, the matrix $A(x_k)$ has full column rank, so that $A(x_k)A(x_k)^T$ is nonsingular. By multiplying (17.14) by $A(x_k)$ and rearranging, we have that

$$\lambda_k = \left[A(x_k)A(x_k)^T\right]^{-1} A(x_k) \left[\nabla f(x_k) - \nabla_x Q(x_k; \mu)\right].$$

Hence by taking the limit as $k \in \mathcal{K}$ goes to ∞, we find that

$$\lim_{k \in \mathcal{K}} \lambda_k = \lambda^* = \left[A(x^*)A(x^*)^T\right]^{-1} A(x^*) \nabla f(x^*).$$

By taking limits in (17.10), we conclude that

$$\nabla f(x^*) - A(x^*)^T \lambda^* = 0,$$

so that λ^* satisfies the first KKT condition (12.30a). We conclude from (12.30) that x^* is a KKT point, with unique Lagrange multiplier vector λ^*. □

When the constraint gradients are linearly dependent, we cannot deduce from (17.12) that $c_i(x^*) = 0$, so the limit point may not be feasible. Even if it *is* feasible, and even if the limiting multiplier value λ^* exists and the KKT conditions (12.30) hold, the necessary conditions for x^* to be a solution of the linear program are still not satisfied because these conditions require linear independence of the constraint gradients (see Theorem 12.1). When the limit x^* is *not* feasible, we have from (17.12) that it is at least a stationary point for the function $\|c(x)\|^2$. Newton-type algorithms can always be attracted to infeasible points of this type. We see the same effect in Chapter 11, in our discussion of methods for nonlinear equations that use the sum-of-squares merit function $\|r(x)\|^2$. Such methods cannot be guaranteed to find a root, but only a stationary point or minimizer of the merit function. In the case in which the nonlinear program (17.1) is infeasible, we will often observe convergence of the quadratic-penalty method to stationary points or minimizers of $\|c(x)\|^2$.

We conclude this section by examining the nature of the ill conditioning (mentioned above) in the Hessian $\nabla_{xx}^2 Q(x; \mu_k)$. An understanding of the properties of this matrix, and the similar Hessians that arise in the other methods of this chapter, is essential in choosing effective algorithms for the minimization problem and for the linear algebra calculations at each iteration.

The Hessian is given by the formula

$$\nabla^2_{xx} Q(x; \mu_k) = \nabla^2 f(x) + \sum_{i \in \mathcal{E}} \frac{c_i(x)}{\mu_k} \nabla^2 c_i(x) + \frac{1}{\mu_k} A(x)^T A(x), \quad (17.15)$$

where we have used the definition (17.13) of $A(x)$. When x is close to the minimizer of $Q(\cdot; \mu_k)$ and the conditions of Theorem 17.2 are satisfied, we have from (17.8) that the first two terms on the right-hand-side of (17.15) are approximately equal to the Hessian of the Lagrangian function defined in (12.28). To be specific, we have

$$\nabla^2_{xx} Q(x; \mu_k) \approx \nabla^2_{xx} \mathcal{L}(x, \lambda^*) + \frac{1}{\mu_k} A(x)^T A(x), \quad (17.16)$$

when x is close to the minimizer of $Q(\cdot; \mu_k)$. We see from this expression that $\nabla^2_{xx} Q(x; \mu_k)$ is approximately equal to the sum of

- a matrix whose elements are independent of μ_k (the Lagrangian term), and
- a matrix of rank $|\mathcal{E}|$ whose nonzero eigenvalues are of order $1/\mu_k$ (the summation term in (17.16)).

The number of constraints $|\mathcal{E}|$ is usually fewer than n. In this case, the summation term is singular, and the overall matrix has some of its eigenvalues approaching a constant, while others are of order $1/\mu_k$. Since μ_k is approaching zero, the increasing ill conditioning of $Q(x; \mu_k)$ is apparent.

One consequence of the ill conditioning is possible inaccuracy in the calculation of the Newton step for $Q(x; \mu_k)$, which is obtained by solving the following system:

$$\nabla^2_{xx} Q(x; \mu) p = -\nabla_x Q(x; \mu). \quad (17.17)$$

Significant roundoff errors will appear in p regardless of the solution technique used, and algorithms will break down as the matrix becomes numerically singular. The presence of roundoff error may not disqualify p from being a good direction of progress for Newton's method, however. As in the Newton/log-barrier step equations (17.28), it could be that the error is concentrated in a direction along which Q does not vary significantly, so that it has little effect on the overall quality of the step.

In any case, we can use an alternative formulation of the equations (17.17) that avoids the ill conditioning. By introducing a "dummy vector" ζ and using the expression (17.15), we see that (17.17) is equivalent to the system

$$\begin{bmatrix} \nabla^2 f(x) + \sum_{i \in \mathcal{E}} \dfrac{c_i(x)}{\mu_k} \nabla^2 c_i(x) & A(x)^T \\ A(x) & -\mu_k I \end{bmatrix} \begin{bmatrix} p \\ \zeta \end{bmatrix} = \begin{bmatrix} -\nabla_x Q(x; \mu) \\ 0 \end{bmatrix}, \quad (17.18)$$

in the sense that the same vector p solves both systems. As $\mu_k \downarrow 0$, however, the coefficient matrix in (17.18) approaches a well-conditioned limit when the second-order sufficient conditions of Theorem 12.6 are satisfied at x^*.

17.2 THE LOGARITHMIC BARRIER METHOD

PROPERTIES OF LOGARITHMIC BARRIER FUNCTIONS

We start by describing the concept of barrier functions in terms of inequality-constrained optimization problems. Given the problem

$$\min_x f(x) \quad \text{subject to } c_i(x) \geq 0, \quad i \in \mathcal{I}, \quad (17.19)$$

the *strictly feasible region* is defined by

$$\mathcal{F}^o \stackrel{\text{def}}{=} \{x \in \mathbf{R}^n \mid c_i(x) > 0 \text{ for all } i \in \mathcal{I}\}; \quad (17.20)$$

we assume that $\mathcal{F}^o$ is nonempty for purposes of this discussion. Barrier functions for this problem have the properties that

- they are infinite everywhere *except* in $\mathcal{F}^o$;
- they are smooth inside $\mathcal{F}^o$;
- their value approaches ∞ as x approaches the boundary of $\mathcal{F}^o$.

The most important barrier function is the *logarithmic barrier function*, which for the constraint set $c_i(x) \geq 0, i \in \mathcal{I}$, has the form

$$-\sum_{i \in \mathcal{I}} \log c_i(x), \quad (17.21)$$

where $\log(\cdot)$ denotes the natural logarithm. For the inequality-constrained optimization problem (17.19), the combined objective/barrier function is given by

$$P(x; \mu) = f(x) - \mu \sum_{i \in \mathcal{I}} \log c_i(x), \quad (17.22)$$

17.2. THE LOGARITHMIC BARRIER METHOD

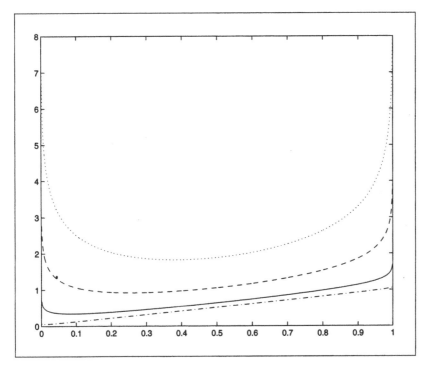

Figure 17.3 Plots of $P(x; \mu)$ for $\mu = 1.0, 0.4, 0.1, 0.01$.

where μ is referred to here as the *barrier parameter*. From now on, we refer to $P(x; \mu)$ itself as the "logarithmic barrier function for the problem (17.19)," or simply the "log barrier function" for short. As for the quadratic penalty function (17.2), we can show that the minimizer of $P(x; \mu)$, which we denote by $x(\mu)$, approaches a solution of (17.19) as $\mu \downarrow 0$, under certain conditions; see Theorems 17.3 and 17.4.

❑ **EXAMPLE 17.2**

Consider the following problem in a single variable x:

$$\min x \quad \text{subject to } x \geq 0, 1 - x \geq 0,$$

for which we have

$$P(x; \mu) = x - \mu \log x - \mu \log(1 - x).$$

We graph this function for different values of μ in Figure 17.3. Naturally, for small values of μ, the function $P(x; \mu)$ is close to the objective f over most of the feasible set; it approaches

∞ only in narrow "boundary layers." (In Figure 17.3, the curve $P(x; 0.01)$ is almost indistinguishable from $f(x)$ to the resolution of our plot, though this function also approaches ∞ when x is very close to the endpoints 0 and 1.) Also, it is clear that as $\mu \downarrow 0$, the minimizer $x(\mu)$ of $P(x; \mu)$ is approaching the solution $x^* = 0$ of the constrained problem.

Since the minimizer $x(\mu)$ of $P(x; \mu)$ lies in the strictly feasible set $\mathcal{F}^o$ (where no constraints are active), we can in principle search for it by using the unconstrained minimization algorithms described in the first part of this book. (These methods need to be modified slightly to keep the iterates inside $\mathcal{F}^o$.) Unfortunately, the minimizer $x(\mu)$ becomes more and more difficult to find as $\mu \downarrow 0$. The scaling of the function $P(x; \mu)$ becomes poorer and poorer, and the quadratic Taylor series approximation (on which Newton-like methods are based) does not adequately capture the behavior of the true function $P(x; \mu)$, except in a small neighborhood of $x(\mu)$. These difficulties can be illustrated by the following two-variable example.

❏ EXAMPLE 17.3

Consider the problem

$$\min (x_1 + 0.5)^2 + (x_2 - 0.5)^2 \quad \text{subject to} \quad x_1 \in [0, 1], \quad x_2 \in [0, 1], \quad (17.23)$$

for which the log-barrier function is

$$P(x; \mu) = (x_1 + 0.5)^2 + (x_2 - 0.5)^2 \qquad (17.24)$$
$$- \mu \left[\log x_1 + \log(1 - x_1) + \log x_2 + \log(1 - x_2) \right].$$

Contours of this function for the values $\mu = 1$, $\mu = 0.1$, and $\mu = 0.01$ are plotted in Figure 17.4. From Figure 17.3 we see that except near the boundary of the feasible region, the contours approach those of the parabolic objective function as $\mu \downarrow 0$. Note that the contours that surround the minimizer in the first two plots are not too eccentric and not too far from being elliptical, indicating that most unconstrained minimization algorithms can be applied successfully to identify the minimizers $x(\mu)$.

For the value $\mu = 0.01$, however, the contours are more elongated and less elliptical as the minimizer $x(\mu)$ is pushed toward the "boundary layer" of the barrier function. (A closeup is shown in Figure 17.5.) As in the case of the quadratic penalty function, the elongated nature of the contours indicates poor scaling, which causes poor performance of unconstrained optimization methods such as quasi-Newton, steepest descent and conjugate gradient. Newton's method is insensitive to the poor scaling, but the nonelliptical property— the contours in Figure 17.5 are almost straight along the left edge while being circular along

17.2. THE LOGARITHMIC BARRIER METHOD   503

the right edge—indicates that the quadratic approximation on which Newton's method is based does not capture well the behavior of the true log-barrier function. Hence, Newton's

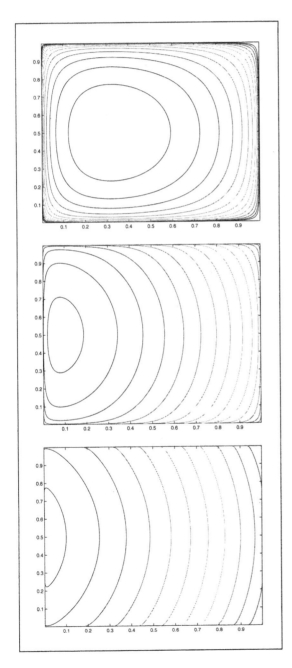

Figure 17.4
Contours of $P(x; \mu)$ from (17.24) for (top to bottom) $\mu = 1, 0.1, 0.01$.

method, too, may not show rapid convergence to $x(\mu)$ except in a small neighborhood of this point.

For more insight into the problem of poor scaling, we examine the structure of the gradient and Hessian of $P(x; \mu)$. We have

$$\nabla_x P(x; \mu) = \nabla f(x) - \sum_{i \in \mathcal{I}} \frac{\mu}{c_i(x)} \nabla c_i(x), \tag{17.25}$$

$$\nabla^2_{xx} P(x; \mu) = \nabla^2 f(x) - \sum_{i \in \mathcal{I}} \frac{\mu}{c_i(x)} \nabla^2 c_i(x) \tag{17.26}$$
$$+ \sum_{i \in \mathcal{I}} \frac{\mu}{c_i^2(x)} \nabla c_i(x) \nabla c_i(x)^T.$$

When x is close to the minimizer $x(\mu)$ and μ is small, we see from Theorem 17.4 and (17.32) that the optimal Lagrange multipliers λ_i^*, $i \in \mathcal{I}$, can be estimated as follows:

$$\lambda_i^* \approx \mu/c_i(x), \quad i \in \mathcal{I}.$$

By substituting into (17.26), and using the definition (12.28) of the Lagrangian $\mathcal{L}(x, \lambda)$, we find that

$$\nabla^2_{xx} P(x; \mu) \approx \nabla^2_{xx} \mathcal{L}(x, \lambda^*) + \sum_{i \in \mathcal{I}} \frac{1}{\mu}(\lambda_i^*)^2 \nabla c_i(x) \nabla c_i(x)^T. \tag{17.27}$$

Note the similarity of this expression to the Hessian of the quadratic penalty function (17.16). A similar analysis of the eigenvalues applies to (17.27) as in that case. If t inequality constraints

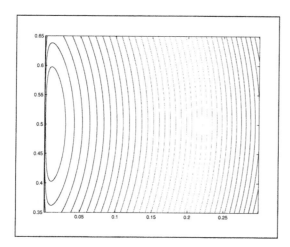

Figure 17.5
Close-up of contours of $P(x; \mu)$ from (17.24) for $\mu = .01$

are active at x^*, and if we assume that the linear independence constraint qualification (LICQ) is satisfied at the minimizer $x(\mu)$, then t of the eigenvalues of $\nabla^2_{xx} P(x;\mu)$ have size of order $1/\mu$, while the remaining $n - t$ are of order 1. Except in the special cases of $t = 0$ and $t = n$, the matrix $\nabla^2_{xx} P(x;\mu)$ becomes increasingly ill conditioned near the minimizer $x(\mu)$, as μ approaches zero.

As in the case of the quadratic penalty function, the properties of the Hessian of $P(x;\mu)$ may lead to complications in solving the Newton step equations, which are

$$\nabla^2_{xx} P(x;\mu) p = -\nabla_x P(x;\mu). \tag{17.28}$$

It is possible to reformulate this system in a similar manner to the reformulation (17.18) of the Newton equations for $Q(x;\mu_k)$, though the presence of inequality constraints rather than equalities makes the reformulation somewhat less obvious. In any case, elimination techniques applied directly to (17.28) yield computed steps p that are good directions of progress for the Newton method, despite the roundoff error caused by the ill conditioning. We discuss this surprising observation further in the Notes and References at the end of the chapter.

ALGORITHMS BASED ON THE LOG-BARRIER FUNCTION

Algorithms based on the log-barrier function aim to identify approximate minima of $P(x;\mu)$ for a decreasing sequence of values of μ. The generic algorithm is very similar to the quadratic penalty function framework, Framework 17.1, and can be written as follows.

Framework 17.2 (Log-Barrier).
  Given $\mu_0 > 0$, tolerance $\tau_0 > 0$, starting point x_0^s;
  for $k = 0, 1, 2, \ldots$
      Find an approximate minimizer x_k of $P(\cdot;\mu_k)$, starting at x_k^s,
          and terminating when $\|\nabla P(x;\mu_k)\| \leq \tau_k$;
      if final convergence test satisfied
          STOP with approximate solution x_k;
      Choose new barrier parameter $\mu_{k+1} \in (0, \mu_k)$;
      Choose new starting point x_{k+1}^s;
  end (for)

As we noted above, the ill conditioning of $\nabla^2_{xx} P(x;\mu)$ for small μ makes Newton's method the only really effective technique for finding the approximate minimizer x_k at each iteration of this framework. Even Newton's method encounters difficulties (discussed above) because of the inadequacy of the quadratic Taylor series approximation on which it is based, but various techniques can be used to ensure that Newton's method enters the domain of convergence for $P(x;\mu_k)$ in just a few steps and thereafter converges rapidly to $x(\mu_k)$. These techniques include the following:

- Enhancing Newton's method with a technique to ensure global convergence, such as a line search or trust-region mechanism. They must ensure that the iterates stay within the domain of $P(x; \mu)$, that is, the strictly feasible region $\mathcal{F}^o$, and they should also take account of the characteristics of the log-barrier function.

- A good starting point x_k^s for minimization of $P(x; \mu_k)$ can be obtained by extrapolating along the path defined by the previous approximate minimizers $x_{k-1}, x_{k-2}, \ldots$. Alternatively, we can calculate the approximate tangent vector to the path $\{x(\mu) \mid \mu > 0\}$ at the point x_{k-1} by performing total differentiation of $\nabla_x P(x; \mu)$ with respect to μ to obtain

$$\nabla_{xx}^2 P(x; \mu) \dot{x} + \frac{\partial}{\partial \mu} \nabla_x P(x; \mu) = 0.$$

By calculating the second term explicitly, we obtain

$$\nabla_{xx}^2 P(x; \mu) \dot{x} - \sum_{i \in \mathcal{I}} \frac{1}{c_i(x)} \nabla c_i(x) = 0. \tag{17.29}$$

By substituting $x = x_{k-1}$ and $\mu = \mu_{k-1}$ into this expression, we obtain the approximate tangent $\dot{x}$, which we can then use to obtain a starting point x_k^s for step k of Framework 17.2 as follows:

$$x_k^s = x_{k-1} + (\mu_k - \mu_{k-1}) \dot{x}. \tag{17.30}$$

Various heuristics have been proposed for the choice of the new barrier parameter μ_{k+1} at each iteration. One common ingredient is that we usually make an ambitious choice of μ_{k+1} ($\mu_{k+1} = 0.2 \mu_k$ or $\mu_{k+1} = 0.1 \mu_k$, say) if the minimization problem for $P(x; \mu_k)$ was not too difficult to solve and if a good starting point x_{k+1}^s can be proposed with some confidence. Since software based on the Framework 17.2 has not made its way into the mainstream (at least not yet), we cannot point to a definitive "best" set of heuristics; determination of reliable, efficient heuristics is a question that requires extensive computational testing with a representative set of nonlinear programming applications.

Despite the difficulties involved in implementing log-barrier methods effectively, interest has revived in recent years partly because of their connection to primal–dual interior-point methods, which have been shown to perform extremely well for large linear programming and convex quadratic programming problems. We explain this connection further below. Remarkably, log-barrier methods remain an active area of research today, almost 45 years after they were originally proposed by Frisch [95] and 30 years after the publication of an authoritative study by Fiacco and McCormick [79].

PROPERTIES OF THE LOG-BARRIER FUNCTION AND FRAMEWORK 17.2

We now describe some of the main properties of the log-barrier function $P(x; \mu)$ defined in (17.22), together with a basic convergence result associated with Framework 17.2.

We start by stating a theorem about the convex programming problem. In this result, we use the following notation:

$\mathcal{M}$ is the set of solutions of (17.19);
f^* is the optimal value of the objective function f.

Theorem 17.3.

Suppose that f and $-c_i$, $i \in \mathcal{I}$, in (17.19) are all convex functions, and that the strictly feasible region $\mathcal{F}^o$ defined by (17.20) is nonempty. Let $\{\mu_k\}$ be a decreasing sequence such that $\mu_k \downarrow 0$, and assume that the solution set $\mathcal{M}$ is nonempty and bounded. Then the following statements are true.

(i) *For any $\mu > 0$, $P(x; \mu)$ is convex in $\mathcal{F}^o$ and attains a minimizer $x(\mu)$ (not necessarily unique) on $\mathcal{F}^o$. Any local minimizer $x(\mu)$ is also a global minimizer of $P(x; \mu)$.*

(ii) *Any sequence of minimizers $x(\mu_k)$ has a convergent subsequence, and all possible limit points of such sequences lie in $\mathcal{M}$.*

(iii) *$f(x(\mu_k)) \to f^*$ and $P(x(\mu_k); \mu_k) \to f^*$, for any sequence of minimizers $\{x(\mu_k)\}$.*

For a proof of this result, see M. Wright [251, Theorem 5].

Note that it is possible to have convex programs for which the set $\mathcal{M}$ is empty (consider $\min x$ subject to $-x \geq 0$) or for which $\mathcal{M}$ is unbounded (consider $\min_{(x_1, x_2)} x_1$ subject to $x_1 \geq 0$), in which cases the conclusions of the theorem do not apply in general.

For general inequality-constrained problems (17.19), the corresponding result is more local in nature. Given a "well-behaved" local solution x^* of the problem (that is, one that satisfies the second-order sufficient conditions of Theorem 12.6 along with strict complementarity and a constraint qualification), the log barrier function $P(x; \mu)$ has a local minimizer close to x^* for all *sufficiently small* values of μ. (We state this result formally below.) However, when the strictly feasible region $\mathcal{F}^o$ is unbounded, the barrier function $P(x; \mu)$ may be unbounded below. It is also possible for a sequence of local minimizers of $P(x; \mu_k)$ to converge to a point that is not a solution of (17.19) as $\mu_k \downarrow 0$. For examples, see the exercises.

There is an important relationship between the minimizer of $P(x; \mu)$ and a point (x, λ) satisfying the KKT conditions for the optimization problem (17.19), which are given in (12.30). At a minimizer $x(\mu)$, the gradient of $P(x; \mu)$ with respect to x is zero, that is,

$$\nabla_x P(x(\mu); \mu) = \nabla f(x(\mu)) - \sum_{i \in \mathcal{I}} \frac{\mu}{c_i(x(\mu))} \nabla c_i(x(\mu)) = 0. \qquad (17.31)$$

If we define a Lagrange multiplier estimate by

$$\lambda_i(\mu) \overset{\text{def}}{=} \frac{\mu}{c_i(x(\mu))}, \quad i \in \mathcal{I}, \tag{17.32}$$

we can write (17.31) as

$$\nabla f(x(\mu)) - \sum_{i \in \mathcal{I}} \lambda_i(\mu) \nabla c_i(x(\mu)) = 0. \tag{17.33}$$

This condition is identical to the first KKT condition $\nabla_x \mathcal{L}(x, \lambda) = 0$ for the problem (17.19), where the Lagrangian function $\mathcal{L}$ is given by

$$\mathcal{L}(x, \lambda) = f(x) - \sum_{i \in \mathcal{I}} \lambda_i c_i(x); \tag{17.34}$$

(see (12.28) for the general definition of $\mathcal{L}$). Let us check the other KKT conditions for the inequality-constrained problem (17.19), which are as follows:

$$c_i(x) \geq 0, \quad \text{for all } i \in \mathcal{I}, \tag{17.35a}$$
$$\lambda_i \geq 0, \quad \text{for all } i \in \mathcal{I}, \tag{17.35b}$$
$$\lambda_i c_i(x) = 0, \quad \text{for all } i \in \mathcal{I}. \tag{17.35c}$$

The nonnegativity conditions (17.35a) and (17.35b) are clearly satisfied by $x = x(\mu)$, $\lambda = \lambda(\mu)$; in fact, the quantities $c_i(x(\mu))$ and $\lambda_i(\mu)$ are strictly positive for all $i \in \mathcal{I}$. The only KKT condition that fails to be satisfied is the complementarity condition (17.35c). By the definition (17.32), we have that

$$\lambda_i(\mu) c_i(x(\mu)) = \mu, \quad \text{for all } i \in \mathcal{I}, \tag{17.36}$$

where μ is strictly positive.

From the observations above, we see that as $\mu \downarrow 0$, a minimizer $x(\mu)$ of $P(x; \mu)$ and its associated Lagrange multiplier estimates $\lambda(\mu)$ come closer and closer to satisfying the KKT conditions of (17.19). In fact, if some additional assumptions are satisfied at the solution x^* of (17.19), we can show that $(x(\mu), \lambda(\mu))$ approaches the optimal primal–dual solution (x^*, λ^*) as $\mu \downarrow 0$.

Theorem 17.4.

Suppose that $\mathcal{F}^o$ is nonempty and that x^* is a local solution of (17.19) at which the KKT conditions are satisfied for some λ^*. Suppose, too, that the linear independence constraint qualification (LICQ) (Definition 12.1), the strict complementarity condition (Definition 12.2), and that the second-order sufficient conditions (Theorem 12.6) hold at (x^*, λ^*). Then the following statements are true.

(i) *There is a unique continuously differentiable vector function $x(\mu)$, defined for all sufficiently small μ by the statement that $x(\mu)$ is a local minimizer of $P(x; \mu)$ in some neighborhood of x^*, such that $\lim_{\mu \downarrow 0} x(\mu) = x^*$.*

(ii) *For the function $x(\mu)$ in (i), the Lagrange multiplier estimates $\lambda(\mu)$ defined by (17.32) converge to λ^* as $\mu \downarrow 0$.*

(iii) *The Hessian $\nabla_{xx}^2 P(x; \mu)$ is positive definite for all μ sufficiently small.*

For a proof of this result, see Fiacco and McCormick [79, Theorem 12] or M. Wright [251, Theorem 8].

The trajectory C_p defined by

$$C_p \stackrel{\text{def}}{=} \{x(\mu) \mid \mu > 0\} \tag{17.37}$$

is often referred to as the *central path* (or sometimes as the *primal central path*, to distinguish it from the primal–dual central path C_{pd} defined below).

HANDLING EQUALITY CONSTRAINTS

We have assumed in the analysis so far that the problem has only inequality constraints and that the strictly feasible region $\mathcal{F}^o$ (17.20) is nonempty. We now discuss one way in which the log-barrier technique can be applied to general nonlinear constrained problems of the form (12.1), which we restate here as follows:

$$\min_x f(x) \quad \text{subject to} \quad \begin{cases} c_i(x) = 0, & i \in \mathcal{E}, \\ c_i(x) \geq 0, & i \in \mathcal{I}. \end{cases} \tag{17.38}$$

We cannot simply replace the equality constraint $c_i(x) = 0$ by two inequalities—namely, $c_i(x) \geq 0$ and $-c_i(x) \geq 0$—to force it into the form (17.19), since then no point x can simultaneously satisfy both inequalities *strictly*. In other words, $\mathcal{F}^o$ will be empty.

One approach for dealing with equality constraints is to include *quadratic* penalty terms for these constraints in the objective. If we assume for simplicity that the coefficient of the quadratic penalty term is $1/\mu$, where μ is the barrier parameter, then the log-barrier/quadratic penalty function to be minimized for each value of μ is

$$B(x; \mu) \stackrel{\text{def}}{=} f(x) - \mu \sum_{i \in \mathcal{I}} \log c_i(x) + \frac{1}{2\mu} \sum_{i \in \mathcal{E}} c_i^2(x). \tag{17.39}$$

As expected, the properties of this function combine those of the log barrier function (17.22) and the quadratic penalty function (17.2). Algorithms can be designed around a framework similar to Frameworks 17.1 and 17.2; that is, successive reductions of the parameter μ alternate with approximate minimizations of the function $B(\cdot; \mu)$. The same issues arise:

the need to drive μ close to zero to obtain an accurate approximation of the true solution x^*; ill-conditioning of the Hessian $\nabla^2_{xx} B(x; \mu)$ when μ is small; the need for careful selection of a starting point for the minimization problem after μ is reduced; and the need for specialized line searches.

Minimization of $B(\cdot; \mu)$ still requires us to identify an initial point that is strictly feasible with respect to the inequality constraints. This task becomes trivial if we introduce slack variables $s_i, i \in \mathcal{I}$, into the formulation (17.38) and apply the method to the equivalent problem

$$\min_{(x,s)} f(x) \quad \text{subject to} \quad \begin{cases} c_i(x) = 0, & i \in \mathcal{E}, \\ c_i(x) - s_i = 0, & i \in \mathcal{I}, \\ s_i \geq 0, & i \in \mathcal{I}. \end{cases} \qquad (17.40)$$

The function corresponding to (17.39) for this problem is

$$f(x) - \mu \sum_{i \in \mathcal{I}} \log s_i + \frac{1}{2\mu} \sum_{i \in \mathcal{E}} c_i^2(x) + \frac{1}{2\mu} \sum_{i \in \mathcal{I}} (c_i(x) - s_i)^2.$$

Any point (x, s) for which $s > 0$ (strict positivity) lies in the domain of this function.

RELATIONSHIP TO PRIMAL–DUAL METHODS

By modifying the log-barrier viewpoint slightly, we can derive a new class of algorithms known as *primal–dual interior-point* methods, which has been the focus of much research activity in the field of optimization since 1984. In primal–dual methods, the Lagrange multipliers are treated as independent variables in the computations, with equal status to the primal variables x. As in the log-barrier approach, however, we still seek the primal–dual pair $(x(\mu), \lambda(\mu))$ that satisfies the approximate conditions (17.33), (17.35a), (17.35b), and (17.36) for successively smaller values of μ.

We explain the basic features of primal–dual methods only for the inequality-constrained formulation (17.19); the extension to handle equality constraints as well is simple (see the exercises). Suppose we rewrite the system of approximate KKT conditions for the problem (17.19) (namely, (17.33), (17.35a), (17.35b), and (17.36)), reformulating slightly by introducing slack variables $s_i, i \in \mathcal{I}$, to obtain

$$\nabla f(x) - \sum_{i \in \mathcal{I}} \lambda_i \nabla c_i(x) = 0, \qquad (17.41a)$$

$$c(x) - s = 0, \qquad (17.41b)$$

$$\lambda_i s_i = \mu, \quad \text{for all } i \in \mathcal{I}, \qquad (17.41c)$$

$$(\lambda, s) \geq 0. \qquad (17.41d)$$

Note that we have gathered the components $c_i(x), i \in \mathcal{I}$, into a vector $c(x)$; similarly for λ_i and $s_i, i \in \mathcal{I}$. As noted above, the pair $(x(\mu), \lambda(\mu))$ defined by (17.31) and (17.32)—

along with $s(\mu)$ defined as $c(x(\mu))$—solve the system (17.41), which we can view as a bound-constrained nonlinear system of equations. We define the *primal–dual central path* $\mathcal{C}_{\text{pd}}$ as

$$\mathcal{C}_{\text{pd}} \stackrel{\text{def}}{=} \{(x(\mu), \lambda(\mu), s(\mu)) \mid \mu > 0\}.$$

Note that the projection of $\mathcal{C}_{\text{pd}}$ onto the space of primal variables x yields the primal central path $\mathcal{C}_p$ defined by (17.37).

The bound constraints (17.41d) play a vital role: Primal–dual points (x, λ, s) that satisfy the first three conditions in (17.41) but violate condition (17.41d) typically lie far from solutions of (17.19). Hence, most primal–dual algorithms require the λ and s components of their iterates to be strictly positive, so that the inequality (17.41d) is satisfied by all iterates. These methods aim to satisfy the other conditions (17.41a), (17.41b), and (17.41c) only in the limit. Primal–dual steps are usually generated by applying a modified Newton method for nonlinear equations to the system formed by the equality conditions (17.41a), (17.41b), and (17.41c). In the log-barrier method, by contrast, we eliminate s and λ directly from the nonlinear system (17.41) (by using (17.41d) and (17.41b)) *before* applying Newton's method.

To show the form of the modified Newton step, we group the first three conditions in (17.41) into a vector function F_μ, that is,

$$F_\mu(x, \lambda, s) \stackrel{\text{def}}{=} \begin{bmatrix} \nabla f(x) - A(x)^T \lambda \\ c(x) - s \\ \Lambda S e - \mu e \end{bmatrix}, \qquad (17.42)$$

where Λ and S are the diagonal matrices whose diagonal elements are λ_i, $i \in \mathcal{I}$, and s_i, $i \in \mathcal{I}$, respectively, e is the vector whose components are all 1, and $A(x)$ is the matrix of constraint gradients, that is,

$$A(x)^T = [\nabla c_i(x)]_{i \in \mathcal{I}}.$$

The modified Newton equations for $F_\mu(x, \lambda, s)$ usually have the form

$$DF_\mu(x, \lambda, s) \begin{bmatrix} \Delta x \\ \Delta \lambda \\ \Delta s \end{bmatrix} = -F_\mu(x, \lambda, s) + \begin{bmatrix} 0 \\ 0 \\ r_{\lambda s} \end{bmatrix},$$

so by substitution from (17.42) we obtain

$$\begin{bmatrix} \nabla^2_{xx} \mathcal{L}(x, \lambda) & -A(x)^T & 0 \\ A(x) & 0 & -I \\ 0 & S & \Lambda \end{bmatrix} \begin{bmatrix} \Delta x \\ \Delta \lambda \\ \Delta s \end{bmatrix} = \begin{bmatrix} -\nabla f(x) + A(x)^T \lambda \\ -c(x) + s \\ -\Lambda S e + \mu e + r_{\lambda s} \end{bmatrix}, \qquad (17.43)$$

where $r_{\lambda s}$ is a modification term that when chosen suitably is key to good theoretical and practical performance of the method. (See Chapter 14 for a discussion of this term in the context of linear programming.)

Note that the structure of the matrix in (17.43) is similar to the matrices in (14.11), (14.20), and (16.53) from Chapters 14 and 16. In fact—if we allow for differences in notation and in the problem formulation—we can see that these formulae are simply specializations of the system (17.43) to the cases of linear and quadratic programming. Those chapters also described the various other ingredients needed to build a complete algorithm around the step formula (17.43). Such ingredients—a strategy for decreasing μ to zero, a way to calculate the step length α along the calculated step, a technique for choosing the modification term $r_{\lambda s}$ to improve the quality of the step, requirements on the conditions that must be satisfied by the iterates (x, λ, s)—are needed also in the nonlinear case, but the nonlinearity gives rise to other requirements for the algorithm as well. The choice of a merit function to govern step selection for the problem of solving $F_\mu(x, \lambda, s) = 0$ is important. In recent work, authors have proposed using the function $B(x; \mu)$ (17.39) in this role, or exact penalty functions that combine the barrier function and constraints. We discuss these issues a little further in the Notes and References at the end of the chapter. We omit details, however, because this topic is a subject of recent and ongoing research, and substantial further efforts in analysis and computation will be needed before clear trends emerge.

We see from (17.41) that we can motivate primal–dual methods independently of log-barrier methods by simply viewing the system (17.41) as a perturbation of the KKT system for (17.19). Just as in other areas of optimization, we are able to motivate a particular approach in a number of different ways, all of which shed a different light on the properties of the method.

17.3 EXACT PENALTY FUNCTIONS

Neither the quadratic penalty function nor the log-barrier function is an *exact* penalty function. By contrast, we gave examples of exact penalty functions in Section 15.3. These are functions with the property that for certain choices of the parameter μ, a *single* minimization yields the exact solution of the nonlinear programming problem (see Definition 15.1). (In Chapter 15 we used the term "merit function" instead of penalty function because we were discussing their use as a measure of the quality of the step, but the terms "merit function" and "penalty function" can be thought of as synonymous.)

An interesting class of algorithms for solving general nonlinear programming problems is based on minimizing the ℓ_1 exact penalty function (15.24), which we restate for convenience here as follows:

$$\phi_1(x; \mu) = f(x) + \frac{1}{\mu} \sum_{i \in \mathcal{E}} |c_i(x)| + \frac{1}{\mu} \sum_{i \in \mathcal{I}} [c_i(x)]^-. \tag{17.44}$$

(As before, we use the notation $[x]^- = \max\{0, -x\}$.) For all sufficiently small, positive values of μ, one minimization of this unconstrained function will yield the solution of the nonlinear program. It is difficult to determine μ a priori in most practical applications, however; rules for adjusting this parameter during the course of the computation must be devised.

Minimization of $\phi_1(x; \mu)$ is made difficult by the fact that it is nonsmooth—the first derivative is not defined at any x for which $c_i(x) = 0$ for some $i \in \mathcal{E} \cup \mathcal{I}$. General techniques for nondifferentiable optimization, such as bundle methods [137], are not efficient. One approach that has proved to be effective in practice is to solve a sequence of subproblems in which the objective function f in (17.44) is replaced by a quadratic approximation whose Hessian is the Hessian of the Lagrangian for the nonlinear programming problem, and in which the constraint terms c_i are replaced by linear approximations about the current iterate x_k. The resulting method turns out to be closely related to sequential quadratic programming methods, and is one of the most powerful techniques for solving constrained optimization problems. For more details, see our discussion in Section 18.5.

Some other algorithms are based on the minimization of exact *differentiable* merit functions related to the augmented Lagrangian; see (15.25), for example. These algorithms also require the solution of a sequence of subproblems, but it has not yet been established whether they can be the basis for reliable and efficient software for nonlinear programming.

17.4 AUGMENTED LAGRANGIAN METHOD

We now discuss an algorithm known as the *method of multipliers* or the *augmented Lagrangian method*. This algorithm is related to the quadratic penalty algorithm of Section 17.1, but it reduces the possibility of ill conditioning of the subproblems that are generated in this approach by introducing explicit Lagrange multiplier estimates at each step into the function to be minimized. It also tends to yield less ill conditioned subproblems than does the log-barrier approach, and it dispenses with the need for iterates to stay strictly feasible with respect to the inequality constraints. It does not introduce nonsmoothness (as do methods based on the ℓ_1 penalty function (17.44), for instance), and implementations can be constructed from standard software for unconstrained or bound-constrained optimization.

The method of multipliers is the basis for a practical implementation of high quality—LANCELOT—which is sketched below.

In this section we use superscripts (usually k and $k+1$) on the Lagrange multiplier estimates to denote iteration index, and subscripts (usually i) to denote the component indices of the vector λ.

MOTIVATION AND ALGORITHM FRAMEWORK

We first consider the equality-constrained problem (17.1). The quadratic penalty function $Q(x; \mu)$ (17.2) penalizes constraint violations by squaring the infeasibilities and scaling them by $1/(2\mu)$. As we see from Theorem 17.2, however, the approximate minimizers x_k of $Q(x; \mu_k)$ do not quite satisfy the feasibility conditions $c_i(x) = 0, i \in \mathcal{E}$. Instead, they are perturbed slightly (see (17.8)) to approximately satisfy

$$c_i(x_k) = -\mu_k \lambda_i^*, \quad \text{for all } i \in \mathcal{E}. \tag{17.45}$$

To be sure, this perturbation vanishes as $\mu_k \downarrow 0$, but one may ask whether we can alter the function $Q(x; \mu_k)$ to avoid this systematic perturbation—that is, to make the approximate minimizers more nearly satisfy the equality constraints $c_i(x) = 0$. By doing so, we may avoid the need to decrease μ to zero, and thereby avoid the ill conditioning and numerical problems associated with $Q(x; \mu)$ for small values of this penalty parameter.

The augmented Lagrangian function $\mathcal{L}_A(x, \lambda; \mu)$ achieves these goals by including an explicit estimate of the Lagrange multipliers λ, based on the formula (17.8), in the objective. From the definition

$$\mathcal{L}_A(x, \lambda; \mu) \stackrel{\text{def}}{=} f(x) - \sum_{i \in \mathcal{E}} \lambda_i c_i(x) + \frac{1}{2\mu} \sum_{i \in \mathcal{E}} c_i^2(x), \tag{17.46}$$

we see that the augmented Lagrangian differs from the (standard) Lagrangian (12.28) for (17.1) by the presence of the squared terms, while it differs from the quadratic penalty function (17.2) in the presence of the summation term involving the λ. In this sense, it is a combination of the Lagrangian and quadratic penalty functions. When we differentiate with respect to x, we obtain

$$\nabla_x \mathcal{L}_A(x, \lambda; \mu) = \nabla f(x) - \sum_{i \in \mathcal{E}} [\lambda_i - c_i(x)/\mu] \nabla c_i(x). \tag{17.47}$$

We now design an algorithm that fixes the barrier parameter μ to some value $\mu_k > 0$ at its kth iteration (as in Frameworks 17.1 and 17.2), fixes λ at the current estimate λ^k, and performs minimization with respect to x. Using x_k to denote the approximate minimizer of $\mathcal{L}_A(x, \lambda; \mu_k)$, we can use the logic in the proof of Theorem 17.2 to deduce that

$$\lambda_i^* \approx \lambda_i^k - c_i(x_k)/\mu_k, \quad \text{for all } i \in \mathcal{E}. \tag{17.48}$$

By rearranging this expression, we have that

$$c_i(x_k) \approx -\mu_k(\lambda_i^* - \lambda_i^k), \quad \text{for all } i \in \mathcal{E},$$

so we conclude that if λ^k is close to the optimal multiplier vector λ^*, the infeasibility in x_k will be much smaller than μ_k, rather than being proportional to μ_k as in (17.45).

How can we update the multiplier estimates λ^k from iteration to iteration, so that they approximate λ^* more and more accurately, based on current information? Equation (17.48) immediately suggests the formula

$$\lambda_i^{k+1} = \lambda_i^k - c_i(x_k)/\mu_k, \quad \text{for all } i \in \mathcal{E}. \tag{17.49}$$

This discussion motivates the following algorithmic framework.

Framework 17.3 (Method of Multipliers–Equality Constraints).
Given $\mu_0 > 0$, tolerance $\tau_0 > 0$, starting points x_0^s and λ^0;
for $k = 0, 1, 2, \ldots$
    Find an approximate minimizer x_k of $\mathcal{L}_A(\cdot, \lambda^k; \mu_k)$, starting at x_k^s,
    and terminating when $\|\nabla_x \mathcal{L}_A(x, \lambda^k; \mu_k)\| \leq \tau_k$;
    if final convergence test satisfied
        STOP with approximate solution x_k;
    Update Lagrange multipliers using (17.49) to obtain λ^{k+1};
    Choose new penalty parameter $\mu_{k+1} \in (0, \mu_k)$;
    Set starting point for the next iteration to $x_{k+1}^s = x_k$;
end (for)

We show below that convergence of this method can be assured without decreasing μ to a very small value. Ill conditioning is therefore less of a problem than in Frameworks 17.1 and 17.2, so the choice of starting point x_{k+1}^s for each value of k in Framework 17.3 is less critical. In Framework 17.3 we simply start the search at iteration $k+1$ from the previous approximate minimizer x_k. Bertsekas [9, p. 347] suggests setting the tolerance τ_k to

$$\tau_k = \min(\epsilon_k, \gamma_k \|c(x_k)\|),$$

where $\{\epsilon_k\}$ and $\{\gamma_k\}$ are two sequences decreasing to 0. An alternative choice is used by the code LANCELOT (see Algorithm 17.4).

❏ **EXAMPLE 17.4**

Consider again problem (17.3), for which the augmented Lagrangian is

$$\mathcal{L}_A(x, \lambda; \mu) = x_1 + x_2 - \lambda(x_1^2 + x_2^2 - 2) + \frac{1}{2\mu}(x_1^2 + x_2^2 - 2)^2. \tag{17.50}$$

The primal solution is $x^* = (-1, -1)^T$, and the optimal Lagrange multiplier is $\lambda^* = -0.5$.
Suppose that at iterate k we have $\mu_k = 1$ (as in Figure 17.1), while the current multiplier estimate is $\lambda^k = -0.4$. Figure 17.6 plots the function $\mathcal{L}_A(x, -0.4; 1)$. Note that the spacing of the contours indicates that the conditioning of this problem is similar to that of the quadratic penalty function $Q(x; 1)$ illustrated in Figure 17.1. However, the minimizing

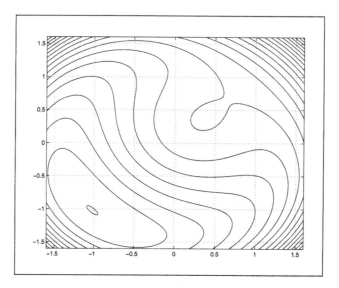

Figure 17.6 Contours of $\mathcal{L}_A(x, \lambda; \mu)$ from (17.4) for $\lambda = -0.4$ and $\mu = 1$, contour spacing 0.5.

value of $x_k \approx (-1.02, -1.02)$ is much closer to the solution $x^* = (-1, -1)^T$ than is the minimizing value of $Q(x; 1)$, which is approximately $(-1.1, -1.1)$. This example shows that the inclusion of the Lagrange multiplier term in the function $\mathcal{L}_A(x, \lambda; \mu)$ can result in a substantial improvement over the quadratic penalty method, as a way to reformulate the constrained optimization problem (17.1).

EXTENSION TO INEQUALITY CONSTRAINTS

When the problem formulation contains *general inequality constraints*—as in the formulation (17.38)—we can convert it to a problem with equality constraints and bound constraints by introducing slack variables s_i and replacing the inequalities $c_i(x) \geq 0, i \in \mathcal{I}$, by

$$c_i(x) - s_i = 0, \quad s_i \geq 0, \quad \text{for all } i \in \mathcal{I}. \tag{17.51}$$

This transformation gives rise to a problem containing equality constraints and bound constraints, and can be solved by the algorithm in LANCELOT, which treats bound constraints explicitly (see (17.66)).

An alternative to handling the bounds directly in the subproblem is to eliminate the slack variables s_i, $i \in \mathcal{I}$, directly from the minimization procedure in a manner that we now describe. Supposing for simplicity that there are no equality constraints ($\mathcal{E} = \emptyset$), the introduction of slack variables transforms the problem to the following form:

$$\min_{x,s} f(x) \quad \text{subject to } c_i(x) - s_i = 0, \quad s_i \geq 0, \quad \text{for all } i \in \mathcal{I}.$$

By defining the augmented Lagrangian in terms of the constraints $c_i(x) - s_i = 0$ and applying the bound constraints $s_i \geq 0$ explicitly, we obtain the following subproblem to be solved at iteration k of Framework 17.3:

$$\min_{x,s} f(x) - \sum_{i \in \mathcal{I}} \lambda_i^k (c_i(x) - s_i) + \frac{1}{2\mu_k} \sum_{i \in \mathcal{I}} (c_i(x) - s_i)^2 \quad (17.52\text{a})$$

$$\text{subject to } s_i \geq 0, \text{ for all } i \in \mathcal{I}. \quad (17.52\text{b})$$

Each s_i occurs in just two terms of (17.52a), which is in fact a convex quadratic function with respect to each of these slack variables. We can therefore perform an explicit minimization in (17.52) with respect to each of the s_i separately. The partial derivative of the subproblem objective (17.52a) with respect to s_i is $(\lambda^k)_i - (1/\mu_k)(c_i(x) - s_i)$. The unconstrained minimizer of (17.52a) with respect to s_i occurs when this partial derivative equals zero, that is,

$$s_i = c_i(x) - \mu \lambda_i^k. \quad (17.53)$$

If this unconstrained minimizer is smaller than the lower bound of 0, then since (17.52a) is convex in s_i, the optimal value of s_i in (17.52) is 0. Summarizing, we find that the optimal values of s_i in (17.52) are

$$s_i = \max(c_i(x) - \mu \lambda_i^k, 0), \quad \text{for all } i \in \mathcal{I}. \quad (17.54)$$

We can use this formula to substitute for s and obtain an equivalent form of the subproblem (17.52) in x alone. By isolating the terms involving s_i, we obtain that

$$-\lambda_i^k (c_i(x) - s_i) + \frac{1}{2\mu}(c_i(x) - s_i)^2$$

$$= \begin{cases} -\lambda_i^k c_i(x) + \dfrac{1}{2\mu} c_i^2(x) & \text{if } c_i(x) - \mu \lambda_i^k \leq 0, \\ -\dfrac{\mu}{2}(\lambda_i^k)^2 & \text{otherwise.} \end{cases} \quad (17.55)$$

By defining the function $\psi(t, \sigma; \mu)$ of the scalar arguments t, σ, and μ to have the form of the right-hand-side in (17.55), that is,

$$\psi(t, \sigma; \mu) \overset{\text{def}}{=} \begin{cases} -\sigma t + \dfrac{1}{2\mu} t^2 & \text{if } t - \mu\sigma \leq 0, \\ -\dfrac{\mu}{2}\sigma^2 & \text{otherwise,} \end{cases} \quad (17.56)$$

we obtain the following transformed subproblem:

$$\min_x \mathcal{L}_A(x, \lambda^k; \mu_k) \stackrel{\text{def}}{=} f(x) + \sum_{i \in \mathcal{I}} \psi(c_i(x), \lambda_i^k; \mu_k). \tag{17.57}$$

This definition of $\mathcal{L}_A$ represents a natural extension of the augmented Lagrangian to the inequality-constrained case. With this extension, we can apply Framework 17.3 to the case of inequality constraints as well, with one further modification: Once the approximate solution x_k is obtained, we use the following formula to update the Lagrange multipliers:

$$\lambda_i^{k+1} = \max(\lambda_i^k - c_i(x_k)/\mu_k, 0), \quad \text{for all } i \in \mathcal{I}. \tag{17.58}$$

This formula follows from the fact that the derivative of $\mathcal{L}_A(x, \lambda; \mu_k)$ in (17.57) with respect to x should be close to zero at the approximate solution x_k. By using (17.55), we obtain

$$\nabla_x \mathcal{L}_A(x_k, \lambda^k; \mu_k) = \nabla f(x_k) - \sum_{i \in \mathcal{I} \mid c_i(x) \leq \mu \lambda_i^k} \left[\lambda_i^k - c_i(x_k)/\mu_k\right] \nabla c_i(x_k) \approx 0.$$

The update formula (17.58) follows by comparing this formula to the first KKT condition (12.30a), which we can state as

$$\nabla f(x^*) - \sum_{i \in \mathcal{I} \mid c_i(x^*) = 0} \lambda_i^* \nabla c_i(x^*) = 0.$$

It makes intuitive sense to maintain nonnegativity of the components of λ, since we know from the KKT conditions (12.30d) that the optimal Lagrange multipliers for the inequality constraints are nonnegative.

Each of the functions $\psi(c_i(x), \lambda_i; \mu)$ is continuously differentiable with respect to x, but there is in general a discontinuity in the second derivative with respect to x wherever $c_i(x) = \mu \lambda_i$ for some $i \in \mathcal{I}$. However, when the strict complementarity condition holds (Definition 12.2), the approximate minimizer x_k of each subproblem (17.57) is usually not close to the nonsmooth regions, so it does not usually interfere with the algorithm that solves the subproblem. To verify this claim, we consider separately the cases in which constraint i is active and inactive. When constraint i is active, then for a sufficiently advanced iterate k we have $c_i(x_k) \approx 0$, while λ_i^k and μ_k are both significantly larger than zero. When the ith constraint is inactive, we have $c_i(x_k) > 0$ while $\lambda_i^k \approx 0$. In both cases, we can expect the point x_k to comfortably avoid the region in which the discontinuity condition $c_i(x) = \mu \lambda_i$ is satisfied.

PROPERTIES OF THE AUGMENTED LAGRANGIAN

We now prove two results that justify the use of the augmented Lagrangian function and the method of multipliers. For simplicity, we confine our attention to the case of equality constraints only, that is, the problem (17.1) for which the augmented Lagrangian is given by (17.46).

The first result validates the approach of Framework 17.3 by showing that when we have knowledge of the exact Lagrange multiplier vector λ^*, the solution x^* of (17.1) is a strict minimizer of $\mathcal{L}_A(x, \lambda^*; \mu)$ for all μ sufficiently small. Although we do not know λ^* exactly in practice, the result and its proof strongly suggest that we can obtain a good estimate of x^* by minimizing $\mathcal{L}_A(x, \lambda; \mu)$ even when μ is not particularly close to zero, provided that λ is a reasonable estimate of λ^*. (We observed these properties already in Example 17.4.)

Theorem 17.5.

Let x^* be a local solution of (17.1) at which the LICQ is satisfied (that is, the gradients $\nabla c_i(x^*)$, $i \in \mathcal{E}$, are linearly independent vectors), and the second-order sufficient conditions specified in Theorem 12.6 are satisfied for $\lambda = \lambda^*$. Then there is a threshold value $\bar{\mu}$ such that for all $\mu \in (0, \bar{\mu}]$, x^* is a strict local minimizer of $\mathcal{L}_A(x, \lambda^*; \mu)$.

PROOF. We prove the result by showing that x^* satisfies the second-order sufficient conditions to be a strict local minimizer of $\mathcal{L}_A(x, \lambda^*; \mu)$ (see Theorem 2.4); that is,

$$\nabla_x \mathcal{L}_A(x^*, \lambda^*; \mu) = 0, \quad \nabla^2_{xx} \mathcal{L}_A(x^*, \lambda^*; \mu) \text{ positive definite}. \tag{17.59}$$

By using the KKT conditions (12.30) and the formula (17.47), we have that

$$\nabla_x \mathcal{L}_A(x^*, \lambda^*; \mu) = \nabla f(x^*) - \sum_{i \in \mathcal{E}} [\lambda_i^* - c_i(x^*)/\mu] \nabla c_i(x^*)$$

$$= \nabla f(x^*) - \sum_{i \in \mathcal{E}} \lambda_i^* \nabla c_i(x^*) = \nabla_x \mathcal{L}(x^*, \lambda^*) = 0,$$

verifying the first part of (17.59) (independently of μ).

We now prove the second part of (17.59) by showing that $u^T \nabla^2_{xx} \mathcal{L}_A(x^*, \lambda^*; \mu) u > 0$ for all $u \in \mathbf{R}^n$ and all $\mu > 0$ sufficiently small. From (17.13), we have

$$A^T = [\nabla c_i^*]_{i \in \mathcal{E}},$$

and note that by LICQ (Definition 12.1), A has full row rank. Note, too, that

$$\nabla^2_{xx} \mathcal{L}_A(x^*, \lambda^*; \mu) = \nabla^2_{xx} \mathcal{L}(x^*, \lambda^*) + \frac{1}{\mu} A^T A. \tag{17.60}$$

By the fundamental theorem of algebra, we can partition any $u \in \mathbf{R}^n$ into components in $\text{Null} A$ and $\text{Range} A^T$, and write

$$u = w + A^T v,$$

where $w \in \text{Null}\, A$ and v is a vector in $\mathbf{R}^{|\mathcal{E}|}$. By using (17.60) and the properties of w and v, we have that

$$u^T \nabla^2_{xx} \mathcal{L}_A(x^*, \lambda^*; \mu)u = w^T \nabla^2_{xx} \mathcal{L}(x^*, \lambda^*)w + 2w^T \nabla^2_{xx}\mathcal{L}(x^*,\lambda^*)A^T v \quad (17.61)$$
$$+ v^T A \nabla^2_{xx}\mathcal{L}(x^*,\lambda^*) A^T v + v^T A \left((1/\mu)A^T A\right) A^T v.$$

We seek bounds on the three terms on the right-hand-side of this expression.

Because of (12.63) and compactness of the unit sphere intersected with $\text{Null}\, A$, there is a scalar $a > 0$ such that

$$w^T \nabla_{xx}\mathcal{L}(x^*, \lambda^*)w \geq a\|w\|^2, \quad \text{for all } w \in \text{Null}\, A,$$

giving us a lower bound on the first right-hand-side term in (17.61). For the second term, define $b \stackrel{\text{def}}{=} \|\nabla^2_{xx}\mathcal{L}(x^*,\lambda^*)A^T\|$, so that

$$2w^T \nabla^2_{xx}\mathcal{L}(x^*,\lambda^*)A^T v \geq -2b\,\|w\|\,\|v\|,$$

giving a lower bound for the second term. For the third term, if we define $c \stackrel{\text{def}}{=} \|A\nabla^2_{xx}\mathcal{L}(x^*,\lambda^*)A^T\|$, we obtain

$$v^T A \nabla^2_{xx}\mathcal{L}(x^*,\lambda^*)A^T v \geq -c\|v\|^2.$$

Finally, if d is the smallest eigenvalue of AA^T, we have that

$$v^T A \left(\frac{1}{\mu}A^T A\right) A^T v \geq \frac{1}{\mu}\|AA^T v\|^2 \geq \frac{d^2}{\mu}\|v\|^2.$$

By substituting these lower bounds into (17.61), we obtain

$$u^T \nabla^2_{xx}\mathcal{L}_A(x^*,\lambda^*;\mu)u \geq a\|w\|^2 - 2b\|v\|\,\|w\| + (d^2/\mu - c)\|v\|^2$$
$$= a\,[\|w\| - (b/a)\|v\|]^2 + (d^2/\mu - c - b^2/a)\|v\|^2. \quad (17.62)$$

The first term on the right-hand-side of this expression is clearly nonnegative. Since $d > 0$ by the full rank of A, the second term is also nonnegative, provided that we choose $\bar{\mu}$ to be any value such that

$$\frac{d^2}{\bar{\mu}} - c - \frac{b^2}{a} > 0$$

and choose $\mu \in (0, \bar{\mu}]$. In fact, the right-hand-side of (17.62) is strictly positive unless $v = 0$ and $w = 0$, which implies that $\nabla^2_{xx}\mathcal{L}_A(x^*, \lambda^*; \mu)$ is positive definite, as required. Hence, we have verified (17.59) and completed the proof. $\square$

The second result, given by Bertsekas [9, Proposition 4.2.3], describes the more realistic situation of $\lambda \neq \lambda^*$. It gives conditions under which there is a minimizer of $\mathcal{L}_A(x, \lambda; \mu)$ that lies close to x^* and gives error bounds on both x_k and the updated multiplier estimate λ^{k+1} obtained from solving the subproblem at iteration k.

Theorem 17.6.

Suppose that the assumptions of Theorem 17.5 are satisfied at x^* and λ^*, and let $\bar{\mu}$ be chosen as in that theorem. Then there exist positive scalars δ, ϵ, and M such that the following claims hold:

(a) *For all λ^k and μ_k satisfying*

$$\|\lambda^k - \lambda^*\| \leq \delta/\mu_k, \quad \mu_k \leq \bar{\mu}, \qquad (17.63)$$

the problem

$$\min_x \mathcal{L}_A(x, \lambda^k; \mu_k) \quad \text{subject to } \|x - x^*\| \leq \epsilon$$

has a unique solution x_k. Moreover, we have

$$\|x_k - x^*\| \leq M\mu_k\|\lambda^k - \lambda^*\|.$$

(b) *For all λ^k and μ_k that satisfy (17.63), we have*

$$\|\lambda^{k+1} - \lambda^*\| \leq M\mu_k\|\lambda^k - \lambda^*\|,$$

where λ^{k+1} is given by the formula (17.49).

(c) *For all λ^k and μ_k that satisfy (17.63), the matrix $\nabla^2_{xx}\mathcal{L}_A(x_k, \lambda^k)$ is positive definite and the constraint gradients $\nabla c_i(x_k)$, $i \in \mathcal{E}$, are linearly independent.*

PRACTICAL IMPLEMENTATION

We now outline a practical algorithm that forms the basis of the LANCELOT implementation of Conn, Gould, and Toint [53]. This program transforms inequalities into equalities by means of slack variables, as in (17.51), and solves problems of the form

$$\min_{x \in \mathbb{R}^n} f(x) \quad \text{subject to} \quad \begin{cases} c_i(x) = 0, & i = 1, 2, \ldots, m, \\ l \leq x \leq u. \end{cases} \qquad (17.64)$$

The slacks are considered to be part of the vector x, and l and u are vectors of lower and upper bounds. (Some components of l may be set to $-\infty$, signifying that there is no lower bound on the components of x in question; similarly for u.) LANCELOT is also designed to be efficient on special cases of (17.64); for example, bound-constrained problems (in which $m = 0$) and unconstrained problems. It can also take advantage of partially separable structure in the objective function and constraints (see Chapter 9).

The augmented Lagrangian function used in LANCELOT [53] incorporates only the equality constraints from (17.64), that is,

$$\mathcal{L}_A(x, \lambda; \mu) = f(x) - \sum_{i=1}^{m} \lambda_i c_i(x) + \frac{1}{2\mu} \sum_{i=1}^{m} c_i^2(x). \tag{17.65}$$

The bound constraints are enforced explicitly in the subproblem, which has the form

$$\min_{x} \mathcal{L}_A(x, \lambda; \mu) \quad \text{subject to } l \leq x \leq u. \tag{17.66}$$

The multiplier estimates λ and penalty parameter μ are held fixed during each subproblem. To find an approximate solution of (17.66), the algorithm forms a quadratic model of $\mathcal{L}_A(x, \lambda; \mu)$ and applies the gradient projection method described in Section 16.6 to obtain a new iterate. The Hessian of the quadratic model uses either exact second derivatives or quasi-Newton estimates of the Hessian of $\mathcal{L}_A$. We refer the reader to [53] and the references therein for details.

By specializing the KKT conditions (12.30) to the problem (17.66) (see the exercises), we find that the first-order necessary condition for x to be a solution of (17.66) is that

$$P_{[l,u]} \nabla \mathcal{L}_A(x, \lambda; \mu) = 0, \tag{17.67}$$

where $P_{[l,u]} g$ is the projection of the vector $g \in \mathbf{R}^n$ onto the rectangular box $[l, u]$ defined by

$$(P_{[l,u]} g)_i = \begin{cases} \min(0, g_i) & \text{if } x_i = l_i, \\ g_i & \text{if } x_i \in (l_i, u_i), \\ \max(0, g_i) & \text{if } x_i = u_i. \end{cases} \quad \text{for all } i = 1, 2, \ldots, n. \tag{17.68}$$

The framework for LANCELOT, stated below, allows the subproblems (17.66) to be solved inexactly and includes procedures for updating the tolerance used for the approximate solution of the subproblem and for updating the penalty parameter μ. We omit details of the gradient-projection procedure for solving the subproblem.

Algorithm 17.4 (LANCELOT–Method of Multipliers).
 Choose positive constants $\bar{\eta}, \bar{\omega}, \bar{\mu} \leq 1, \tau < 1, \bar{\gamma} < 1, \alpha_\omega, \beta_\omega, \alpha_\eta, \beta_\eta, \alpha_*, \beta_*$
  satisfying $\alpha_\eta < \min(1, \alpha_\omega), \beta_\eta < \min(1, \beta_\omega)$;
 Choose $\lambda^0 \in R^m$;
 Set $\mu_0 = \bar{\mu}, \alpha_0 = \min(\mu_0, \bar{\gamma}), \omega_0 = \bar{\omega}(\alpha_0)^{\alpha_\omega}, \eta_0 = \bar{\eta}(\alpha_0)^{\alpha_\eta}$;
 for $k = 0, 1, 2, \ldots$

Find an approximate solution x_k of the subproblem (17.66) such that

$$\|P_{[l,u]}\nabla\mathcal{L}_A(x_k,\lambda^k;\mu_k)\| \leq \omega_k,$$

where the projection operator $P_{[l,u]}$ is defined as in (17.68).

if $\|c(x_k)\| \leq \eta_k$
    (* test for convergence *)
    if $\|c(x_k)\| \leq \eta_*$ and $\|P_{[l,u]}\nabla\mathcal{L}_A(x_k,\lambda^k;\mu_k)\| \leq \omega_*$
        STOP with approximate solution x_k;
    end (if)
    (* update multipliers, tighten tolerances *)
    $\lambda^{k+1} = \lambda^k - c(x_k)/\mu_k$;
    $\mu_{k+1} = \mu_k$;
    $\alpha_{k+1} = \mu_{k+1}$;
    $\eta_{k+1} = \eta_k \alpha_{k+1}^{\beta_\eta}$;
    $\omega_{k+1} = \omega_k \alpha_{k+1}^{\beta_\omega}$;
else
    (* decrease penalty parameter, tighten tolerances *)
    $\lambda^{k+1} = \lambda^k$;
    $\mu_{k+1} = \tau\mu_k$;
    $\alpha_{k+1} = \mu_{k+1}\bar{\gamma}$;
    $\eta_{k+1} = \bar{\eta}\alpha_{k+1}^{\beta_\eta}$;
    $\omega_{k+1} = \bar{\omega}\alpha_{k+1}^{\beta_\omega}$;
end (if)
end (for)

The main branch in the algorithm tests the norm of $c(x_k)$ against the tolerance η_k. If $\|c(x_k)\| > \eta_k$ (the *else* branch), we decrease the penalty parameter μ_k to ensure that the next subproblem will place more emphasis on decreasing the constraint violations. If $\|c(x_k)\| \leq \eta_k$ (the *then* branch), we decide that the current value of μ_k is doing a good job of maintaining near-feasibility of the iterate x_k, so we update the multipliers according to formula (17.49) without decreasing μ. In both cases, we decrease the tolerances ω_k and η_k to force the subsequent primal iterates $x_{k+1}, x_{k+2}, \ldots$ to be increasingly accurate solutions of the subproblem.

17.5 SEQUENTIAL LINEARLY CONSTRAINED METHODS

The principal idea behind *sequential linearly constrained* (SLC), or *reduced Lagrangian*, methods is to generate a step by minimizing the Lagrangian subject to linearizations of the constraints. In contrast to the sequential quadratic programming methods described in

Chapter 18, which minimize a quadratic approximation of the Lagrangian subject to linear constraints, the subproblem in SLC methods uses a nonlinear objective function.

Let us first consider the equality-constrained problem (17.1). The SLC subproblem in this case takes the form

$$\min_x F_k(x) \quad \text{subject to} \quad \nabla c_i(x_k)^T (x - x_k) + c_i(x_k) = 0, \quad \text{for all } i \in \mathcal{E}. \quad (17.69)$$

There are several possible choices for $F_k(x)$. Early SLC methods defined

$$F_k(x) = f(x) - \sum_{i \in \mathcal{E}} \lambda_i^k \bar{c}_i^k(x), \quad (17.70)$$

where λ^k is the current Lagrange multiplier estimate and $\bar{c}_i^k(x)$ is the difference between $c_i(x)$ and its linearization at x_k, that is,

$$\bar{c}_i^k(x) = c_i(x) - c_i(x_k) - \nabla c_i(x_k)^T (x - x_k). \quad (17.71)$$

One can show that as x_k converges to a solution x^*, the Lagrange multiplier associated with (17.69) converges to the optimal multiplier. Therefore, we take the multiplier estimate λ^k in (17.70) to be the multiplier of the subproblem (17.69) at the previous iteration.

To obtain reliable convergence from remote starting points, the most popular SLC method defines F_k to be the augmented Lagrangian function

$$F_k(x) = f(x) - \sum_{i \in \mathcal{E}} \lambda_i^k \bar{c}_i^k(x) + \frac{1}{2\mu} \sum_{i \in \mathcal{E}} [\bar{c}_i^k(x)]^2, \quad (17.72)$$

where μ is a positive penalty parameter. Note that this definition differs from (17.46) in that the original constraints $c_i(x)$ have been replaced by $\bar{c}_i^k(x)$ from (17.71). The new definition is reasonable, however, since all feasible points x must satisfy $\bar{c}_i^k(x) = 0$ whenever x_k is a solution of the equality-constrained problem (17.1).

SLC methods are mainly used to solve large problems, and the model function (17.72) is incorporated in the very successful implementation MINOS [175]. Since the step computation requires the minimization of the nonlinear subproblem (17.69), several iterations—and thus several function and constraint evaluations—are required to generate the new iterate. Few "outer" iterations are required, however, in comparison with most other optimization methods.

Newton or quasi-Newton iterations can be used to solve the linearly constrained problem (17.69), and the sparse elimination techniques of Chapter 15 are used to exploit sparsity in the constraint gradients.

For nonlinear optimization problems with inequality constraints (17.19), we can introduce slacks where necessary and transform all inequalities into equalities together with bound constraints. The subproblem is then given by (17.69) with the additional bounds

$$l \leq x \leq u,$$

where x now contains the variables and slacks, and l and u are suitable vectors of bounds. It can be shown that in a neighborhood of a solution satisfying the second-order sufficiency conditions, this subproblem will identify the optimal active set.

A drawback of SLC methods is that the subproblem must be solved quite accurately to ensure that the Lagrange multiplier estimates are of good quality, so that a large number of function evaluations may be required per outer iteration. Even though it is widely believed that SQP methods are preferable to SLC methods, the now classical MINOS package implementing an SLC method still ranks among the most effective implementations for large nonlinear programming.

NOTES AND REFERENCES

The quadratic penalty function was first proposed by Courant [60]. Gould [118] addresses the issue of stable determination of the Newton step for $Q(x; \mu_k)$. His formula (2.2) differs from our formula (17.17) in the right-hand-side, but both systems give rise to the same p component. A reformulation of the log-barrier system (17.28) is also discussed in [118], but it requires an estimate of the active set to be made. Gould [119] also discusses the use of an extrapolation technique for finding a good starting point x_k^s for each major iterate of Framework 17.1. Analogous techniques for the log-barrier function are discussed by Conn, Gould, and Toint [55] and Dussault [78].

For a discussion of the history of barrier function methods, see Nash [178]. The term "interior-point," which is widely used in connection with both log-barrier and primal–dual methods that insist on strict satisfaction of the inequality constraints at each iteration, appears to originate in the book of Fiacco and McCormick [79]. As mentioned above, this book has remained the standard reference in the area of log-barrier functions, and its importance only increased after 1984 when interior-point methods for nonlinear programming became a topic of intense research following the publication of Karmarkar's paper [140]. Fiacco and McCormick [79] also introduce the barrier/quadratic penalty function (17.39) and the central path $\mathcal{C}_p$.

Reformulation of the problem with a linear objective, which causes the Newton step for $P(\cdot; \mu_{k+1})$ taken from the previous approximate minimizer x_k to eventually pass close to $x(\mu_{k+1})$, is discussed by Wright and Jarre [258]. Analysis of the Newton/log-barrier method, including the size of the region of convergence for the minimizer of $P(\cdot; \mu)$ and of the possibility of superlinear convergence, can be found in S. Wright [254].

Standard error analysis indicates that when the solution of the system (17.28) is computed in a finite-precision environment, the ill conditioning in $\nabla_{xx}^2 P(x; \mu)$ leads to large errors in the step p. Remedies based on exploiting the structure in $\nabla_{xx}^2 P(x; \mu)$ exposed by (17.27) have been proposed, but more detailed analysis has shown that because of the special structure of both the gradient and the Hessian of $P(x; \mu)$, the errors in the computed step p are less significant than a naive analysis would indicate, so that the computed p remains a good search direction for the Newton algorithm until μ becomes quite small. For details, see M. Wright [252] and S. Wright [256]. (These papers actually deal with primal–dual algorithms for the problem (17.19), but their major conclusions can be applied to the system (17.28) as well.) Specialized line search techniques that are suitable for use with barrier functions are described by Murray and Wright [174] and Fletcher and McCann [87].

Techniques based on the log-barrier/quadratic penalty function defined in (17.39) have been investigated by Gould [119] and Dussault [78].

Primal–dual methods have been investigated in a number of recent papers. Forsgren and Gill [91] use the function $B(x; \mu)$ (17.39) as a merit function for steps generated by a primal–dual approach for solving the system $F_\mu(x, \lambda, s) = 0$. Gay, Overton, and Wright [99] add a Lagrangian term $\sum_{i \in \mathcal{I} \cup \mathcal{E}} \lambda_i c_i(x)$ to $B(x; \mu)$ to obtain their primary merit function. Byrd, Hribar, and Nocedal [34] use a nondifferentiable ℓ_2 merit function for the barrier problem. When the problem (12.1) is a convex program, algorithms more closely related to the linear programming primal–dual algorithms of Chapter 14 than the ones described above may be appropriate. For example, Ralph and Wright [211] describe a method with good global and local convergence properties.

The method of multipliers was proposed by Hestenes [134] and Powell [195], and the definitive work in this area is the book of Bertsekas [8]. Chapters 1–3 of Bertsekas's book contains a thorough motivation of the method that outlines its connections to other approaches. Other introductory discussions are given by Fletcher [83, Section 12.2], and Polak [193, Section 2.8]. The extension to inequality constraints was described by Rockafellar [216] and Powell [198].

SLC methods were proposed by Robinson [215] and Rosen and Kreuser [218]. The MINOS implementation is due to Murtagh and Saunders [175]. Our discussion of SLC methods is based on that of Gill et al. [112].

✎ EXERCISES

✎ 17.1 For $z \in \mathbb{R}$, show that the function $\min(0, z)^2$ has a discontinuous second derivative at $z = 0$. (It follows that quadratic penalty function (17.5) may not have continuous second derivatives even when f and c_i, $i \in \mathcal{E} \cup \mathcal{I}$, in (12.1) are all twice continuously differentiable.)

✎ 17.2 Consider the scalar minimization problem:

$$\min_x \frac{1}{1+x^2}, \quad \text{subject to } x \geq 1.$$

Write down $P(x; \mu)$ for this problem, and show that $P(x; \mu)$ is unbounded below for any positive value of μ. (See Powell [197] and M. Wright [251].)

17.5. SEQUENTIAL LINEARLY CONSTRAINED METHODS

⌾ **17.3** Consider the scalar minimization problem

$$\min x \quad \text{subject to } x^2 \geq 0, \, x + 1 \geq 0,$$

for which the solution is $x^* = -1$. Write down $P(x; \mu)$ for this problem and find its local minimizers. Show that for any sequence $\{\mu_k\}$ such that $\mu_k \downarrow 0$, there exists a corresponding sequence of local minimizers $x(\mu_k)$ that converges to 0.

⌾ **17.4** By building on (17.26), find the third-derivative tensor for $P(x; \mu)$. Use the approximation $\lambda_i^* \approx \mu/c_i(x)$, $i \in \mathcal{I}$, to identify the largest term in this expression, and estimate its size. (The large size of this tensor is an indicator of the inadequacy of the Taylor series quadratic approximation to $P(x; \mu)$ discussed in the text.)

⌾ **17.5** Suppose that the current point x_k is the exact minimizer of $P(x; \mu_k)$. Write down the formula for the Newton step for $P(x; \mu_{k+1})$ from the current point x_k, using the fact that $\nabla P(x_k; \mu_k) = 0$ to eliminate the term $\nabla f(x_k)$ from the right-hand-side. How does the resulting step differ from the predictor step (17.29), (17.30) obtained by extrapolating along the central path?

⌾ **17.6** Modify the system of nonlinear equations (17.41) that is used as the basis of primal–dual methods for the case in which equality constraints are also present (that is, the problem has the form (17.38)).

⌾ **17.7** Verify that the KKT conditions for the bound-constrained problem

$$\min_{x \in \mathbf{R}^n} \phi(x) \quad \text{subject to } l \leq x \leq u$$

are equivalent to the compactly stated condition

$$P_{[l,u]} \nabla \phi(x) = 0,$$

where the projection operator $P_{[l,u]}$ onto the rectangular box $[l, u]$ is defined in (17.68).

⌾ **17.8** Show that the function $\psi(t, \sigma; \mu)$ defined in (17.56) has a discontinuity in its second derivative with respect to t when $t = \mu\sigma$. Assuming that $c_i : \mathbf{R}^n \to \mathbf{R}$ is twice continuously differentiable, write down the second partial derivative matrix of $\psi(c_i(x), \lambda_i; \mu)$ with respect to x for the two cases $c_i(x) < \mu\lambda_i$ and $c_i(x) \geq \mu\lambda_i$.

Chapter 18

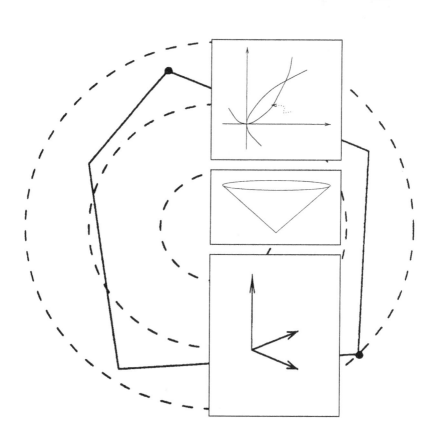

Sequential Quadratic Programming

One of the most effective methods for nonlinearly constrained optimization generates steps by solving quadratic subproblems. This sequential quadratic programming (SQP) approach can be used both in line search and trust-region frameworks, and it is appropriate for small or large problems. Unlike sequential linearly constrained methods (Chapter 17), which are effective when most of the constraints are linear, SQP methods show their strength when solving problems with significant nonlinearities.

Our development of SQP methods will be done in two stages. First we will present a local algorithm that motivates the SQP approach and that allows us to introduce the step computation and Hessian approximation techniques in a simple setting. We then consider practical line search and trust-region methods that achieve convergence from remote starting points.

18.1 LOCAL SQP METHOD

Let us begin by considering the equality-constrained problem

$$\min f(x) \tag{18.1a}$$
$$\text{subject to } c(x) = 0, \tag{18.1b}$$

where $f : \mathbf{R}^n \to \mathbf{R}$ and $c : \mathbf{R}^n \to \mathbf{R}^m$ are smooth functions. Problems containing only equality constraints are not very common in practice, but an understanding of (18.1) is crucial in the design of SQP methods for problems with general constraints.

The essential idea of SQP is to model (18.1) at the current iterate x_k by a quadratic programming subproblem and to use the minimizer of this subproblem to define a new iterate x_{k+1}. The challenge is to design the quadratic subproblem so that it yields a good step for the underlying constrained optimization problem and so that the overall SQP algorithm has good convergence properties and good practical performance. Perhaps the simplest derivation of SQP methods, which we now present, views them as an application of Newton's method to the KKT optimality conditions for (18.1).

From (12.28) we know that the Lagrangian function for this problem is $\mathcal{L}(x, \lambda) = f(x) - \lambda^T c(x)$. We use $A(x)$ to denote the Jacobian matrix of the constraints, that is,

$$A(x)^T = [\nabla c_1(x), \nabla c_2(x), \ldots, \nabla c_m(x)], \tag{18.2}$$

where $c_i(x)$ is the ith component of the vector $c(x)$. By specializing the first-order (KKT) conditions (12.30) to the equality-constrained case, we obtain a system of $n + m$ equations in the $n + m$ unknowns x and λ:

$$F(x, \lambda) = \begin{bmatrix} \nabla f(x) - A(x)^T \lambda \\ c(x) \end{bmatrix} = 0. \tag{18.3}$$

If A_* has full rank, any solution (x^*, λ^*) of the equality-constrained problem (18.1) satisfies (18.3). One approach that suggests itself is to solve the nonlinear equations (18.3) by using Newton's method, as described in Chapter 11.

The Jacobian of (18.3) is given by

$$\begin{bmatrix} W(x, \lambda) & -A(x)^T \\ A(x) & 0 \end{bmatrix}, \tag{18.4}$$

where W denotes the Hessian of the Lagrangian,

$$W(x, \lambda) = \nabla^2_{xx} \mathcal{L}(x, \lambda). \tag{18.5}$$

The Newton step from the iterate (x_k, λ_k) is thus given by

$$\begin{bmatrix} x_{k+1} \\ \lambda_{k+1} \end{bmatrix} = \begin{bmatrix} x_k \\ \lambda_k \end{bmatrix} + \begin{bmatrix} p_k \\ p_\lambda \end{bmatrix}, \quad (18.6)$$

where p_k and p_λ solve the KKT system

$$\begin{bmatrix} W_k & -A_k^T \\ A_k & 0 \end{bmatrix} \begin{bmatrix} p_k \\ p_\lambda \end{bmatrix} = \begin{bmatrix} -\nabla f_k + A_k^T \lambda_k \\ -c_k \end{bmatrix}. \quad (18.7)$$

This iteration, which is sometimes called the *Newton–Lagrange method*, is well-defined when the KKT matrix is nonsingular. We saw in Chapter 16 that nonsingularity is a consequence of the following conditions.

Assumption 18.1.

(a) The constraint Jacobian A_k has full row rank.

(b) The matrix W_k is positive definite on the tangent space of the constraints, i.e., $d^T W_k d > 0$ for all $d \neq 0$ such that $A_k d = 0$.

The first assumption is the linear independence constraint qualification discussed in Chapter 12 (see Definition 12.1), which we assume throughout this chapter. The second condition holds whenever (x, λ) is close to the optimum (x^*, λ^*) and the second-order sufficient condition is satisfied at the solution (see Theorem 12.6). The Newton iteration (18.6), (18.7) can be shown to be quadratically convergent under these assumptions and constitutes an excellent algorithm for solving equality-constrained problems, provided that the starting point is close enough to x^*.

SQP FRAMEWORK

There is an alternative way to view the iteration (18.6), (18.7). Suppose that at the iterate (x_k, λ_k) we define the quadratic program

$$\min_p \tfrac{1}{2} p^T W_k p + \nabla f_k^T p \quad (18.8a)$$

$$\text{subject to} \quad A_k p + c_k = 0. \quad (18.8b)$$

If Assumptions 18.1 hold, this problem has a unique solution (p_k, μ_k) that satisfies

$$W_k p_k + \nabla f_k - A_k^T \mu_k = 0, \quad (18.9a)$$

$$A_k p_k + c_k = 0. \quad (18.9b)$$

A key observation is that p_k and μ_k can be identified with the solution of the Newton equations (18.7). If we subtract $A_k^T \lambda_k$ from both sides of the first equation in (18.7), we obtain

$$\begin{bmatrix} W_k & -A_k^T \\ A_k & 0 \end{bmatrix} \begin{bmatrix} p_k \\ \lambda_{k+1} \end{bmatrix} = \begin{bmatrix} -\nabla f_k \\ -c_k \end{bmatrix}. \tag{18.10}$$

Hence, by nonsingularity of the coefficient matrix, we have that $p = p_k$ and $\lambda_{k+1} = \mu_k$.

We refer to this interesting relationship as *the equivalence between SQP and Newton's method*: If Assumptions 18.1 hold at x_k, then the new iterate (x_{k+1}, λ_{k+1}) can be defined either as the solution of the quadratic program (18.8) or as the iterate generated by Newton's method (18.6), (18.7) applied to the optimality conditions of the problem. These alternative interpretations are quite useful: The Newton point of view facilitates the analysis, whereas the SQP framework enables us to derive practical algorithms and to extend the technique to the inequality-constrained case.

We now state the SQP method in its simplest form.

Algorithm 18.1 (Local SQP Algorithm).
   Choose an initial pair (x_0, λ_0);
      for $k = 0, 1, 2, \ldots$
           Evaluate f_k, ∇f_k, $W_k = W(x_k, \lambda_k)$, c_k, and A_k;
           Solve (18.8) to obtain p_k and μ_k;
           $x_{k+1} \leftarrow x_k + p_k$; $\lambda_{k+1} \leftarrow \mu_k$;
           if convergence test satisfied
                 STOP with approximate solution (x_{k+1}, λ_{k+1});
   end (for).

It is straightforward to establish a local convergence result for this algorithm, since we know that it is equivalent to Newton's method applied to the nonlinear system $F(x, \lambda) = 0$. More specifically, if Assumptions 18.1 hold at a solution (x^*, λ^*) of (18.1), if f and c are twice differentiable with Lipschitz continuous second derivatives, and if the initial point (x_0, λ_0) is sufficiently close to (x^*, λ^*), then the iterates generated by Algorithm 18.1 converge quadratically to (x^*, λ^*) (see Section 18.10).

We should note in passing that in the objective (18.8a) of the quadratic program we could replace the linear term $\nabla f_k^T p$ by $\nabla_x \mathcal{L}(x_k, \lambda_k)^T p$, since the constraint (18.8b) makes the two choices equivalent. In this case, (18.8a) is a quadratic approximation of the Lagrangian function, and this leads to an alternative motivation of the SQP method. We replace the nonlinear program (18.1) by the problem of minimizing the Lagrangian subject to the equality constraints (18.1b). By making a quadratic approximation of the Lagrangian and a linear approximation of the constraints we obtain (18.8).

INEQUALITY CONSTRAINTS

The SQP framework can be extended easily to the general nonlinear programming problem

$$\min \quad f(x) \quad (18.11a)$$
$$\text{subject to} \quad c_i(x) = 0, \quad i \in \mathcal{E}, \quad (18.11b)$$
$$c_i(x) \geq 0, \quad i \in \mathcal{I}. \quad (18.11c)$$

To define the subproblem we now linearize both the inequality and equality constraints to obtain

$$\min \tfrac{1}{2} p^T W_k p + \nabla f_k^T p \quad (18.12a)$$
$$\text{subject to} \quad \nabla c_i(x_k)^T p + c_i(x_k) = 0, \quad i \in \mathcal{E}, \quad (18.12b)$$
$$\nabla c_i(x_k)^T p + c_i(x_k) \geq 0, \quad i \in \mathcal{I}. \quad (18.12c)$$

We can use one of the algorithms for quadratic programming described in Chapter 16 to solve this problem.

A local SQP method for (18.11) follows from Algorithm 18.1 above, with one modification: The step p_k and the new multiplier estimate λ_{k+1} are defined as the solution and the corresponding Lagrange multiplier of (18.12). The following result shows that this approach eventually identifies the optimal active set for the inequality-constrained problem (18.11). Recall that strict complementarity is said to hold at a solution pair (x^*, λ^*) if there is no index $i \in \mathcal{I}$ such that $\lambda_i^* = c_i(x^*) = 0$.

Theorem 18.1.

Suppose that x^* is a solution point of (18.11). Assume that the Jacobian A_* of the active constraints at x^* has full row rank, that $d^T W_* d > 0$ for all $d \neq 0$ such that $A_* d = 0$, and that strict complementarity holds. Then if (x_k, λ_k) is sufficiently close to (x^*, λ^*), there is a local solution of the subproblem (18.12) whose active set $\mathcal{A}_k$ is the same as the active set $\mathcal{A}(x^*)$ of the nonlinear program (18.11) at x^*.

As the iterates of the SQP method approach a minimizer satisfying the conditions given in the theorem, the active set will remain fixed. The subproblem (18.12) behaves like an equality-constrained quadratic program, since we can eventually ignore the inequality constraints that do not fall into the active set $\mathcal{A}(x^*)$, while treating the active constraints as equality constraints.

IQP VS. EQP

There are two ways of implementing the SQP method for solving the general nonlinear programming problem (18.11). The first approach solves at every iteration the quadratic subprogram (18.12), taking the active set at the solution of this subproblem as a guess of the optimal active set. This approach is referred to as the IQP (inequality-constrained QP) approach, and has proved to be quite successful in practice. Its main drawback is the expense of solving the general quadratic program (18.12), which can be high when the problem is large. As the iterates of the SQP method converge to the solution, however, solving the quadratic subproblem becomes very economical if we carry information from the previous iteration to make a good guess of the optimal solution of the current subproblem. This *hot-start* strategy is described later.

The second approach selects a subset of constraints at each iteration to be the so-called working set, and solves only equality-constrained subproblems of the form (18.8), where the constraints in the working sets are imposed as equalities and all other constraints are ignored. The working set is updated at every iteration by rules based on the Lagrange multiplier estimates, or by solving an auxiliary subproblem. This EQP approach has the advantage that the equality-constrained quadratic subproblems are less expensive to solve than (18.12) and require less sophisticated software.

An example of an EQP method is the gradient projection method described in Section 16.6. In this method, the working set is determined by minimizing the quadratic model along the path obtained by projecting the steepest descent direction onto the feasible region. Another variant of the EQP method makes use of the method of *successive linear programming*. This approach obtains a linear program by omitting the quadratic term $p^T W_k p$ from (18.12a), applying a trust-region constraint $\|p\| \leq \Delta_k$ on the step p (see (18.45c)), and taking the active set of this subproblem to be the working set for the current iteration. It then fixes the constraints in the working set and solves an equality-constrained quadratic program (with the term $p^T W_k p$ reinserted) to obtain the step.

18.2 PREVIEW OF PRACTICAL SQP METHODS

To be practical, an SQP method must be able to converge from remote starting points and on nonconvex problems. We now outline how the local SQP strategy can be adapted to meet these goals.

We begin by drawing an analogy with unconstrained optimization. In its simplest form, the Newton iteration for minimizing a function f takes a step to the minimum of the quadratic model

$$m_k(p) = f_k + \nabla f_k^T p + \tfrac{1}{2} p^T \nabla^2 f_k p.$$

This framework is useful near the solution where the Hessian $\nabla^2 f(x_k)$ is normally positive definite and the quadratic model has a well-defined minimizer. When x_k is not close to the solution, however, the model function m_k may not be convex. Trust-region methods ensure that the new iterate is always well-defined and useful by restricting the candidate step p_k to some neighborhood of the origin. Line search methods modify the Hessian in $m_k(p)$ to make it positive definite (possibly replacing it by a quasi-Newton approximation B_k), to ensure that p_k is a descent direction for the objective function f.

Similar strategies are used to globalize SQP methods. If W_k is positive definite on the tangent space of the constraints, the quadratic subproblem (18.8) has a unique solution. When W_k does not have this property, line search methods either replace it by a positive definite approximation B_k or modify W_k directly during the process of matrix factorization. A third possibility is to define W_k as the Hessian of an *augmented Lagrangian* function having certain convexity properties. In all these cases, the subproblem (18.8) will be well-defined.

Trust-region SQP methods add a constraint to the subproblem, limiting the step to a region where the model (18.8) is considered to be reliable. Because they impose a trust-region bound on the step, they are able to use Hessians W_k that fail to satisfy the convexity properties. Complications may arise, however, because the inclusion of the trust region may cause the subproblem to become infeasible. At some iterations, it is necessary to relax the constraints, which complicates the algorithm and increases its computational cost. Due to these tradeoffs, neither one of the two SQP approaches—line search or trust region—can be regarded as clearly superior to the other.

The technique used to solve the line search and trust-region subproblems has a great impact in the efficiency and robustness of SQP methods, particularly for large problems. For line search methods, the quadratic programming algorithms described in Chapter 16 can be used, whereas trust-region methods require special techniques.

Another important question in the development of SQP methods is the choice of a merit function that guides the algorithms toward the solution. In unconstrained optimization the merit function is simply the objective f, and it is fixed throughout the course of the minimization. SQP methods can use any of the merit functions discussed in Chapter 15, but the parameters of these merit functions may need to be adjusted on some iterations to ensure that the direction obtained from the subproblem is indeed a descent direction with respect to this function. The rules for updating these parameters require careful consideration, since they have a strong influence on the practical behavior of SQP methods.

In the remainder of this chapter we expand on these ideas to produce practical SQP algorithms. We first discuss a variety of techniques for solving the quadratic subproblems (18.8) and (18.12), and observe the effect that they have in the form of the SQP iteration. We then consider various formulations of the quadratic model that ensure their adequacy in a line search context, studying the case in which W_k is the exact Lagrangian Hessian and also the case in which it is a quasi-Newton approximation. We present two important merit functions and show that the SQP directions are descent directions for these functions. This discussion will set the stage for our presentation of practical line search and trust-region SQP methods.

18.3 STEP COMPUTATION

EQUALITY CONSTRAINTS

We begin by considering the equality-constrained quadratic program (18.8). We have assumed so far that the Hessian of this model is defined as the Hessian of the Lagrangian $W_k = \nabla_{xx}^2 \mathcal{L}(x_k \lambda_k)$ (as in (18.5)). In later sections we will replace this matrix by a quasi-Newton approximation B_k, but the discussion in this section is independent of this choice.

We have noted already that if Assumptions 18.1 hold, the solution of (18.8) is given by the KKT system (18.10). In Chapter 16 we described a variety of techniques for solving this system. We review these techniques here, pointing out some simplifications that can be made when we know that the system is derived from an SQP subproblem, rather than from a general QP.

Direct Solution of the KKT System.

The first alternative is to solve the full $(n+m) \times (n+m)$ KKT system (18.10) with a symmetric indefinite factorization. In this approach, the KKT matrix is factored as LDL^T, where D is a block diagonal matrix with blocks of size 1×1 and 2×2 and L is unit lower triangular. Two popular algorithms for computing this factorization are the methods of Bunch and Kaufman [30] (for the dense case), and the method of Duff and Reid [75] (for the sparse case). This approach is often called the *augmented system approach*. It provides us directly with a step p_k in the primal variables, and a new Lagrange multiplier estimate λ_{k+1}.

The KKT system (18.10) can also be solved by iterative methods such as the QMR and LSQR iterations; see the Notes and References in Chapter 16. Early termination of iteration is a delicate matter, since the resulting search direction may not lead toward a minimizer. Due to this difficulty, iterative methods for the full KKT system are not yet commonly employed in practical implementations of the SQP method. The topics of termination conditions and preconditioning of the KKT system are subjects of current research.

Dual or Range-Space Approach.

If W_k is positive definite, we can decouple the KKT system (18.10) and solve the following two systems in sequence to obtain λ_{k+1} and p_k:

$$(A_k W_k^{-1} A_k^T) \lambda_{k+1} = A_k W_k^{-1} \nabla f_k - c_k, \tag{18.13a}$$

$$W_k p_k = -\nabla f_k + A_k^T \lambda_{k+1}. \tag{18.13b}$$

This approach is particularly effective for quasi-Newton methods that explicitly maintain positive definite approximations H_k to the matrix W_k^{-1}. If we are using a direct method to solve (18.13a), we simply form the product $A_k W_k^{-1} A_k^T$ each time W_k^{-1} is updated. If we are using an iterative method such as conjugate gradient, the explicit availability of W_k^{-1} makes it easy to calculate matrix–vector products involving this matrix.

18.3. STEP COMPUTATION

Null-Space Method.

The null-space approach is at the heart of many SQP methods. It requires knowledge of the matrices Y_k and Z_k that span the range space of A^T and null space of A_k, respectively; see Chapter 15. By writing

$$p_k = Y_k p_Y + Z_k p_Z,$$

we can substitute into (18.10) to obtain the following systems to be solved for for p_Y and p_Z:

$$(A_k Y_k) p_Y = -c_k, \tag{18.14a}$$
$$\left(Z_k^T W_k Z_k\right) p_Z = -Z_k^T W_k Y_k p_Y - Z_k^T \nabla f_k. \tag{18.14b}$$

The Lagrange multipliers λ_{k+1}, which we refer to as *QP multipliers* for reasons that become clear below, can be obtained by solving

$$(A_k Y_k)^T \lambda_{k+1} = Y_k^T (\nabla f_k + W_k p_k). \tag{18.15}$$

The only assumption we make on W_k is that the reduced Hessian $Z_k^T W_k Z_k$ is positive definite. See Chapter 16 for a more extensive discussion of the null-space method.

There are several variants of this approach in which we compute an approximation of the pure KKT step p_k. The first variant is simply to delete the term involving p_k from the right-hand-side of (18.15), thereby decoupling the computations of p_k and λ_{k+1}. This simplification can be justified by observing that p_k converges to zero as we approach the solution, whereas ∇f_k normally does not. Therefore, the new multipliers will be good estimates of the QP multipliers near the solution. If we happen to choose $Y_k = A_k^T$ (which is a valid choice for Y_k when A_k has full row rank; see Chapter 15), we obtain

$$\hat{\lambda}_{k+1} = (A_k A_k^T)^{-1} A_k \nabla f_k. \tag{18.16}$$

These are called the *least-squares multipliers* because they can also be derived by solving the problem

$$\min_{\lambda} \left\| \nabla f_k - A_k^T \lambda \right\|_2^2. \tag{18.17}$$

It is clear from this relation that the least-squares multipliers are useful even when the current iterate is far from the solution, because they seek to satisfy the first-order optimality condition in (18.3) as closely as possible. In practice, it is desirable to use the most current information when computing least-squares multipliers. Therefore, in Algorithm 18.3 of Section 18.6 we first compute the step p_k using (18.14), evaluate the gradients of the functions and constraints at the new point x_{k+1}, and define λ_{k+1} by (18.16) but with the terms on the right-hand side evaluated at x_{k+1}.

Conceptually, the use of least-squares multipliers transforms the SQP method from an iteration in x and λ to a purely primal iteration in the x variable alone.

A second set of variants, known as *reduced-Hessian methods*, go one step further and remove the cross term $Z_k^T W_k Y_k p_Y$ in (18.14b), thereby yielding the system

$$(Z_k^T W_k Z_k) p_Z = -Z_k^T \nabla f_k. \tag{18.18}$$

This variant has the advantage that we need to store (or approximate) and factorize only the matrix $Z_k^T W_k Z_k$, not the cross-term matrix $Z_k^T W_k Y_k$. As we show later, dropping the cross term is justified because the normal component p_Y usually converges to zero faster than the tangential component p_Z, thereby making (18.18) a good approximation to (18.14b).

INEQUALITY CONSTRAINTS

The search direction in *line search* SQP methods for inequality-constrained optimization is obtained by solving the subproblem (18.12). We can do this by means of the active-set method for quadratic programming (Algorithm 16.1) described in Chapter 16. If W_k is not convex, then we need to introduce the variations for indefinite quadratic programming described in that chapter.

We can make significant savings in the solution of the quadratic subproblem by so-called *hot-start* procedures. Specifically, we can use the solution $\tilde{p}$ of the previous quadratic subproblem as the starting point for the current subproblem. Two phase-I iterations that can take advantage of a good starting point were described in Chapter 16. We can also initialize the working set for each QP subproblem to be the final active set from the previous SQP iteration. Additionally, it is sometimes possible—particularly when the problem contains only linear constraints—to reuse or update certain matrix factorizations from the previous iteration. Hot-start procedures are crucial to the efficiency of line search SQP methods.

A common difficulty in line search SQP methods is that the linearizations (18.12b), (18.12c) of the nonlinear constraints may give rise to an infeasible subproblem. Consider, for example, the case where $n = 1$ and where the constraints are $x \leq 1$ and $x^2 \geq 0$. When we linearize these constraints at $x_k = 3$, we obtain the inequalities

$$3 + p \leq 1 \quad \text{and} \quad 9 + 6p \geq 0,$$

which are inconsistent.

To overcome this difficulty, we can define a relaxation of the SQP subproblem that is guaranteed to be feasible. For example, the SNOPT program [108] for large-scale optimization first attempts to solve (18.12a)–(18.12c), but if this quadratic program is found to be infeasible, it solves the auxiliary problem

$$\min \quad f(x) + \gamma e^T (v + w) \tag{18.19a}$$
$$\text{subject to} \quad c_i(x) - v_i + w_i = 0, \quad i \in \mathcal{E} \tag{18.19b}$$

$$c_i(x) - v_i + w_i \geq 0, \quad i \in \mathcal{I}, \tag{18.19c}$$

$$v \geq 0, \quad w \geq 0. \tag{18.19d}$$

Here γ is a nonnegative penalty parameter, and $e^T = (1, \ldots, 1)$. If the nonlinear problem (18.11) has a feasible solution and γ is sufficiently large, the solutions to (18.19) and (18.11) are identical. If, on the other hand, there is no feasible solution to the nonlinear problem and γ is large enough, then the auxiliary problem (18.19) usually determines a "good" infeasible point. The choice of γ requires heuristics; SNOPT uses the value $\gamma = 100\|\nabla f(x_s)\|$, where x_s is the first iterate at which inconsistent linearized constraints were detected. If a subproblem is found to be infeasible, then all subsequent iterates are computed by solving the subproblem (18.19).

An alternative approach for solving the quadratic subproblem (18.12) is to use an interior-point method; see Section 16.7. For general problems this approach is competitive with active-set methods only on early iterations, when the active sets change substantially from iteration to iteration and not much is to be gained from hot-start information. (Interior-point methods cannot take advantage of prior information about the solution or active set to the same extent as active-set methods.) On some problems with special structure (for example, certain applications in control) interior-point methods are better able to exploit the structure than active-set methods, and therefore become competitive.

The step computation in *trust-region* SQP methods adds a trust-region bound to the subproblem, and also reformulates the constraints in (18.12a)–(18.12c) to ensure feasibility of the subproblem. We defer detailed discussion to Section 18.8, but mention here that the $S\ell_1 QP$ method formulates a subproblem in which the linearized constraints are moved to the objective of the quadratic program in the form of a penalty term. The method described in Section 18.9 indirectly introduces a relaxation of the constraints by first computing a normal step by means of (18.47).

18.4 THE HESSIAN OF THE QUADRATIC MODEL

Let us now consider the choice of the matrix W_k in the quadratic model (18.8a). For simplicity, we focus first on the equality-constrained optimization problem (18.1).

The equivalence between SQP and Newton's method applied to the optimality conditions (18.3) is based on the choice of W_k as the Hessian of the Lagrangian, $\nabla_{xx}^2 \mathcal{L}(x_k, \lambda_k)$ (see (18.5)). This choice leads to a quadratic rate of convergence under reasonable assumptions and often also produces rapid progress when the iterates are distant from the solution. However, this matrix is made up of second derivatives of the objective function and constraints, which may not be easy to compute. Moreover, it may not always be positive definite on the constraint null space. In this section we discuss various alternative choices for W_k.

FULL QUASI-NEWTON APPROXIMATIONS

The first idea that comes to mind is to maintain a quasi-Newton approximation B_k to the full Lagrangian Hessian $\nabla_{xx}^2 \mathcal{L}(x_k, \lambda_k)$. Since the BFGS formula has proved to be very successful in the context of unconstrained optimization, we can try to employ it in this case as well.

The update for B_k that results from the step from iterate k to iterate $k+1$ will make use of the vectors s_k and y_k, which can be defined as follows:

$$s_k = x_{k+1} - x_k, \qquad y_k = \nabla_x \mathcal{L}(x_{k+1}, \lambda_{k+1}) - \nabla_x \mathcal{L}(x_k, \lambda_{k+1}). \tag{18.20}$$

We then compute the new approximation B_{k+1} using the BFGS formula (8.19). We can view this as the application of quasi-Newton updating to the case where the objective function is given by the Lagrangian $\mathcal{L}(x, \lambda)$ (with λ fixed). These definitions immediately reveal the strengths and weaknesses of this approach.

If $\nabla_{xx}^2 \mathcal{L}$ is positive definite in the region where the minimization takes place, the quasi-Newton approximations $\{B_k\}$ will reflect some of the curvature information of the problem, and the iteration will converge robustly and rapidly, just as in the unconstrained BFGS method. If, however, $\nabla_{xx}^2 \mathcal{L}$ contains negative eigenvalues, then the BFGS approach of approximating it with a positive definite matrix may be ineffective. In fact, BFGS updating requires that s_k and y_k satisfy the curvature condition $s_k^T y_k > 0$, which may not hold when s_k and y_k are defined by (18.20), even when the iterates are close to the solution.

To overcome this difficulty, we could *skip* the BFGS update if the condition

$$s_k^T y_k \geq \theta s_k^T B_k s_k \tag{18.21}$$

is not satisfied, where θ is a positive parameter (10^{-2}, say). This skipping strategy has been used in some SQP implementations, and it has performed well on many problems. On other problems, however, it yields poor performance or even failure, so it cannot be regarded as adequate for general-purpose algorithms.

A more effective modification ensures that the update is always well-defined by modifying the definition of y_k.

Procedure 18.2 (Damped BFGS Updating for SQP).
Define s_k and y_k as in (18.20) and set

$$r_k = \theta_k y_k + (1 - \theta_k) B_k s_k,$$

where the scalar θ_k is defined as

$$\theta_k = \begin{cases} 1 & \text{if } s_k^T y_k \geq 0.2 s_k^T B_k s_k, \\ (0.8 s_k^T B_k s_k)/(s_k^T B_k s_k - s_k^T y_k) & \text{if } s_k^T y_k < 0.2 s_k^T B_k s_k. \end{cases} \tag{18.22}$$

18.4. THE HESSIAN OF THE QUADRATIC MODEL

Update B_k as follows:

$$B_{k+1} = B_k - \frac{B_k s_k s_k^T B_k}{s_k^T B_k s_k} + \frac{r_k r_k^T}{s_k^T r_k}. \qquad (18.23)$$

The formula (18.23) is simply the standard BFGS update formula, with y_k replaced by r_k. It guarantees that B_{k+1} is positive definite, since it is easy to show that when $\theta_k \neq 1$ we have

$$s_k^T r_k = 0.2 s_k^T B_k s_k > 0. \qquad (18.24)$$

To gain more insight into this strategy, note that when $\theta_k = 0$ we have $B_{k+1} = B_k$, and that $\theta_k = 1$ gives the (possibly indefinite) matrix produced by the unmodified BFGS update. A value $\theta_k \in (0, 1)$ thus produces a matrix that interpolates the current approximation B_k and the one produced by the unmodified BFGS formula. The choice of θ_k ensures that the new approximation stays close enough to the current approximation B_k to ensure positive definiteness.

Damped BFGS updating has been incorporated in various SQP programs and has performed well on many problems. Nevertheless, numerical experiments show that it, too, can behave poorly on difficult problems. It still fails to address the underlying problem that the Lagrangian Hessian may not be positive definite. In this setting, SR1 updating (8.24) is more appropriate, and is indeed a good choice for trust-region SQP methods. Line search methods, however, cannot accept indefinite Hessian approximations and would therefore need to modify the SR1 formula, which is not totally desirable.

The strategy described next takes a different approach. It modifies the Lagrangian Hessian directly by adding terms to the Lagrangian function, the effect of which is to ensure positive definiteness.

HESSIAN OF AUGMENTED LAGRANGIAN

Let us consider the augmented Lagrangian function defined by

$$\mathcal{L}_A(x, \lambda; \mu) = f(x) - \lambda^T c(x) + \frac{1}{2\mu} \|c(x)\|^2, \qquad (18.25)$$

for some positive scalar μ. We have shown in Chapter 17 that at a minimizer (x^*, λ^*) satisfying the second-order sufficiency condition, the Hessian of this function, which is

$$\nabla_{xx}^2 \mathcal{L}_A = \nabla_{xx}^2 \mathcal{L}(x^*, \lambda^*) + \mu^{-1} A(x^*)^T A(x^*), \qquad (18.26)$$

is positive definite for all μ smaller than a certain threshold value μ^*. Note that the last term in (18.25) adds positive curvature to the Lagrangian on the space spanned by the columns of $A(x)^T$ while leaving the curvature on the null space of $A(x)$ unchanged. We could now choose the matrix W_k in the quadratic subproblem (18.8) to be $\nabla_{xx}^2 \mathcal{L}_A(x_k, \lambda_k; \mu)$, or some quasi-Newton approximation B_k to this matrix. For sufficiently small choices of μ, this Hessian will always be positive definite and can be used directly in line search SQP methods.

This strategy is not without its difficulties. The threshold value μ^* depends on quantities that are normally not known, such as bounds on the second derivatives of the problem functions, so it can be difficult to choose an appropriate value for μ. Choices of μ that are too small can result in the last term in (18.26) dominating the original Lagrangian Hessian, sometimes leading to poor practical performance. If μ is too large, the Hessian of the augmented Lagrangian may not be positive definite and $\mathcal{L}_A$ could be nonconvex, so that the curvature condition of $y_k^T s$ may not be satisfied.

A variant of this approach is based on a vector y_k^A defined by

$$y_k^A \equiv \nabla_x \mathcal{L}_A(x_{k+1}, \lambda_{k+1}; \mu) - \nabla_x \mathcal{L}_A(x_k, \lambda_{k+1}; \mu)$$
$$= y_k + \mu^{-1} A_{k+1}^T c_{k+1},$$

where we have used the definition (18.20) to derive the second equality. One can show that near the solution there is a maximum value of μ that guarantees uniform positiveness of $(y_k^A)^T s$. One can then design an algorithm that selects μ adaptively to satisfy a positiveness criterion and replaces y_k by y_k^A in the BFGS update formula. There is as yet insufficient numerical experience to know whether this approach leads to robust and efficient SQP methods.

REDUCED-HESSIAN APPROXIMATIONS

The two approaches just mentioned compute or approximate a full $n \times n$ Hessian. An alternative is to approximate only the reduced Hessian of the Lagrangian $Z_k^T \nabla_{xx}^2 \mathcal{L}(x_k, \lambda_k) Z_k$, a matrix of smaller dimension that under the standard assumptions is positive definite in a neighborhood of the solution. This approach is used in *reduced-Hessian quasi-Newton methods*, which compute the search direction from the formulae (18.14), (18.16) and a quasi-Newton variant of (18.18). We restate these systems as follows:

$$\lambda_k = (A_k A_k^T)^{-1} A_k \nabla f_k, \tag{18.27a}$$
$$(A_k Y_k) p_Y = -c_k, \tag{18.27b}$$
$$M_k p_Z = -Z_k^T \nabla f_k. \tag{18.27c}$$

We use M_k to distinguish the reduced-Hessian approximation from the full Hessian approximation B_k. Note also that we have now defined the least-squares multipliers λ_k so as to use the most recent gradient information.

We now discuss how the quasi-Newton approximations M_k to the reduced Hessian $Z_k^T \nabla_{xx}^2 \mathcal{L}(x_k, \lambda_k) Z_k$ can be constructed. As before, let us define $W_k = \nabla_{xx}^2 \mathcal{L}(x_k, \lambda_k)$, and suppose that we have just taken a step $\alpha_k p_Z$ from (x_k, λ_k) to (x_{k+1}, λ_{k+1}). By Taylor's theorem,

18.4. THE HESSIAN OF THE QUADRATIC MODEL

we have

$$W_{k+1}\alpha_k p_k \approx [\nabla_x \mathcal{L}(x_k + \alpha_k p_k, \lambda_{k+1}) - \nabla_x \mathcal{L}(x_k, \lambda_{k+1})],$$

where $p_k = x_{k+1} - x_k = Z_k p_z + Y_k p_Y$. By premultiplying by Z_k^T, we have

$$Z_k^T W_{k+1} Z_k \alpha_k p_z \quad (18.28)$$
$$\approx -Z_k^T W_{k+1} Y_k \alpha_k p_Y + Z_k^T [\nabla_x \mathcal{L}(x_k + \alpha_k p_k, \lambda_{k+1}) - \nabla_x \mathcal{L}(x_k, \lambda_{k+1})].$$

The secant equation for M_k is obtained by dropping the cross term $Z_k^T W_{k+1} Y_k \alpha_k p_Y$ (using the rationale mentioned in the previous section), which yields

$$M_{k+1} s_k = y_k, \quad (18.29)$$

where s_k and y_k are defined by

$$s_k = \alpha_k p_z, \quad (18.30a)$$
$$y_k = Z_k^T [\nabla_x \mathcal{L}(x_k + \alpha_k p_k, \lambda_{k+1}) - \nabla_x \mathcal{L}(x_k, \lambda_{k+1})]. \quad (18.30b)$$

We can therefore apply the BFGS formula (8.19) using these definitions for the correction vectors s_k and y_k to define the new approximation M_{k+1}. In Section 18.7 we discuss procedures for ensuring that the curvature condition $s_k^T y_k > 0$ holds.

One aspect of this description requires clarification. In the left-hand-side of (18.28) we have $Z_k^T W_{k+1} Z_k$ rather than $Z_{k+1}^T W_{k+1} Z_{k+1}$. We could have used Z_{k+1} in (18.28), avoiding an inconsistency of indices, and by a longer argument show that y_k can be defined by (18.30b) with Z_k replaced by Z_{k+1}. But this more complex argument is not necessary, since the convergence properties of the simpler algorithm we described above are just as strong.

There are several variations of (18.30b) that have similar properties. One such formula is

$$y_k = Z_k^T [\nabla f(x_{k+1}) - \nabla f(x_k)]. \quad (18.31)$$

A more satisfying approach, proposed by Coleman and Conn [45], is to use curvature information gathered along the tangent space of the constraints, not along the full steps of the algorithm. In this approach, y_k is defined by

$$y_k = Z_k^T [\nabla_x \mathcal{L}(x_k + Z_k p_z, \lambda_{k+1}) - \nabla_x \mathcal{L}(x_k, \lambda_{k+1})], \quad (18.32)$$

which requires an additional evaluation of the gradient of the function and constraints at the intermediate point $x_k + Z_k p_z$. One can show (see the exercises) that this definition of y_k guarantees positiveness of $y_k^T s_k$ in a neighborhood of a solution point, so that BFGS updating can be safely applied once we are close enough to a solution. Since the cost of the additional gradients at the intermediate point is significant, practical implementations apply (18.32) only at some iterations and in the rest of the iterations use (18.30b).

More details on reduced-Hessian methods will be given in Section 18.7.

18.5 MERIT FUNCTIONS AND DESCENT

To ensure that the SQP method converges from remote starting points it is common to use a merit function ϕ to control the size of the steps (in line search methods) or to determine whether a step is acceptable and whether the trust-region radius needs to be modified (in trust-region methods). It plays the role of the objective function in unconstrained optimization, since we insist that each step provide a sufficient reduction in it. A variety of merit functions have been used in conjunction with SQP methods. Here we focus on the non-differentiable ℓ_1 merit function and on Fletcher's exact and differentiable function. These two merit functions are representative of most of those used in practice. To simplify the discussion, we focus our attention on the equality-constrained problem (18.1).

Although the merit function is needed to induce global convergence, we do not want it to interfere with "good" steps—those that make progress toward a solution. In this section we discuss the conditions on the problem and on the merit functions that ensure that the functions show a decrease on a step generated by the SQP method.

Let us begin with the ℓ_1 *merit function* (see (15.24)), which for the equality-constrained problem (18.1) is defined as

$$\phi_1(x; \mu) = f(x) + \frac{1}{\mu}\|c(x)\|_1, \qquad (18.33)$$

where $\mu > 0$ is called the penalty parameter. This function is not differentiable everywhere; specifically, at points x for which one or more components of $c(x)$ are zero, the gradient is not defined. However, it always has a directional derivative (see (A.14) in the Appendix for background on directional derivatives). The following result describes the directional derivative along the direction p_k generated by the SQP subproblem.

Lemma 18.2.

Let p_k and λ_{k+1} be generated by the SQP iteration (18.10). Then the directional derivative of ϕ_1 in the direction p_k satisfies

$$D(\phi_1(x_k; \mu); p_k) \leq -p_k^T W_k p_k - (\mu^{-1} - \|\lambda_{k+1}\|_\infty)\|c_k\|_1. \qquad (18.34)$$

PROOF. By applying Taylor's theorem (see (2.5)) to f and $c_i, i = 1, 2, \ldots, m$, we obtain

$$\phi_1(x_k + \alpha p; \mu) - \phi_1(x_k; \mu) = f(x_k + \alpha p) - f_k + \mu^{-1}\|c(x_k + \alpha p)\|_1 - \mu^{-1}\|c_k\|_1$$
$$\leq \alpha \nabla f_k^T p + \gamma \alpha^2 \|p\|^2 + \mu^{-1}\|c_k + \alpha A_k p\|_1 - \mu^{-1}\|c_k\|_1,$$

where the positive constant γ bounds the second-derivative terms in f and c. If $p = p_k$ is given by (18.10), we have that $A_k p_k = -c_k$, so for $\alpha \leq 1$ we have that

$$\phi_1(x_k + \alpha p_k; \mu) - \phi_1(x_k; \mu) \leq \alpha[\nabla f_k^T p_k - \mu^{-1}\|c_k\|_1] + \alpha^2\gamma\|p_k\|^2.$$

By arguing similarly, we also obtain the following lower bound:

$$\phi_1(x_k + \alpha p_k; \mu) - \phi_1(x_k; \mu) \geq \alpha[\nabla f_k^T p_k - \mu^{-1}\|c_k\|_1] - \alpha^2\gamma\|p_k\|^2.$$

Taking limits, we conclude that the directional derivative of ϕ_1 in the direction p_k is given by

$$D(\phi_1(x_k; \mu); p_k) = \nabla f_k^T p_k - \mu^{-1}\|c_k\|_1.$$

The fact that p_k satisfies the first equation in (18.10) implies that

$$D(\phi_1(x_k; \mu); p_k) = -p_k^T W_k p_k + p_k^T A_k^T \lambda_{k+1} - \mu^{-1}\|c_k\|_1.$$

From the second equation in (18.10), we can replace the term $p_k^T A_k^T \lambda_{k+1}$ in this expression by $-c_k^T \lambda_{k+1}$. By making this substitution in the expression above and invoking the inequality

$$-c_k^T \lambda_{k+1} \leq \|c_k\|_1 \|\lambda_{k+1}\|_\infty,$$

we obtain (18.34). $\square$

It follows from (18.34) that p_k will be a descent direction for ϕ_1 if W_k is positive definite and μ is sufficiently small. A more detailed analysis shows that this assumption on W_k can be relaxed, and that all that is necessary is that W_k be positive definite on the tangent space of the constraints. A suitable value for the penalty parameter is obtained by choosing a constant $\bar{\delta} > 0$ and defining μ at every iteration to be

$$\mu^{-1} = \|\lambda_{k+1}\|_\infty + \bar{\delta}. \tag{18.35}$$

We now consider *Fletcher's augmented Lagrangian merit function* (15.25), which is defined to be

$$\phi_F(x; \mu) \stackrel{\text{def}}{=} f(x) - \lambda(x)^T c(x) + \frac{1}{2\mu}\|c(x)\|^2, \tag{18.36}$$

where $\mu > 0$ is the penalty parameter, $\|\cdot\|$ denotes the ℓ_2 norm, and

$$\lambda(x) = [A(x)A(x)^T]^{-1}A(x)\nabla f(x) \tag{18.37}$$

are the least-squares multiplier estimates (18.17).

Fletcher's merit function (15.25) is differentiable with gradient

$$\nabla \phi_F(x_k; \mu) = \nabla f_k - A_k^T \lambda_k - (\lambda_k')^T c_k + \mu^{-1} A_k^T c_k,$$

where λ_k' is the $m \times n$ Jacobian of $\lambda(x)$ evaluated at x_k. If p_k satisfies the SQP equation (18.10), we have

$$\nabla \phi_F(x_k; \mu)^T p_k = \nabla f_k^T p_k + \lambda_k^T c_k - c_k^T \lambda_k' p_k - \mu^{-1} \|c_k\|^2.$$

Let us write $p_k = Z_k p_z + A_k^T p_Y$, where Z_k is a basis for the null-space of A_k, and where we have defined $Y_k = A_k^T$. Recalling (18.27b), we have that

$$A_k^T p_Y = -A_k^T [A_k A_k^T]^{-1} c_k,$$

which, when combined with (18.37) evaluated at $x = x_k$, implies that $\nabla f_k^T A_k^T p_Y = -\lambda_k^T c_k$. We thus obtain

$$\nabla \phi_F(x_k; \mu)^T p_k = \nabla f_k^T Z_k p_z + \nabla f_k^T A_k^T p_Y + \lambda_k^T c_k - c_k^T \lambda_k' p_k - \mu^{-1} \|c_k\|^2$$
$$= \nabla f_k^T Z_k p_z - c_k^T \lambda_k' p_k - \mu^{-1} \|c_k\|^2.$$

We note from (18.10) that $W_k p_k = W_k Z_k p_z + W_k A_k^T p_Y = -\nabla f_k$. By using this we obtain

$$\nabla \phi_F(x_k; \mu)^T p_k = -p_z^T (Z_k^T W_k Z_k) p_z - p_Y^T A_k W_k Z_k p_z \qquad (18.38)$$
$$- c_k^T \lambda_k' p_k - \mu^{-1} \|c_k\|^2.$$

Thus p_k will be a descent direction for Fletcher's merit function if the reduced Hessian of the Lagrangian $Z_k^T W_k Z_k$ is positive definite and if μ satisfies the condition

$$\mu^{-1} > \left[\frac{-\frac{1}{2} p_z^T Z_k^T W_k Z_k p_z - p_Y^T A_k W_k Z_k p_z - c_k^T \lambda_k' p_k}{\|c_k\|^2} \right] + \bar{\delta}, \qquad (18.39)$$

for some positive constant $\bar{\delta}$. (If $\|c_k\| = 0$, the descent property holds for any value of μ.) The factor $\frac{1}{2}$ is somewhat arbitrary, and other values can be used; its objective is to ensure that the directional derivative is at least as large as a fraction of $-p_z^T Z_k^T W_k Z_k p_z$. We note that the formula (18.39) for μ is not as simple as the corresponding formula for the ℓ_1 merit function, and its value will depend on the singular values of both A_k and $Z_k^T W_k Z_k$. However, this choice of μ is practical, and it allows us to establish global convergence results.

We summarize the results of this section, which couple the step-generation procedure with the choice of merit function in SQP methods.

Theorem 18.3.
 Suppose that x_k is not a stationary point of the equality-constrained nonlinear problem (18.1), and that the reduced Hessian $Z_k^T W_k Z_k$ is positive definite. Then the search direction p_k generated by the pure SQP iteration (18.10) is a descent direction for the ℓ_1 merit function ϕ_1 if μ is given by (18.35). It is a descent direction for Fletcher's function ϕ_F if μ satisfies (18.39).

In practice, as well as for the purposes of the analysis, it is desirable for the merit function parameters $\{\mu_k\}$ to eventually stay the same as the iterates converge to the solution. To make this scenario more likely, we use an update rule that leaves the current value of μ unchanged whenever its value seems to be adequate, and that decreases it by a significant amount otherwise. Choose a constant $\delta > 0$ and define

$$\mu_k = \begin{cases} \mu_{k-1} & \text{if } \mu_{k-1}^{-1} \geq \gamma + \delta, \\ (\gamma + 2\delta)^{-1} & \text{otherwise,} \end{cases} \quad (18.40)$$

where we set γ to $\|\lambda_{k+1}\|_\infty$ in the case of the ℓ_1 merit function, and to the term inside square brackets in (18.39) for Fletcher's function ϕ_F.

The rule (18.40) forces the penalty parameter to be monotonically decreasing, but this is not always desirable in practice. In some cases, μ_k takes on a very small value in the early iterations, and as a result the constraints are highly penalized during the rest of the run. Therefore, some SQP implementations include heuristics that allow μ_k to increase at certain iterations, without interfering with the global convergence properties of the iteration.

The descent properties described above assume that the step is obtained by means of the pure SQP iteration (18.10). This assumption holds in many line search algorithms, but other variants of SQP such as reduced-Hessian methods and trust-region approaches may compute the search direction differently. In all such cases, however, the descent properties can be analyzed by some adaptation of the analysis above.

18.6 A LINE SEARCH SQP METHOD

From the discussion in the previous sections, we can see that there is a wide variety of line search SQP methods that differ in the way the Hessian approximation is computed, in the choice of the merit function, and in the step-computation procedure. We now incorporate the ideas discussed so far into a practical quasi-Newton algorithm for solving the nonlinear programming problem (18.11) with equality and inequality constraints.

Algorithm 18.3 (SQP Algorithm for Nonlinear Programming).
 Choose parameters $\eta \in (0, 0.5)$, $\tau \in (0, 1)$; Choose an initial pair (x_0, λ_0);
 Choose an initial $n \times n$ symmetric positive definite Hessian
  approximation B_0;

Evaluate $f_0, \nabla f_0, c_0, A_0$;
for $k = 0, 1, 2, \ldots$
    if termination test is satisfied
        STOP;
    Compute p_k by solving (18.12)
    Choose μ_k such that p_k is a descent direction for ϕ at x_k;
    Set $\alpha_k = 1$;
    while $\phi(x_k + \alpha_k p_k; \mu_k) > \phi(x_k, \mu_k) + \eta \alpha_k D\phi(x_k; p_k)$
        Reset $\alpha_k \leftarrow \tau_\alpha \alpha_k$ for some $\tau_\alpha \in (0, \tau)$;
    end (while)
    Set $x_{k+1} = x_k + \alpha_k p_k$;
    Evaluate $f_{k+1}, \nabla f_{k+1}, c_{k+1}, A_{k+1}$;
    Compute λ_{k+1} by solving

$$\lambda_{k+1} = -[A_{k+1} A_{k+1}^T]^{-1} A_{k+1} \nabla f_{k+1};$$

    Set

$$s_k = \alpha_k p_k, \quad y_k = \nabla_x \mathcal{L}(x_{k+1}, \lambda_{k+1}) - \nabla_x \mathcal{L}(x_k, \lambda_{k+1});$$

    Obtain B_{k+1} by updating B_k using a quasi-Newton formula;
end (for)

A version of this algorithm that uses second-derivative approximation is easily obtained by replacing B_k by the Hessian of the Lagrangian W_k. As mentioned in Section 18.1, precautions should be taken so that the constraints in (18.12) are consistent. We have left freedom in the choice of the quasi-Newton approximation, and this algorithm requires only that it be positive definite. Therefore, one option is to define B_k by means of the damped BFGS update discussed in the previous section. We have also left the choice of the merit function unspecified.

18.7 REDUCED-HESSIAN SQP METHODS

We now return to the case where it is advantageous to approximate only the reduced Hessian of the Lagrangian (as opposed to the full Hessian, as in the algorithm given in the previous section). We have mentioned that methods that follow this approach are called reduced-Hessian methods, and they have proved to be very effective in many areas of application, such as optimal control.

18.7. REDUCED-HESSIAN SQP METHODS

Reduced-Hessian quasi-Newton methods are designed for solving problems in which second derivatives are difficult to compute, and in which the number of degrees of freedom in the problem, $(n-m)$, is small. They update an $(n-m) \times (n-m)$ approximation M_k of $Z_k^T W_k Z_k$, and are required to satisfy (18.29). The strength of these methods is that when $n-m$ is small, M_k is of high quality, and computation of the null-space component p_z of the step via (18.27c) is inexpensive. Another advantage, compared to full-Hessian quasi-Newton approximations, is that the reduced Hessian is much more likely to be positive definite, even when the current iterate is some distance from the solution, so that the safeguarding mechanism in the quasi-Newton update will be required less often in line search implementations. In this section we discuss some of the salient properties of reduced-Hessian methods and then present a practical implementation.

SOME PROPERTIES OF REDUCED-HESSIAN METHODS

Let us consider first the equality-constrained case and review the argument that leads to the tangential step (18.18).

If we were to retain p_Y and compute the step p_k by using a quasi-Newton approximation to the Hessian W_k, we would note that the p_Y component is based on a Newton-like iteration applied to the constraints $c(x) = 0$ (see (18.27b)), so that normally we can expect p_Y to converge to zero quadratically. By contrast, the tangential component p_Z converges only superlinearly, since a quasi-Newton approximation to the Hessian is used in its computation. Hence, it is common to observe in practice that

$$\frac{\|p_Y\|}{\|p_Z\|} \to 0, \qquad (18.41)$$

a behavior that is commonly known as *tangential convergence*. We deduce that p_Y is ultimately less significant, so we are justified in dropping it from the computation (that is, setting $p_Y = 0$). By doing so, we eliminate the need to maintain an approximation to $Z_k^T W_k Y_k$. The price we pay is that the total step p_k is no longer a solution of the KKT system (18.10), but rather an approximation to it. As a result, the overall convergence rate drops from 1-step superlinear to 2-step superlinear. This difference does not appear to be significant in practice. In fact, as we discuss in Section 18.10, practical methods often achieve one-step superlinear convergence in any case.

In the Coleman–Conn method based on (18.32), the step computation (18.27c), (18.27b) takes the form

$$M_k p_z = -Z_k^T \nabla f_k, \qquad (18.42a)$$

$$A_k Y_k p_Y = -c(x_k + Z_k p_z). \qquad (18.42b)$$

Note that the constraints are evaluated at the intermediate iterate $x_k + Z_k p_z$. One can show that this variant achieves a one-step superlinear convergence rate and avoids the Maratos effect described in Section 18.11.

UPDATE CRITERIA FOR REDUCED-HESSIAN UPDATING

Let us now consider the global behavior of reduced-Hessian methods and examine the question of whether the curvature condition $y_k^T s_k > 0$ can be expected to hold near the solution. Let us define the "averaged" Lagrangian Hessian $\bar{W}_k$ over the step p_k to be

$$\bar{W}_k = \int_0^1 \nabla_{xx}^2 \mathcal{L}(x_k + \tau p_k, \lambda_{k+1}) d\tau. \tag{18.43}$$

Then for our first definition (18.30b) of y_k, we have by Taylor's theorem that

$$y_k = \bar{W}_k p_k = \bar{W}_k Z_k p_z + \bar{W}_k Y_k p_Y.$$

If we take s_k to be the full step p_k, we thus obtain

$$y_k^T s_k = p_z^T Z_k^T \bar{W}_k Z_k p_z + p_z Z_k^T \bar{W}_k Y_k p_Y. \tag{18.44}$$

Near the solution, the first term on the right-hand-side is positive under our assumption of second-order sufficiency, but the last term is of uncertain sign. Since reduced-Hessian quasi-Newton methods normally exhibit tangential convergence (18.41), p_Y converges to zero faster than does p_z. Therefore, the first term on the right-hand-side of (18.44) will eventually dominate the second term, resulting in a positive value for $y_k^T s_k$. However, we cannot guarantee this property, even arbitrarily close to a solution, so a safeguarding mechanism is needed to ensure robustness.

To prevent a bad quasi-Newton update from taking place we could simply *skip* the update if a condition like (18.21) does not hold. Skipping the BFGS update is more justifiable in the case of reduced-Hessian methods than for full-Hessian methods. For one thing, it occurs less often; the discussion of the previous paragraph indicates that $y_k^T s_k$ is usually significantly positive near the solution. In situations in which the update is skipped, the second term in (18.44) outweighs the first term, suggesting that the step component p_Y is not small relative to p_z. Since the normal component p_Y is determined by first-derivative information (see (18.27b)), which is known accurately, it generally makes good progress toward the solution. The behavior of the step component p_z, which depends on the reduced-Hessian approximation, is therefore of less concern in this particular situation, so we can afford to skip an update.

This discussion motivates the following skipping rule, which explicitly monitors the magnitudes of the normal and tangential components of the step.

18.7. REDUCED-HESSIAN SQP METHODS

Procedure 18.4 (Update–Skip).
Given a sequence of positive numbers γ_k with $\sum_{k=1}^{\infty} \gamma_k < \infty$;
if $y_k^T s_k > 0$ and $\|p_Y\| \le \gamma_k \|p_Z\|$;
    Compute s_k and y_k from (18.30);
    Update M_k with the BFGS formula to obtain M_{k+1};
else
    Set $B_{k+1} = B_k$;
end (if).

The choice of the sequence γ_k is somewhat arbitrary; values such as $\gamma_k = 0.1k^{-1.1}$ have been used in practice.

In the Coleman–Conn method, y_k is given by (18.32). In this case, we modify the definition of the averaged Lagrangian Hessian $\bar{W}_k$ by replacing the term τp_k inside the integral (18.43) by τp_z. We then obtain for $y_k^T s_k$ that

$$y_k^T s_k = p_z^T Z_k^T \bar{W}_k Z_k p_z,$$

guaranteeing that $y_k^T s_k$ will be positive near the solution under our second-order sufficient assumptions. Away from the solution, however, this term can be negative or very small. Recall that in the unconstrained case, we used a line search strategy to ensure that $y_k^T s_k$ is sufficiently large. A similar effect can be achieved in the constrained case by performing a curved line search that roughly follows the constraints [101].

CHANGES OF BASES

All the discussion in this section has been framed in the context of an equality-constrained problem because this simplifies the presentation and because the concepts pertained to a single step of the algorithm. We now consider the effects of changes in the working set, and this highlights one of the potential weaknesses of reduced-Hessian quasi-Newton methods.

Since the working set (the set of indices that represents our best guess of the optimal active set) changes from iteration k to iteration $k+1$, so does the size of the reduced-Hessian matrix. If one or more indices are *added* to the active set at a given iteration, the active constraint gradients can be used to project the current reduced-Hessian approximation M_k onto a smaller approximation M_{k+1} at the new iterate. If, on the other hand, one or more indices are *deleted* from the active set, the matrix M_{k+1} has larger dimension than M_k. It is not obvious how the new rows and columns of M_{k+1} can be initialized; some iterations may be required before these new entries accumulate enough information to start contributing to the quality of the steps p_k. Thus if constraints are added and deleted frequently, the quasi-Newton approximations will not be of high quality. Fortunately, these difficulties will disappear as the iterates converge to the solution, since the working set tends to settle down.

Several procedures have been proposed to cope with this case where the Hessian approximation increases in dimension, but none of them is completely satisfying. Therefore, for simplicity, we will focus our specification of a practical reduced-Hessian method given below on the equality-constrained problem (18.1) and refer the reader to [108] for a detailed treatment of inequality constraints.

Even when the dimension of the working set does not change, there may be an abrupt change in the null-space basis Z from one step to the next due to a different definition of basic variables in (15.14). In addition, if the selection of the basic variables is unfortunate, Z may be unnecessarily badly conditioned, which can introduce roundoff errors in the step computation. All of these potential pitfalls need to be addressed in robust reduced-Hessian methods, showing that their implementations can be quite sophisticated; see [108], [86], [146].

A PRACTICAL REDUCED-HESSIAN METHOD

The discussion above shows that there is a great deal of complexity in a practical reduced-Hessian method, especially if we intend it to solve large problems. Nevertheless, the framework for this type of method, for the case of equality constraints, can be specified as follows.

Algorithm 18.5 (Reduced-Hessian Method—Equality).
Choose parameters $\eta \in (0, 0.5)$, $\tau \in (0, 1)$;
Choose an initial pair (x_0, λ_0);
Choose an initial $(n - m) \times (n - m)$ symmetric positive definite
    reduced-Hessian approximation M_0;
Evaluate $f_0, \nabla f_0, c_0, A_0$;
Compute Y_0 and Z_0 that span the range space of A_0^T and
    null space of A_0, respectively;
for $k = 0, 1, 2, \ldots$
    if termination test is satisfied
        STOP;
    Compute p_Y and p_Z by solving

$$(A_k Y_k) p_Y = -c_k, \qquad M_k p_Z = -Z_k^T \nabla f_k;$$

    Set $p_k = Y_k p_Y + Z_k p_Z$;
    Choose μ_k such that p_k is a descent direction for ϕ at x_k;
    Set $\alpha_k = 1$;
    while $\phi(x_k + \alpha_k p_k; \mu_k) > \phi(x_k, \mu_k) + \eta \alpha_k D\phi(x_k; p_k)$
        Reset $\alpha_k \leftarrow \tau_\alpha \alpha_k$ for some $\tau_\alpha \in (0, \tau)$;
    end (while)

Set $x_{k+1} = x_k + \alpha_k p_k$;
Evaluate $f_{k+1}, \nabla f_{k+1}, c_{k+1}, A_{k+1}$;
Compute Y_{k+1} and Z_{k+1} that span the range space of A_{k+1}^T and
    null space of A_{k+1}, respectively;
Compute λ_{k+1} by solving

$$\lambda_{k+1} = -[Y_{k+1}^T A_{k+1}^T]^{-1} Y_{k+1}^T \nabla f_{k+1};$$

Set

$$s_k = \alpha_k p_z, \quad y_k = Z_k^T [\nabla_x \mathcal{L}(x_{k+1}, \lambda_{k+1}) - \nabla_x \mathcal{L}(x_k, \lambda_{k+1})];$$

if update criterion is satisfied
    obtain M_{k+1} by updating M_k via formula (8.19);
else
    $M_{k+1} \leftarrow M_k$;
end (if)
end (for)

The matrix M_k should be a quasi-Newton approximation of the reduced Hessian of the Lagrangian, modified if necessary so that it is sufficiently positive definite.

18.8 TRUST-REGION SQP METHODS

Trust-region implementations of the SQP approach have several attractive properties. Among them are the fact that they provide a means for treating the case where active constraint gradients are linearly dependent, and that they can make direct use of second-derivative information. Nevertheless, some of the algorithms described in this section have only recently been proposed and have been fully developed only for the equality-constrained case.

Therefore, we begin by focusing on the equality-constrained problem (18.1), for which the quadratic subproblem is (18.8). By adding a trust-region constraint, we obtain the new model

$$\min_p \frac{1}{2} p^T W_k p + \nabla f_k^T p \qquad (18.45a)$$

$$\text{subject to } A_k p + c_k = 0, \qquad (18.45b)$$

$$\|p\| \leq \Delta_k. \qquad (18.45c)$$

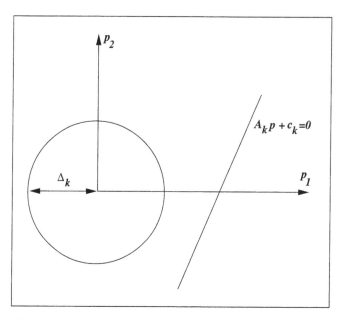

Figure 18.1 Inconsistent constraints in trust-region model.

We assume for the moment that $\|\cdot\|$ denotes the ℓ_2 norm, even though in practice it is common to work with a scaled trust region of the form $\|Sp\| \leq \Delta_k$. The trust-region radius Δ_k will be updated depending on how the predicted reduction in the merit function compares to the actual reduction. If there is good agreement, the trust-region radius is unaltered or increased, whereas if the agreement is poor, the radius is decreased.

The inclusion of the ℓ_2 trust-region constraint makes the subproblem (18.45) considerably more difficult to solve than the quadratic program (18.8). Moreover, the new model may not always have a solution because the constraints (18.45b), (18.45c) can be inconsistent, as illustrated in Figure 18.1. In this example, any step p that satisfies the linearized constraints must lie outside the trust region.

To resolve the possible conflict between satisfying the linear constraints (18.45b) and the trust-region constraint (18.45c), it is not appropriate simply to increase Δ_k until the set of steps p satisfying the linear constraints (18.45b) intersects the trust region. This approach would defeat the purpose of using the trust region in the first place as a way to define a region within which we trust the model (18.45a), (18.45b) to accurately reflect the behavior of the true objective and constraint functions in (18.1), and it would harm the convergence properties of the algorithm. A more appropriate viewpoint is that there is no reason to try to satisfy the equality constraints exactly at every step; rather, we should aim to improve the feasibility of these constraints at each step and to satisfy them exactly only in the limit. This point of view leads to three different techniques for reformulating the trust-region subproblem, which we describe in turn.

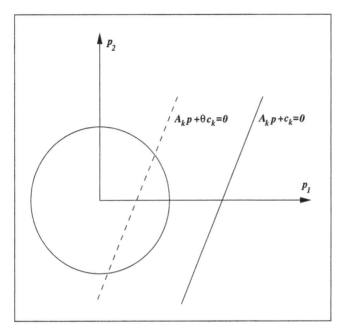

Figure 18.2 Relaxed constraints becoming consistent.

APPROACH I: SHIFTING THE CONSTRAINTS

In this approach we replace (18.45b) by the shifted constraint

$$A_k p + \theta c_k = 0, \tag{18.46}$$

where $\theta \in (0, 1]$ is chosen small enough that the constraint set (18.45b), (18.45c) is feasible. It is not difficult to find values of θ that achieve this goal. In the example of Figure 18.1, we see that θ has the effect of shifting the line corresponding to the linearized constraints to a parallel line that intersects the trust region; see Figure 18.2.

Though it is easy to find a value of θ_k that makes the constraint set (18.45b), (18.45c) feasible, it is not a simple matter to make a choice of this parameter that ensures good practical performance of the algorithm. The value of θ_k strongly influences performance because it controls whether the computed step p tends more to satisfy constraint feasibility or to minimize the objective function. If we always choose θ_k to be much smaller than its maximum acceptable value (that is, if the shifted constraint line in Figure 18.2 almost passes through the current iterate x_k), then the step computed in the subproblem (18.45) will tend to do little to improve the feasibility of the constraints $c(x) = 0$, focusing instead on reduction of the objective function f.

The following implementation of the constraint-shifting approach overcomes this difficulty by computing the step in two stages. In the first stage, we ignore the objective function altogether, and investigate instead just how close we can come to satisfying the linearized constraints (18.45b) while staying well inside the trust region (18.45c). To be precise, we solve the following "normal subproblem":

$$\min_{v} \|A_k v + c_k\|_2 \tag{18.47a}$$

$$\text{subject to } \|v\|_2 \leq \zeta \Delta_k, \tag{18.47b}$$

where ζ is some parameter in the interval $(0, 1)$; a typical value is $\zeta = 0.8$. We call the solution v_k of this subproblem the *normal step*. We now demand that the total step p_k of the trust-region method give as much reduction toward satisfying the constraints as the normal step. This is achieved by replacing θc_k by $-A_k v_k$ in (18.46), yielding the new subproblem

$$\min_{p} \frac{1}{2} p^T W_k p + \nabla f_k^T p \tag{18.48a}$$

$$\text{subject to } A_k p = A_k v_k, \tag{18.48b}$$

$$\|p\|_2 \leq \Delta_k. \tag{18.48c}$$

It is clear that the constraints (18.48b), (18.48c) of this subproblem are consistent, since the choice $p = v_k$ satisfies both of these constraints.

The new subproblem (18.48) differs from the trust-region subproblems for unconstrained optimization described in Chapter 4 because of the presence of the equality constraints (18.48b). However, as we show in the next section, where we outline a practical implementation of this approach, we can eliminate these constraints and obtain a subproblem that can be solved by the trust-region techniques described in Chapter 4.

APPROACH II: TWO ELLIPTICAL CONSTRAINTS

Another modification of the approach is to reformulate the SQP subproblem as

$$\min_{p} \frac{1}{2} p^T W_k p + \nabla f_k^T p \tag{18.49a}$$

$$\text{subject to } \|A_k p + c_k\|_2 \leq \pi_k, \tag{18.49b}$$

$$\|p\| \leq \Delta_k. \tag{18.49c}$$

(Note that this problem reduces to (18.45) when $\pi_k = 0$.) There are several ways to choose the bound π_k. The first is to demand that the step p_k be at least as good toward satisfying the constraints as a steepest descent step. We define

$$\pi_k = \|A_k p^c + c_k\|_2,$$

where p^c is the Cauchy point for the problem

$$\min_{v} m(v) = \|A_k v + c_k\|_2^2 \quad \text{subject to } \|v\|_2 \leq \Delta_k. \tag{18.50}$$

(Recall from Chapter 4 that the Cauchy point is the minimizer of m along the steepest descent direction $-\nabla m(0)$ subject to the trust-region constraint.) It is clear that this choice of π_k makes the constraints (18.49b), (18.49c) consistent, since p^c satisfies both constraints. Since this choice forces the step obtained by solving (18.50) to make at least as much progress toward feasibility as the Cauchy point, one can show that the iteration has good global convergence properties.

An alternative way to select π_k is to define it to be any number satisfying

$$\min_{\|p\|_2 \leq b_1 \Delta_k} \|A_k p + c_k\|_2^2 \leq \pi_k \leq \min_{\|p\|_2 \leq b_2 \Delta_k} \|A_k p + c_k\|_2^2,$$

where b_1 and b_2 are two constants such that $0 < b_2 \leq b_1 < 1$. For the standard choices of b_1 and b_2, this is a more stringent condition that forces greater progress toward satisfying feasibility, and it may provide better algorithm performance.

Regardless of the value of π_k, subproblem (18.49) is more difficult to solve than a standard trust-region problem. Various techniques for finding exact or approximate solutions have been proposed, and all are satisfactory for the case in which n is small or A_k and W_k are dense. However, efficient algorithms for this subproblem are still not established for the large-scale case. For this reason, we will not give further details on this method.

APPROACH III: Sℓ_1QP (SEQUENTIAL ℓ_1 QUADRATIC PROGRAMMING)

The two approaches above were developed specifically with equality-constrained optimization in mind, and it is not trivial to extend them to inequality-constrained problems. The approach to be described here, however, handles inequality constraints in a straightforward way, so we describe it in terms of the general problem (18.11).

The SQP quadratic programming subproblem is now given by (18.12). As before, we face the difficulty that addition of a trust-region bound to this model may cause the constraint set for this subproblem to become infeasible. To avoid this difficulty, the Sℓ_1QP approach moves the linearized constraints (18.12b), (18.12c) into the objective of the quadratic program, in the form of an ℓ_1 penalty term, leaving only the trust region as a constraint. This strategy yields the following subproblem:

$$\min_{p} \nabla f_k^T p + \frac{1}{2} p^T W_k p + \frac{1}{\mu_k} \sum_{i \in \mathcal{E}} |c_i(x_k) + \nabla c_i(x_k)^T p|$$

$$+ \frac{1}{\mu_k} \sum_{i \in \mathcal{I}} [c_i(x_k) + \nabla c_i(x_k)^T p]^- \tag{18.51}$$

$$\text{subject to } \|p\|_\infty \leq \Delta_k,$$

where we use the notation $[x]^- = \max\{0, -x\}$. The positive scalar μ_k is the penalty parameter, which defines the weight that we assign to constraint satisfaction relative to minimization of the objective.

The trust region uses the ℓ_∞ norm because it makes the subproblem easier to solve. Indeed, by introducing slacks and artificial variables into (18.51), we can formulate this problem as a quadratic program and solve it with standard algorithms for quadratic programming. These algorithms can be modified to take advantage of the special structure in (18.51), and they should make use of so-called hot-start information. This consists of good initial guesses for the initial point and optimal active set for (18.51), reuse of matrix factorizations, and so on.

We use the ℓ_1 merit function ϕ_1,

$$\phi_1(x; \mu) = f(x) + \frac{1}{\mu} \sum_{i \in \mathcal{E}} |c_i(x)| + \frac{1}{\mu} \sum_{i \in \mathcal{I}} [c_i(x)]^-, \tag{18.52}$$

to decide on the usefulness of the step generated by solving (18.51). In fact, the subproblem (18.51) can be viewed as an approximation to (18.52) in which we set $\mu = \mu_k$, replace each constraint function c_i by its linearization, and replace the "smooth part" f by a quadratic function whose curvature term includes information from both objective and constraint functions. The $S\ell_1QP$ algorithm accepts the approximate solution of (18.51) as a step if the ratio of actual to predicted reduction in ϕ_1 is not too small. Otherwise, it decreases the trust-region radius Δ_k and re-solves the subproblem to obtain a new candidate step. Other details of implementation follow those of trust-region methods for unconstrained optimization.

This approach has many attractive properties. Not only does the formulation (18.51) overcome the possible inconsistency between the linearized constraints and the trust region, it also allows relaxation of the regularity assumption on A_k. In the subproblem (18.51), the matrix W_k can be defined as the exact Hessian of the Lagrangian or a quasi-Newton approximation, and there is no requirement that it be positive definite. Most importantly, when μ_k (equivalently, μ in (18.52)) is chosen to be sufficiently large, local minimizers of the merit function ϕ_1 normally correspond to local solutions of (18.11), so that the algorithm has good global convergence properties.

It has been conjectured, but not fully established, that the $S\ell_1QP$ algorithm can be very sensitive to the choice of penalty parameter μ_k. As noted above, it should be chosen small enough to ensure that local minimizers of ϕ_1 (with $\mu = \mu_k$) correspond to local solutions of (18.11). There exist simple examples for which a value of μ_k that is just above the required threshold can produce an objective in (18.51) that is unbounded below, leading to erratic behavior of the algorithm. On the other hand, as in any penalty function approach (Chapter 17), very small values of μ will cause the constraint violation terms to "swamp" the objective function term, possibly leading to behavior similar to unmodified SQP (at best) and to termination at a point that is feasible but not optimal (at worst).

There is another complication with this basic $S\ell_1QP$ approach, which is known as the *Maratos effect*. In this phenomenon, steps that make good progress toward the solution of

(18.11) are rejected because they cause an increase in the merit function ϕ_1. Left unattended, the Maratos effect can cause the algorithm to repeatedly reject good steps, and therefore to become extremely slow on some problems. Fortunately, the Maratos effect can be overcome by applying the techniques described in Section 18.11. For concreteness we now describe how one of these techniques—the *second-order correction*—can be applied in this context.

We first solve (18.51) to obtain a step p_k. If this step gives rise to an increase in the merit function ϕ_1, then this may be an indication that our linear approximations to the constraints are not sufficiently accurate. To overcome this we could resolve (18.51) with the linear terms $c_i(x_k) + \nabla c_i(x_k)^T p$ replaced by quadratic approximations,

$$c_i(x_k) + \nabla c_i(x_k)^T p + \tfrac{1}{2} p^T \nabla^2 c_i(x_k) p. \tag{18.53}$$

But this is not practical, even if the Hessians of the constraints are individually available, because the subproblem becomes very hard to solve. Instead, we evaluate the constraint values at the new point $x_k + p_k$ and make use of the following approximations. Ignoring third-order terms, we have

$$c_i(x_k + p_k) = c_i(x_k) + \nabla c_i(x_k)^T p_k + \tfrac{1}{2} p_k^T \nabla^2 c_i(x)^T p_k. \tag{18.54}$$

Even though we don't know what the step p would be if we used the quadratic approximations (18.53) in the subproblem, we will assume that

$$p^T \nabla^2 c_i(x)^T p = p_k^T \nabla^2 c_i(x)^T p_k. \tag{18.55}$$

Making this substitution in (18.53) and using (18.54) yields the second-order correction subproblem

$$\min_p \nabla f_k^T p + \tfrac{1}{2} p^T W_k p + \frac{1}{\mu_k} \sum_{i \in \mathcal{E}} |d_i + \nabla c_i(x_k)^T p| + \frac{1}{\mu_k} \sum_{i \in \mathcal{I}} [d_i + \nabla c_i(x_k)^T p]^-$$
$$\text{subject to } \|p\|_\infty \leq \Delta_k, \tag{18.56}$$

where

$$d_i = c_i(x_k + p_k) - \nabla c_i(x_k)^T p_k.$$

This problem has a similar form to (18.51) and can be solved with the same techniques. Since the active set at the solution of (18.56) is often the same as, or similar to, that of the original problem (18.51), hot-start techniques can be used to solve (18.56) with little additional effort.

The second-order correction step requires evaluation of the constraints $c_i(x_k + p_k)$ for $i \in \mathcal{E} \cup \mathcal{I}$, and therefore it is preferable not to apply it every time that the merit function increases. Moreover, the crucial assumption (18.55) will normally not hold if the step is large, and we need to make sure that the trust region is asymptotically large enough not to interfere with the correction step. As a result it is necessary to use sophisticated heuristics to maximize the benefit of the second-order correction step; see Fletcher [83, Chapter 14].

18.9 A PRACTICAL TRUST-REGION SQP ALGORITHM

In the previous section we have described three practical frameworks for trust-region SQP methods. We will now elaborate on Approach I, which is based on (18.47) and (18.48), and give a detailed algorithm. There is a considerable amount of flexibility in the computation of the normal step v and the total step p, particularly since exact solutions of the normal subproblem (18.47) or the main trust-region subproblem (18.48) are not needed; approximate solutions often suffice.

The normal subproblem (18.47) can be approximately solved by the conjugate gradient method (see Algorithm 4.3 in Chapter 4) or by means of the dogleg method. The latter cannot be implemented exactly as in Chapter 4, because the Newton step is not uniquely defined: Singularity of the Hessian in (18.47a) ensures that there are infinitely many "Newton steps" that minimize the quadratic model. If we assume that A_k has full rank and choose our particular "Newton step" in the dogleg method to be the vector of minimum Euclidean norm that satisfies $A_k v + c_k = 0$, we have that

$$p^{\text{B}} = -A_k^T [A_k A_k^T]^{-1} c_k.$$

As defined in Chapter 4, the Cauchy point is the minimizing point for the objective (18.47a) along the direction $-A_k^T c$, so it, too, lies in the range space of A_k^T. Since the dogleg step v_k is a linear combination of these two directions, it, too, lies in the range space of A_k^T.

Let us now consider the solution of the main subproblem (18.48). An efficient technique, which is suitable for small or large problems, is to use a reduced-Hessian approach. Since the normal step v_k is in the range space of A_k^T, we express the total step p_k as

$$p_k = v_k + Z_k u_k, \quad \text{for some} \quad u \in \mathbf{R}^{n-m}, \tag{18.57}$$

where Z_k is a basis for the null space of A_k^T. That is, we fix the component of p_k in the range space of A_k^T to be exactly v_k, the normal-step vector, and allow no other movement in this direction. By substituting this form of p_k into (18.48a), (18.48b), we obtain the following *tangential subproblem* in the reduced variable u:

$$\min_u \; m_k(u) \stackrel{\text{def}}{=} (\nabla f_k + W_k v_k)^T Z_k u + \frac{1}{2} u^T Z_k^T W_k Z_k u \tag{18.58a}$$

$$\text{subject to} \quad \|Z_k u\|_2 \leq \sqrt{\Delta_k^2 - \|v_k\|_2^2}. \tag{18.58b}$$

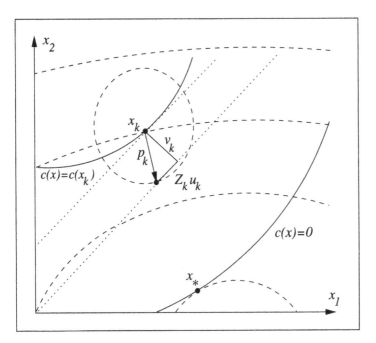

Figure 18.3 The step $p_k = v_k + Z_k u_k$ of the trust-region method.

(Note that we have dropped all terms in the objective (18.58a) that are independent of u, since they do not affect the solution.) The simple form (18.58b) of the trust-region constraint is a consequence of the fact that the components v_k and $Z_k u_k$ in (18.57) are orthogonal. We cannot treat (18.58b) as a scaled trust-region constraint, since the matrix Z_k is not even square, in general. Nevertheless, we can write (18.58b) as

$$u^T S_k^2 u \leq \Delta_k^2, \quad \text{with } S_k^2 \equiv Z_k^T Z_k,$$

where S_k is positive definite, since Z_k has linearly independent columns.

We call the solution $Z_k u_k$ of (18.58) the *tangent step*. It cannot be computed by the dogleg method, unless we have an assurance that the reduced Hessian $Z_k^T W_k Z_k$ is positive definite. But the conjugate gradient (CG) method, as given in Algorithm 4.3 of Chapter 4, can always be applied to this subproblem. Effective ways of preconditioning the CG iteration in this case are the subject of continuing investigation.

Figure 18.3 illustrates the normal and tangent steps for a simple problem with two unknowns and one nonlinear equality constraint. The dashed elliptical lines represent level curves of the objective $f(x)$, with the minimum in the lower right part of the picture. The equality constraint is represented as a solid line, and the broken circle is the trust-region constraint. The tangent space of the constraints is shown as a dotted line through the point x_k (Z_k has just a single column in this example), and the normal step is perpendicular to this space. This same manifold translated to the point $x_k + v_k$ defines the possible set of tangential steps, as shown by the lower dotted line. The final step p_k in this case reaches all the way to the trust-region constraint.

A merit function that fits well with this approach is the nondifferentiable ℓ_2 function

$$\phi(x, \mu) = f(x) + \frac{1}{\mu}\|c(x)\|_2. \tag{18.59}$$

(Note that the ℓ_2 norm is not squared). The use of the ℓ_2 norm is consistent here and in the normal subproblem, but other merit functions can also be used. The actual reduction in this function at the candidate step p_k is defined as

$$\mathbf{ared} = \phi(x_k, \mu_k) - \phi(x_k + p_k, \mu_k),$$

while the predicted reduction is

$$\mathbf{pred} = m_k(0) - m_k(u) + \mu_k^{-1}\mathbf{vpred},$$

where m_k is defined in (18.58a), and **vpred** is the reduction in the model (18.47a) obtained by the normal step v_k, and is given by

$$\mathbf{vpred} = \|c(x_k)\| - \|c(x_k) + A_k v_k\|.$$

We require that μ_k be small enough that **pred** be positive and proportional to **vpred**, i.e.,

$$\mathbf{pred} \geq \rho \mu_k^{-1} \mathbf{vpred}, \tag{18.60}$$

where $0 < \rho < 1$ (for example, $\rho = 0.3$). We can enforce this inequality by choosing the penalty parameter μ_k such that

$$\mu_k^{-1} \geq \frac{m_k(0) - m_k(u)}{(1 - \rho)\mathbf{vpred}}. \tag{18.61}$$

We can now give a precise description of this trust-region SQP method for the equality-constrained optimization problem (18.1).

Algorithm 18.6 (Trust-Region SQP).
  Choose constants $\epsilon > 0$ and $\eta, \zeta, \gamma \in (0, 1)$;
  Choose starting point x_0, initial trust region $\Delta_0 > 0$;
  for $k = 0, 1, 2, \ldots$
    Compute $f_k, c_k, \nabla f_k, A_k$;
    Compute multiplier estimates $\hat{\lambda}_k$ by (18.16);
    if $\|\nabla f_k - A_k^T \hat{\lambda}_k\|_\infty < \epsilon$ **and** $\|c_k\|_\infty < \epsilon$
      STOP with approximate solution x_k;

Solve the normal subproblem (18.47) for v_k;
Compute matrix Z_k whose columns span the null space of A_k;
Compute or update W_k;
Solve (18.58) for u_k;
Set $p_k = v_k + Z_k u_k$;
Compute $\rho_k = $ **ared**/**pred**;
if $\rho_k > \eta$
    Set $x_{k+1} = x_k + p_k$;
    Choose Δ_{k+1} to satisfy $\Delta_{k+1} \geq \Delta_k$;
else
    Set $x_{k+1} = x_k$;
    Choose Δ_{k+1} to satisfy $\Delta_{k+1} \leq \gamma \| p_k \|$;
end (for).

The constant ζ that defines the smaller trust region for the normal problem (18.47) is commonly chosen to be 0.8, but this value is not critical to the performance of the method.

Algorithm 18.6 can be extended to problems with inequality constraints in two different ways. One is by using an active-set-type approach, and the other is in an interior-point framework, where an adaptation of Algorithm 18.6 is used to solve barrier problems; see the Notes and References. As in any null-space method, the drawback of defining the step p_k in terms of two components v_k and $Z_k u$ is the need to compute the null-space basis Z_k. Nevertheless, one can rearrange the CG iteration to bypass the computation of Z altogether (see the exercises).

Unfortunately, the merit function (18.59) suffers from the Maratos effect, and it is therefore appropriate to include in Algorithm 18.6 the watchdog technique or the second-order correction term (18.67) described in Section 18.11. If we use the latter, it is not efficient to apply this correction after every step that increases the merit function, as in the Sℓ_1QP method. A heuristic that has proved to be effective in practice is to apply the second-order correction only if the normal step is significantly smaller than both the tangential step and the trust-region radius.

18.10 RATE OF CONVERGENCE

We now derive conditions that guarantee the local convergence of SQP methods, as well as conditions that ensure a superlinear rate of convergence. For simplicity, we limit our discussion to Algorithm 18.1 given in Section 18.1 for equality-constrained optimization, but we will consider both exact Hessian and quasi-Newton versions of it. The results presented here can be applied to algorithms for inequality-constrained problems, once the active set has settled (see Theorem 18.1), and they can also be invoked in studying global algorithms, since in the vicinity of a solution that satisfies the assumptions below, the globalization strategies normally do not interfere with the good convergence behavior of the local algorithm. (An exception to this comment is provided by the Maratos effect, which we discuss in the next section.)

We begin by listing a set of assumptions on the problem and the quasi-Newton approximations B_k that will be useful in this section. Each of the results below makes use of one or both of these conditions.

Assumption 18.2.

(a) At the solution point x^* with optimal Lagrange multipliers λ^*, the constraint Jacobian A_* has full row rank, and the Hessian of the Lagrangian $\nabla^2_{xx}\mathcal{L}(x^*, \lambda^*)$ is positive definite on the tangent space of the constraints.

(b) The sequences $\{B_k\}$ and $\{B_k^{-1}\}$ are bounded, that is, there exists a constant β_2 such that

$$\|B_k\| \leq \beta_2 \quad \|B_k^{-1}\| \leq \beta_2, \quad \text{for all } k.$$

We first consider a Newton SQP method that uses exact second derivatives.

Theorem 18.4.

Suppose that Assumption 18.2(a) holds and that f and c are twice differentiable, with Lipschitz continuous second derivatives, in a neighborhood of (x^*, λ^*). Then if x_0 and λ_0 are sufficiently close to x^* and λ^*, the pairs (x_k, λ_k) generated by Algorithm 18.1 with W_k defined as the Hessian of the Lagrangian converge quadratically to (x^*, λ^*).

The proof follows directly from Theorem 11.2.

We turn now to quasi-Newton variants of Algorithm 18.1, in which the augmented Lagrangian Hessian $W_k = \nabla^2_{xx}\mathcal{L}(x_k, \lambda_k)$ is replaced by a quasi-Newton approximation B_k. We will make use of the projection matrix P_k defined for each iterate k by

$$P_k = I - A_k^T \left[A_k A_k^T\right]^{-1} A_k = Z_k Z_k^T,$$

where Z_k has orthonormal columns. This matrix projects any vector in $\mathbf{R}^n$ into the null space of the constraint gradients.

The *projected Hessian* $P_k W_k$ plays a crucial role in the analysis. To motivate this, we multiply the first equation of the KKT system (18.10) by P_k, and use the fact that $P_k A_k^T = 0$ to obtain

$$P_k W_k p_k = -P_k \nabla f_k.$$

This equation shows that the step is completely determined by the one-sided projection of W_k and is independent of the rest of W_k. It also suggests that a quasi-Newton method will be locally convergent if $P_k B_k$ is a reasonable approximation of $P_k W_k$, and that it will be superlinearly convergent if $P_k B_k$ approximates $P_k W_k$ accurately. To make the second

18.10. RATE OF CONVERGENCE

statement more precise, we present a result that quantifies the quality of this approximation in a way that can be viewed as an extension of the Dennis and Moré characterization (3.5) of superlinear convergence from the unconstrained case to the equality-constrained case.

Theorem 18.5 (Boggs, Tolle, and Wang [24]).

Suppose that Assumption 18.2(a) holds and that the iterates x_k generated by Algorithm 18.1 with quasi-Newton approximate Hessians B_k converge to x^*. Then x_k converges superlinearly if and only if the Hessian approximation B_k satisfies

$$\lim_{k \to \infty} \frac{\|P_k(B_k - W_*)(x_{k+1} - x_k)\|}{\|x_{k+1} - x_k\|} = 0. \qquad (18.62)$$

Let us apply these results to the quasi-Newton updating schemes discussed earlier in this chapter. We begin with the full BFGS approximation based on (18.20). To guarantee that it is always well-defined we make the (strong) assumption that the Hessian of the Lagrangian is positive definite at the solution.

Theorem 18.6.

Suppose that W_* and B_0 are symmetric and positive definite, and that Assumptions 18.2(a), (b) hold. If $\|x_0 - x^*\|$ and $\|B_0 - W_*\|$ are sufficiently small, then the iterates x_k generated by Algorithm 18.1 with BFGS Hessian approximations B_k defined by (18.20) and (18.23) (with $r_k = s_k$) satisfy the limit (18.62). Therefore, the iterates x_k converge superlinearly to x^*.

For the damped BFGS updating strategy given in Procedure 18.2, a weaker result has been obtained. If we assume that the iterates x_k converge to a solution point x^*, we can show that the rate of convergence is R-superlinear (not the usual Q-superlinear; see Chapter 2). Similar results can be obtained for the updating strategies based on the Hessian of the augmented Lagrangian (18.25).

CONVERGENCE RATE OF REDUCED-HESSIAN METHODS

Reduced-Hessian SQP methods update an approximation M_k to $Z_k^T W_k Z_k$. From the definition of P_k we see that $Z_k M_k Z_k^T$ can be considered as an approximation to the two-sided projection $P_k W_k P_k$. Therefore, since reduced-Hessian methods do not approximate the one-sided projection $P_k W_k$, we cannot expect (18.62) to hold. For these methods we can derive a condition for superlinear convergence by writing (18.62) as

$$\lim_{k \to \infty} \left[\frac{P_k(B_k - W_*)P_k(x_{k+1} - x_k)}{\|x_{k+1} - x_k\|} \right. \qquad (18.63)$$
$$\left. + \frac{P_k(B_k - W_*)(I - P_k)(x_{k+1} - x_k)}{\|x_{k+1} - x_k\|} \right] = 0.$$

To use this limit, we define the $n \times n$ matrix $B_k = Z_k M_k Z_k^T$. The following result shows that it is necessary only for the first term in (18.63) to go to zero to obtain a weaker form of superlinear convergence, namely, two-step superlinear convergence.

Theorem 18.7.

Suppose that Assumption 18.2(a) holds, and that the matrices B_k are bounded. Assume also that the iterates x_k generated by Algorithm 18.1 with approximate Hessians B_k converge to x^*, and that

$$\lim_{k \to \infty} \frac{\|P_k(B_k - W_*)P_k(x_{k+1} - x_k)\|}{\|x_{k+1} - x_k\|} = 0. \qquad (18.64)$$

Then the sequence $\{x_k\}$ converges to x^* two-step superlinearly, i.e.,

$$\lim_{k \to \infty} \frac{\|x_{k+2} - x^*\|}{\|x_k - x^*\|} = 0.$$

We have mentioned earlier that reduced-Hessian methods often exhibit *tangential convergence*, meaning that the normal component of the step is much smaller than the tangential component; see (18.41). In other words, the ratio

$$\frac{\|(I - P_k)(x_{k+1} - x_k)\|}{\|x_{k+1} - x_k\|}$$

tends to zero. Then it follows from (18.63) and (18.64) that the rate of convergence is actually one-step superlinear. This rate will occur even when the term $P_k(B_k - W_*)(I - P_k)$ is not small, as is the case when we ignore this cross term (see (18.18)).

Let us consider a local reduced-Hessian method using BFGS updating. The iteration is $x_{k+1} = x_k + Y_k p_Y + Z_k p_Z$, where p_Y and p_Z are given by (18.27). The reduced-Hessian approximation M_k is updated by the BFGS formula using the correction vectors (18.30), and the initial approximation M_0 is symmetric and positive definite. If we make the assumption that the null space bases Z_k used to define the correction vectors (18.30) vary smoothly, i.e.,

$$\|Z_k - Z_*\| = O(\|x_k - x^*\|), \qquad (18.65)$$

then we can establish the following result that makes use of Theorem (18.7).

Theorem 18.8.

Suppose that Assumption 18.2(a) holds. Let x_k be generated by the reduced-Hessian method just described. If the sequence $\{x_k\}$ converges to x^* R-linearly, then $\{M_k\}$ and $\{M_k^{-1}\}$ are uniformly bounded and x_k converges two-step superlinearly.

It is necessary to make use of the smoothness assumption (18.65) on Z_k for the following reason. There are many null-space bases for the matrix of constraint gradients A_k (Chapter 15). If the choice of this null-space basis changes too much from one iterate to the next, superlinear convergence will be impeded because the quasi-Newton update will be based on correction vectors s_k and y_k that exhibit jumps. Indeed, in many cases, any procedure for computing Z_k *as a function of A_k alone* will have discontinuities, even for full-rank A_k. Before computing a basis Z_k we need to take into consideration the choice made during the previous iteration.

Two procedures have been proposed to cope with this difficulty. One is to obtain Z_k by computing a QR factorization of A_k (see (A.54)) in which the inherent arbitrary choices in the factorization algorithm are made, near the solution, in the same way as in computing Z_{k-1} from A_{k-1}. As a result, when the matrices A_k are close to A_*, the same choice of sign will be made at each step. Another procedure consists in applying the orthogonal factor of the QR factorization of A_{k-1} to A_k, and then computing the QR factorization of $Q_{k-1}^T A_k$ to obtain Q_k, and thus Z_k. One can show that either of these two procedures satisfies (18.65).

18.11 THE MARATOS EFFECT

We noted in Section 18.9 that some merit functions can impede the rapid convergence behavior of SQP methods by rejecting steps that make good progress toward a solution. This undesirable phenomenon is often called the *Maratos effect*, because it was first observed by Maratos [156]. It is illustrated by the following example, in which the SQP steps p_k give rise to a quadratic convergence rate but cause an increase both in the objective function value and the constraint norm. As a result, these steps will be rejected by many merit functions.

❏ **EXAMPLE 18.1** (POWELL [209])

Consider the problem

$$\min\ f(x_1, x_2) = 2(x_1^2 + x_2^2 - 1) - x_1, \quad \text{subject to} \quad x_1^2 + x_2^2 - 1 = 0.$$

It is easy to verify (see Figure 18.4) that the optimal solution is $x^* = (1, 0)^T$, that the corresponding Lagrange multiplier is $\lambda^* = \frac{3}{2}$, and that $\nabla_{xx}^2 \mathcal{L}(x^*, \lambda^*) = I$.

Let us consider an iterate x_k of the form $x_k = (\cos\theta, \sin\theta)^T$, which is feasible for any value of θ. We now generate a search direction p_k by solving the subproblem (18.8) with $B_k = \nabla_{xx}^2 \mathcal{L}(x^*, \lambda^*) = I$. Since

$$f(x_k) = -\cos\theta, \quad \nabla f(x_k) = \begin{pmatrix} 4\cos\theta - 1 \\ 4\sin\theta \end{pmatrix}, \quad A(x_k)^T = \begin{pmatrix} 2\cos\theta \\ 2\sin\theta \end{pmatrix},$$

568 CHAPTER 18. SEQUENTIAL QUADRATIC PROGRAMMING

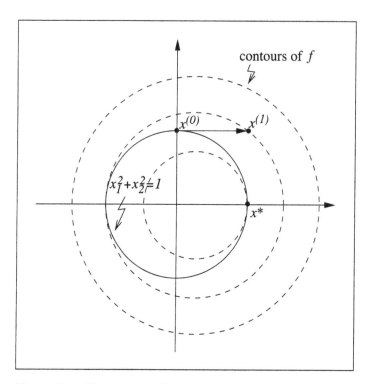

Figure 18.4 The Maratos effect: Example 18.1.

the quadratic subproblem (18.8) takes the form

$$\min -\cos\theta + (4\cos\theta - 1)p_1 + 4\sin\theta p_2 + \frac{1}{2}p_1^2 + \frac{1}{2}p_2^2$$
$$\text{subject to } p_2 = -\cot\theta p_1.$$

By solving this subproblem, we obtain

$$p_k = \begin{pmatrix} \sin^2\theta \\ -\sin\theta\cos\theta \end{pmatrix}, \tag{18.66}$$

which yields a new trial point

$$x_k + p_k = \begin{pmatrix} \cos\theta + \sin^2\theta \\ \sin\theta(1-\cos\theta) \end{pmatrix}.$$

If $\sin\theta \neq 0$, we have that

$$\|x_k + p_k - x^*\|_2 = 2\sin^2(\theta/2), \qquad \|x_k - x^*\|_2 = 2|\sin(\theta/2)|,$$

and therefore

$$\frac{\|x_k + p_k - x^*\|_2}{\|x_k - x^*\|_2^2} = \frac{1}{2}.$$

Hence, this step approaches the solution at a rate consistent with Q-quadratic convergence. Note, however, that

$$f(x_k + p_k) = \sin^2\theta - \cos\theta > -\cos\theta = f(x_k),$$
$$c(x_k + p_k) = \sin^2\theta > c(x_k) = 0,$$

so that both the objective function value and constraint violation increase over this step.

In Figure 18.4 we illustrate the case where $\theta = \pi/2$ and the SQP method moves from $x^{(0)} = (1, 0)$ to $x^{(1)} = (1, 1)$. We have chosen a large value of θ for clarity, but note that we have shown that the step will be rejected for any nonzero value of θ. □

This example shows that any merit function of the form

$$\Psi(x; \mu) = f(x) + \frac{1}{\mu} h(c(x))$$

(where $h(\cdot)$ is a nonnegative function satisfying $h(0) = 0$) will reject the step (18.66), so that any algorithm based on a merit function of this type will suffer from the Maratos effect. Examples of such merit functions include the smooth ℓ_2 merit function $f(x) + \mu^{-1}\|c(x)\|_2^2$ and the nondifferentiable ℓ_1 function $f(x) + \mu^{-1}\|c(x)\|_1$.

If no measures are taken, the Maratos effect can dramatically slow down SQP methods. Not only does it interfere with good steps away from the solution, but it can also prevent superlinear convergence from taking place. Techniques for avoiding the Maratos effect include the following.

- We can use a merit function that does not suffer from the Maratos effect. An example is Fletcher's augmented Lagrangian function (18.36), for which one can show that steps generated by Algorithm 18.1 will be accepted in a neighborhood of a solution point that satisfies the second-order sufficient conditions.

- We can use a second-order correction in which we add to p_k a step p'_k, which is computed at $c(x_k + p_k)$ and which provides sufficient decrease in the constraints. See, for example, (18.56), which is the second-order correction subproblem for the Sℓ_1QP method.

- We may allow the merit function Ψ to *increase* on certain iterations, that is, we can use a nonmonotone strategy. An example is the watchdog technique discussed below.

We now discuss the last two approaches.

SECOND-ORDER CORRECTION

By adding a correction term that provides further decrease in the constraints, the SQP iteration overcomes the difficulties associated with the Maratos effect. We have already described this second-order correction technique in Section 18.8 in connection with the $S\ell_1 QP$ algorithm, which is a trust-region algorithm based on the nondifferentiable ℓ_1 merit function. Here, we outline the way in which the same idea can be applied to a line search algorithm, also based on the ℓ_1 merit function.

Given an SQP step p_k, the second-order correction step w_k is defined to be

$$w_k = -A_k^T (A_k A_k^T)^{-1} c(x_k + p_k). \tag{18.67}$$

Note that w_k has the property that is satisfies a linearization of the constraints c at the point $x_k + p_k$, that is,

$$A_k w_k + c(x_k + p_k) = 0.$$

In fact, w_k is the minimum-norm solution of this equation. The motivation is similar to that used in (18.53) and (18.54), which gives the corresponding strategy for the trust-region $S\ell_1 QP$ algorithm. The effect of the correction step w_k, which is normal to the constraints, is to decrease the quantity $\|c(x)\|$ to the order of $\|x_k - x^*\|^3$. By doing so, we guarantee that the step from x_k to $x_k + p_k + w_k$ will decrease the merit function, at least near the solution. The price we pay is an additional evaluation of the constraint function c at $x_k + p_k$.

Following is a detailed implementation of the second-order correction step in a line search SQP method in which the Hessian approximation is updated by a quasi-Newton formula. As before, we denote the directional derivative of the merit function ϕ_1 in the direction p_k by $D(\phi_1(x_k; \mu); p_k)$. It is understood that the penalty parameter μ is held fixed until a successful step is computed.

Algorithm 18.7 (SQP Algorithm with Second-Order Correction).
  Choose parameters $\eta \in (0, 0.5)$ and τ_1, τ_2 with $0 < \tau_1 < \tau_2 < 1$;
  Choose initial point x_0 and approximate Hessian B_0;
  for $k = 0, 1, 2, \ldots$
    Evaluate $f_k, \nabla f_k, c_k, A_k$;
    if termination test is satisfied
      STOP;
    Compute the SQP step p_k;
    Set $\alpha_k \leftarrow 1$, newpoint $\leftarrow$ false;
    while not newpoint
      if $\phi_1(x_k + \alpha_k p_k) \leq \phi_1(x_k) + \eta \alpha_k D\phi_1(x_k; p_k)$
        Set $x_{k+1} \leftarrow x_k + \alpha_k p_k$;
        Set newpoint $\leftarrow$ true;
      else if $\alpha_k = 1$

            Compute w_k from (18.67);
            if $\phi_1(x_k + p_k + w_k) \leq \phi_1(x_k) + \eta D\phi_1(x_k; p_k)$
                Set $x_{k+1} \leftarrow x_k + p_k + w_k$;
                Set newpoint $\leftarrow$ true;
            end (if)
            Choose new α_k in $[\tau_1 \alpha_k, \tau_2 \alpha_k]$;
        else
            Choose new α_k in $[\tau_1 \alpha_k, \tau_2 \alpha_k]$;
        end (if)
    end (while)
    Update B_k using a quasi-Newton formula to obtain B_{k+1};
end (for)

One can show that after a finite number of iterations of Algorithm 18.7, the value $\alpha_k = 1$ always produces a new iterate x_{k+1}, which can have the form either $x_{k+1} = x_k + p_k$ or $x_{k+1} = x_k + p_k + w_k$. Therefore, the merit function does not interfere with the iteration, and superlinear convergence is obtained, as for the local algorithm.

Note that the step w_k can be defined more generally in terms of any matrix Y_k whose columns are a basis for the subspace spanned by the columns of A_k.

The second-order correction strategy is effective in practice. The additional cost of performing the extra constraint function evaluation in (18.67) is outweighed by added robustness and efficiency.

WATCHDOG (NONMONOTONE) STRATEGY

The inefficiencies caused by the Maratos effect can also be avoided by occasionally accepting steps that increase the merit function. (Such steps are called "relaxed steps.") There is a limit to our liberality, however: If a sufficient reduction of the merit function has not been obtained within a certain number of iterates of the relaxed step ($\hat{t}$ iterates, say), then we return to the iterate before the relaxed step and perform a normal step, using a line search or some other technique to force a reduction in the ℓ_1 merit function.

Our hope is, of course, that the relaxed step is a good step in the sense of making progress toward the solution, even though it increases the merit function. The step taken immediately after the relaxed step serves a similar purpose to the second-order correction step above; that is, it corrects the SQP step for its not taking sufficient account of curvature information for the constraint functions in its formulation of the SQP subproblem.

We now describe a particular instance of this technique, which is often called the watchdog strategy. We set $\hat{t} = 1$, so that we allow the merit function to increase on just a single step before insisting on a sufficient decrease of some type. As above, we focus our discussion on a line search SQP algorithm that uses the ℓ_1 merit function. We assume that the penalty parameter μ_k is not changed until a successful cycle has been completed. For simplicity we omit the details of the calculation of step length α_k and the termination test.

Algorithm 18.8 (Watchdog).
Choose constant $\eta \in (0, 0.5)$;
Choose initial point x_0 and approximate Hessian B_0;
Set $k \leftarrow 0, \mathcal{S} \leftarrow \{0\}$;
repeat
    Evaluate $f_k, \nabla f_k, c_k, A_k$;
    if termination test satisfied
        STOP with approximate solution x_k;
    Compute the SQP step p_k;
    Set $x_{k+1} \leftarrow x_k + p_k$;
    Update B_k using a quasi-Newton formula to obtain B_{k+1};
    if $\phi_1(x_{k+1}) \leq \phi_1(x_k) + \eta D\phi_1(x_k; p_k)$
        $k \leftarrow k + 1$;
        $\mathcal{S} \leftarrow \mathcal{S} \cup \{k\}$;
    else
        Compute the SQP step p_{k+1};
        Find α_{k+1} such that
            $\phi_1(x_{k+2}) \leq \phi_1(x_{k+1}) + \eta\alpha_{k+1} D\phi_1(x_{k+1}; p_{k+1})$;
        Set $x_{k+2} \leftarrow x_{k+1} + \alpha_{k+1} p_{k+1}$;
        Update B_{k+1} using a quasi-Newton formula to obtain B_{k+2};
        if $\phi_1(x_{k+1}) \leq \phi(x_k)$ **or** $\phi_1(x_{k+2}) \leq \phi_1(x_k) + \eta D\phi_1(x_k; p_k)$
            $k \leftarrow k + 2$;
            $\mathcal{S} \leftarrow \mathcal{S} \cup \{k\}$;
        else if $\phi_1(x_{k+2}) > \phi_1(x_k)$
            Find α_k such that $\phi_1(x_{k+3}) \leq \phi_1(x_k) + \eta\alpha_k D\phi_1(x_k; p_k)$;
            Compute $x_{k+3} = x_k + \alpha_k p_k$;
        else
            Compute the SQP step p_{k+2};
            Find α_{k+2} such that
                $\phi_1(x_{k+3}) \leq \phi_1(x_{k+2}) + \eta\alpha_{k+2} D\phi_1(x_{k+2}; p_{k+2})$;
            Set $x_{k+3} \leftarrow x_{k+2} + \alpha_{k+2} p_{k+2}$;
            Update B_{k+2} using a quasi-Newton formula to obtain B_{k+3};
            $k \leftarrow k + 3$;
            $\mathcal{S} \leftarrow \mathcal{S} \cup \{k\}$;
        end (if)
    end (if)
end (repeat)

A quasi-Newton update at x_{k+1} is always performed using information from the immediately preceding step $x_{k+1} - x_k$. The set $\mathcal{S}$ is not required by the algorithm and is introduced only to identify the iterates for which a sufficient merit function reduction was obtained. Note that at least one third of the iterates have their indices in $\mathcal{S}$. By using this fact, one can show that the SQP method using the watchdog technique is locally convergent, that for all sufficiently large k the step length is $\alpha_k = 1$, and that the rate of convergence is superlinear.

In practice, it may be advantageous to allow increases in the merit function for more than 1 iteration. Values of $\hat{t}$ such as 5 or 10 are typical. For simplicity, we have assumed that the matrix is updated at each iterate along the direction moved to reach that iterate, even though in practice it may be preferable not to do so at certain iterates that will be rejected.

As this discussion indicates, careful implementations of the watchdog technique have a certain degree of complexity, but the added complexity is worthwhile because the approach has shown good practical performance. A potential advantage of the watchdog technique over the second-order correction strategy is that it may require fewer evaluations of the constraints. Indeed, in the best case, most of the steps will be full SQP steps, and there will rarely be a need to return to an earlier point. There is insufficient numerical experience, however, to allow us to make broad claims about the relative merits of watchdog and second-order correction strategies.

NOTES AND REFERENCES

SQP methods were first proposed in 1963 by Wilson [245] and were developed in the 1970s (Garcia-Palomares and Mangasarian [97], Han [131, 132], Powell [202, 205, 204]). See Boggs and Tolle [23] for a literature survey. The successive linear programming approach outlined at the end of Section 18.1 is described by Fletcher and Sainz de la Maza [89]. The relation between SQP and augmented Lagrangian methods is explored by Tapia [234]

Another method that depends on x alone can be derived by applying Newton's method to the reduced form of the first-order KKT equations,

$$\mathcal{F}(x) \equiv \begin{bmatrix} Z(x)^T g(x) \\ c(x) \end{bmatrix} = 0. \tag{18.68}$$

This system of n nonlinear equations in n unknowns does not involve any Lagrange multiplier estimates and is an excellent example of a primal method for nonlinear optimization. The Jacobian of this system requires the computation of the derivative $Z'(x)$, which is not directly available. However, when we drop terms that involve Z', we obtain a scheme that is related to SQP methods (Goodman [117]).

A procedure for coping with inconsistent linearized constraints that is different from the one presented in Section 18.3 is described by Powell [202].

Reduced-Hessian SQP methods were perhaps first proposed by Murray and Wright and Coleman and Conn; see [23] for a literature survey. Since then, many papers have been written on the subject, and the algorithms have proved to be useful in applications such as chemical process control and trajectory optimization.

Some analysis shows that several—but not all—of the good properties of BFGS updating are preserved by damped BFGS updating. Numerical experiments exposing the weakness of the approach are reported by Powell [208].

Most SQP methods are normally *infeasible* methods, meaning that neither the initial point nor any of the subsequent iterates need to be feasible. This can be advantageous when the problem contains significantly nonlinear constraints, because it can be computationally expensive to stay inside the feasible region. In some applications, however, the functions that make up the problem are not always defined outside of the feasible region. To cope with this phenomenon, *feasible SQP methods* have been developed; see, for example, Panier and Tits [189].

Many SQP codes treat linear constraints differently from nonlinear constraints. There is a first phase in which the iterates become feasible with respect to all linear constraints, and all subsequent iterates remain feasible with respect to them. This is the case in NPSOL [111] and SNOPT [108]. To deal with inconsistent or nearly dependent constraints, these codes make use of a flexible mode in which the constraints of the problem are relaxed.

The interior-point method for nonlinear programming described by Byrd, Hribar, and Nocedal [34] uses Algorithm 18.6 to solve the equality-constrained subproblems arising in barrier methods.

Second-order correction strategies were proposed by Coleman and Conn [44], Fletcher [82], Gabay [96], and Mayne and Polak [161]. The watchdog technique was proposed by Chamberlain et al. [40].

Procedures for computing a smoothly varying sequence of null-space matrices Z_k are described by Coleman and Sorensen [50] and Gill et al. [109].

More realistic convergence results have been proved for reduced-Hessian quasi-Newton methods; see Byrd and Nocedal [36].

Exercises

18.1 Prove Theorem 18.4.

18.2 Write a program that implements Algorithm 18.1. Use it to solve the problem

$$\min e^{x_1 x_2 x_3 x_4 x_5} - \tfrac{1}{2}(x_1^3 + x_2^3 + 1)^2$$
$$\text{subject to } x_1^2 + x_2^2 + x_3^2 + x_4^2 + x_5^2 - 10 = 0,$$
$$x_2 x_3 - 5 x_4 x_5 = 0,$$
$$x_1^3 + x_2^3 + 1 = 0.$$

Use the starting point $x_0 = (-1.71, 1.59, 1.82, -0.763, -0.763)$. The solution is $x^* = (-1.8, 1.7, 1.9, -0.8, -0.8)$.

18.3 Show that the damped BFGS updating satisfies (18.24).

18.4 Suppose that x^* is a solution of problem (18.1) where the second-order sufficiency conditions hold. Show that $y_k^T s_k > 0$ in a neighborhood of this point when y_k and s_k

are defined by (18.30a), (18.32). What assumption do you need to make on the multiplier estimates?

✏ **18.5** Prove (18.38).

✏ **18.6** Consider the constraint $x_1^2 + x_2^2 = 1$. Write the linearized constraints (18.8b) at the following points: $(0, 0)$, $(0, 1)$, $(0.1, 0.02)$, $-(0.1, 0.02)$.

✏ **18.7** Write a program that implements the reduced-Hessian method (EQ) and use it to solve the problem given in Exercise 3.

✏ **18.8** Show that the normal problem (18.47a)–(18.47b) always has a solution v_k lying in the range space of A_k^T. Hint: First show that if the trust-region constraint (18.47b) is active, then v_k lies in the range space of A_k^T; then show that if the trust region is inactive, then the minimum-norm solution of (18.47a) lies in the range space of A_k^T.

✏ **18.9** Rewrite (18.51) as a quadratic program.

✏ **18.10** Let us consider the solution of the tangential problem (18.58a)–(18.58b) by means of the conjugate gradient method. We perform the change of variables $\tilde{u} = Z_k u$ that transforms (18.58b) into a spherical trust region. Applying CG to the transformed problem is nothing but a preconditioned CG iteration with preconditioner $[Z_k^T Z_k]^{-1}$. Noting that the solution u of this problem needs to be multiplied by Z_k to obtain the tangent step $Z_k u_k$, we see that it is convenient to perform this multiplication by Z_k directly in the CG iteration. Show that by doing this the null-space basis Z_k enters in the CG iteration only through products of the form

$$Z_k [Z_k^T Z_k]^{-1} Z_k^T w,$$

where w is a vector. Use this to show that this multiplication can be performed as

$$A_k^T [A_k A_k^T]^{-1} A_k w. \qquad (18.69)$$

(This procedure shows that the Algorithm 18.6 can be implemented without computing Z_k.) Finally, show that the computation (18.69) can also be performed by solving

$$\begin{bmatrix} I & A_k^T \\ A_k & 0 \end{bmatrix}.$$

Appendix A

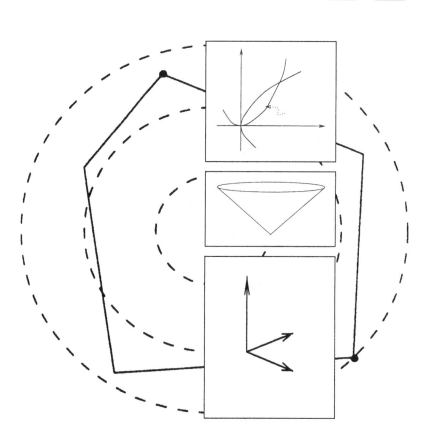

Background Material

A.1 ELEMENTS OF ANALYSIS, GEOMETRY, TOPOLOGY

TOPOLOGY OF THE EUCLIDEAN SPACE $\mathbf{R}^n$

Let $\mathcal{F}$ be a subset of $\mathbf{R}^n$, and suppose that $\{x_k\}$ is a sequence of points belonging to $\mathcal{F}$. We say that a sequence $\{x_k\}$ *converges* to some point x, written $\lim_{k\to\infty} x_k = x$, if for any $\epsilon > 0$, there is an index K such that

$$\|x_k - x\| \le \epsilon, \quad \text{for all } k \ge K.$$

For example, the sequence $\{x_k\}$ defined by $x_k = (1 - 2^{-k}, 1/k^2)^T$ converges to $(1, 0)^T$. Other examples of convergent sequences are given in Chapter 2, where the *rate of convergence* is also discussed.

We say that $\hat{x} \in \mathbf{R}^n$ is an *accumulation point* or *limit point* for $\{x_k\}$ if there is some infinite subsequence of indices $k_1, k_2, k_3, \ldots$ such that

$$\lim_{i\to\infty} x_{k_i} = \hat{x}.$$

Another way to define a limit point is to say that for any $\epsilon > 0$ and all positive integers K, we have

$$\|x_k - x\| \leq \epsilon, \quad \text{for some } k \geq K.$$

An example is given by the sequence

$$\begin{bmatrix} 1 \\ 1 \end{bmatrix}, \begin{bmatrix} 1/2 \\ 1/2 \end{bmatrix}, \begin{bmatrix} 1 \\ 1 \end{bmatrix}, \begin{bmatrix} 1/4 \\ 1/4 \end{bmatrix}, \begin{bmatrix} 1 \\ 1 \end{bmatrix}, \begin{bmatrix} 1/8 \\ 1/8 \end{bmatrix}, \ldots, \quad (A.1)$$

which has exactly two limit points: $\hat{x} = (0, 0)^T$ and $\hat{x} = (1, 1)^T$. A sequence can even have an infinite number of limit points. An example is the sequence $x_k = \sin k$, for which every point in the interval $[-1, 1]$ is a limit point.

If $\{t_k\}$ is a sequence of *real* numbers, we can define two other concepts: the lim inf and lim sup. The lim inf is the smallest accumulation point, while the lim sup is the largest. That is, we write $\hat{t} = \lim \sup_{k \to \infty} t_k$ if the following two conditions hold:

(i) there is an infinite subsequence $k_1, k_2, \ldots$ such that $\lim_{i \to \infty} t_{k_i} = \hat{t}$, and

(ii) there is no other limit point t such that $t > \hat{t}$.

The sequence $1, \frac{1}{2}, 1, \frac{1}{4}, 1, \frac{1}{8}, \ldots$ has a lim inf of 0 and a lim sup of 1.

The set $\mathcal{F}$ is *bounded* if there is some real number $M > 0$ such that

$$\|x\| \leq M, \quad \text{for all } x \in \mathcal{F}.$$

A subset $\mathcal{F} \subset \mathbf{R}^n$ is *open* if for every $x \in \mathcal{F}$, we can find a positive number $\epsilon > 0$ such that the ball of radius ϵ around x is contained in $\mathcal{F}$; that is,

$$\{y \in \mathbf{R}^n \mid \|y - x\| \leq \epsilon\} \subset \mathcal{F}.$$

The set $\mathcal{F}$ is *closed* if for all possible sequences of points $\{x_k\}$ in $\mathcal{F}$, all limit points of $\{x_k\}$ are elements of $\mathcal{F}$. For instance, the set $\mathcal{F} = (0, 1) \cup (2, 10)$ is an open subset of $\mathbf{R}$, while $\mathcal{F} = [0, 1] \cup [2, 5]$ is a closed subset of $\mathbf{R}$. The set $\mathcal{F} = (0, 1]$ is a subset of $\mathbf{R}$ that is neither open nor closed.

The *interior* of a set $\mathcal{F}$, denoted by int$\mathcal{F}$, is the largest open set contained in $\mathcal{F}$. The *closure* of $\mathcal{F}$, denoted by cl$\mathcal{F}$, is the smallest closed set containing $\mathcal{F}$. In other words, we have

$$x \in \text{cl}\mathcal{F} \quad \text{if } \lim_{k \to \infty} x_k = x \text{ for some sequence } \{x_k\} \text{ of points in } \mathcal{F}.$$

If $\mathcal{F} = (-1, 1] \cup [2, 4)$, then

$$\text{cl}\mathcal{F} = [-1, 1] \cup [2, 4], \quad \text{int}\mathcal{F} = (-1, 1) \cup (2, 4).$$

Note that if $\mathcal{F}$ is open, then int$\mathcal{F} = \mathcal{F}$, while if $\mathcal{F}$ is closed, then cl$\mathcal{F} = \mathcal{F}$.

The set $\mathcal{F}$ is *compact* if every sequence $\{x^k\}$ of points in $\mathcal{F}$ has at least one limit point, and all such limit points are in $\mathcal{F}$. (This definition is equivalent to the more formal one involving covers of $\mathcal{F}$.) The following is a central result in topology:

$\mathcal{F} \in \mathbf{R}^n$ is closed and bounded $\Rightarrow$ $\mathcal{F}$ is compact.

Given a point $x \in \mathbf{R}^n$, we call $\mathcal{N} \in \mathbf{R}^n$ a *neighborhood of* x if it is an open set containing x. An especially useful neighborhood is the *open ball of radius ϵ around x*, which is denoted by $\mathbf{B}(x, \epsilon)$; that is,

$$\mathbf{B}(x, \epsilon) = \{y \mid \|y - x\| < \epsilon\}.$$

A *cone* is a set $\mathcal{F}$ with the property that for all $x \in \mathcal{F}$ we have

$$x \in \mathcal{F} \Rightarrow \alpha x \in \mathcal{F}, \quad \text{for all } \alpha \geq 0. \tag{A.2}$$

For instance, the set $\mathcal{F} \subset \mathbf{R}^2$ defined by

$$\{(x_1, x_2) \mid x_1 > 0, \ x_2 \geq 0\}$$

is a cone in $\mathbf{R}^2$.

Finally, we define the affine hull and relative interior of a set. Given $\mathcal{F} \subset \mathbf{R}^n$, the affine hull is the smallest affine set containing $\mathcal{F}$, that is,

$$\text{aff}\,\mathcal{F} = \{x \mid x \text{ is a linear combination of vectors in } \mathcal{F}\}. \tag{A.3}$$

For instance, when $\mathcal{F}$ is the "ice-cream cone" defined as below by (A.26), we have aff$\,\mathcal{F} = \mathbf{R}^3$, while if $\mathcal{F}$ is the set of two isolated points $\mathcal{F} = \{(1, 0, 0), (0, 2, 0)\}$, we have

$$\text{aff}\,\mathcal{F} = \{(x_1, x_2, 0) \mid \text{for all } x_1 \text{ and } x_2\}.$$

The *relative interior* ri$\mathcal{F}$ of the set S is its *interior relative to* aff$\,\mathcal{F}$. If $x \in \mathcal{F}$, then $x \in $ ri$\mathcal{F}$ if there is an $\epsilon > 0$ such that

$$(x + \epsilon B) \cap \text{aff}\,\mathcal{F} \subset \mathcal{F}.$$

Referring again to the ice-cream cone (A.26), we have that

$$\text{ri}\mathcal{F} = \left\{x \in \mathbf{R}^3 \ \middle| \ x_3 > 2\sqrt{x_1^2 + x_2^2}\right\}.$$

For the set of two isolated points $\mathcal{F} = \{(1, 0, 0), (0, 2, 0)\}$, we have $\operatorname{ri}\mathcal{F} = \emptyset$. For the set $\mathcal{F}$ defined by

$$\mathcal{F} \stackrel{\text{def}}{=} \{x \in \mathbf{R}^3 \mid x_1 \in [0, 1],\ x_2 \in [0, 1],\ x_3 = 0\},$$

we have that

$$\operatorname{aff}\mathcal{F} = \mathbf{R} \times \mathbf{R} \times \{0\}, \qquad \operatorname{ri}\mathcal{F} = \{x \in \mathbf{R}^3 \mid x_1 \in (0, 1),\ x_2 \in (0, 1),\ x_3 = 0\}.$$

CONTINUITY AND LIMITS

Let f be a function that maps some domain $\mathcal{D} \subset \mathbf{R}^n$ to the space $\mathbf{R}^m$. For some point $x_0 \in \operatorname{cl}\mathcal{D}$, we write

$$\lim_{x \to x_0} f(x) = f_0 \tag{A.4}$$

(spoken "the limit of $f(x)$ as x approaches x_0 is f_0") if for all $\epsilon > 0$, there is a value $\delta > 0$ such that

$$\|x - x_0\| < \delta \text{ and } x \in \mathcal{D} \;\Rightarrow\; \|f(x) - f_0\| < \epsilon.$$

We say that f is *continuous* at x_0 if $x_0 \in \mathcal{D}$ and the expression (A.4) holds with $f_0 = f(x_0)$. We say that f is continuous on its domain $\mathcal{D}$ if f is continuous for all $x_0 \in \mathcal{D}$.

An example is provided by the function

$$f(x) = \begin{cases} -x & \text{if } x \in [-1, 1],\ x \neq 0, \\ 5 & \text{for all other } x \in [-10, 10]. \end{cases} \tag{A.5}$$

This function is defined on the domain $[-10, 10]$ and is continuous at all points of the domain except the points $x = 0$, $x = 1$, and $x = -1$. At $x = 0$, the expression (A.4) holds with $f_0 = 0$, but the function is not continuous at this point because $f_0 \neq f(0) = 5$. At $x = -1$, the limit (A.4) is not defined, because the function values in the neighborhood of this point are close to both 5 and -1, depending on whether x is slightly smaller or slightly larger than -1. Hence, the function is certainly not continuous at this point. The same comments apply to the point $x = 1$.

In the special case of $n = 1$ (that is, the argument of f is a real scalar), we can also define the *one-sided limit*. Given $x_0 \in \operatorname{cl}\mathcal{D}$, We write

$$\lim_{x \downarrow x_0} f(x) = f_0 \tag{A.6}$$

(spoken "the limit of $f(x)$ as x approaches x_0 from above is f_0") if for all $\epsilon > 0$, there is a value $\delta > 0$ such that

$$x_0 < x < x_0 + \delta \text{ and } x \in \mathcal{D} \implies \|f(x) - f_0\| < \epsilon.$$

Similarly, we write

$$\lim_{x \uparrow x_0} f(x) = f_0 \tag{A.7}$$

(spoken "the limit of $f(x)$ as x approaches x_0 from below is f_0") if for all $\epsilon > 0$, there is a value $\delta > 0$ such that

$$x_0 - \delta < x < x_0 \text{ and } x \in \mathcal{D} \implies \|f(x) - f_0\| < \epsilon.$$

For the function defined in (A.5), we have that

$$\lim_{x \downarrow 1} f(x) = 5, \quad \lim_{x \uparrow 1} f(x) = 1.$$

The function f is said to be *Lipschitz continuous* if there is a constant $M > 0$ such that for any two points x_0, x_1 in $\mathcal{D}$, we have

$$\|f(x_1) - f(x_0)\| \leq M \|x_1 - x_0\|. \tag{A.8}$$

The function f is *locally Lipschitz continuous* at a point $x_0 \in \text{int}\mathcal{D}$ if there is some neighborhood $\mathcal{N}$ with $x_0 \in \mathcal{N} \subset \mathcal{D}$ such that the property (A.8) holds for all x_0 and x_1 in $\mathcal{N}$.

DERIVATIVES

Let $\phi : \mathbf{R} \to \mathbf{R}$ be a real-valued function of a real variable (sometimes known as a *univariate* function). The first derivative $\phi'(\alpha)$ is defined by

$$\frac{d\phi}{d\alpha} = \phi'(\alpha) \stackrel{\text{def}}{=} \lim_{\epsilon \to 0} \frac{\phi(\alpha + \epsilon) - \phi(\alpha)}{\epsilon}. \tag{A.9}$$

The second derivative is obtained by substituting ϕ by ϕ' in this same formula; that is,

$$\frac{d^2\phi}{d\alpha^2} = \phi''(\alpha) \stackrel{\text{def}}{=} \lim_{\epsilon \to 0} \frac{\phi'(\alpha + \epsilon) - \phi'(\alpha)}{\epsilon}. \tag{A.10}$$

(We assume implicitly that ϕ is smooth enough that these limits exist.) Suppose now that α in turn depends on another quantity β (we denote this dependence by writing $\alpha = \alpha(\beta)$).

We can use the *chain rule* to calculate the derivative of ϕ with respect to β:

$$\frac{d\phi(\alpha(\beta))}{d\beta} = \frac{d\phi}{d\alpha}\frac{d\alpha}{d\beta} = \phi'(\alpha)\alpha'(\beta). \tag{A.11}$$

Consider now the function $f : \mathbf{R}^n \to \mathbf{R}$, which is a real-valued function of n independent variables. We typically gather the variables into a vector x, which we can write componentwise as $x = (x_1, x_2, \ldots, x_n)$. The n-vector of first derivatives of f—the *gradient*—is defined as

$$\nabla f(x) = \begin{bmatrix} \frac{\partial f}{\partial x_1} \\ \vdots \\ \frac{\partial f}{\partial x_n} \end{bmatrix}. \tag{A.12}$$

(The notation "∇" is used frequently in the optimization literature to denote the first derivative. In cases of ambiguity, a subscript such as "x" or "t" is added to ∇ to indicate the variable with respect to which the differentiation takes place.) Each partial derivative $\partial f/\partial x_i$ measures the sensitivity of the function to just one of the components of x; that is,

$$\partial f/\partial x_i$$
$$\stackrel{\text{def}}{=} \lim_{\epsilon \to 0} \frac{f(x_1, \ldots, x_{i-1}, x_i + \epsilon, x_{i+1}, \ldots, x_n) - f(x_1, \ldots, x_{i-1}, x_i, x_{i+1}, \ldots, x_n)}{\epsilon}$$
$$= \frac{f(x + \epsilon e_i) - f(x)}{\epsilon},$$

where e_i is the vector $(0, 0, \ldots, 0, 1, 0, \ldots, 0)$, where the 1 appears in the ith position. The matrix of second partial derivatives of f is known as the *Hessian*, and is defined as

$$\nabla^2 f(x) = \begin{bmatrix} \frac{\partial^2 f}{\partial x_1^2} & \frac{\partial^2 f}{\partial x_1 \partial x_2} & \cdots & \frac{\partial^2 f}{\partial x_1 \partial x_n} \\ \frac{\partial^2 f}{\partial x_2 \partial x_1} & \frac{\partial^2 f}{\partial x_2^2} & \cdots & \frac{\partial^2 f}{\partial x_2 \partial x_n} \\ \vdots & \vdots & & \vdots \\ \frac{\partial^2 f}{\partial x_n \partial x_1} & \frac{\partial^2 f}{\partial x_n \partial x_2} & \cdots & \frac{\partial^2 f}{\partial x_n^2} \end{bmatrix}.$$

We say that f is *differentiable* if all first partial derivatives of f exist, and *continuously differentiable* if in addition these derivatives are continuous functions of x. Similarly, f is *twice differentiable* if all second partial derivatives of f exist and *twice continuously differentiable* if they are also continuous. Note that when f is twice continuously differentiable, the Hessian is a symmetric matrix, since

$$\frac{\partial^2 f}{\partial x_i \partial x_j} = \frac{\partial^2 f}{\partial x_j \partial x_i}, \quad \text{for all } i, j = 1, 2, \ldots, n.$$

When the vector x in turn depends on another vector t (that is, $x = x(t)$), the chain rule (A.11) for the univariate function can be extended as follows:

$$\nabla_t f(x(t)) = \sum_{i=1}^{n} \frac{\partial f}{\partial x_i} \nabla x_i(t). \tag{A.13}$$

DIRECTIONAL DERIVATIVES

If f is continuously differentiable and $p \in \mathbf{R}^n$, then the *directional derivative* of f in the direction p is given by

$$D(f(x); p) \stackrel{\text{def}}{=} \lim_{\epsilon \to 0} \frac{f(x + \epsilon p) - f(x)}{\epsilon} = \nabla f(x)^T p. \tag{A.14}$$

To verify this formula, we define the function

$$\phi(\alpha) = f(x + \alpha p) = f(y(\alpha)), \tag{A.15}$$

where $y(\alpha) = x + \alpha p$. Note that

$$\lim_{\epsilon \to 0} \frac{f(x + \epsilon p) - f(x)}{\epsilon} = \lim_{\epsilon \to 0} \frac{\phi(\epsilon) - \phi(0)}{\epsilon} = \phi'(0).$$

By applying the chain rule (A.13) to $f(y(\alpha))$, we obtain

$$\phi'(\alpha) = \sum_{i=1}^{n} \frac{\partial f(y(\alpha))}{\partial y_i} \nabla y_i(\alpha) \tag{A.16}$$

$$= \sum_{i=1}^{n} \frac{\partial f(y(\alpha))}{\partial y_i} p_i = \nabla f(y(\alpha))^T p = \nabla f(x + \alpha p)^T p.$$

We obtain (A.14) by setting $\alpha = 0$ and comparing the last two expressions.

The directional derivative is sometimes defined even when the function f itself is not differentiable. For instance, if $f(x) = \|x\|_1$, we have from the definition (A.14) that

$$D(\|x\|_1; p) = \lim_{\epsilon \to 0} \frac{\|x + \epsilon p\|_1 - \|x\|_1}{\epsilon} = \lim_{\epsilon \to 0} \frac{\sum_{i=1}^{n} |x_i + \epsilon p_i| - \sum_{i=1}^{n} |x_i|}{\epsilon}.$$

If $x_i > 0$, we have $|x_i + \epsilon p_i| = |x_i| + \epsilon p_i$ for all ϵ sufficiently small. If $x_i < 0$, we have $|x_i + \epsilon p_i| = |x_i| - \epsilon p_i$, while if $x_i = 0$, we have $|x_i + \epsilon p_i| = \epsilon |p_i|$. Therefore, we have

$$D(\|x\|_1; p) = \sum_{i | x_i < 0} -p_i + \sum_{i | x_i > 0} p_i + \sum_{i | x_i = 0} |p_i|,$$

so the directional derivative of this function exists for any x and p. Its first derivative does *not* exist, however, whenever any of the components of x are zero.

❑ EXAMPLE A.1

Let $f : \mathbf{R}^2 \to \mathbf{R}$ be defined by $f(x_1, x_2) = x_1^2 + x_1 x_2$, where $x_1 = \sin t_1 + t_2^2$ and $x_2 = (t_1 + t_2)^2$. The chain rule (A.13) yields

$$\nabla_t f(x(t))$$

$$= \sum_{i=1}^{n} \frac{\partial f}{\partial x_i} \nabla x_i(t)$$

$$= (2x_1 + x_2) \begin{bmatrix} \cos t_1 \\ 2t_2 \end{bmatrix} + x_1 \begin{bmatrix} 2(t_1 + t_2) \\ 2(t_1 + t_2) \end{bmatrix}$$

$$= \left(2 \left(\sin t_1 + t_2^2\right) + (t_1 + t_2)^2\right) \begin{bmatrix} \cos t_1 \\ 2t_2 \end{bmatrix} + \left(\sin t_1 + t_2^2\right) \begin{bmatrix} 2(t_1 + t_2) \\ 2(t_1 + t_2) \end{bmatrix}.$$

If, on the other hand, we substitute directly for x into the definition of f, we obtain

$$f(x(t)) = \left(\sin t_1 + t_2^2\right)^2 + \left(\sin t_1 + t_2^2\right)(t_1 + t_2)^2.$$

The reader should verify that the gradient of this expression is identical to the one obtained above by applying the chain rule.

❑

MEAN VALUE THEOREM

We now recall the mean value theorem for univariate functions. Given a continuously differentiable function $\phi : \mathbf{R} \to \mathbf{R}$ and two real numbers α_0 and α_1 that satisfy $\alpha_1 > \alpha_0$, we have that

$$\phi(\alpha_1) = \phi(\alpha_0) + \phi'(\xi)(\alpha_1 - \alpha_0) \tag{A.17}$$

for some $\xi \in (\alpha_0, \alpha_1)$. An extension of this result to a multivariate function $f : \mathbf{R}^n \to \mathbf{R}$ is that for any vector p we have

$$f(x+p) = f(x) + \nabla f(x + \alpha p)^T p, \tag{A.18}$$

for some $\alpha \in (0, 1)$. (This result can be proved by defining $\phi(\alpha) = f(x + \alpha p)$, $\alpha_0 = 0$, and $\alpha_1 = 1$ and applying the chain rule, as above.)

❏ EXAMPLE A.2

Consider $f : \mathbf{R}^2 \to \mathbf{R}$ defined by $f(x) = x_1^3 + 3x_1 x_2^2$, and let $x = (0, 0)^T$ and $p = (1, 2)^T$. It is easy to verify that $f(x) = 0$ and $f(x+p) = 13$. Since

$$\nabla f(x + \alpha p) = \begin{bmatrix} 3(x_1 + \alpha p_1)^2 + 3(x_2 + \alpha p_2)^2 \\ 6(x_1 + \alpha p_1)(x_2 + \alpha p_2) \end{bmatrix} = \begin{bmatrix} 15\alpha^2 \\ 12\alpha^2 \end{bmatrix},$$

we have that $\nabla f(x + \alpha p)^T p = 39\alpha^2$. Hence the relation (A.18) holds when we set $\alpha = 1/\sqrt{13}$, which lies in the open interval $(0, 1)$, as claimed.

❏

An alternative expression to (A.18) can be stated for twice differentiable functions: We have

$$f(x+p) = f(x) + \nabla f(x)^T p + \frac{1}{2} p^T \nabla^2 f(x + \alpha p)^T p, \tag{A.19}$$

for some $\alpha \in (0, 1)$. In fact, this expression is one form of Taylor's theorem, Theorem 2.1 in Chapter 2, to which we refer throughout the book.

IMPLICIT FUNCTION THEOREM

The implicit function theorem lies behind a number of important results in local convergence theory of optimization algorithms and in the characterization of optimality (see Chapter 12). We present a brief outline here based on the discussion in Lang [147, p. 131].

Theorem A.1 (Implicit Function Theorem).
Let $h : \mathbf{R}^n \times \mathbf{R}^m \to \mathbf{R}^n$ be a function such that

(i) $h(z^*, 0) = 0$ for some $z^* \in \mathbf{R}^n$,

(ii) the function $h(\cdot, \cdot)$ is Lipschitz continuously differentiable in some neighborhood of $(z^*, 0)$, and

(iii) $\nabla_z h(z, t)$ is nonsingular at the point $(z, t) = (z^*, 0)$.

Then the function $z : \mathbf{R}^m \to \mathbf{R}^n$ defined implicitly by $h(z(t), t) = 0$ is well-defined and Lipschitz continuous for $t \in \mathbf{R}^m$ in some neighborhood of the origin.

This theorem is frequently applied to parametrized systems of linear equations, in which z is obtained as the solution of

$$M(t)z = g(t),$$

where $M(\cdot) \in \mathbf{R}^{n \times n}$ has $M(0)$ nonsingular, and $g(\cdot) \in \mathbf{R}^n$. To apply the theorem, we define

$$h(z, t) = M(t)z - g(t).$$

If $M(\cdot)$ and $g(\cdot)$ are Lipschitz continuously differentiable in some neighborhood of 0, the theorem implies that $z(t) = M(t)^{-1} g(t)$ is a Lipschitz continuous function of t for all t in some neighborhood of 0.

GEOMETRY OF FEASIBLE SETS

In describing the theory of constrained optimization, we assume for the most part that the set of feasible points is described by algebraic equations. That is, we aim to minimize a function f over the set Ω defined by

$$\Omega = \{x \in \mathbf{R}^n \mid c_i(x) = 0, \ i \in \mathcal{E}; \ c_i(x) \geq 0, \ i \in \mathcal{I}\}, \tag{A.20}$$

where $\mathcal{E}$ and $\mathcal{I}$ are the index sets of equality and inequality constraints, respectively (see also (12.2)). In some situations, however, it is useful to abandon the algebraic description and merely consider the set Ω on its own merits, as a general subset of Ω. By doing this, we no longer tie Ω down to any particular algebraic description. (Note that for any given set Ω, there may be infinitely many collections of constraint functions $c_i, i \in \mathcal{E} \cup \mathcal{I}$, such that the identification (A.20) holds.)

There are advantages and disadvantages to taking the purely geometric viewpoint. By not tying ourselves to a particular algebraic description, we avoid the theoretical complications caused by redundant, linearly dependent, or insufficiently smooth constraints. We also avoid some practical problems such as poor scaling. On the other hand, most of the development of theory, algorithms, and software for constrained optimization assumes that an algebraic description of Ω is available. If we choose instead to work explicitly with Ω, we must reformulate the optimality conditions and algorithms in terms of this set, without reference to its algebraic description. Many such tools are available, and they have been applied successfully in a number of applications, including optimal control. (See, for example, Clarke [42], Dunn [77].) We describe a few of the relevant concepts here.

Given a constrained optimization problem formulated as

$$\min\ f(x)\ \text{subject to}\ x \in \Omega, \tag{A.21}$$

where Ω is a closed subset of $\mathbf{R}^n$, most of the theory revolves around *tangent cones* and *normal cones* to the set Ω at various feasible points $x \in \Omega$. A definition of a tangent vector is given by Clarke [42, Theorem 2.4.5].

Definition A.1 (Tangent).

A vector $w \in \mathbf{R}^n$ is tangent to Ω at $x \in \Omega$ if for all vector sequences $\{x_i\}$ with $x_i \to x$ and $x_i \in \Omega$, and all positive scalar sequences $t_i \downarrow 0$, there is a sequence $w_i \to w$ such that $x_i + t_i w_i \in \Omega$ for all i.

The *tangent cone* $T_\Omega(x)$ is the collection of all tangent vectors to Ω at x. The *normal cone* $N_\Omega(x)$ is simply the orthogonal complement to the tangent cone; that is,

$$N_\Omega(x) = \{v \mid v^T w \le 0 \text{ for all } w \in T_\Omega(x)\}. \tag{A.22}$$

Note that the zero vector belongs to both T_Ω and N_Ω.

It is not difficult to verify that both T_Ω and N_Ω are indeed cones according to the definition (A.2). To prove this for T_Ω, we take $w \in T_\Omega(x)$ and show that $\alpha w \in T_\Omega(x)$ for some given $\alpha \ge 0$. Let the sequence $\{x_i\}$ and $\{t_i\}$ be given, as in Definition A.1, and define the sequence $\{\bar{t}_i\}$ by $\bar{t}_i = t_i/\alpha$. Since $\{\bar{t}_i\}$ is also a valid sequence of scalars according to Definition A.1, there is a sequence of vectors $\{\bar{w}_i\}$ such that $x_i + \bar{t}_i \bar{w}_i \in \Omega$ and $\bar{w}_i \to w$. By defining $w_i = \alpha \bar{w}_i$, we have that $x_i + t_i w_i = x_i + \bar{t}_i \bar{w}_i \in \Omega$. Hence we have identified a sequence $\{w_i\}$ satisfying the conditions of Definition A.1 with the property that $w_i \to \alpha w$, so we conclude that $\alpha w \in T_\Omega(x)$, as required.

As an example, consider the set defined by a single equality constraint:

$$\Omega = \{x \in \mathbf{R}^n \mid h(x) = 0\}, \tag{A.23}$$

where $h : \mathbf{R}^n \to \mathbf{R}$ is smooth, and let x be a particular point for which $\nabla h(x) \ne 0$. If $w \in T_\Omega(x)$ for some $x \in \Omega$, we have from Definition A.1 that there is a sequence $w_i \to w$ such that $x + t_i w_i \in \Omega$ for any sequence $t_i \downarrow 0$. (We have made the legal choice $x_i \equiv x$ in the definition.) Therefore, we have

$$h(x + t_i w_i)\ \text{for all}\ i, \qquad h(x) = 0.$$

By using this fact together with a Taylor series expansion and smoothness of h, we find that

$$0 = \frac{1}{t_i}[h(x + t_i w_i) - h(x)] = w_i^T \nabla h(x) + o(t_i)/t_i \to w^T \nabla h(x).$$

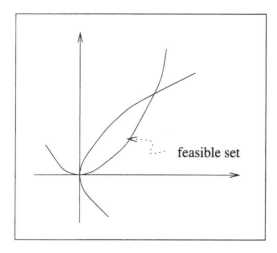

Figure A.1
Feasible set defined by $x_1 \geq x_2^2$, $x_2 \geq x_1^2$.

An argument based on the implicit function theorem can be used to prove the converse, which is that $w^T \nabla h(x) = 0 \Rightarrow w \in T_\Omega(x)$. Hence we have

$$T_\Omega(x) = \{w \mid w^T \nabla h(x) = 0\} = \text{Null}\left(\nabla h(x)^T\right). \tag{A.24}$$

It follows from the definition (A.22) that the normal cone is

$$N_\Omega(x) = \{\alpha \nabla h(x) \mid \alpha \in \mathbf{R}\} = \text{Range}(\nabla h(x)). \tag{A.25}$$

For the next example, consider a set Ω defined by two parabolic constraints:

$$\Omega = \{x \in \mathbf{R}^2 \mid x_1 \geq x_2^2,\ x_2 \geq x_1^2\}$$

(see Figure A.1). The tangent and normal cones at the most interesting point—the origin— are given by

$$T_\Omega(0, 0) = \{w \in \mathbf{R}^2 \mid w_1 \geq 0,\ w_2 \geq 0\},$$
$$N_\Omega(0, 0) = \{v \in \mathbf{R}^2 \mid v_1 \leq 0,\ v_2 \leq 0\}.$$

To show that, for instance, the tangent vector $(0, 1)$ fits into Definition A.1, suppose we are given the feasible sequence $x_i = (1/i^2, 1/i) \to (0, 0)$ and the positive scalar sequence $t_i = 1/i \downarrow 0$. If we define $w_i = \left(\frac{1}{3i}, 1\right)$, it is easy to check that $x_i + t_i w_i$ is feasible for each i and that $w_i \to (0, 1)$.

A three-dimensional example is given by the "ice-cream cone" defined by

$$\Omega = \left\{ x \in \mathbf{R}^3 \mid x_3 \geq 2\sqrt{x_1^2 + x_2^2} \right\} \tag{A.26}$$

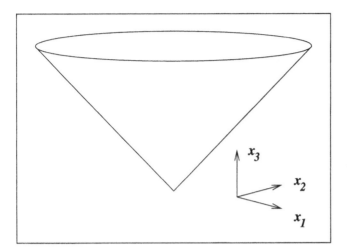

Figure A.2 "Ice-cream cone" set.

(see Figure A.2). At the origin $x = 0$, the tangent cone coincides with the feasible set: $T_\Omega(0) = \Omega$. The normal cone is defined as

$$N_\Omega(0) = \left\{ v \mid v_3 \leq -\frac{1}{2}\sqrt{v_1^2 + v_2^2} \right\}.$$

To verify that $v^T w \leq 0$ for all $v \in N_\Omega(0)$, $w \in T_\Omega(0)$, we have

$$\begin{aligned}
v^T w &= v_1 w_1 + v_2 w_2 + v_3 w_3 \\
&\leq v_1 w_1 + v_2 w_2 - \frac{1}{2}(v_1^2 + v_2^2)^{1/2}(2)(w_1^2 + w_2^2)^{1/2} \\
&= \sqrt{(v_1 w_1 + v_2 w_2)^2} - \sqrt{(v_1^2 + v_2^2)(w_1^2 + w_2^2)} \\
&= \sqrt{v_1^2 w_1^2 + v_2^2 w_2^2 + 2 v_1 v_2 w_1 w_2} - \sqrt{v_1^2 w_1^2 + v_2^2 w_2^2 + v_2^2 w_1^2 + v_1^2 w_2^2}.
\end{aligned}$$

Since

$$v_1^2 w_2^2 + v_2^2 w_1^2 \geq 2 v_1 v_2 w_2 w_2,$$

the second term outweighs the first, so we have $v^T w \leq 0$, as required.

A final example is given by the two-dimensional subset

$$\Omega = \left\{ x \in \mathbf{R}^2 \mid x_2 \geq 0,\ x_2 \leq x_1^3 \right\}, \tag{A.27}$$

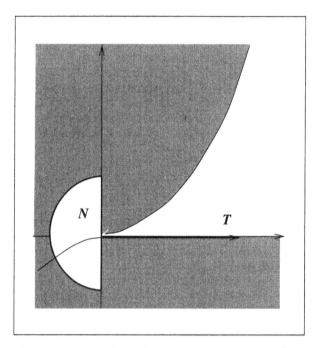

Figure A.3 Feasible set defined by $x_2 \geq 0$, $x_2 \leq x_1^3$, showing tangent and normal cones.

illustrated in Figure A.3. The tangent cone at the origin is

$$T_\Omega(0) = \{(w_1, 0) \mid w_1 \geq 0\}, \tag{A.28}$$

that is, all positive multiples of the vector $(1, 0)$. The normal cone is

$$N_\Omega(0) = \{v \mid v_1 \leq 0\}. \tag{A.29}$$

Finally, we recall the definitions of affine hull aff$\mathcal{F}$ and relative interior ri$\mathcal{F}$ from earlier in the chapter. For the set Ω defined by (A.27), we have that

$$\text{ri}\,\Omega = \{x \mid x_2 > 0, \ x_2 < x_1^3\},$$

$$\text{ri}\,T_\Omega(0) = \{(w_1, 0) \mid w_1 > 0\},$$

$$\text{ri}\,N_\Omega(0) = \{(v_1, v_2) \mid v_1 < 0\}.$$

ORDER NOTATION

In much of our analysis we are concerned with how the members of a sequence behave *eventually*, that is, when we get far enough along in the sequence. For instance, we might ask whether the elements of the sequence are bounded, or whether they are similar in size to the elements of a corresponding sequence, or whether they are decreasing and, if so, how rapidly. Order notation is useful shorthand to use when questions like these are being examined. It saves us defining many constants that clutter up the argument and the analysis.

We will use three varieties of order notation: $O(\cdot)$, $o(\cdot)$, and $\Omega(\cdot)$. Given two nonnegative infinite sequences of scalars $\{\eta_k\}$ and $\{\nu_k\}$, we write

$$\eta_k = O(\nu_k)$$

if there is a positive constant C such that

$$|\eta_k| \leq C|\nu_k|$$

for all k sufficiently large. We write

$$\eta_k = o(\nu_k)$$

if the sequence of ratios $\{\eta_k/\nu_k\}$ approaches zero, that is,

$$\lim_{k \to \infty} \frac{\eta_k}{\nu_k} = 0.$$

Finally, we write

$$\eta_k = \Omega(\nu_k)$$

if there are two constants C_0 and C_1 with $0 < C_0 \leq C_1 < \infty$ such that

$$C_0|\nu_k| \leq |\eta_k| \leq C_1|\nu_k|,$$

that is, the corresponding elements of both sequences stay in the same ballpark for all k. This definition is equivalent to saying that $\eta_k = O(\nu_k)$ and $\nu_k = O(\eta_k)$.

The same notation is often used in the context of quantities that depend continuously on each other as well. For instance, if $\eta(\cdot)$ is a function that maps $\mathbf{R}$ to $\mathbf{R}$, we write

$$\eta(\nu) = O(\nu)$$

if there is a constant C such that $|\eta(\nu)| \leq C|\nu|$ for all $\nu \in \mathbf{R}$. (Typically, we are interested only in values of ν that are either very large or very close to zero; this should be clear from the context. Similarly, we use

$$\eta(\nu) = o(\nu) \tag{A.30}$$

to indicate that the ratio $\eta(\nu)/\nu$ approaches zero either as $\nu \to 0$ or $\nu \to \infty$. (Again, the precise meaning should be clear from the context.)

As a slight variant on the definitions above, we write

$$\eta_k = O(1)$$

to indicate that there is a constant C such that $|\eta_k| \leq C$ for all k, while

$$\eta_k = o(1)$$

indicates that $\lim_{k \to \infty} \eta_k = 0$. We sometimes use vector and matrix quantities as arguments, and in these cases the definitions above are intended to apply to the norms of these quantities. For instance, if $f : \mathbf{R}^n \to \mathbf{R}^n$, we write $f(x) = O(\|x\|)$ if there is a constant $C > 0$ such that $\|f(x)\| \leq C\|x\|$ for all x in the domain of f. Typically, as above, we are interested only in some subdomain of f, usually a small neighborhood of 0. As before, the precise meaning should be clear from the context.

ROOT-FINDING FOR SCALAR EQUATIONS

In Chapter 11 we discussed methods for finding solutions of nonlinear systems of equations $F(x) = 0$, where $F : \mathbf{R}^n \to \mathbf{R}^n$. Here we discuss briefly the case of scalar equations ($n = 1$), for which the algorithm is easy to illustrate. Scalar root-finding is needed in the trust-region algorithms of Chapter 4, for instance. Of course, the general theorems of Chapter 11 can be applied to derive rigorous convergence results for this special case.

The basic step of Newton's method (Algorithm Newton of Chapter 11) in the scalar case is simply

$$p_k = -F(x_k)/F'(x_k), \quad x_{k+1} \leftarrow x_k + p_k \tag{A.31}$$

(cf. (11.9)). Graphically, such a step involves taking the tangent to the graph of F at the point x_k and taking the next iterate to be the intersection of this tangent with the x axis (see Figure A.4). Clearly, if the function F is nearly linear, the tangent will be quite a good approximation to F itself, so the Newton iterate will be quite close to the true root of F.

The secant method for scalar equations can be viewed as the specialization of Broyden's method to the case of $n = 1$. The issues are simpler in this case, however, since the secant equation (11.26) completely determines the value of the 1×1 approximate Hessian B_k. That is, we do not need to apply extra conditions to ensure that B_k is fully determined. By

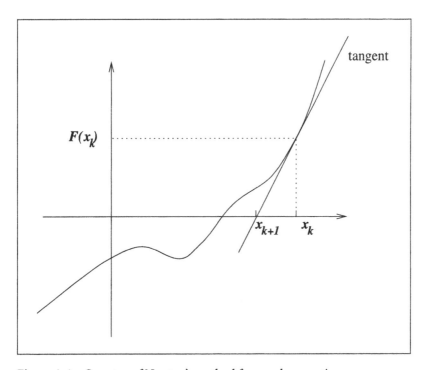

Figure A.4 One step of Newton's method for a scalar equation.

combining (11.24) with (11.26), we find that the secant method for the case of $n = 1$ is defined by

$$B_k = (F(x_k) - F(x_{k-1}))/(x_k - x_{k-1}), \qquad (A.32a)$$
$$p_k = -F(x_k)/B_k, \qquad x_{k+1} = x_k + p_k. \qquad (A.32b)$$

By illustrating this algorithm, we see the origin of the term "secant." B_k approximates the slope of the function at x_k by taking the secant through the points $(x_{k-1}, F(x_{k-1}))$ and $(x_k, F(x_k))$, and x_{k+1} is obtained by finding the intersection of this secant with the x axis. The method is illustrated in Figure A.5.

A.2 ELEMENTS OF LINEAR ALGEBRA

VECTORS AND MATRICES

In this book we work exclusively with vectors and matrices whose components are real numbers. Vectors are usually denoted by lowercase roman characters, and matrices by

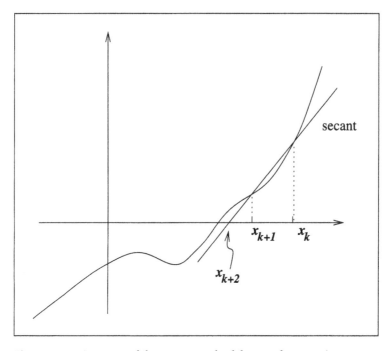

Figure A.5 One step of the secant method for a scalar equation.

uppercase roman characters. The space of real vectors of length n is denoted by $\mathbf{R}^n$, while the space of real $m \times n$ matrices is denoted by $\mathbf{R}^{m \times n}$.

We say that a matrix $A \in \mathbf{R}^{n \times n}$ is *symmetric* if $A = A^T$. A symmetric matrix A is *positive definite* if there is a positive constant α such that

$$x^T A x \geq \alpha \|x\|^2, \qquad \text{for all } x \in \mathbf{R}^n. \tag{A.33}$$

It is *positive semidefinite* if the relationship (A.33) holds with $\alpha = 0$, that is, $x^T A x \geq 0$ for all $x \in \mathbf{R}^n$.

NORMS

For a vector $x \in \mathbf{R}^n$, we define the following norms:

$$\|x\|_1 \stackrel{\text{def}}{=} \sum_{i=1}^n |x_i|, \tag{A.34a}$$

$$\|x\|_2 \stackrel{\text{def}}{=} \left(\sum_{i=1}^n x_i^2 \right)^{1/2} = (x^T x)^{1/2}, \tag{A.34b}$$

$$\|x\|_\infty \stackrel{\text{def}}{=} \max_{i=1,\ldots,n} |x_i|. \tag{A.34c}$$

The norm $\|\cdot\|_2$ is often called the *Euclidean norm*. All these norms measure the length of the vector in some sense, and they are equivalent in the sense that each one is bounded above and below by a multiple of the other. To be precise, we have

$$\|x\|_\infty \le \|x\|_2 \le \sqrt{n}\|x\|_\infty, \quad \|x\|_\infty \le \|x\|_1 \le n\|x\|_\infty, \tag{A.35}$$

and so on. In general, a norm is any mapping $\|\cdot\|$ from $\mathbf{R}^n$ to the nonnegative real numbers that satisfies the following properties:

$$\|x + z\| \le \|x\| + \|z\|, \quad \text{for all } x, z \in \mathbf{R}^n; \tag{A.36a}$$
$$\|x\| = 0 \Rightarrow x = 0; \tag{A.36b}$$
$$\|\alpha x\| = |\alpha|\|x\|, \quad \text{for all } \alpha \in \mathbf{R} \text{ and } x \in \mathbf{R}^n. \tag{A.36c}$$

Equality holds in (A.36a) if and only if one of the vectors x and z is a nonnegative scalar multiple of the other.

Another interesting property that holds for the Euclidean norm $\|\cdot\| = \|\cdot\|_2$ is the Hölder inequality, which states that

$$|x^T z| \le \|x\|\,\|z\|, \tag{A.37}$$

with equality if and only if one of these vectors is a nonnegative multiple of the other. We can prove this result as follows:

$$0 \le \|\alpha x + z\|^2 = \alpha^2 \|x\|^2 + 2\alpha x^T z + \|z\|^2.$$

The right-hand-side is a convex function of α, and it satisfies the required nonnegativity property only if there exist fewer than 2 distinct real roots, that is,

$$(2x^T z)^2 \le 4\|x\|^2 \|z\|^2,$$

proving (A.37). Equality occurs when the quadratic α has exactly one real root (that is, $|x^T z| = \|x\|\,\|z\|$) and when $\alpha x + z = 0$ for some α, as claimed.

We can derive definitions for certain matrix norms from these vector norm definitions. If we let $\|\cdot\|$ be generic notation for the three norms listed in (A.34), we define the corresponding matrix norm as

$$\|A\| \stackrel{\text{def}}{=} \sup_{x \ne 0} \frac{\|Ax\|}{\|x\|}. \tag{A.38}$$

The matrix norms defined in this way are said to be *consistent* with the vector norms (A.34). Explicit formulae for these norms are as follows:

$$\|A\|_1 = \max_{j=1,\ldots,n} \sum_{i=1}^{m} |A_{ij}|, \tag{A.39a}$$

$$\|A\|_2 = \lambda_1(A^T A)^{1/2}, \quad \text{where } \lambda_1(\cdot) \text{ denotes the largest eigenvalue,} \tag{A.39b}$$

$$\|A\|_\infty = \max_{i=1,\ldots,m} \sum_{j=1}^{n} |A_{ij}|. \tag{A.39c}$$

The Frobenius norm $\|A\|_F$ of the matrix A is defined by

$$\|A\|_F = \left(\sum_{i=1}^{m} \sum_{j=1}^{n} a_{ij}^2 \right)^{1/2}. \tag{A.40}$$

This norm is useful for many purposes, but it is not consistent with any vector norm. Once again, these various matrix norms are equivalent with each other in a sense similar to (A.35).
    For the Euclidean norm $\|\cdot\| = \|\cdot\|_2$, the following property holds:

$$\|AB\| \le \|A\| \, \|B\|, \tag{A.41}$$

for all matrices A and B with consistent dimensions.
    The *condition number* of a nonsingular matrix is defined as

$$\kappa(A) = \|A\| \, \|A^{-1}\|, \tag{A.42}$$

where any matrix norm can be used in the definition. We distinguish the different norms by the use of a subscript: $\kappa_1(\cdot)$, $\kappa_2(\cdot)$, and $\kappa_\infty(\cdot)$, respectively. (Of course, different norm definitions give different measures of condition number, in general.)
    Norms also have a meaning for scalar, vector, and matrix-valued functions that are defined on a particular domain. In these cases, we can define Hilbert spaces of functions for which the inner product and norm are defined in terms of an integral over the domain. We omit details, since all the development of this book takes place in the space $\mathbf{R}^n$, though many of the algorithms can be extended to more general Hilbert spaces. However, we mention for purposes of Newton analysis that the following inequality holds for functions of the type that we consider in this book:

$$\left\| \int_a^b F(t) \right\| \le \int_a^b \|F(t)\| \, dt,$$

where F is a scalar-, vector-, or matrix-valued function on the interval $[a, b]$.

SUBSPACES

Given the Euclidean space $\mathbf{R}^n$, the subset $S \subset \mathbf{R}^n$ is a *subspace of* $\mathbf{R}^n$ if the following property holds: If x and y are any two elements of S, then

$$\alpha x + \beta y \in S, \quad \text{for all } \alpha, \beta \in \mathbf{R}.$$

For instance, S is a subspace of $\mathbf{R}^2$ if it consists of (i) the whole space $\mathbf{R}^n$; (ii) any line passing through the origin; (iii) the origin alone; or (iv) the empty set.

Given any set of vectors $a_i \in \mathbf{R}^n$, $i = 1, 2, \ldots, m$, the set

$$S = \{w \in \mathbf{R}^n \mid a_i^T w = 0, \ i = 1, 2, \ldots, m\} \tag{A.43}$$

is a subspace. However, the set

$$\{w \in \mathbf{R}^n \mid a_i^T w \geq 0, \ i = 1, 2, \ldots, m\} \tag{A.44}$$

is not in general a subspace. For example, if we have $n = 2$, $m = 1$, and $a_1 = (1, 0)^T$, this set would consist of all vectors $(w_1, w_2)^T$ with $w_1 \geq 0$, but then given two vectors $x = (1, 0)^T$ and $y = (2, 3)$ in this set, it is easy to choose multiples α and β such that $\alpha x + \beta y$ has a negative first component, and so lies outside the set.

Sets of the forms (A.43) and (A.44) arise in the discussion of second-order optimality conditions for constrained optimization.

For any subspace S of the Euclidean space $\mathbf{R}^n$, the set of vectors $s_1, s_2, \ldots, s_m$ in S is called a *linearly independent set* if there are no real numbers $\alpha_1, \alpha_2, \ldots, \alpha_m$ such that

$$\alpha_1 s_2 + \alpha_2 s_2 + \cdots + \alpha_m s_m,$$

unless we make the trivial choice $\alpha_1 = \alpha_2 = \cdots = \alpha_m = 0$. Another way to define linear independence is to say that none of the vectors $s_1, s_2, \ldots, s_m$ can be written as a linear combination of the other vectors in this set. We say that this set of vectors is a *spanning set* for S if *any* vector $s \in S$ can be written as

$$s = \alpha_1 s_2 + \alpha_2 s_2 + \cdots + \alpha_m s_m,$$

for some particular choice of the coefficients $\alpha_1, \alpha_2, \ldots, \alpha_m$.

If the vectors $s_1, s_2, \ldots, s_m$ are both linearly independent and a spanning set for S, we call them a *basis*. In this case, m (the number of elements in the basis) is referred to as the *dimension* of S. Notationally, we write $\dim(S)$ to denote the dimension of S. Note that there are many ways to choose a basis of S in general, but that all bases do, in fact, contain the same number of vectors.

If A is any real matrix, the *null space* is the subspace

$$\text{Null}(A) = \{w \mid Aw = 0\},$$

while the *range space* is

$$\text{Range}(A) = \{w \mid w = Av \text{ for some vector } v\}.$$

The *fundamental theorem of linear algebra* states that

$$\text{Null}(A) \oplus \text{Range}(A^T) = \mathbf{R}^n,$$

where n is the number of columns in A.

EIGENVALUES, EIGENVECTORS, AND THE SINGULAR-VALUE DECOMPOSITION

A scalar value λ is an *eigenvalue* of the $n \times n$ matrix A if there is a nonzero vector x such that

$$Aq = \lambda q.$$

The vector q is called an *eigenvector* of A. The matrix A is nonsingular if none of its eigenvalues are zero. The eigenvalues of symmetric matrices are all real numbers, while non symmetric matrices may have imaginary eigenvalues. If the matrix is positive definite as well as symmetric, its eigenvalues are all positive real numbers.

All matrices $A \in \mathbf{R}^{m \times n}$ can be decomposed as a product of three matrices with special properties, as follows:

$$A = USV^T. \tag{A.45}$$

Here, U and V are orthogonal matrices of dimension $m \times m$ and $n \times n$, respectively, which means that they satisfy the relations $U^T U = UU^T = I$ and $V^T V = VV^T = I$. The matrix S is a diagonal matrix of dimension $m \times n$, wih diagonal elements $\sigma_i, i = 1, 2, \ldots, \min(m, n)$, that satisfy

$$\sigma_1 \geq \sigma_2 \geq \cdots \geq \sigma_{\min(m,n)} \geq 0.$$

These diagonal values are called the singular values of A, and (A.45) is called the singular-value decomposition.

When A is symmetric, its n real eigenvalues $\lambda_1, \lambda_2, \ldots, \lambda_n$ and their associated eigenvectors $q_1, q_2, \ldots, q_n$ can be used to write a *spectral decomposition* of A as follows:

$$A = \sum_{i=1}^{n} \lambda_i q_i q_i^T.$$

This decomposition can be restated in matrix form by defining

$$\Lambda = \text{diag}(\lambda_1, \lambda_2, \cdots, \lambda_n), \qquad Q = [q_1 | q_2 | \ldots | q_n],$$

and writing

$$A = Q \Lambda Q^T. \tag{A.46}$$

In fact, when A is positive definite as well as symmetric, this decomposition is identical to the singular-value decomposition (A.45), where we define $U = V = Q$ and $S = \Lambda$. Note that the singular values σ_i, $i = 1, 2, \ldots, m$, and the eigenvalues λ_i, $i = 1, 2, \ldots, m$, coincide in this case.

In the case of the Euclidean norm (A.39b), we have for symmetric positive definite matrices A that the singular values and eigenvalues of A coincide, and that

$$\|A\| = \sigma_1(A) = \text{largest eigenvalue of } A,$$
$$\|A^{-1}\| = \sigma_n(A)^{-1} = \text{inverse of smallest eigenvalue of } A.$$

Hence, we have for all $x \in \mathbb{R}^n$ that

$$\sigma_n(A)\|x\|^2 = \|x\|^2 / \|A^{-1}\| \leq x^T A x \leq \|A\| \|x\|^2 = \sigma_1(A) \|x\|^2.$$

For an orthogonal matrix Q, we have for the Euclidean norm that

$$\|Qx\| = \|x\|,$$

and that all the singular values of this matrix are equal to 1.

DETERMINANT AND TRACE

The trace of an $n \times n$ matrix A is defined by

$$\text{trace}(A) = \sum_{i=1}^{n} A_{ii}. \tag{A.47}$$

If the eigenvalues of A are denoted by $\lambda_1, \lambda_2, \ldots, \lambda_n$, it can be shown that

$$\text{trace}(A) = \sum_{i=1}^{n} \lambda_i, \tag{A.48}$$

that is, the trace of the matrix is the sum of its eigenvalues.

The determinant of an $n \times n$ matrix A is the product of its eigenvalues; that is,

$$\det A = \prod_{i=1}^{n} \lambda_i. \tag{A.49}$$

The determinant has several appealing (and revealing) properties. For instance,

$\det A = 0$ if and only if A is singular;

$\det AB = (\det A)(\det B)$;

$\det A^{-1} = 1/\det A$.

Recall that any orthogonal matrix A has the property that $QQ^T = Q^TQ = I$, so that $Q^{-1} = Q^T$. It follows from the property of the determinant that $\det Q = \det Q^T = \pm 1$. The properties above are used in the analysis of Chapters 6 and 8.

MATRIX FACTORIZATIONS: CHOLESKY, LU, QR

Matrix factorizations are important both in the design of algorithms and in their analysis. One such factorization is the singular-value decomposition defined above in (A.45). Here we define the other important factorizations.

All the factorization algorithms described below make use of *permutation matrices*. Suppose that we wish to exchange the first and fourth rows of a matrix A. We can perform this operation by premultiplying A by a permutation matrix P, which is constructed by interchanging the first and fourth rows of an identity matrix that contains the same number of rows as A. Suppose, for example, that A is a 5×5 matrix. The appropriate choice of P would be

$$P = \begin{bmatrix} 0 & 0 & 0 & 1 & 0 \\ 0 & 1 & 0 & 0 & 0 \\ 0 & 0 & 1 & 0 & 0 \\ 1 & 0 & 0 & 0 & 0 \\ 0 & 0 & 0 & 0 & 1 \end{bmatrix}.$$

A similar technique is used to to find a permutation matrix P that exchanges columns of a matrix.

The LU factorization of a matrix $A \in \mathbf{R}^{n\times n}$ is defined as

$$PA = LU, \tag{A.50}$$

where

P is an $n \times n$ permutation matrix (that is, it is obtained by rearranging the rows of the $n \times n$ identity matrix),

L is unit lower triangular (that is, lower triangular with diagonal elements equal to 1, and

U is upper triangular.

This factorization can be used to solve a linear system of the form $Ax = b$ efficiently by the following three-step process:

form $\tilde{b} = Pb$ by permuting the elements of b;

solve $Lz = \tilde{b}$ by performing triangular forward-substitution, to obtain the vector z;

solve $Ux = z$ by performing triangular back-substitution, to obtain the solution vector x.

The factorization (A.50) can be found by using Gaussian elimination with row partial pivoting, an algorithm that requires approximately $2n^3/3$ floating-point operations when A is dense. Standard software that implements this algorithm (notably, LAPACK [4]) is readily available. The method can be stated as follows:

Algorithm A.1 (Gaussian Elimination with Row Partial Pivoting).
   Given $A \in \mathbf{R}^{n\times n}$;
   Set $P \leftarrow I$, $L \leftarrow 0$;
   for $i = 1, 2, \ldots, n$
       find the index $j \in \{i, i+1, \ldots, n\}$ such that $|A_{ji}| = \max_{k=i,i+1,\ldots,n} |A_{ki}|$;
       if $A_{ij} = 0$
           stop; (* matrix A is singular *)
       if $i \neq j$
           swap rows i and j of matrices A and L;
       (* elimination step*)
       $L_{ii} \leftarrow 1$;
       for $k = i+1, i+2, \ldots, n$
           $L_{ki} \leftarrow A_{ki}/A_{ii}$;
           for $l = i+1, i+2, \ldots, n$
               $A_{kl} \leftarrow A_{kl} - L_{ki} A_{il}$;
           end (for)
       end (if)

end (for)
$U \leftarrow$ upper triangular part of A.

Variants of the basic algorithm allow for rearrangement of the columns as well as the rows during the factorization, but these do not add to the practical stability properties of the algorithm. Column pivoting may, however, improve the performance of Gaussian elimination when the matrix A is sparse, by ensuring that the factors L and U are also reasonably sparse.

Gaussian elimination can be applied also to the case in which A is not square. When A is $m \times n$, with $m > n$, the standard row pivoting algorithm produces a factorization of the form (A.50), where $L \in \mathbf{R}^{m \times n}$ is unit lower triangular and $U \in \mathbf{R}^{n \times n}$ is upper triangular. When $m < n$, we can find an LU factorization of A^T rather than A, that is, we obtain

$$PA^T = \begin{bmatrix} L_1 \\ L_2 \end{bmatrix} U, \tag{A.51}$$

where L_1 is $m \times m$ (square) unit lower triangular, U is $m \times m$ upper triangular, and L_2 is a general $(n-m) \times m$ matrix. If A has full row rank, we can use this factorization to calculate its null space explicitly as the space spanned by the columns of the matrix

$$M = P^T \begin{bmatrix} L_1^{-T} L_2^T \\ -I \end{bmatrix} U^{-T}. \tag{A.52}$$

(It is easy to check that M has dimensions $n \times (n-m)$ and that $AM = 0$; we leave this an exercise.)

When $A \in \mathbf{R}^{n \times n}$ is symmetric positive definite, it is possible to compute a similar but more specialized factorization at about half the cost—about $n^3/3$ operations. This factorization, known as the Cholesky factorization, produces a matrix L such that

$$A = LL^T. \tag{A.53}$$

(If we require L to have positive diagonal elements, it is uniquely defined by this formula.) The algorithm can be specified as follows.

Algorithm A.2 (Cholesky Factorization).
Given $A \in \mathbf{R}^{n \times n}$ symmetric positive definite;
for $i = 1, 2, \ldots, n$;
$\quad L_{ii} \leftarrow \sqrt{A_{ii}}$;
$\quad$ for $j = i+1, i+2, \ldots, n$
$\quad\quad L_{ji} \leftarrow A_{ji}/L_{ii}$;
$\quad\quad$ for $k = i+1, i+2, \ldots, j$
$\quad\quad\quad A_{jk} \leftarrow A_{jk} - L_{ji} L_{ki}$;

            end (for)
        end (for)
    end (for)

Note that this algorithm references only the lower triangular elements of A; in fact, it is only necessary to store these elements in any case, since by symmetry they are simply duplicated in the upper triangular positions.

Unlike the case of Gaussian elimination, the Cholesky algorithm can produce a valid factorization of a symmetric positive definite matrix without swapping any rows or columns. However, symmetric permutation (that is, reordering the rows and columns in the same way) can be used to improve the sparsity of the factor L. In this case, the algorithm produces a permutation of the form

$$P^T A P = L L^T$$

for some permutation matrix P.

The Cholesky factorization can be used to compute solutions of the system $Ax = b$ by performing triangular forward- and back-substitutions with L and L^T, respectively, as in the case of L and U factors produced by Gaussian elimination.

Another useful factorization of rectangular matrices $A \in \mathbf{R}^{m \times n}$ has the form

$$AP = QR, \qquad (A.54)$$

where

P is an $n \times n$ permutation matrix,

A is $m \times m$ orthogonal, and

R is $m \times n$ upper triangular.

In the case of a square matrix $m = n$, this factorization can be used to compute solutions of linear systems of the form $Ax = b$ via the following procedure:

set $\tilde{b} = Q^T b$;

solve $Rz = \tilde{b}$ for z by performing back-substitution;

set $x = P^T z$ by rearranging the elements of x.

For a dense matrix A, the cost of computing the QR factorization is about $4m^2 n/3$ operations. In the case of a square matrix, this is about twice as many operations as required to compute an LU factorization via Gaussian elimination. Unlike the case of Gaussian elimination, the QR factorization procedure cannot be modified in general to ensure efficiency on sparse matrices. That is, no matter how the column permutation matrix P is chosen, the factors Q

and R will be dense in general. (This remains true even if we allow row pivoting as well as column pivoting.)

Algorithms to perform QR factorization are almost as simple as algorithms for Gaussian elimination and for Cholesky factorization. The most widely used algorithms work by applying a sequence of special orthogonal matrices to A, known either as Householder transformations or Givens rotations, depending on the algorithm. We omit the details, and refer instead to Golub and Van Loan [115, Chapter 5] for a complete description.

In the case of a rectangular matrix A with $m < n$, we can use the QR factorization of A^T to find a matrix whose columns span the null space of A. To be specific, we write

$$A^T P = QR = [\ Q_1 \quad Q_2\ ] R,$$

where Q_1 consists of the first m columns of Q, and Q_2 contains the last $n - m$ columns. It is easy to show that columns of the matrix Q_2 span the null space of A. This procedure yields a more satisfactory basis matrix for the null space than the Gaussian elimination procedure (A.52), because the columns of Q_2 are orthogonal to each other and have unit length. It may be more expensive to compute, however, particularly in the case in which A is sparse.

When A has full column rank, we can make an identification between the R factor in (A.54) and the Cholesky factorization. By multiplying the formula (A.54) by its transpose, we obtain

$$P^T A^T A P = R^T Q^T Q R = R^T R,$$

and by comparison with (A.53), we see that R^T is simply the Cholesky factor of the symmetric positive definite matrix $P^T A^T A P$. Recalling that L is uniquely defined when we restrict its diagonal elements to be positive, this observation implies that R is also uniquely defined for a given choice of permutation matrix P, provided that we enforce positiveness of the diagonals of R. Note, too, that since we can rearrange (A.54) to read $APR^{-1} = Q$, we can conclude that Q is also uniquely defined under these conditions.

Note that by definition of the Euclidean norm and the property (A.41), and the fact that the Euclidean norms of the matrices P and Q in (A.54) are both 1, we have that

$$\|A\| = \|QRP^T\| \leq \|Q\|\,\|R\|\,\|P^T\| = \|R\|,$$

while

$$\|R\| = \|Q^T A P\| \leq \|Q^T\|\,\|A\|\,\|P\| = \|A\|.$$

We conclude from these two inequalities that $\|A\| = \|R\|$. When A is square, we have by a similar argument that $\|A^{-1}\| = \|R^{-1}\|$. Hence the Euclidean-norm condition number of A can be estimated by substituting R for A in the expression (A.42). This observation is significant because various techniques are available for estimating the condition number of triangular matrices R; see Golub and Van Loan [115, pp. 128–130] for a discussion.

SHERMAN–MORRISON–WOODBURY FORMULA

If the square nonsingular matrix A undergoes a rank-one update to become

$$\bar{A} = A + ab^T,$$

where $a, b \in \mathbf{R}^n$, then if $\bar{A}$ is nonsingular, we have

$$\bar{A}^{-1} = A^{-1} - \frac{A^{-1}ab^T A^{-1}}{1 + b^T A^{-1}a}. \tag{A.55}$$

It is easy to verify this formula: Simply multiply the definitions of $\bar{A}$ and $\bar{A}^{-1}$ together and check that they produce the identity.

This formula can be extended to higher-rank updates. Let U and V be matrices in $\mathbf{R}^{n \times p}$ for some p between 1 and n. If we define

$$\hat{A} = A + UV^T,$$

then

$$\hat{A}^{-1} = A^{-1} - A^{-1}U(I + V^T A^{-1}U)^{-1}V^T A^{-1}. \tag{A.56}$$

We can use this formula to solve linear systems of the form $\bar{A}x = d$. Since

$$x = \hat{A}^{-1}d = A^{-1}d - A^{-1}U(I + V^T A^{-1}U)^{-1}V^T A^{-1}d,$$

we see that x can be found by solving $p + 1$ linear systems with the matrix A (to obtain $A^{-1}d$ and $A^{-1}U$), inverting the $p \times p$ matrix $I + V^T A^{-1}U$, and performing some elementary matrix algebra. Inversion of the $p \times p$ matrix $I + V^T A^{-1}U$ is inexpensive when $p \ll n$.

INTERLACING EIGENVALUE THEOREM

The following result is proved in Golub and Van Loan [115, Theorem 8.1.8].

Theorem A.2 (Interlacing Eigenvalue Theorem).
Let $A \in \mathbf{R}^{n \times n}$ be a symmetric matrix with eigenvalues $\lambda_1, \lambda_2, \ldots, \lambda_n$ satisfying

$$\lambda_1 \geq \lambda_2 \geq \cdots \geq \lambda_n,$$

and let $z \in \mathbf{R}^n$ be a vector with $\|z\| = 1$, and $\alpha \in \mathbf{R}$ be a scalar. Then if we denote the eigenvalues of $A + \alpha zz^T$ by $\xi_1, \xi_2, \ldots, \xi_n$ (in decreasing order), we have for $\alpha > 0$ that

$$\xi_1 \geq \lambda_1 \geq \xi_2 \geq \lambda_2 \geq \xi_3 \geq \cdots \geq \xi_n \geq \lambda_n,$$

with

$$\sum_{i=1}^{n} \xi_i - \lambda_i = \alpha. \tag{A.57}$$

If $\alpha < 0$, we have that

$$\lambda_1 \geq \xi_1 \geq \lambda_2 \geq \xi_2 \geq \lambda_3 \geq \cdots \geq \lambda_n \geq \xi_n,$$

where the relationship (A.57) is again satisfied.

Informally stated, the eigenvalues of the modified matrix "interlace" the eigenvalues of the original matrix, with nonnegative adjustments if the coefficient α is positive, and nonpositive adjustments if α is negative. The total magnitude of the adjustments equals α, whose magnitude is identical to the Euclidean norm $\|\alpha zz^T\|_2$ of the modification.

ERROR ANALYSIS AND FLOATING-POINT ARITHMETIC

In most of this book our algorithms and analysis deal with real numbers. Modern digital computers, however, cannot store or compute with general real numbers. Instead, they work with a subset known as *floating-point numbers*. Any quantities that are stored on the computer, whether they are read directly from a file or program or arise as the intermediate result of a computation, must be approximated by a floating-point number. In general, then, the numbers that are produced by practical computation differ from those that would be produced if the arithmetic were exact. Of course, we try to perform our computations in such a way that these differences are as tiny as possible.

Discussion of errors requires us to distinguish between *absolute error* and *relative error*. If x is some exact quantity (scalar, vector, matrix) and $\tilde{x}$ is its approximate value, the absolute error is the norm of the difference, namely, $\|x - \tilde{x}\|$. (In general, any of the norms (A.34a), (A.34b), and (A.34c) can be used in this definition.) The relative error is the ratio of the absolute error to the size of the exact quantity, that is,

$$\frac{\|x - \tilde{x}\|}{\|x\|}.$$

When this ratio is significantly less than one, we can replace the denominator by the size of the approximate quantity—that is, $\|\tilde{x}\|$—without affecting its value very much.

Most computations associated with optimization algorithms are performed in double-precision arithmetic. Double-precision numbers are stored in words of length 64 bits. Most of these bits (say t) are devoted to storing the *fractional part*, while the remainder encode the *exponent* e and other information, such as the sign of the number, or an indication of whether it is zero or "undefined." Typically, the fractional part has the form

$$.d_1 d_2 \ldots d_t,$$

where each $d_i, i = 1, 2, \ldots, t$, is either zero or one. (In some systems d_1 is implicitly assumed to be 1 and is not stored.) The value of the floating-point number is then

$$\sum_{i=1}^{t} d_i 2^{-i} \times 2^e.$$

The value 2^{-t} is known as *unit roundoff* and is denoted by **u**. Any real number whose absolute value lies in the range $[2^L, 2^U]$ (where L and U are lower and upper bounds on the value of the exponent e) can be approximated to within a relative accuracy of **u** by a floating-point number, that is,

$$\text{fl}(x) = x(1 + \epsilon), \qquad \text{where } |\epsilon| \leq \mathbf{u}, \tag{A.58}$$

where $\text{fl}(\cdot)$ denotes floating-point approximation. The value of **u** for double-precision computations is typically about 10^{-15}. In other words, if the real number x and its floating-point approximation are both written as base-10 numbers (the usual fashion), they agree to at least 15 digits.

For further information on floating-point computations, see Golub and Van Loan [115, Section 2.4] and Higham [136].

When an arithmetic operation is performed with one or two floating-point numbers, the result must also be stored as a floating-point number. This process introduces a small *roundoff error*, whose size can be quantified in terms of the size of the arguments. If x and y are two floating-point numbers, we have that

$$|\text{fl}(x * y) - x * y| \leq \mathbf{u}|x * y|, \tag{A.59}$$

where $*$ denotes any of the operations $+, -, \times, \div$.

Although the error in a single floating-point operation appears benign, more significant errors may occur when the arguments x and y are floating-point approximations of two *real* numbers, or when a sequence of computations are performed in succession. Suppose, for instance, that x and y are large real numbers whose values are very similar. When we store them in a computer, we approximate them with floating-point numbers $\text{fl}(x)$ and $\text{fl}(y)$

that satisfy

$$\mathrm{fl}(x) = x + \epsilon_x, \quad \mathrm{fl}(y) = y + \epsilon_y, \quad \text{where } |\epsilon_x| \leq \mathbf{u}|x|, |\epsilon_y| \leq \mathbf{u}|y|.$$

If we take the difference of the two stored numbers, we obtain a final result $\mathrm{fl}(\mathrm{fl}(x) - \mathrm{fl}(y))$ that satisfies

$$\mathrm{fl}(\mathrm{fl}(x) - \mathrm{fl}(y)) = (\mathrm{fl}(x) - \mathrm{fl}(y))(1 + \epsilon_{xy}), \quad \text{where } |\epsilon_{xy}| \leq \mathbf{u}.$$

By combining these expressions, we find that the difference between this result and the true value $x - y$ may be as large as

$$\epsilon_x + \epsilon_y + \epsilon_{xy},$$

which is bounded by $\mathbf{u}(|x| + |y| + |x - y|)$. Hence, since x and y are large and close together, the relative error is approximately $2\mathbf{u}|x|/|x - y|$, which may be quite large, since $|x| \gg |x - y|$.

This phenomenon is known as *cancellation*. It can also be explained (less formally) by noting that if both x and y are accurate to k digits, and if they agree in the first $\bar{k}$ digits, then their difference will contain only about $k - \bar{k}$ significant digits—the first $\bar{k}$ digits cancel each other out. This observation is the reason for the well-known adage of numerical computing—that one should avoid taking the difference of two similar numbers if at all possible.

CONDITIONING AND STABILITY

Conditioning and *stability* are two terms that are used frequently in connection with numerical computations. Unfortunately, their meaning sometimes varies from author to author, but the general definitions below are widely accepted, and we adhere to them in this book.

Conditioning is a property of the numerical problem at hand (whether it is a linear algebra problem, an optimization problem, a differential equations problem, or whatever). A problem is said to be *well conditioned* if its solution is not affected greatly by small perturbations to the data that define the problem. Otherwise, it is said to be *ill conditioned*.

A simple example is given by the following 2×2 system of linear equations:

$$\begin{bmatrix} 1 & 2 \\ 1 & 1 \end{bmatrix} \begin{bmatrix} x_1 \\ x_2 \end{bmatrix} = \begin{bmatrix} 3 \\ 2 \end{bmatrix}.$$

By computing the inverse of the coefficient matrix, we find that the solution is simply

$$\begin{bmatrix} x_1 \\ x_2 \end{bmatrix} = \begin{bmatrix} -1 & 2 \\ 1 & -1 \end{bmatrix} \begin{bmatrix} 3 \\ 2 \end{bmatrix} = \begin{bmatrix} 1 \\ 1 \end{bmatrix}.$$

If we replace the first right-hand-side element by 3.00001, the solution becomes $(x_1, x_2)^T = (0.99999, 1.00001)^T$, which is only slightly different from its exact value $(1, 1)^T$. We would note similar insensitivity if we were to perturb the other elements of the right-hand-side or elements of the coefficient matrix. We conclude that this problem is well conditioned. On the other hand, the problem

$$\begin{bmatrix} 1.00001 & 1 \\ 1 & 1 \end{bmatrix} \begin{bmatrix} x_1 \\ x_2 \end{bmatrix} = \begin{bmatrix} 2.00001 \\ 2 \end{bmatrix}$$

is ill conditioned. Its exact solution is $x = (1, 1)^T$, but if we change the first element of the right-hand-side from 2.00001 to 2, the solution would change drastically to $x = (0, 2)^T$.

For general square linear systems $Ax = b$ where $A \in \mathbf{R}^{n \times n}$, the condition number of the matrix (defined in (A.42)) can be used to quantify the conditioning. Specifically, if we perturb A to $\tilde{A}$ and b to $\tilde{b}$ and take $\tilde{x}$ to be the solution of the perturbed system $\tilde{A}\tilde{x} = \tilde{b}$, it can be shown that

$$\frac{\|x - \tilde{x}\|}{\|x\|} \approx \kappa(A) \left[\frac{\|A - \tilde{A}\|}{\|A\|} + \frac{\|b - \tilde{b}\|}{\|b\|} \right]$$

(see, for instance, Golub and Van Loan [115, Section 2.7]). Hence, a large condition number $\kappa(A)$ indicates that the problem $Ax = b$ is ill conditioned, while a modest value indicates well conditioning.

Note that the concept of conditioning has nothing to do with the particular algorithm that is used to solve the problem, only with the numerical problem itself.

Stability, on the other hand, is a property of the algorithm. An algorithm is stable if it is guaranteed to produce accurate answers to all well-conditioned problems in its class, even when floating-point arithmetic is used.

As an example, consider again the linear equations $Ax = b$. We can show that Algorithm A.1, in combination with triangular substitution, yields a computed solution $\tilde{x}$ whose relative error is approximately

$$\frac{\|x - \tilde{x}\|}{\|x\|} \approx \kappa(A) \frac{\text{growth}(A)}{\|A\|} \mathbf{u}, \tag{A.60}$$

where growth(A) is the size of the largest element that arises in A during execution of Algorithm A.1. In the worst case, we can show that growth(A)/$\|A\|$ may be around 2^{n-1}, which indicates that Algorithm A.1 is an unstable algorithm, since even for modest n (say,

$n = 200$), the right-hand-side of (A.60) may be large even when $\kappa(A)$ is modest. In practice, however, large growth factors are rarely observed, so we conclude that Algorithm A.1 is stable for all practical purposes.

Gaussian elimination without pivoting, on the other hand, is definitely unstable. If we omit the possible exchange of rows in Algorithm A.1, the algorithm will fail to produce a factorization even of some well-conditioned matrices, such as

$$A = \begin{bmatrix} 0 & 1 \\ 1 & 2 \end{bmatrix}.$$

For systems $Ax = b$ in which A is symmetric positive definite, the Cholesky factorization in combination with triangular substitution constitutes a stable algorithm for producing a solution x.

References

[1] R. K. AHUJA, T. L. MAGNANTI, AND J. B. ORLIN, *Network Flows: Theory, Algorithms, and Applications*, Prentice-Hall, Englewood Cliffs, N.J., 1993.

[2] H. AKAIKE, *On a successive transformation of probability distribution and its application to the analysis of the optimum gradient method*, Annals of the Institute of Statistical Mathematics, 11 (1959), pp. 1–17.

[3] M. AL-BAALI, *Descent property and global convergence of the Fletcher-Reeves method with inexact line search*, I.M.A. Journal on Numerical Analysis, 5 (1985), pp. 121–124.

[4] E. ANDERSON, Z. BAI, C. BISCHOF, J. DEMMEL, J. DONGARRA, J. DU CROZ, A. GREENBAUM, S. HAMMARLING, A. MCKENNEY, S. OSTROUCHOV, AND D. SORENSEN, *LAPACK User's Guide*, SIAM, Philadelphia, 1992.

[5] B. M. AVERICK, R. G. CARTER, J. J. MORÉ, AND G. XUE, *The MINPACK-2 test problem collection*, Preprint MCS–P153–0692, Mathematics and Computer Science Division, Argonne National Laboratory, Argonne, Ill., 1992.

[6] P. BAPTIST AND J. STOER, *On the relation between quadratic termination and convergence properties of minimization algorithms, Part II: Applications*, Numerische Mathematik, 28 (1977), pp. 367–392.

[7] M. BAZARAA, H. SHERALI, AND C. SHETTY, *Nonlinear Programming, Theory and Applications.*, John Wiley & Sons, New York, second ed., 1993.

[8] D. P. BERTSEKAS, *Constrained Optimization and Lagrange Multiplier Methods*, Academic Press, New York, 1982.

[9] ———, *Nonlinear Programming*, Athena Scientific, Belmont, Mass., 1995.

[10] M. BERZ, C. BISCHOF, C. F. CORLISS, AND A. GRIEWANK, eds., *Computational Differentiation: Techniques, Applications, and Tools*, SIAM Publications, Philadelphia, Penn., 1996.

[11] J. R. BIRGE AND F. LOUVEAUX, *Introduction to Stochastic Programming*, Springer-Verlag, New York, 1997.

[12] C. BISCHOF, A. BOUARICHA, P. KHADEMI, AND J. J. MORÉ, *Computing gradients in large-scale optimization using automatic differentiation*, INFORMS Journal on Computing, 9 (1997), pp. 185–194.

[13] C. BISCHOF, A. CARLE, P. KHADEMI, AND A. MAUER, *ADIFOR 2.0: Automatic differentiation of FORTRAN 77 programs*, IEEE Computational Science & Engineering, 3 (1996), pp. 18–32.

[14] C. BISCHOF, G. CORLISS, AND A. GRIEWANK, *Structured second- and higher-order derivatives through univariate Taylor series*, Optimization Methods and Software, 2 (1993), pp. 211–232.

[15] C. BISCHOF AND M. R. HAGHIGHAT, *On hierarchical differentiation*, in Computational Differentiation: Techniques, Applications, and Tools, M. Berz, C. Bischof, G. Corliss, and A. Griewank, eds., SIAM, Philadelphia, 1996, pp. 83–94.

[16] C. BISCHOF, P. KHADEMI, A. BOUARICHA, AND A. CARLE, *Efficient computation of gradients and Jacobians by transparent exploitation of sparsity in automatic differentiation*, Optimization Methods and Software, 7 (1996), pp. 1–39.

[17] C. BISCHOF, L. ROH, AND A. MAUER, *ADIC: An extensible automatic differentiation tool for ANSI-C*, Software—Practice and Experience, 27 (1997), pp. 1427–1456.

[18] Å. BJÖRCK, *Least squares methods*, in Handbook of Numerical Analysis, P. G. Ciarlet and J. L. Lions, eds., Elsevier/North-Holland, Amsterdam, The Netherlands, 1990.

[19] ———, *Numerical Methods for Least Squares Problems*, SIAM Publications, Philadelphia, Penn., 1996.

[20] P. T. BOGGS, R. H. BYRD, AND R. B. SCHNABEL, *A stable and efficient algorithm for nonlinear orthogonal distance regression*, SIAM Journal on Scientific and Statistical Computing, 8 (1987), pp. 1052–1078.

[21] P. T. BOGGS, J. R. DONALDSON, R. H. BYRD, AND R. B. SCHNABEL, *ODRPACK—Software for weighted orthogonal distance regression*, ACM Transactions on Mathematical Software, 15 (1981), pp. 348–364.

[22] P. T. BOGGS AND J. W. TOLLE, *Convergence properties of a class of rank-two updates*, SIAM Journal on Optimization, 4 (1994), pp. 262–287.

[23] ———, *Sequential quadratic programming*, Acta Numerica, 4 (1996), pp. 1–51.

[24] P. T. BOGGS, J. W. TOLLE, AND P. WANG, *On the local convergence of quasi-Newton methods for constrained optimization*, SIAM Journal on Control and Optimization, 20 (1982), pp. 161–171.

[25] I. BONGARTZ, A. R. CONN, N. I. M. GOULD, AND P. L. TOINT, *CUTE: Constrained and unconstrained testing environment*, Research Report, IBM T.J. Watson Research Center, Yorktown Heights, NY, 1993.

[26] S. BOYD, L. EL GHAOUI, E. FERON, AND V. BALAKRISHNAN, *Linear Matrix Inequalities in Systems and Control Theory*, SIAM Publications, Phildelphia, 1994.

[27] R. P. BRENT, *Algorithms for minimization without derivatives*, Prentice Hall, Englewood Cliffs, N.J., 1973.

[28] A. BUCKLEY AND A. LENIR, *QN-like variable storage conjugate gradients*, Mathematical Programming, 27 (1983), pp. 155–175.

[29] R. BULIRSCH AND J. STOER, *Introduction to Numerical Analysis*, Springer-Verlag, New York, 1980.

[30] J. R. BUNCH AND L. KAUFMAN, *Some stable methods for calculating inertia and solving symmetric linear systems*, Mathematics of Computation, 31 (1977), pp. 163–179.

[31] J. R. BUNCH AND B. N. PARLETT, *Direct methods for solving symmetric indefinite systems of linear equations*, SIAM Journal on Numerical Analysis, 8 (1971), pp. 639–655.

[32] J. V. BURKE AND J. J. MORÉ, *Exposing constraints*, SIAM Journal on Optimization, 4 (1994), pp. 573–595.

[33] W. BURMEISTER, *Die Konvergenzordnung des Fletcher-Powell Algorithmus*, Z. Angew. Math. Mech., 53 (1973), pp. 693–699.

[34] R. H. BYRD, M. E. HRIBAR, AND J. NOCEDAL, *An interior-point algorithm for large-scale nonlinear programming*, Technical Report 97/05, Optimization Technology Center, Argonne National Laboratory and Northwestern University, July 1997.

[35] R. H. BYRD, H. F. KHALFAN, AND R. B. SCHNABEL, *Analysis of a symmetric rank-one trust region method*, SIAM Journal on Optimization, 6 (1996), pp. 1025–1039.

[36] R. H. BYRD AND J. NOCEDAL, *An analysis of reduced Hessian methods for constrained optimization*, Mathematical Programming, 49 (1991), pp. 285–323.

[37] R. H. BYRD, J. NOCEDAL, AND R. B. SCHNABEL, *Representations of quasi-Newton matrices and their use in limited-memory methods*, Mathematical Programming, Series A, 63 (1994), pp. 129–156.

[38] R. H. BYRD, J. NOCEDAL, AND Y. YUAN, *Global convergence of a class of quasi-Newton methods on convex problems*, SIAM Journal on Numerical Analysis, 24 (1987), pp. 1171–1190.

[39] R. H. BYRD, R. B. SCHNABEL, AND G. A. SCHULTZ, *Approximate solution of the trust regions problem by minimization over two-dimensional subspaces*, Mathematical Programming, 40 (1988), pp. 247–263.

[40] R. CHAMBERLAIN, C. LEMARÉCHAL, H. C. PEDERSEN, AND M. J. D. POWELL, *The watchdog technique for forcing convergence in algorithms for constrained optimization*, Mathematical Programming, 16 (1982), pp. 1–17.

[41] S. H. CHENG AND H. J. HIGHAM, *A modified Cholesky algorithm based on a symmetric indefinite factorization*, technical report, University of Manchester, 1996.

[42] F. H. CLARKE, *Optimization and Nonsmooth Analysis*, John Wiley & Sons, New York, 1983.

[43] A. COHEN, *Rate of convergence of several conjugate gradient algorithms*, SIAM Journal on Numerical Analysis, 9 (1972), pp. 248–259.

[44] T. F. COLEMAN AND A. R. CONN, *Non-linear programming via an exact penalty-function: Asymptotic analysis*, Mathematical Programming, 24 (1982), pp. 123–136.

[45] T. F. COLEMAN AND A. R. CONN, *On the local convergence of a quasi-Newton method for the nonlinear programming problem*, SIAM Journal on Numerical Analysis, 21 (1984), pp. 755–769.

[46] T. F. COLEMAN, B. GARBOW, AND J. J. MORÉ, *Software for estimating sparse Jacobian matrices*, ACM Transactions on Mathematical Software, 10 (1984), pp. 329–345.

[47] ———, *Software for estimating sparse Hessian matrices*, ACM Transactions on Mathematical Software, 11 (1985), pp. 363–377.

[48] T. F. COLEMAN AND J. J. MORÉ, *Estimation of sparse Jacobian matrices and graph coloring problems*, SIAM Journal on Numerical Analysis, 20 (1983), pp. 187–209.

[49] ———, *Estimation of sparse Hessian matrices and graph coloring problems*, Mathematical Programming, 28 (1984), pp. 243–270.

[50] T. F. COLEMAN AND D. C. SORENSEN, *A note on the computation of an orthonormal basis for the null space of a matrix*, Mathematical Programming, 29 (1984), pp. 234–242.

[51] A. R. CONN, N. I. M. GOULD, AND P. L. TOINT, *Testing a class of algorithms for solving minimzation problems with simple bounds on the variables*, Mathematics of Computation, 50 (1988), pp. 399–430.

[52] ——, *Convergence of quasi-Newton matrices generated by the symmetric rank one update*, Mathematical Programming, 50 (1991), pp. 177–195.

[53] ——, *LANCELOT: a FORTRAN package for large-scale nonlinear optimization (Release A)*, no. 17 in Springer Series in Computational Mathematics, Springer-Verlag, New York, 1992.

[54] ——, *Numerical experiments with the LANCELOT package (Release A) for large-scale nonlinear optimization*, Report 92/16, Department of Mathematics, University of Namur, Belgium, 1992.

[55] ——, *A note on using alternative second-order models for the subproblems arising in barrier function methods for minimization*, Numerische Mathematik, 68 (1994), pp. 17–33.

[56] W. J. COOK, W. H. CUNNINGHAM, W. R. PULLEYBLANK, AND A. SCHRIJVER, *Combinatorial Optimization*, John Wiley & Sons, New York, 1997.

[57] B. F. CORLISS AND L. B. RALL, *An introduction to automatic differentiation*, in Computational Differentiation: Techniques, Applications, and Tools, M. Berz, C. Bischof, G. F. Corliss, and A. Griewank, eds., SIAM Publications, Philadelphia, Penn., 1996, ch. 1.

[58] T. H. CORMEN, C. E. LEISSERSON, AND R. L. RIVEST, *Introduction to Algorithms*, MIT Press, 1990.

[59] R. W. COTTLE, J.-S. PANG, AND R. E. STONE, *The Linear Complementarity Problem*, Academic Press, San Diego, 1992.

[60] R. COURANT, *Variational methods for the solution of problems with equilibrium and vibration*, Bull. Amer. Math. Soc., 49 (1943), pp. 1–23.

[61] H. P. CROWDER AND P. WOLFE, *Linear convergence of the conjugate gradient method*, IBM Journal of Research and Development, 16 (1972), pp. 431–433.

[62] A. CURTIS, M. J. D. POWELL, AND J. REID, *On the estimation of sparse Jacobian matrices*, Journal of the Institute of Mathematics and its Applications, 13 (1974), pp. 117–120.

[63] G. B. DANTZIG, *Linear Programming and Extensions*, Princeton University Press, Princeton, New Jersey, 1963.

[64] W. C. DAVIDON, *Variable metric method for minimization*, Technical Report ANL–5990 (revised), Argonne National Laboratory, Argonne, Il, 1959.

[65] ——, *Variable metric method for minimization*, SIAM Journal on Optimization, 1 (1991), pp. 1–17.

[66] R. S. DEMBO, S. C. EISENSTAT, AND T. STEIHAUG, *Inexact Newton methods*, SIAM Journal on Numerical Analysis, 19 (1982), pp. 400–408.

[67] J. E. DENNIS, D. M. GAY, AND R. E. WELSCH, *Algorithm 573 — NL2SOL, An adaptive nonlinear least-squares algorithm*, ACM Transactions on Mathematical Software, 7 (1981), pp. 348–368.

[68] J. E. DENNIS AND J. J. MORÉ, *Quasi-Newton methods, motivation and theory*, SIAM Review, 19 (1977), pp. 46–89.

[69] J. E. DENNIS AND R. B. SCHNABEL, *Numerical Methods for Unconstrained Optimization*, Prentice-Hall, Englewood Cliffs, NJ, 1983. Reprinted by SIAM Publications, 1993.

[70] J. E. DENNIS AND R. B. SCHNABEL, *A view of unconstrained optimization*, in Optimization, vol. 1 of Handbooks in Operations Research and Management, Elsevier Science Publishers, Amsterdam, the Netherlands, 1989, pp. 1–72.

[71] P. DEUFLHARD, R. W. FREUND, AND A. WALTER, *Fast secant methods for the iterative solution of large nonsymmetric linear systems*, Impact of Computing in Science and Engineering, 2 (1990), pp. 244–276.

[72] I. I. DIKIN, *Iterative solution of problems of linear and quadratic programming*, Soviet Mathematics-Doklady, 8 (1967), pp. 674–675.

[73] I. S. DUFF, J. NOCEDAL, AND J. K. REID, *The use of linear programming for the solution of sparse sets of nonlinear equations*, SIAM Journal on Scientific and Statistical Computing, 8 (1987), pp. 99–108.

[74] I. S. DUFF AND J. K. REID, *The multifrontal solution of indefinite sparse symmetric linear equations*, ACM Transactions on Mathematical Software, 9 (1983), pp. 302–325.

[75] ———, *The design of MA48, a code for direct solution of sparse unsymmetric linear systems of equations*, ACM Transactions on Mathematical Software, 22 (1996), pp. 187–226.

[76] I. S. DUFF, J. K. REID, N. MUNKSGAARD, AND H. B. NEILSEN, *Direct solution of sets of linear equations whose matrix is sparse symmetric and indefinite*, Journal of the Institute of Mathematics and its Applications, 23 (1979), pp. 235–250.

[77] J. C. DUNN, *A projected Newton method for minimization problems with nonlinear inequality constraints*, Numerische Mathematik, 53 (1988), pp. 377–409.

[78] J. DUSSAULT, *Numerical stability and efficiency of penalty algorithms*, SIAM Journal on Numerical Analysis, 32 (1995), pp. 296–317.

[79] A. V. FIACCO AND G. P. MCCORMICK, *Nonlinear Programming: Sequential Unconstrained Minimization Techniques*, John Wiley & Sons, New York, NY, 1968. reprinted by SIAM Publications, 1990.

[80] R. FLETCHER, *A general quadratic programming algorithm*, Journal of the Institute of Mathematics and its Applications, 7 (1971), pp. 76–91.

[81] ———, *A class of methods for nonlinear programming, iii: Rates of convergence*, in Numerical Methods for Non-Linear Optimization, F. A. Lootsma, ed., Academic Press, London and New York, 1972, pp. 371–382.

[82] ———, *Second order corrections for non-differentiable optimization*, in Numerical Analysis, D. Griffiths, ed., Springer Verlag, 1982, pp. 85–114. Proceedings Dundee 1981.

[83] ———, *Practical Methods of Optimization*, John Wiley & Sons, New York, second ed., 1987.

[84] ———, *An optimal positive definite update for sparse Hessian matrices*, SIAM Journal on Optimization, 5 (1995), pp. 192–218.

[85] R. FLETCHER, A. GROTHEY, AND S. LEYFFER, *Computing sparse Hessian and Jacobian approximations with optimal hereditary properties*, technical report, Department of Mathematics, University of Dundee, 1996.

[86] R. FLETCHER AND S. LEYFFER, *Nonlinear programming without a penalty function*, Tech. Rep. NA/171, Department of Mathematics, University of Dundee, September 1997.

[87] R. FLETCHER AND A. P. MCCANN, *Acceleration techniques for nonlinear programming*, in Optimization, R. Fletcher, ed., Academic Press, London, 1969, pp. 203–214.

[88] R. FLETCHER AND C. M. REEVES, *Function minimization by conjugate gradients*, Computer Journal, 7 (1964), pp. 149–154.

[89] R. FLETCHER AND E. SAINZ DE LA MAZA, *Nonlinear programming and nonsmooth optimization by successive linear programming*, Mathematical Programming, (1989), pp. 235–256.

[90] C. FLOUDAS AND P. PARDALOS, eds., *Recent Advances in Global Optimization*, Princeton University Press, Princeton, NJ, 1992.

[91] A. FORSGREN AND P. E. GILL, *Primal-dual interior methods for nonconvex nonlinear programming*, SIAM Journal on Optimization, 8 (1998), pp. 1132–1152.

[92] R. FOURER, D. M. GAY, AND B. W. KERNIGHAN, *AMPL: A Modeling Language for Mathematical Programming*, The Scientific Press, South San Francisco, Calif., 1993.

[93] R. FOURER AND S. MEHROTRA, *Solving symmetric indefinite systems in an interior-point method for linear programming*, Mathematical Programming, 62 (1993), pp. 15–39.

[94] R. FREUND AND N. NACHTIGAL, *QMR: A quasi-minimal residual method for non-Hermitian linear systems*, Numerische Mathematik, 60 (1991), pp. 315–339.

[95] K. R. FRISCH, *The logarithmic potential method of convex programming*, Technical Report, University Institute of Economics, Oslo, Norway, 1955.

[96] D. GABAY, *Reduced quasi-Newton methods with feasibility improvement for nonlinearly constrained optimization*, Mathematical Programming Studies, 16 (1982), pp. 18–44.

[97] U. M. GARCIA-PALOMARES AND O. L. MANGASARIAN, *Superlinearly convergent quasi-Newton methods for nonlinearly constrained optimization problems*, Mathematical Programming, 11 (1976), pp. 1–13.

[98] D. M. GAY, *More AD of nonlinear AMPL models: computing Hessian information and exploiting partial separability*. To appear in the Proceedings of the Second International Workshop on Computational Differentiation, 1996.

[99] D. M. GAY, M. L. OVERTON, AND M. H. WRIGHT, *A primal-dual interior method for nonconvex nonlinear programming*, Technical Report 97-4-08, Computing Sciences Research Center, Bell Laboratories, Murray Hill, N.J., July 1997.

[100] R.-P. GE AND M. J. D. POWELL, *The convergence of variable metric matrices in unconstrained optimization*, Mathematical Programming, 27 (1983), pp. 123–143.

[101] J. GILBERT, *Maintaining the positive definiteness of the matrices in reduced secant methods for equality constrained optimization*, Mathematical Programming, 50 (1991), pp. 1–28.

[102] J. GILBERT AND C. LEMARÉCHAL, *Some numerical experiments with variable-storage quasi-Newton algorithms*, Mathematical Programming, Series B, 45 (1989), pp. 407–435.

[103] J. GILBERT AND J. NOCEDAL, *Global convergence properties of conjugate gradient methods for optimization*, SIAM Journal on Optimization, 2 (1992), pp. 21–42.

[104] P. GILL, W. MURRAY, AND M. H. WRIGHT, *Practical Optimization*, Academic Press, 1981.

[105] P. E. GILL, G. H. GOLUB, W. MURRAY, AND M. A. SAUNDERS, *Methods for modifying matrix factorizations*, Mathematics of Computation, 28 (1974), pp. 505–535.

[106] P. E. GILL AND M. W. LEONARD, *Limited-memory reduced-Hessian methods for unconstrained optimization*, Numerical Analysis Report NA 97-1, University of California, San Diego, 1997.

[107] P. E. GILL AND W. MURRAY, *Numerically stable methods for quadratic programming*, Mathematical Programming, 14 (1978), pp. 349–372.

[108] P. E. GILL, W. MURRAY, AND M. A. SAUNDERS, *User's guide for SNOPT (Version 5.3): A FORTRAN package for large-scale nonlinear programming*, Technical Report NA 97-4, Department of Mathematics, University of California, San Diego, 1997.

[109] P. E. GILL, W. MURRAY, M. A. SAUNDERS, G. W. STEWART, AND M. H. WRIGHT, *Properties of a representation of a basis for the null space*, Mathematical Programming, 33 (1985), pp. 172–186.

[110] P. E. GILL, W. MURRAY, M. A. SAUNDERS, AND M. H. WRIGHT, *User's guide for SOL/QPSOL*, Technical Report SOL84-6, Department of Operations Research, Stanford University, Stanford, California, 1984.

[111] ———, *User's guide for NPSOL (Version 4.0): A FORTRAN package for nonlinear programming*, Technical Report SOL 86-2, Department of Operations Research, Stanford University, Stanford, CA, 1986.

[112] P. E. GILL, W. MURRAY, M. A. SAUNDERS, AND M. H. WRIGHT, *Constrained nonlinear programming*, in Optimization, vol. 1 of Handbooks in Operations Research and Management, Elsevier Science Publishers, Amsterdam, the Netherlands, 1989, pp. 171–210.

[113] D. GOLDFARB, *Curvilinear path steplength algorithms for minimization which use directions of negative curvature*, Mathematical Programming, 18 (1980), pp. 31–40.

[114] D. GOLDFARB AND J. FORREST, *Steepest edge simplex algorithms for linear programming*, Mathematical Programming, 57 (1992), pp. 341–374.

[115] G. H. GOLUB AND C. F. VAN LOAN, *Matrix Computations*, The Johns Hopkins University Press, Baltimore, 3rd ed., 1996.

[116] J. GONDZIO, *Multiple centrality corrections in a primal-dual method for linear programming*, Computational Optimization and Applications, 6 (1996), pp. 137–156.

[117] J. GOODMAN, *Newton's method for constrained optimization*, Mathematical Programming, 33 (1985), pp. 162–171.

[118] N. I. M. GOULD, *On the accurate determination of search directions for simple differentiable penalty functions*, I.M.A. Journal on Numerical Analysis, 6 (1986), pp. 357–372.

[119] ———, *On the convergence of a sequential penalty function method for constrained minimization*, SIAM Journal on Numerical Analysis, 26 (1989), pp. 107–128.

[120] ———, *An algorithm for large scale quadratic programming*, I.M.A. Journal on Numerical Analysis, 11 (1991), pp. 299–324.

[121] N. I. M. GOULD, S. LUCIDI, M. ROMA, AND P. L. TOINT, *Solving the trust-region subproblem using the Lanczos method*. To appear in SIAM Journal on Optimization, 1998.

[122] A. GRIEWANK, *Achieving logarithmic growth of temporal and spatial complexity in reverse automatic differentiation*, Optimization Methods and Software, 1 (1992), pp. 35–54.

[123] ———, *Automatic directional differentiation of nonsmooth composite functions*, in Seventh French-German Conference on Optimization, 1994.

[124] ———, *Computational Differentiation and Optimization*, in Mathematical Programming: State of the Art 1994, J. R. Birge and K. G. Murty, eds., The University of Michigan, Michigan, USA, 1994, pp. 102–131.

[125] A. GRIEWANK AND G. F. CORLISS, eds., *Automatic Differentition of Algorithms*, SIAM Publications, Philadelphia, Penn., 1991.

[126] A. GRIEWANK, D. JUEDES, AND J. UTKE, *ADOL-C, A package for the automatic differentiation of algorithms written in C/C++*, ACM Transactions on Mathematical Software, 22 (1996), pp. 131–167.

[127] A. GRIEWANK AND P. L. TOINT, *Local convergence analysis of partitioned quasi-Newton updates*, Numerische Mathematik, 39 (1982), pp. 429–448.

[128] ———, *On the unconstrained optimization of partially separable objective functions*, in Nonlinear Optimization 1981, M. J. D. Powell, ed., Academic Press, London, 1982, pp. 301–312.

[129] ———, *Partitioned variable metric updates for large structured optimization problems*, Numerische Mathematik, 39 (1982), pp. 119–137.

[130] J. GRIMM, L. POTTIER, AND N. ROSTAING-SCHMIDT, *Optimal time and minimum space time product for reversing a certain class of programs*, in Computational Differentiation, Techniques, Applications, and Tools, M. Berz, C. Bischof, G. Corliss, and A. Griewank, eds., SIAM, Philadelphia, 1996, pp. 95–106.

[131] S. P. HAN, *Superlinearly convergent variable metric algorithms for general nonlinear programming problems*, Mathematical Programming, 11 (1976), pp. 263–282.

[132] ———, *A globally convergent method for nonlinear programming*, Journal of Optimization Theory and Applications, 22 (1977), pp. 297–309.

[133] *Harwell Subroutine Library, Release 10*, Advanced Computing Department, AEA Industrial Technology, Harwell Laboratory, Oxfordshire, United Kingdom, 1990.

[134] M. R. HESTENES, *Multiplier and gradient methods*, Journal of Optimization Theory and Applications, 4 (1969), pp. 303–320.

[135] M. R. HESTENES AND E. STIEFEL, *Methods of conjugate gradients for solving linear systems*, Journal of Research of the National Bureau of Standards, 49 (1952), pp. 409–436.

[136] N. J. HIGHAM, *Accuracy and Stability of Numerical Algorithms*, SIAM Publications, Philadelphia, 1996.

[137] J.-B. HIRIART-URRUTY AND C. LEMARÉCHAL, *Convex Analysis and Minimization Algorithms*, Springer-Verlag, Berlin, New York, 1993.

[138] J. E. DENNIS, JR. AND R. B. SCHNABEL, *Numerical Methods for Unconstrained Optimization and Nonlinear Equations*, Prentice-Hall, Englewood Cliffs, N.J., 1983.

[139] P. KALL AND S. W. WALLACE, *Stochastic Programming*, John Wiley & Sons, New York, 1994.

[140] N. KARMARKAR, *A new polynomial-time algorithm for linear programming*, Combinatorics, 4 (1984), pp. 373–395.

[141] C. T. KELLEY, *Iterative Methods for Linear and Nonlinear Equations*, SIAM Publications, Philadelphia, Penn., 1995.

[142] L. G. KHACHIYAN, *A polynomial algorithm in linear programming*, Soviet Mathematics Doklady, 20 (1979), pp. 191–194.

[143] H. F. KHALFAN, R. H. BYRD, AND R. B. SCHNABEL, *A theoretical and experimental study of the symmetric rank one update*, SIAM Journal on Optimization, 3 (1993), pp. 1–24.

[144] V. KLEE AND G. J. MINTY, *How good is the simplex algorithm?* in Inequalities, O. Shisha, ed., Academic Press, New York, 1972, pp. 159–175.

[145] H. W. KUHN AND A. W. TUCKER, *Nonlinear programming*, in Proceedings of the Second Berkeley Symposium on Mathematical Statistics and Probability, J. Neyman, ed., Berkeley, CA, 1951, University of California Press, pp. 481–492.

[146] M. LALEE, J. NOCEDAL, AND T. PLANTENGA, *On the implementation of an algorithm for large-scale equality constrained optimization*, SIAM Journal on Optimization, (1998), pp. 682–706.

[147] S. LANG, *Real Analysis*, Addison-Wesley, Reading, MA, second ed., 1983.

[148] C. L. LAWSON AND R. J. HANSON, *Solving Least Squares Problems*, Prentice-Hall, Englewood Cliffs, NJ, 1974.

[149] C. LEMARÉCHAL, *A view of line searches*, in Optimization and Optimal Control, W. Oettli and J. Stoer, eds., no. 30 in Lecture Notes in Control and Information Science, Springer-Verlag, 1981, pp. 59–78.

[150] K. LEVENBERG, *A method for the solution of certain non-linear problems in least squares*, Quarterly of Applied Mathematics, 2 (1944), pp. 164–168.

[151] D. C. LIU AND J. NOCEDAL, *On the limited-memory BFGS method for large scale optimization*, Mathematical Programming, 45 (1989), pp. 503–528.

[152] D. LUENBERGER, *Introduction to Linear and Nonlinear Programming*, Addison Wesley, second ed., 1984.

[153] *Macsyma User's Guide*, second ed., 1996.

[154] O. L. MANGASARIAN, *Nonlinear Programming*, McGraw-Hill, New York, 1969. Reprinted by SIAM Publications, 1995.

[155] O. L. MANGASARIAN AND L. L. SCHUMAKER, *Discrete splines via mathematical programming*, SIAM Journal on Control, 9 (1971), pp. 174–183.

[156] N. MARATOS, *Exact penalty function algorithms for finite dimensional and control optimization problems*, Ph.D. thesis, University of London, 1978.

[157] H. M. MARKOWITZ, *Portfolio selection*, Journal of Finance, 8 (1952), pp. 77–91.

[158] ———, *The elimination form of the inverse and its application to linear programming*, Management Science, 3 (1957), pp. 255–269.

[159] ———, *Portfolio Selection: Efficient Diversification of Investments*, Basil Blackwell, Cambridge, Mass., 1991.

[160] D. W. MARQUARDT, *An algorithm for least squares estimation of non-linear parameters*, SIAM Journal, 11 (1963), pp. 431–441.

[161] D. Q. MAYNE AND E. POLAK, *A superlinearly convergent algorithm for constrained optimization problems*, Mathematical Programming Studies, 16 (1982), pp. 45–61.

[162] L. MCLINDEN, *An analogue of Moreau's proximation theorem, with applications to the nonlinear complementarity problem*, Pacific Journal of Mathematics, 88 (1980), pp. 101–161.

[163] N. MEGIDDO, *Pathways to the optimal set in linear programming*, in Progress in Mathematical Programming: Interior-Point and Related Methods, N. Megiddo, ed., Springer-Verlag, New York, N.Y., 1989, ch. 8, pp. 131–158.

[164] S. MEHROTRA, *On the implementation of a primal-dual interior point method*, SIAM Journal on Optimization, 2 (1992), pp. 575–601.

[165] S. MIZUNO, M. TODD, AND Y. YE, *On adaptive step primal-dual interior-point algorithms for linear programming*, Mathematics of Operations Research, 18 (1993), pp. 964–981.

[166] J. J. MORÉ, *The Levenberg-Marquardt algorithm: Implementation and theory*, in Lecture Notes in Mathematics, No. 630–Numerical Analysis, G. Watson, ed., Springer-Verlag, 1978, pp. 105–116.

[167] ———, *Recent developments in algorithms and software for trust region methods*, in Mathematical Programming: The State of the Art, Springer-Verlag, Berlin, 1983, pp. 258–287.

[168] ———, *A collection of nonlinear model problems*, in Computational Solution of Nonlinear Systems of Equations, vol. 26 of Lectures in Applied Mathematics, American Mathematical Society, Providence, R.I., 1990, pp. 723–762.

[169] J. J. MORÉ AND D. C. SORENSEN, *On the use of directions of negative curvature in a modified Newton method*, Mathematical Programming, 16 (1979), pp. 1–20.

[170] ———, *Computing a trust region step*, SIAM Journal on Scientific and Statistical Computing, 4 (1983), pp. 553–572.

[171] ———, *Newton's method*, in Studies in Numerical Analysis, vol. 24 of MAA Studies in Mathematics, The Mathematical Association of America, 1984, pp. 29–82.

[172] J. J. MORÉ AND D. J. THUENTE, *Line search algorithms with guaranteed sufficient decrease*, ACM Transactions on Mathematical Software, 20 (1994), pp. 286–307.

[173] J. J. MORÉ AND S. J. WRIGHT, *Optimization Software Guide*, SIAM Publications, Philadelphia, 1993.

[174] W. MURRAY AND M. H. WRIGHT, *Line search procedures for the logarithmic barrier function*, SIAM Journal on Optimization, 4 (1994), pp. 229–246.

[175] B. A. MURTAGH AND M. A. SAUNDERS, *MINOS 5.1 User's guide*, Technical Report SOL-83-20R, Stanford University, 1987.

[176] K. G. MURTY AND S. N. KABADI, *Some NP-complete problems in quadratic and nonlinear programming*, Mathematical Programming, 19 (1987), pp. 200–212.

[177] S. G. NASH, *Newton-type minimization via the Lanczos method*, SIAM Journal on Numerical Analysis, 21 (1984), pp. 553–572.

[178] ———, *SUMT (Revisited)*, Operations Research, 46 (1998), pp. 763–775.

[179] G. L. NEMHAUSER AND L. A. WOLSEY, *Integer and Combinatorial Optimization*, John Wiley & Sons, New York, 1988.

[180] A. S. NEMIROVSKII AND D. B. YUDIN, *Problem complexity and method efficiency*, John Wiley & Sons, New York, 1983.

[181] Y. E. NESTEROV AND A. S. NEMIROVSKII, *Interior Point Polynomial Methods in Convex Programming*, SIAM Publications, Philadelphia, 1994.

[182] G. N. NEWSAM AND J. D. RAMSDELL, *Estimation of sparse Jacobian matrices*, SIAM Journal on Algebraic and Discrete Methods, 4 (1983), pp. 404–418.

[183] J. NOCEDAL, *Updating quasi-Newton matrices with limited storage*, Mathematics of Computation, 35 (1980), pp. 773–782.

[184] ———, *Theory of algorithms for unconstrained optimization*, Acta Numerica, 1 (1992), pp. 199–242.

[185] J. M. ORTEGA AND W. C. RHEINBOLDT, *Iterative solution of nonlinear equations in several variables*, Academic Press, New York and London, 1970.

[186] M. R. OSBORNE, *Nonlinear least squares—the Levenberg algorithm revisited*, Journal of the Australian Mathematical Society, Series B, 19 (1976), pp. 343–357.

[187] ———, *Finite Algorithms in Optimization and Data Analysis*, John Wiley & Sons, 1985.

[188] C. C. PAIGE AND M. A. SAUNDERS, *LSQR: An algorithm for sparse linear equations and sparse least squares*, ACM Transactions on Mathematical Software, 8 (1982), pp. 43–71.

[189] E. R. PANIER AND A. L. TITS, *On combining feasibility, descent and superlinear convergence in inequality constrained optimization*, Mathematical Programming, 59 (1993), pp. 261–276.

[190] C. H. PAPADIMITRIOU AND K. STEIGLITZ, *Combinatorial Optimization: Algorithms and Complexity*, Prentice Hall, Englewood Cliffs, NJ, 1982.

[191] R. J. PLEMMONS, *Least squares computations for geodetic and related problems*, in High Speed Computing, R. Williamson, ed., University of Illinois Press, 1989, pp. 198–200.

[192] R. J. PLEMMONS AND R. WHITE, *Substructuring methods for computing the nullspace of equilibrium matrices*, SIAM Journal on Matrix Analysis and Applications, 11 (1990), pp. 1–22.

[193] E. POLAK, *Optimization: Algorithms and Consistent Approximations*, no. 124 in Applied Mathematical Sciences, Springer, 1997.

[194] E. POLAK AND G. RIBIÈRE, *Note sur la convergence de méthodes de directions conjuguées*, Revue Française d'Informatique et de Recherche Opérationnelle, 16 (1969), pp. 35–43.

[195] M. J. D. POWELL, *A method for nonlinear constraints in minimization problems*, in Optimization, R. Fletcher, ed., Academic Press, New York, NY, 1969, pp. 283–298.

[196] ———, *A hybrid method for nonlinear equations*, in Numerical Methods for Nonlinear Algebraic Equations, P. Rabinowitz, ed., Gordon & Breach, London, 1970, pp. 87–114.

[197] ———, *Problems related to unconstrained optimization*, in Numerical Methods for Unconstrained Optimization, W. Murray, ed., Academic Press, 1972, pp. 29–55.

[198] ———, *On search directions for minimization algorithms*, Mathematical Programming, 4 (1973), pp. 193–201.

[199] ———, *Convergence properties of a class of minimization algorithms*, in Nonlinear Programming 2, O. L. Mangasarian, R. R. Meyer, and S. M. Robinson, eds., Academic Press, New York, 1975, pp. 1–27.

[200] ———, *Some convergence properties of the conjugate gradient method*, Mathematical Programming, 11 (1976), pp. 42–49.

[201] ———, *Some global convergence properties of a variable metric algorithm for minimization without exact line searches*, in Nonlinear Programming, SIAM-AMS Proceedings, Vol. IX, R. W. Cottle and C. E. Lemke, eds., SIAM Publications, 1976, pp. 53–72.

[202] ———, *A fast algorithm for nonlinearly constrained optimization calculations*, in Numerical Analysis Dundee 1977, G. A. Watson, ed., Springer Verlag, Berlin, 1977, pp. 144–157.

[203] ———, *Restart procedures for the conjugate gradient method*, Mathematical Programming, 12 (1977), pp. 241–254.

[204] ———, *Algorithms for nonlinear constraints that use Lagrangian functions*, Mathematical Programming, 14 (1978), pp. 224–248.

[205] ———, *The convergence of variable metric methods for nonlinearly constrained optimization calculations*, in Nonlinear Programming 3, Academic Press, New York and London, 1978, pp. 27–63.

[206] ———, *On the rate of convergence of variable metric algorithms for unconstrained optimization*, Technical Report DAMTP 1983/NA7, Department of Applied Mathematics and Theoretical Physics, Cambridge University, 1983.

[207] ———, *Nonconvex minimization calculations and the conjugate gradient method*, Lecture Notes in Mathematics, 1066 (1984), pp. 122–141.

[208] ———, *The performance of two subroutines for constrained optimization on some difficult test problems*, in Numerical Optimization, P. T. Boggs, R. H. Byrd, and R. B. Schnabel, eds., SIAM Publications, Philadelphia, 1984.

[209] ———, *Convergence properties of algorithms for nonlinear optimization*, SIAM Review, 28 (1986), pp. 487–500.

[210] M. J. D. POWELL AND P. L. TOINT, *On the estimation of sparse Hessian matrices*, SIAM Journal on Numerical Analysis, 16 (1979), pp. 1060–1074.

[211] D. RALPH AND S. J. WRIGHT, *Superlinear convergence of an interior-point method for monotone variational inequalities*, in Complementarity and Variational Problems: State of the Art, SIAM Publications, Philadelphia, Penn., 1997, pp. 345–385.

[212] Z. REN AND K. MOFFATT, *Quantitative analysis of synchotron Laue diffraction patterns in macromolecular crystallography*, Journal of Applied Crystallography, 28 (1995), pp. 461–481.

[213] J. M. RESTREPO, G. K. LEAF, AND A. GRIEWANK, *Circumventing storage limitations in variational data assimilation studies*, Preprint ANL/MCS-P515-0595, Mathematics and Computer Science Division, Argonne National Laboratory, Argonne, Ill., 1995.

[214] K. RITTER, *On the rate of superlinear convergence of a class of variable metric methods*, Numerische Mathematik, 35 (1980), pp. 293–313.

[215] S. M. ROBINSON, *A quadratically convergent algorithm for general nonlinear programming problems*, Mathematical Programming, (1972), pp. 145–156.

[216] R. T. ROCKAFELLAR, *The multiplier method of Hestenes and Powell applied to convex programming*, Journal of Optimization Theory and Applications, 12 (1973), pp. 555–562.

[217] ———, *Lagrange multipliers and optimality*, SIAM Review, 35 (1993), pp. 183–238.

[218] J. B. ROSEN AND J. KREUSER, *A gradient projection algorithm for nonlinear constraints*, in Numerical Methods for Non-Linear Optimization, F. A. Lootsma, ed., Academic Press, London and New York, 1972, pp. 297–300.

[219] N. ROSTAING, S. DALMAS, AND A. GALLIGO, *Automatic differentiation in Odyssee*, Tellus, 45a (1993), pp. 558–568.

[220] Y. SAAD, *Iterative Methods for Sparse Linear Systems*, PWS Publishing Company, 1996.

[221] Y. SAAD AND M. SCHULTZ, *GMRES: A generalized minimal residual algorithm for solving nonsymmetric linear systems*, SIAM Journal on Scientific and Statistical Computing, 7 (1986), pp. 856–869.

[222] K. SCHITTKOWSKI, *The nonlinear programming method of Wilson, Han and Powell with an augmented Lagrangian type line search function*, Numerische Mathematik, (1981), pp. 83–114.

[223] R. B. SCHNABEL AND E. ESKOW, *A new modified Cholesky factorization*, SIAM Journal on Scientific Computing, 11 (1991), pp. 1136–1158.

[224] R. B. SCHNABEL AND P. D. FRANK, *Tensor methods for nonlinear equations*, SIAM Journal on Numerical Analysis, 21 (1984), pp. 815–843.

[225] G. SCHULLER, *On the order of convergence of certain quasi-Newton methods*, Numerische Mathematik, 23 (1974), pp. 181–192.

[226] G. A. SCHULTZ, R. B. SCHNABEL, AND R. H. BYRD, *A family of trust-region-based algorithms for unconstrained minimization with strong global convergence properties*, SIAM Journal on Numerical Analysis, 22 (1985), pp. 47–67.

[227] G. A. F. SEBER AND C. J. WILD, *Nonlinear Regression*, John Wiley & Sons, New York, 1989.

[228] D. F. SHANNO AND K. H. PHUA, *Remark on Algorithm 500: Minimization of unconstrained multivariate functions*, ACM Transactions on Mathematical Software, 6 (1980), pp. 618–622.

[229] A. SHERMAN, *On Newton-iterative methods for the solution of systems of nonlinear equations*, SIAM Journal on Numerical Analysis, 15 (1978), pp. 755–771.

[230] D. D. SIEGEL, *Implementing and modifying Broyden class updates for large scale optimization,*, Technical Report AMTP 1992/NA12, Department of Applied Mathematics and Theoretical Physics, University of Cambridge, 1992.

[231] T. STEIHAUG, *The conjugate gradient method and trust regions in large scale optimization*, SIAM Journal on Numerical Analysis, 20 (1983), pp. 626–637.

[232] J. STOER, *On the relation between quadratic termination and convergence properties of minimization algorithms. Part I: Theory*, Numerische Mathematik, 28 (1977), pp. 343–366.

[233] K. TANABE, *Centered Newton method for mathematical programming*, in System Modeling and Optimization: Proceedings of the 13th IFIP conference, vol. 113 of Lecture Notes in Control and Information Systems, Berlin, 1988, Springer-Verlag, pp. 197–206.

[234] R. A. TAPIA, *Quasi-Newton methods for equality constrained optimization: Equivalence of existing methods and a new implementation*, in Nonlinear Programming 3 (O. Mangasarian, R. Meyer, and S. Robinson, eds), Academic Press, New York, NY, (1978) pp. 125–164.

[235] M. J. TODD, *Potential reduction methods in mathematical programming*, Mathematical Programming, Series B, 76 (1997), pp. 3–45.

[236] M. J. TODD AND Y. YE, *A centered projective algorithm for linear programming*, Mathematics of Operations Research, 15 (1990), pp. 508–529.

[237] P. L. TOINT, *On sparse and symmetric matrix updating subject to a linear equation*, Mathematics of Computation, 31 (1977), pp. 954–961.

[238] ———, *Towards an efficient sparsity exploiting Newton method for minimization*, in Sparse Matrices and Their Uses, Academic Press, New York, 1981, pp. 57–87.

[239] ———, *On large-scale nonlinear least squares calculations*, SIAM Journal on Scientific and Statistical Computing, 8 (1987), pp. 416–435.

[240] L. VANDENBERGHE AND S. BOYD, *Semidefinite programming*, SIAM Review, 38 (1996), pp. 49–95.

[241] S. A. VAVASIS, *Nonlinear Optimization*, Oxford University Press, New York and Oxford, 1991.

[242] H. WALKER, *Implementation of the GMRES method using Householder transformations*, SIAM Journal on Scientific and Statistical Computing, 9 (1989), pp. 815–825.

[243] WATERLOO MAPLE SOFTWARE, INC, *Maple V software package*, 1994.

[244] L. T. WATSON, *Numerical linear algebra aspects of globally convergent homotopy methods*, SIAM Review, 28 (1986), pp. 529–545.

[245] R. B. WILSON, *A simplicial algorithm for concave programming*, Ph.D. thesis, Graduate School of Business Administration, Harvard University, 1963.

[246] W. L. WINSTON, *Operations Research*, Wadsworth Publishing Co., 3rd ed., 1997.

[247] P. WOLFE, *The composite simplex algorithm*, SIAM Review, 7 (1965), pp. 42–54.

[248] S. WOLFRAM, *The Mathematica Book*, Cambridge University Press and Wolfram Media, Inc., third ed., 1996.

[249] L. A. WOLSEY, *Integer Programming*, Wiley–Interscience Series in Discrete Mathematics and Optimization, John Wiley & Sons, New York, NY, 1998.

[250] M. H. WRIGHT, *Numerical Methods for Nonlinearly Constrained Optimization*, Ph.D. thesis, Stanford University, Stanford University, CA, 1976.

[251] ———, *Interior methods for constrained optimization*, in Acta Numerica 1992, Cambridge University Press, 1992, pp. 341–407.

[252] ———, *Ill-conditioning and computational error in interior methods for nonlinear programming*, SIAM Journal on Optimization, 9 (1999), pp. 84–111.

[253] S. J. WRIGHT, *Applying new optimization algorithms to model predictive control*, in Chemical Process Control-V, J. C. Kantor, ed., CACHE, 1997.

[254] ———, *On the convergence of the Newton/log-barrier method*, Preprint ANL/MCS-P681-0897, Mathematics and Computer Science Division, Argonne National Laboratory, Argonne, Ill., August 1997.

[255] ———, *Primal-Dual Interior-Point Methods*, SIAM Publications, Philadelphia, Pa, 1997.

[256] ———, *Effects of finite-precision arithmetic on interior-point methods for nonlinear programming*, Preprint MCS-P705-0198, Mathematics and Computer Science Division, Argonne National Laboratory, Argonne, Ill., January 1998.

[257] S. J. WRIGHT AND J. N. HOLT, *An inexact Levenberg-Marquardt method for large sparse nonlinear least squares problems*, Journal of the Australian Mathematical Society, Series B, 26 (1985), pp. 387–403.

[258] S. J. WRIGHT AND F. JARRE, *The role of linear objective functions in barrier methods*, Preprint MCS-P485-1294, Mathematics and Computer Science Division, Argonne National Laboratory, Argonne, Ill., 1994. Revised 1998. To appear in Mathematical Programming, Series A.

[259] C. ZHU, R. H. BYRD, P. LU, AND J. NOCEDAL, *Algorithm 778: L-BFGS-B, FORTRAN subroutines for large scale bound constrained optimization*, ACM Transactions on Mathematical Software, 23 (1997), pp. 550–560.

Index

Accumulation point, *see* Limit point
Active set, 331, 336, 345, 347, 353, 422
    definition of, 327
Affine scaling
    direction, 398, 400, 405, 409
    method, 417
Alternating variables method, 53, 104
Angle test, 47
Applications
    design optimization, 1
    finance, 7, 342
    portfolio optimization, 1
    transportation, 4, 7
Armijo line search, 38, 139, 141
Augmented Lagrangian function, 424–425
    as merit function, 437
    definition, 514
    exactness of, 519–521
    example, 515

Augmented Lagrangian method, 424, 491, 495, 526
    convergence, 521
    framework for, 515, 518
    inequality constraints, 516–518
    LANCELOT code, 513, 515, 516, 521–523
    motivation, 513–515
Automatic differentiation, 136, 140, 165
    adjoint variables, 180, 181
    and graph-coloring algorithms, 184, 188–190
    basis in elementary arithmetic, 176
    checkpointing, 182
    common expressions, 184
    computational graph, 177–178, 180, 182, 183, 185, 187
    computational requirements, 178–179, 182, 186, 188, 190
    forward mode, 178–179, 285

Automatic (*cont.*)
  forward sweep, 177, 180, 182, 185–187, 190
  foundations in elementary arithmetic, 166
  Hessian calculation
    forward mode, 185–187
    interpolation formulae, 186–187
    reverse mode, 187–188
  intermediate variables, 177–180, 184, 190
  Jacobian calculation, 183–185
    forward mode, 184
    reverse mode, 184–185
  limitations of, 188–189
  nonlinear least-squares and, 252
  reverse mode, 179–182
  reverse sweep, 180–182, 190
  seed vectors, 178, 179, 184, 185, 188
  software, 166, 182, 189

Backtracking, 41
Barrier functions, 500
Barrier methods, 424
Barrier parameter, 424
Basic variables, 428
Basis, *see* Subspace, basis
Basis matrix, 428–430
BFGS method, 25, 31, 194–201
  damping, 540
  implementation, 200–201
  properties, 199–200, 219
  self-correction, 200
  skipping, 201, 540, 550
Bound-constrained optimization, 96, 422, 476, 513
Boundary layer, 502
Boundedness
  of sets, 578
Broyden class, 207
Broyden's method, 279, 280, 292, 311, 592
  derivation of, 286–288
  limited-memory variants, 290
  rate of convergence, 288–289
  statement of algorithm, 288

Calculus of variations, 9

Cancellation error, *see* Floating-point arithmetic, cancellation
Cauchy point, 68–77, 99, 159, 160, 263, 477
  calculation of, 69–70, 95–96
  for nonlinear equations, 299–300
  role in global convergence, 87–89
Cauchy–Schwarz inequality, 98
Central path, 399–400, 402, 417, 509, 511
  neighborhoods of, 402–404, 409, 412, 415, 484
Chain rule, 31, 166, 176, 178–180, 185, 243, 582, 583
Choleksy factorization
  modified, 155
Cholesky factorization, 82, 199, 201, 219, 256–257, 264, 296, 300, 602–604, 610
  incomplete, 158
  modified, 141, 144–150, 162, 163
    bounded modified factorization property, 141
    Gershgorin modification, 150
  sparse, 408–409
  stability of, 147, 610
Complementarity, 399
Complementarity condition, 78, 323, 328, 344, 508
  strict, 328, 330, 348–350, 508, 518
Complexity of algorithms, 392, 395, 396, 410, 415–417
Conditioning, *see also* Matrix, condition number, 426, 429, 430, 432, 608–609
  ill conditioned, 31, 495, 505, 510, 514, 608, 609
  well conditioned, 608, 609
Cone, 579
Cone of feasible directions, *see* Tangent cone
Conjugacy, 102
Conjugate direction method, 102
  expanding subspace minimization, 106
  termination of, 103
Conjugate gradient method, 75, 101–132, 135, 139, 154, 163, 270, 285
  n-step quadratic convergence, 132

relation to limited memory, 227
VA14, 229
clustering, 115
condition number, 117
CONMIN, 124, 229
expanding subspace, 76, 112
Fletcher–Reeves, *see* Fletcher–Reeves method
global convergence, 46
Hestenes–Stiefel, 122
Krylov subspace, 112
nonlinear, 26, 120–131
numerical performance, 124
optimal polynomial, 113
optimal process, 112
Polak–Ribière, *see* Polak–Ribière method
practical version, 111
preconditioned, 118, 154
rate of convergence, 112
restarts, 122
superlinear convergence, 132
superquadratic, 132
termination, 114, 122
Constrained optimization, 6
linear, 422
nonlinear, 6, 184, 363, 421, 422, 491, 492, 494, 509
Constraint qualifications, 328, 336–339, 345, 351–353, 357
linear independence (LICQ), 328, 329, 336, 337, 339, 341, 343, 348, 351, 353, 355, 359, 360, 366, 454, 497, 498, 505, 508, 519
Mangasarian–Fromovitz (MFCQ), 353, 359
Constraints, 1, 2, 319, 421
bounds, 434, 511, 516, 522
equality, 315
hard and soft, 423
inequality, 315
Continuation methods for nonlinear equations, 280, 312
convergence of, 308–310
formulation as initial-value ODE, 306–307
motivation, 304–305

predictor–corrector method, 307–308
zero path, 305–307, 309, 310, 312
divergence of, 309–310
tangent, 306–307, 309
turning point, 305, 308
Convergence, rate of, 28–30, 162
n-step quadratic, 132
linear, 28, 29, 137, 267
quadratic, 24, 29, 30, 32, 138, 142, 262
sublinear, 32
superlinear, 8, 24, 29, 30, 32, 71, 132, 138, 198, 200, 218, 220, 267, 269, 311, 409, 525
superquadratic convergence, 132
Convex programming, 6, 8, 347, 350–351, 422, 507
Convexity, 8
of functions, 8, 17, 31, 135, 163, 256, 350
of sets, 8, 31, 350, 358
Coordinate descent method, 53, *see* Alternating variables method
Coordinate relaxation step, 430

Data-fitting problems, 12–13, 254
Degeneracy, 328, 455
of linear program, 373
Dennis and Moré characterization, 50
Descent direction, 22, 31, 35
DFP method, 197
Differential equations
ordinary, 307
partial, 188, 310
Directional derivative, 178, 179, 583–584
Discrete optimization, 4–5, 423
Dual variables, *see also* Lagrange multipliers, 437
Duality, 357
in linear programming, 367–370

Eigenvalues, 79, 258, 349, 598, 605
negative, 74, 94
of symmetric matrix, 504, 599
Eigenvectors, 79, 258, 598
Element function, 235
Elimination of Variables, 425
linear equality constraints, 427–434
nonlinear, 426–427

Elimination (*cont.*)
  when inequality constraints are present, 434
Ellipsoid algorithm, 392, 395, 416
Error
  absolute, 606
  relative, 167, 257, 258, 606, 609
  truncation, 188
Errors-in-variables models, 271

Feasible sequences, 332–339, 343–345, 347
  limiting directions of, 332–339, 341–342, 345–346, 351
Feasible set, 3, 315, 317, 350, 351, 491
  geometric properties of, 351, 353–357, 359, 586–590
  primal, 365
  primal–dual, 397, 398, 402, 409, 415
  strictly, 500, 507, 509
Finite differencing, 136, 140, 165–176, 188, 274, 285
  and graph-coloring algorithms, 172–176
  central-difference formula, 165, 168–169, 173, 174, 189
  forward-difference formula, 167, 168, 173, 174, 189
  gradient approximation, 166–169
  Hessian approximation, 173–176
  Jacobian approximation, 169–173, 290
First-order feasible descent direction, 322–327
First-order optimality conditions, *see also* Karush–Kuhn–Tucker (KKT) conditions, 86, 282, 319–342, 351, 354, 358, 498
  derivation of, 331–342
  examples, 319–327, 329–330, 332–335
  fundamental principle, 335
  unconstrained optimization, 15–16, 437
Fixed-regressor model, 253
Fletcher–Reeves method, 101, 120–131
  convergence of, 124
  numerical performance, 124
Floating-point arithmetic, 188, 607–609
  cancellation, 430, 607–608
  double-precision, 607
  roundoff error, 167, 189, 257, 499, 505, 525, 607

unit roundoff, 167, 189, 607
Floating-point numbers, 606
  exponent, 607
  fractional part, 607
Forcing sequence, *see* Newton's method, inexact, forcing sequence
Function
  continuous, 580
  continuously differentiable, 582
  derivatives of, 581–585
  differentiable, 582
  Lipschitz continuous, 581, 586
  Lipschitz continuously differentiable, 585, 586
  locally Lipschitz continuous, 581
  one-sided limit, 580
  univariate, 581, 583
Fundamental theorem of algebra, 520, 598

Gauss–Newton method, 259–263, 267, 269, 272, 274, 282
  connection to linear least squares, 260
  line search in, 259, 260
  performance on large-residual problems, 267
Gaussian elimination, 144, 429, 603, 604
  sparse, 136, 429, 434
  stability of, 610
  with row partial pivoting, 601–602, 609–610
Global convergence, 87–94, 263, 280
Global minimizer, 13–14, 17, 422, 496, 507
Global optimization, 6, 8
Global solution, *see also* Global minimizer, 6, 78, 84–87, 316, 347, 350, 358
GMRES algorithm, 285
Goldstein condition, 41, 139, 141
Gradient, 582
  generalized, 18
Gradient–projection method, 453, 476–481, 486, 522
Group partial separability, *see* Partially separable function, group partially separable

Hölder inequality, 595
Harwell subroutine library
  VA14, 123

Hessian, 15, 20, 24, 26, 582
  average, 196, 197
Homotopy map, 304
Homotopy methods, *see* Continuation methods for nonlinear equations

Implicit function theorem, 337, 338, 355, 585–586, 588
Inertia of a matrix, 151, 447, 475
Inexact Newton method, *see* Newton's method, inexact
Integer programming, 5
  branch-and-bound algorithm, 5
Integral equations, 310
Interior-point methods, *see* Primal–dual interior-point methods, 392
Interlacing eigenvalue theorem, 605–606
Invariant subspace, *see* Partially separable optimization, invariant subspace

Jacobian, 252, 256, 260, 261, 274, 338, 398

Karmarkar's algorithm, 392, 396, 416
Karush–Kuhn–Tucker (KKT) conditions, 342, 343, 345, 347, 348, 353, 357, 359, 360, 422, 475, 497, 498, 507–510, 512, 518, 519, 522, 527
  for general constrained problem, 328
  for linear programming, 366–367, 374, 375, 397, 399, 402, 410, 411
Krylov subspace, 108

Lagrange multipliers, 321, 322, 330–331, 339, 342, 345, 348, 349, 353, 359, 366–368, 419, 422, 510
  estimates of, 497, 504, 508, 509, 513, 514, 521, 524
Lagrangian function, 86, 321, 323, 325, 342, 343, 347, 508
  for constrained optimization, 327
  for linear program, 366, 367
  Hessian of, 342–345, 347, 348, 366
  projected Hessian of, 349
LANCELOT, *see* Augmented Lagrangian method, LANCELOT code
Lanczos method, 74
LAPACK, 601
Least-squares problems, linear, 256–259

  applications of, 273
  LSQR algorithm, 270
  normal equations, 256–257, 264, 269, 275, 408
  solution via QR factorization, 257
  solution via SVD, 257–258
Least-squares problems, nonlinear, 13, 183
  applications of, 251, 253–254
  Dennis–Gay–Welsch algorithm, 267–269
  Fletcher–Xu algorithm, 267
  large-residual problems, 266–269
  large-scale problems, 269–270
  Levenberg–Marquardt method, *see* Levenberg–Marquardt method
  scaling of, 266
  software for, 268, 274
  statistical justification of, 255
  structure of objective, 252, 259
Least-squares problems, total, 271
Level set, 94, 263
Levenberg–Marquardt method, 262–266, 269, 272
  as trust-region method, 262–264, 300
  for nonlinear equations, 300
  implementation via orthogonal transformations, 264–266
  inexact, 270
  local convergence of, 266
  performance on large-residual problems, 267
lim inf, lim sup, 578
Limit point, 31, 89, 94, 98, 496, 497, 507, 577–578
Limited memory method, 25, 224–233, 247
  compact representation, 230–232
  for nonlinear equations, 247
  L-BFGS, 224–233
  L-BFGS algorithm, 226
  memoryless BFGS method, 227
  performance of, 227
  relation to CG, 227
  scaling, 226
  SR1, 232
  two-loop recursion, 225
Line search, *see also* Step length selection

Line (cont.)
  Armijo, 38
  backtracking, 41
  curvature condition, 38
  for log-barrier function, 506, 510, 526
  Goldstein, 41
  inexact, 37
  Newton's method with, 22–24
  Nonlinear conjugate gradient methods with, 26
  quasi-Newton methods with, 24–25
  search directions, 21–26
  strong Wolfe, 39
  sufficient decrease, 37
  Wolfe conditions, 37
Line search method, 19–20, 35–55, 65, 66, 69, 252
  for nonlinear equations, 278, 293–298
    global convergence of, 294–297
    local convergence of, 298
    poor performance of, 296
Linear complementarity problem, 410–411
Linear programming, 4, 6, 9, 301, 351, 421, 491, 512
  artificial variables, 370, 386–389
  basic feasible points, 370–374
  dual problem, 367–370
  feasible polytope, 364
    vertices of, 372–373
  fundamental theorem of, 371–372
  primal solution set, 364
  slack/surplus variables, 365, 368, 370, 388, 411
  splitting variables, 365, 368
  standard form, 364–365, 411
Linearly dependent, 350, 586
Linearly independent, 353, 431, 497, 498, 519, 521
Lipschitz continuity, *see also* Function, Lipschitz continuous, 281–284, 286, 294, 295, 298, 301, 302, 311
Local minimizer, 13, 15, 280, 527
  isolated, 14, 31
  strict, 13, 15, 16, 31, 519
  weak, 13

Local solution, *see also* Local minimizer, 316–317, 330, 332, 335, 341, 343, 345, 350, 354, 358, 437
  isolated, 317
  strict, 317, 345, 347, 348
  strong, 317
Log-barrier function, 417, 424, 525–527
  definition, 500–501
  difficulty of minimizing, 502, 510
  examples, 501–504
  ill conditioned Hessian of, 504–505
  possible unboundedness, 507
  properties, 507–509
Log-barrier method, 491, 505–506
  convergence of, 508–509
  extrapolation, 506
  modification for equality constraints, 509–510
  relationship to primal–dual interior-point methods, 506, 510–512
LSQR method, 449, 485, 536
LU factorization, 601–602

Maratos effect, 436, 550, 558, 567–573
  example of, 567
  remedies, 569
Matrix
  condition number, 257, 596, 604, 609
  determinant, 600
  diagonal, 258, 408, 429
  full-rank, 306, 308, 309, 498, 604
  indefinite, 73, 74
  lower triangular, 601, 602
  nonsingular, 338, 350, 596, 605
  null space, 306, 337, 348, 430, 431, 598, 602, 604
  orthogonal, 257, 258, 350, 432, 598, 603
  permutation, 257, 429, 601
  positive definite, 16, 23, 30, 67, 74, 75, 349, 594
  positive semidefinite, 16, 78, 349, 410, 411, 594
  range space, 430, 598
  rank-deficient, 259
  rank-one, 25
  rank-two, 25
  singular, 350

symmetric, 25, 67, 408, 411, 594
symmetric indefinite, 409
symmetric positive definite, 602
trace, 599–600
upper triangular, 257, 350, 601–603
Matrix, sparse, 408, 409, 602, 603
 Cholesky factorization, 409
Maximum likelihood estimate, 255
Merit function, *see also* Penalty function, 422, 425, 434–438
 ℓ_1, 301, 435–437, 544–547, 558
  choice of parameter, 547
 ℓ_2, 526
 exact, 435–437
  definition of, 435
  nonsmoothness of, 436–437
  smooth, 513
 Fletcher's augmented Lagrangian, 435–436, 544–547
  choice of parameter, 547
 for feasible methods, 434
 for nonlinear equations, 279, 292–294, 296, 298, 299, 301, 304, 310–312, 498
 for primal–dual interior-point methods, 512
 for SQP, 544–547
Method of multipliers, *see* Augmented Lagrangian methods
Minimum surface problem, 238, 244, 246
MINOS, *see* Sequential linearly constrained methods, MINOS
Modeling, 1–2, 9, 12, 253–255

Negative curvature direction, 73, 75, 76, 139–143, 156, 157, 161–163, 470, 471, 480
Neighborhood, 13, 15, 31, 579
Network optimization, 365
Newton's method, 26, 252, 259, 261, 267
 for log-barrier function, 502, 505, 525, 526
 for nonlinear equations, 278, 280–284, 288, 290, 292, 293, 295–298, 302, 304, 308, 311
  cycling, 292
  inexact, 284–286, 297
 for quadratic penalty function, 495, 499

 global convergence, 45
 Hessian-free, 140, 156
 in one variable, 78, 81, 93, 592
 inexact, 136–138, 156, 162, 185
  forcing sequence, 136–138, 140, 156, 162, 285
 Lanczos, 157
 large scale, 135–162
  LANCELOT, 159
  line search method, 139–142
  MINPACK-2, 159
  trust-region method, 154
 modified, 141–142
  adding a multiple of I, 144
  eigenvalue modification, 143–144
 Newton–CG, 136, 139–141, 156–159, 161–163, 173
  preconditioned, 157–159
 rate of convergence, 29, 51, 137–138, 155, 159, 282–284, 288–289, 298
 scale invariance, 27
Newton–Lagrange method, *see* Sequential quadratic programming
Nondifferentiable optimization, 513
Nonlinear complementarity problems, 417
Nonlinear equations, 169, 183, 185, 423, 592
 degenerate solution, 281, 283, 290, 292, 311
 examples of, 278–279, 296, 310
 merit function, *see* Merit function, for nonlinear equations
 multiple solutions, 279–280
 primal–dual interior-point methods, relation to, 398, 511
 quasi-Newton methods, *see* Broyden's method
 relationship to least squares, 278–279, 282, 298, 300–301, 311
 relationship to optimization, 278
 solution, 278
 statement of problem, 277–278
Nonlinear least-squares, *see* Least-squares problem, nonlinear
Nonlinear programming, *see* Constrained optimization, nonlinear, 301
Nonmonotone algorithms, 19

Nonnegative orthant, 97
Nonsmooth functions, 6, 18, 317, 318, 358
Norm
   Euclidean, 26, 144, 257, 288, 311, 595, 596, 599, 604
   Frobenius, 144, 196, 197, 596
   matrix, 595–596
      consistent with vector norms, 596
   vector, 594–595
Normal cone, 354–357, 587–590
Normal distribution, 255
Normal step, 556
Null space, *see* Matrix, null space
Numerical analysis, 363

Objective function, 1, 2, 11, 315, 421
One-dimensional minimization, 19, 55
Optimal control, 586
Optimality conditions, *see also* First-order optimality conditions, Second-order optimality conditions, 2, 8, 315–316
   for unconstrained local minimizer, 15–17
Order notation, 591–592
Orthogonal distance regression, 271–273
   contrast with least squares, 271–272
   structure of objective, 272–273
Orthogonal transformations, 257, 264–266
   Givens, 264, 604
   Householder, 264, 604
Outliers, 267

Partially separable function, 25, 183, 235–237, 270
   automatic detection, 183, 241
   definition, 183, 241
   group partially separable, 243
   vs. sparsity, 242
Partially separable optimization, 235–247
   BFGS, 246
   compactifying matrix, 236
   element variables, 236
   internal variables, 237, 239
   invariant subspace, 239, 240
   Newton's method, 244
   quasi-Newton method, 237, 245
   SR1, 246

Penalty function, *see also* Merit function, 492
   exact, 424, 491, 512–513
   ℓ_1, 512–513
   quadratic, 424, 492–494, 504, 505, 509, 525, 526
      difficulty of minimizing, 495
      Hessian of, 498–499
      relationship to augmented Lagrangian, 513
Penalty methods, 424
   exterior, 492
Penalty parameter, 435, 492, 514, 522–524
Pivoting, 257, 610
Polak–Ribière method, 121
   convergence of, 130
Polak–Ribière method
   numerical performance, 124
Portfolio optimization, 442–443, 485
Preconditioners, 118–120
   banded, 119
   incomplete Cholesky, 119
   SSOR, 119
Primal–dual interior-point methods, 475, 491, 525, 527
   centering parameter, 400, 401, 403–405, 409
   complexity of, 396, 410, 415–416
   contrasts with simplex method, 364, 396
   convex quadratic programs, 410–411
   corrector step, 404–406, 409
   duality measure, 400
   infeasible-interior-point algorithms, 401–404
   linear algebra issues, 408–409
   Mehrotra's predictor–corrector algorithm, 396, 404–408, 483
   nonlinear programs, 411, 510–512, 526
   path-following algorithms, 402–409, 484
      long-step, 411–416
      predictor–corrector (Mizuno-Todd-Ye) algorithm, 409
      short-step, 409
   potential function, 410
      Tanabe–Todd–Ye, 410, 484

potential-reduction algorithms, 409–410, 484
predictor step, 405–406, 409
quadratic programming, 481–484
relationship to Newton's method, 397, 398, 402
software, 417
Probability density function, 255
Projected Hessian, 564
two-sided, 565

QMR method, 449, 485, 536
QR factorization, 257, 264, 297, 300, 307, 349, 432, 434, 603–605
cost of, 603
relationship to Cholesky factorization, 604
Quadratic penalty method, 491, 494–495, 513
convergence of, 495–500
Quadratic programming, 422, 425, 441–486
active set methods, 457–476
big M method, 463
blocking constraint, 459
convex, 491
cycling, 467
duality, 484
indefinite, 470–476
inertia controlling methods, 470–474
inertia controlling method, 486
initial working set, 465
interior-point method, 481–484
null-space method, 450–452
optimal active set, 457
optimality conditions, 454
phase I, 462
pseudo-constraint, 471
range-space method, 449–450
termination, 466
updating factorizations, 467
working set, 457–467
Quasi-Newton approximate Hessian, 24, 25, 71, 522, 592
Quasi-Newton method, *see also* Limited-memory method, 26, 252, 267, 495, 502
BFGS, *see* BFGS method, 267

bounded deterioration, 219
Broyden class, *see* Broyden class
curvature condition, 195
DFP, *see* DFP method, 247, 269
for nonlinear equations, *see* Broyden's method
for partially separable functions, 25
global convergence, 45
large-scale, 223–247
limited memory, *see* Limited memory method
rate of convergence, 29, 49
secant equation, 24, 25, 195, 197, 268–269, 287, 592
sparse, *see* Sparse quasi-Newton method
SR1, *see* SR1 method
symmetric-rank-one (SR1), 25

Range space, *see* Matrix, range space
Relative interior, 590
Residuals, 12, 251, 260, 266–269, 274
vector of, 18, 169, 252
Robustness, 7
Root, *see* Nonlinear equations, solution
Root-finding algorithm, 265
Rootfinding algorithm, *see also* Newton's method, in one variable, 264, 592–593
for trust-region subproblem, 78–83
Rosenbrock function
extended, 248
Roundoff error, *see* Floating-point arithmetic, roundoff error
Row echelon form, 429

$S\ell_1$QP method, 301, 557–560
Saddle point, 30, 94
Scale invariance, 196, 199
Scaling, 27–28, 94–97, 331, 502, 504
example of poor scaling, 27
matrix, 95
scale invariance, 28
Schur complement, 152
Secant method, *see also* Quasi-Newton method, 287, 592–593
Second-order correction, 558, 569–571
Second-order optimality conditions, 330, 342–350, 597

Second-order (*cont.*)
   necessary, 94, 343–349
   sufficient, 345–349, 508, 519
   unconstrained optimization, 16–17
Semidefinite programming, 411, 491
Sensitivity, 582, 609
Sensitivity analysis, 2, 166, 258, 330–331, 357, 369
Separable problem, 235
Sequential linearly constrained methods, 424, 425, 491, 523–526
   MINOS, 524–526
Sequential quadratic programming, 425, 475, 513, 524, 529–573
   augmented Lagrangian Hessian, 541
   Coleman–Conn method, 543, 549, 551
   derivation, 530–533
   full quasi-Newton Hessian, 540
   identification of optimal active set, 533
   IQP vs. EQP, 534
   KKT system, 282, 531
   least-squares multipliers, 537
   line search algorithm, 547
   local algorithm, 532
   QP multipliers, 537
   rate of convergence, 563–567
   reduced-Hessian approximation, 542–544
   reduced-Hessian method, 538, 548–553
      properties, 549
   $S\ell_1$QP method, *see* $S\ell_1$QP method
   shifting constraints, 555
   step computation, 536–539
      direct, 536
      for inequalities, 538
      iterative, 536
      null-space, 537
      range-space, 536
   tangential convergence, 549
   trust-region method, 553–563
   two elliptical constraints, 556
Set
   affine hull of, 579
   closed, 578
   closure of, 578
   compact, 579
   interior of, 578
   open, 578
   relative interior of, 579
Sherman–Morrison–Woodbury formula, 197, 198, 202, 220, 290, 385, 605
Simplex method
   as active-set method, 391–392
   basic index set $\mathcal{B}$, 370–375, 386
   complexity of, 392
   cycling, 389
      avoidance of, 389–391
   degenerate steps, 378
   description of single iteration, 374–378
   discovery of, 363
   entering index, 375–376, 378, 383–386
   finite termination of, 377–378
   initialization, 386–389
   leaving index, 375, 376, 378
   linear algebra issues, 378–383
   Phase I/Phase II, 386–389
   pricing, 375, 383–384
      multiple, 384
      partial, 383
   steepest-edge rule, 384–386
Singular values, 260, 598
Singular-value decomposition (SVD), 258, 274, 275, 311, 598
Slack variables, *see also* Linear programming, slack/surplus variables, 510, 516, 517, 521
Smooth functions, 11, 15, 317–319, 342
SNOPT, 538
Sparse quasi-Newton method, 233–235, 247
SR1 method, 202, 219
   algorithm, 204
   properties, 205
   safeguarding, 203
   skipping, 203, 218
Stability, 608–610
Starting point, 19
Stationary point, 8, 16, 30, 296, 437, 498
Steepest descent direction, 21, 22, 68, 72, 145
Steepest descent method, 22, 26, 27, 35, 71, 94, 502
   rate of convergence, 29, 47, 49
Step length, 19, 35

unit, 23, 31
Step length selection, *see also* Line search, 55–61
  bracketing phase, 55
  cubic interpolation, 57
  for Wolfe conditions, 58
  initial step length, 58
  interpolation in, 56
  selection phase, 55
Stochastic optimization, 7
Strict complementarity, *see* Complementarity condition, strict
Subgradient, 18
Subspace, 349
  basis, 430, 597
    orthonormal, 433
  dimension, 597
  linearly independent set, 597
  spanning set, 597
Successive linear programming, 534
Sufficient reduction, 69, 70, 89
Sum of absolute values, 254
Sum of squares, *see* Least-squares problem, nonlinear
Symbolic differentiation, 166
Symmetric indefinite factorization, 448, 475, 476
  Bunch–Kaufman, 153
  Bunch–Parlett, 152
  modified, 151–154, 162
  sparse, 153
Symmetric rank-one update, *see* SR1 method

Tangent, 506
Tangent cone, 339, 354–357, 587–590
Tangential step, 560
Taylor series, 16, 23, 30, 31, 66, 67, 281, 320, 342, 344, 345, 495, 502, 505, 527, 587
Taylor's theorem, 16, 21, 22, 24, 90, 122, 137, 138, 165–169, 173, 174, 196, 281, 287, 302, 335, 337, 338, 344, 346, 354–356, 585
  statement of, 15
Tensor methods, 280
  computational results, 292
  derivation, 290–292

performance on degenerate problems, 292
Termination criterion, 93
Triangular substitution, 433, 601, 603, 609, 610
Truncated Newton method, *see* Newton's method, Newton–CG
Trust region
  boundary, 68, 72, 73, 76, 77, 94
  box-shaped, 20, 301
  choice of size for, 65–66, 91
  elliptical, 20, 66, 95, 96, 99
  radius, 20, 26, 67–69, 71, 262, 302
  spherical, 94, 262
Trust-region method, 19–20, 68, 81–82, 87, 89, 90, 92–94, 252, 262, 592
  contrast with line search method, 20, 65
  dogleg method, 68, 71–74, 77, 78, 87, 89, 93, 98, 154–155, 161, 299–301
  double-dogleg method, 98
  for log-barrier function, 506
  for nonlinear equations, 278, 279, 293, 298–304, 311
    global convergence of, 300–302
    local convergence of, 302–304
  global convergence, 69, 70, 74, 76, 77, 87–94, 155, 156
  local convergence, 159–162
  Newton variant, 26–27, 67, 77, 94
  software, 97
  Steihaug's approach, 68, 75–77, 87, 89, 93, 480
  strategy for adjusting radius, 68
  subproblem, 20, 26–27, 67, 70, 71, 74, 87, 93, 95–97, 262, 263
    approximate solution of, 67, 69
    exact solution of, 77–78
    hard case, 82–84
    nearly exact solution of, 68–69, 74, 78–87, 89, 97, 156, 162, 300–301
  two-dimensional subspace minimization, 68, 74, 78, 87, 89, 97, 99, 154–155

Unconstrained optimization, 6, 11–30, 350, 358, 426, 432, 493, 495, 502, 513
Unit ball, 86, 520

Unit roundoff, *see* Floating-point arithmetic, unit roundoff

Variable metric method, *see* Quasi-Newton method
Variable storage method, *see* Limited memory method

Watchdog technique, 569–573

Weakly active, 331
Wolfe conditions, 37–41, 87, 131, 139, 141, 194, 195, 198–201, 204, 218, 226, 260, 294, 295, 298
  scale invariance of, 41
  strong, 39, 40, 121, 122, 124–128, 195, 200, 220, 226

Zoutendijk condition, 43–46, 127, 142, 214, 295

《国外数学名著系列》(影印版)

(按初版出版时间排序)

1. 拓扑学 I：总论　S. P. Novikov (Ed.)　2006.1
2. 代数学基础　Igor R. Shafarevich　2006.1
3. 现代数论导引 (第二版)　Yu. I. Manin　A. A. Panchishkin　2006.1
4. 现代概率论基础 (第二版)　Olav Kallenberg　2006.1
5. 数值数学　Alfio Quarteroni　Riccardo Sacco　Fausto Saleri　2006.1
6. 数值最优化　Jorge Nocedal　Stephen J. Wright　2006.1
7. 动力系统　Jürgen Jost　2006.1
8. 复杂性理论　Ingo Wegener　2006.1
9. 计算流体力学原理　Pieter Wesseling　2006.1
10. 计算统计学基础　James E. Gentle　2006.1
11. 非线性时间序列　Jianqing Fan　Qiwei Yao　2006.1
12. 函数型数据分析 (第二版)　J. O. Ramsay　B. W. Silverman　2006.1
13. 矩阵迭代分析 (第二版)　Richard S. Varga　2006.1
14. 偏微分方程的并行算法　Petter Bjørstad　Mitchell Luskin(Eds.)　2006.1
15. 非线性问题的牛顿法　Peter Deuflhard　2006.1
16. 区域分解算法：算法与理论　A. Toselli　O. Widlund　2006.1
17. 常微分方程的解法 I：非刚性问题 (第二版)　E. Hairer　S. P. Nørsett　G. Wanner　2006.1
18. 常微分方程的解法 II：刚性与微分代数问题 (第二版)　E. Hairer　G. Wanner　2006.1
19. 偏微分方程与数值方法　Stig Larsson　Vidar Thomée　2006.1
20. 椭圆型微分方程的理论与数值处理　W. Hackbusch　2006.1
21. 几何拓扑：局部性、周期性和伽罗瓦对称性　Dennis P. Sullivan　2006.1
22. 图论编程：分类树算法　Victor N. Kasyanov　Vladimir A. Evstigneev　2006.1
23. 经济、生态与环境科学中的数学模型　Natali Hritonenko　Yuri Yatsenko　2006.1
24. 代数数论　Jürgen Neukirch　2007.1
25. 代数复杂性理论　Peter Bürgisser　Michael Clausen　M. Amin Shokrollahi　2007.1
26. 一致双曲性之外的动力学：一种整体的几何学的与概率论的观点　Christian Bonatti　Lorenzo J. Díaz　Marcelo Viana　2007.1
27. 算子代数理论 I　Masamichi Takesaki　2007.1
28. 离散几何中的研究问题　Peter Brass　William Moser　János Pach　2007.1
29. 数论中未解决的问题 (第三版)　Richard K. Guy　2007.1
30. 黎曼几何 (第二版)　Peter Petersen　2007.1
31. 递归可枚举集和图灵度：可计算函数与可计算生成集研究　Robert I. Soare　2007.1
32. 模型论引论　David Marker　2007.1

33. 线性微分方程的伽罗瓦理论 Marius van der Put Michael F. Singer 2007.1
34. 代数几何 II：代数簇的上同调，代数曲面 I. R. Shafarevich (Ed.) 2007.1
35. 伯克利数学问题集 (第三版) Paulo Ney de Souza Jorge-Nuno Silva 2007.1
36. 陶伯理论：百年进展 Jacob Korevaar 2007.1

37. 同调代数方法 (第二版) Sergei I. Gelfand Yuri I. Manin 2009.1
38. 图像处理与分析：变分，PDE，小波及随机方法 Tony F. Chan Jianhong Shen 2009.1
39. 稀疏线性系统的迭代方法 Yousef Saad 2009.1
40. 模型参数估计的反问题理论与方法 Albert Tarantola 2009.1
41. 常微分方程和微分代数方程的计算机方法 Uri M. Ascher Linda R. Petzold 2009.1
42. 无约束最优化与非线性方程的数值方法 J. E. Dennis Jr. Robert B. Schnabel 2009.1
43. 代数几何 I：代数曲线，代数流形与概型 I. R. Shafarevich (Ed.) 2009.1
44. 代数几何 III：复代数簇，代数曲线及雅可比行列式 A. N. Parshin I. R. Shafarevich (Eds.) 2009.1
45. 代数几何 IV：线性代数群，不变量理论 A. N. Parshin I. R. Shafarevich (Eds.) 2009.1
46. 代数几何 V：Fano 簇 A. N. Parshin I. R. Shafarevich (Eds.) 2009.1
47. 交换调和分析 I：总论，古典问题 V. P. Khavin N. K. Nikol'skij (Eds.) 2009.1
48. 复分析 I：整函数与亚纯函数，多解析函数及其广义性 A. A. Gonchar V. P. Havin N. K. Nikolski (Eds.) 2009.1
49. 计算不变量理论 Harm Derksen Gregor Kemper 2009.1
50. 动力系统 V：分歧理论和突变理论 V. I. Arnol'd (Ed.) 2009.1
51. 动力系统 VII：可积系统，不完整动力系统 V. I. Arnol'd, S. P. Novikov (Eds.) 2009.1
52. 动力系统 VIII：奇异系统 II：应用 V. I. Arnol'd (Ed.) 2009.1
53. 动力系统 IX：带有双曲性的动力系统 D. V. Anosov (Ed.) 2009.1
54. 动力系统 X：旋涡的一般理论 V. V. Kozlov 2009.1
55. 几何 I：微分几何基本思想与概念 R. V. Gamkrelidze (Ed.) 2009.1
56. 几何 II：常曲率空间 E. B. Vinberg (Ed.) 2009.1
57. 几何 III：曲面理论 Yu. D. Burago V. A. Zalgaller (Eds.) 2009.1
58. 几何 IV：非正规黎曼几何 Yu. G. Reshetnyak (Ed.) 2009.1
59. 几何 V：最小曲面 R. Osserman (Ed.) 2009.1
60. 几何 VI：黎曼几何 M. M. Postnikov 2009.1
61. 李群与李代数 I：李理论基础，李交换群 A. L. Onishchik (Ed.) 2009.1
62. 李群与李代数 II：李群的离散子群,李群与李代数的上同调 A. L. Onishchik E. B. Vinberg (Eds.) 2009.1
63. 李群与李代数 III：李群与李代数的结构 A. L. Onishchik E. B. Vinberg (Eds.) 2009.1
64. 经典力学与天体力学中的数学问题 Vladimir I. Arnold Valery V. Kozlov Anatoly I. Neishtadt 2009.1
65. 数论 IV：超越数 A. N. Parshin I. R. Shafarevich (Eds.) 2009.1
66. 偏微分方程 IV：微局部分析和双曲型方程 Yu. V. Egorov M. A. Shubin (Eds.) 2009.1
67. 拓扑学 II：同伦与同调，经典流形 S. P. Novikov V. A. Rokhlin (Eds.) 2009.1